Advanced Materials
for Building Construction

Advanced Materials for Building Construction

Edited by
Jonathan Hunter

www.statesacademicpress.com

Published by States Academic Press,
109 South 5th Street,
Brooklyn, NY 11249, USA

ISBN: 978-1-63989-017-0

Cataloging-in-Publication Data

Advanced materials for building construction / edited by Jonathan Hunter.
p. cm.
Includes bibliographical references and index.
ISBN 978-1-63989-017-0
1. Building. 2. Building materials. 3. Structural engineering. I. Hunter, Jonathan.
TA403 .A28 2022
620.11--dc23

For information on all States Academic Press publications
visit our website at www.statesacademicpress.com

Contents

Preface

Construction refers to the act of building structures and infrastructure. It is primarily associated with the planning, designing and financing of the construction projects. Construction is applied in the three major sectors – buildings, infrastructure and industry. Building construction involves residential and non-residential institutions. Infrastructure includes dams, bridges, highways, etc. Industrial sector involves refineries, mills and manufacturing plants. Some of the steps that fall under the domain of construction of buildings are developing floor plans, excavating the foundation and building the main load bearing structure. Most of the topics introduced in this book cover new techniques and the applications of advanced materials for building construction. It is compiled in such a manner, that it will provide an in-depth knowledge about the theory and practice of the subject. A number of latest researches have been included to keep the readers up-to-date with the global concepts in this area of study.

After months of intensive research and writing, this book is the end result of all who devoted their time and efforts in the initiation and progress of this book. It will surely be a source of reference in enhancing the required knowledge of the new developments in the area. During the course of developing this book, certain measures such as accuracy, authenticity and research focused analytical studies were given preference in order to produce a comprehensive book in the area of study.

This book would not have been possible without the efforts of the authors and the publisher. I extend my sincere thanks to them. Secondly, I express my gratitude to my family and well-wishers. And most importantly, I thank my students for constantly expressing their willingness and curiosity in enhancing their knowledge in the field, which encourages me to take up further research projects for the advancement of the area.

Editor

Effect of Adjusting for Particle-Size Distribution of Cement on Strength Development of Concrete

Daegeon Kim ⓘ

Architecture Engineering, Dongseo University, Busan, Republic of Korea

Correspondence should be addressed to Daegeon Kim; gun43@gdsu.dongseo.ac.kr

Academic Editor: Guoqiang Xie

The massive construction projects such as nuclear power plants and gas storage plants operate their own batch plants for appropriate supply of a massive amount of ASTM Type I cement. The batch plants have difficulty responding to diversifying consumer requirements such as early strength development of concrete under the current limited production line of cement. In order to respond to the needs, this study collected several sizes of cement particles from different filters at a cement plant, and conducted the adjustment of cement particle-size distribution to enhance the properties of concrete based on ASTM Type I cement. This paper shows the chemical properties and physical tests such as setting time and compression tests considering the effects of the distribution of cement particles. Also, the classification based on cement particle-size distribution was defined as FMC (fineness modulus of cement). As FMC increased, compressive strength was relatively low at early age but the difference became smaller for later age. The test results show the effects of the adjusted mix proportion considering particle-size distribution of cement on strength development of concrete. Therefore, it is possible to see that customized cement or concrete can be manufactured in response to various consumer requirements by introducing such method that by-passes and collects fine or coarse cement in the cement-crushing process and then re-mixes them with OPC ASTM Type I cement depending on FMC.

1. Introduction

It is known that the quality and type of cement are significant factors to determine the strength development of concrete, which is specifically affected by the fineness and mineral composition of cement. The fineness of cement also affects its reactivity with water. Generally, the finer the cement, the more rapidly it will react [1, 2]. However, the cost of grinding and the heat evolved on hydration set some limits on fineness.

It is reported in the study that the rate of reactivity and the strength development can be enhanced by finer grinding of cements. It is generally agreed that cement particles larger than 45×10^{-6} m are difficult to hydrate and those larger than 75×10^{-6} m could never hydrate completely. However, an estimate of the relative rates of reactivity for similar cement composition cannot be made without knowing the complete particle-size distribution by sedimentation methods [3–5].

The concrete of early strength development attributable to more contents of fineness cement particles is used for construction sites which is sensitive to the duration and cost of construction completion. Sometimes, customers require early concrete strength of over 5 MPa in fourteen hours, even if the only production line of normal cement is available in the cement plant, or if the limited number of batch plants for normal cement in the construction sites is operated for the massive construction projects. In order to address those needs, various approaches such as the use of different types of cement and the change of distribution of cement particle sizes can be provided, which the former increases its cost. As considering economy, the cement resulted in the way of adjusting the particle-size distribution through separately collecting the fineness cement in the regular cement-grinding process of normal cement production line without additional grinding process can be equatable with the cement of ASTM Type III in terms of early strength development.

Han et al. [6] reported how the grinding process in cement plant and cement fineness affects in mortar exclusive of the effect of aggregates. According to the process of the general production line of Type I cement, which is called the normal cement, in South Korea, when a clinker and other mineral additives are provided and grinded in the cement ball mill, the grinded material is transferred to the cyclone separator. At this stage, the insufficiently grinded particles are going to be fed back and go through the regrinding process in the cement ball mill so that fine particles are sent to cement silos and made into cement mixtures. In the process of air classification, excessively grinded particles are flown into and collected in the separator bag filter along with discharged air and then are mixed with properly grinded particles together with the fine particles that have been collected in the main bag filter, and finally are inserted into the silo through the bucket elevator for the production of regular Type I cement products. Also, the research conducted several tests to evaluate the properties of cement mortar containing the finer cement particles collected from the bag filter and the coarser cement particles collected from the tube mill than Type I cement. It was reported that the effect of the fineness cement particles on the strength development of concrete was investigated. And it was concluded that the mortar including different sizes of fineness cement particles behaves close to concrete properties of ASTM Type III.

The effect of cement fineness on mortar was also investigated by Hu et al. [7]. It was reported that cement fineness and water-to-cement ratio (W/C) affected the heat of hydration and set times. The initial and final set times of coarser cement were found to be late compared to finer cement. It was concluded that the decreased surface area of coarser cement delayed the rate of hydration.

Sajedi and Razak [8] reported that the surface area of cement particles was associated with strength development and hydration. The authors showed that the rate of hydration depended on the fineness of cement particles, and high fineness was required for a rapid development of strength.

The review of the literatures indicates that the fineness of cement has been studied due to the rapid development of concrete strength. Also, it is indicated that the appropriate distribution of fine particles of cement can be contributable to the decrease of heat of hydration. But, the studies were confined to investigation of only mortar exclusive of effects of aggregates. It is thought that the limited results may be attributable to not much quantity of fineness cement collected from plant equipment. This paper focused on investigating the properties of fresh and hardened concrete based on the different contents of the very fine particles adding coarse aggregates to apply the mixed concrete to real construction in terms of strength development. The mix proportion in previous researches was newly required considering the effects of coarse aggregates in this study.

This paper aims to analyze the effect of the various classifications of particle-size distribution of cement through re-mixing to enable customized production of cement for consumers in the limited Type I cement production facilities. The main objectives of this study are thus to investigate the properties of fresh concrete including finer cement than Type I cement, to investigate the flowability and strength development of concrete with respect to the various classifications of particle-size distribution of cement, and to develop the relationship of compressive strength using the various classifications of particle-size distribution of cement and concrete ages.

2. Experimental Plan and Approach

2.1. Material.
Chemical composition of the material used in this study is shown in Table 1. Table 1 shows that the ignition loss (LOI) of fineness cement (FC) is high comparing to ASTM Type I cement (OPC) and coarse cement (CC). Cement with a density of $3.15 \, g/cm^3$ was used in this study, and the Blaine fineness for FC, OPC, and CC is given in Table 2. River sand and crushed sand were mixed in a proportion of $50:50$ for fine aggregates, whereas coarse aggregates were a mixture of 5 to 10 mm and 10 to 25 mm aggregates in a proportion of $35:65$. Other physical properties of aggregates are shown in Table 3. And the properties of admixtures are shown in Table 4.

2.2. Approach.
To evaluate the effect of the classifications of particle-size distribution of cement, the mix proportion was newly required. Nine mix proportions by the amount of different sizes of cement were classified based on the accumulated remains denoted as R at $10 \, \mu m$ (R10), $20 \, \mu m$ (R20), $40 \, \mu m$ (R40), and $80 \, \mu m$ (R80) using an Alpine air jet sieve shaker, as shown in Table 5, and the fineness modulus (FMC) of the distributed cement particle sizes as integers shown in (1) was used. Generally, a lower value of FMC has the effect of reducing the size of cement particles which is advantageous to the increase of initial strength, and on the contrary, as the higher value of FMC increases the size of cement particles, it is more favorable in the reduction of hydration heat.

$$\text{FMC} = \frac{(R80 + R40 + R20 + R10)}{100}. \quad (1)$$

The experimental variable related to the change of FMC is finest cement (FMC: 0.50), which are particles excessively grinded during the production process of Type I cement (OPC) that have flown into and collected in the separator bag filter together with discharged air, and relatively coarse cement (FMC: 1.69) primarily grinded in the tube mill were input proportionally to substitute Type I cement (OPC) of FMC with 1.11 for manufacturing. Table 6 shows the results of fineness distribution using the sieves of $1 \, \mu m$, $24 \, \mu m$, $64 \, \mu m$, and $200 \, \mu m$.

2.3. Experimental Plan.
The scope of tests on the properties of concrete using fineness cement consisted of the fundamental tests with fresh concrete, slump, air contents, and setting time [9–11]. The experiment was planned with the following targets; slump of $120 \, mm \pm 25 \, mm$, water-to-cement ratio of 0.5, and air content of $4.5\% \pm 1.5\%$. As for experimental items, unit volume mass [12] and setting time were measured for fresh concrete, and compressive strength [13] of up to 91 days of age was to be measured for hardened concrete. The density of all the types of cement was planned to be 3.15.

The mixture proportion for the experiment was designed by FMC with nine levels through proportional substitution of fineness cement shown in Table 1. Data on the concrete mix fixed at this stage is shown in Table 7.

TABLE 1: Chemical composition of cement.

Name	Chemical element (%)								LSF	SM	IM
	LOI	SiO$_2$	Al$_2$O$_3$	Fe$_2$O$_3$	CaO	MgO	SO$_3$	K$_2$O			
FC	0.54	21.24	5.01	3.68	62.91	2.00	3.17	1.31	89.55	2.44	1.36
FO1	0.50	21.40	5.01	3.68	63.23	2.01	2.84	1.22	89.78	2.46	1.37
FO2	0.45	21.56	5.02	3.67	63.55	2.01	2.50	1.12	90.00	2.48	1.37
FO3	0.40	21.72	5.02	3.67	63.87	2.01	2.17	1.02	90.22	2.50	1.37
OPC	0.35	21.88	5.02	3.66	64.18	2.01	1.83	0.92	90.44	2.52	1.37
CO1	0.31	21.96	5.06	3.70	64.25	2.02	1.66	0.91	90.34	2.51	1.37
CO2	0.27	22.03	5.10	3.74	64.31	2.03	1.49	0.90	90.24	2.50	1.36
CO3	0.23	22.11	5.14	3.78	64.38	2.04	1.32	0.89	90.14	2.49	1.36
CC	0.18	22.18	5.17	3.82	64.44	2.05	1.14	0.87	90.03	2.47	1.35

LSF, lime saturation factor; SM, silica modulus; IM, iron modulus. FO1, FO2, and FO3 consist of the mixture of FC and OPC by the different value of FMC. CO1, CO2, and CO3 consist of the mixture of CC and OPC by the different value of FMC.

TABLE 2: Blaine fineness of cement (cm^2/g).

FC	FO1	FO2	FO3	OPC	CO1	CO2	CO3	CC
6,479	5,752	5,024	4,174	3,324	2,920	2,515	2,164	1,813

TABLE 3: Physical properties of aggregate.

Classification	Density (g/cm^3)	Fineness modulus	Absorption rate (%)	Amount of pass through 0.08 mm strainer (%)
Fine aggregate	2.59	2.67	1.11	1.12
Coarse aggregate	2.71	7.01	1.18	0.11

TABLE 4: Physical properties of admixture.

Classification	Form	Property	Color	pH	Density (g/cm^3)
SP agent	Liquid	Polycarbonate	Ivory white	6.5	1.06
AE agent	Liquid	Negative ion	Ivory white	—	1.04

SP, superplasticizer; AE agent, air-entrained agent.

3. Experiment Results and Analysis

3.1. Fresh Concrete. Figure 1 shows the fine aggregate ratio as well as the unit water contents according to the change of FMC. As FMC increased, fine aggregate ratio had to be gradually decreased to secure target fluidity, and unit water content had to be first decreased and then increased centering on OPC of which FMC is 1.11. This is considered to be related with the ratio increase of 200 μm size of which the particle diameter is relatively big as can be seen in Table 6.

Figure 2 illustrates the usage amount of SP and AE agents following the change of FMC. The usage amount of the SP agent had to be reduced, whereas the usage amount of the AE agent did not have any special effect. As FMC increases, the size of cement particles increases, which leads to smaller surface area that can contact water and lower

viscosity. To secure viscosity, the unit volume had to be increased with regard to unit cement volume, and as FMC increased by 0.1, the average diameter of cement grew by approximately 1.7 μm, which eventually is the effect of increased fluidity due to the decrease in the number of cement particles, having to reduce the usage amount of the SP agent.

Figure 3 and Table 8 demonstrate penetration resistance according to progressed time by FMC. As FMC increased, the setting time was delayed in proportion, which is considered to be a result of delayed hydration due to smaller surface that can contact water as there is more distribution of coarse particles as FMC increased. As FMC increased by 0.1, the final set was delayed by about 0.42 hours. But, slump results did not show any tendency for the various FMC values.

3.2. Hardened Concrete. Table 9 shows the experiment results of relationship between compressive strength and FMC according to age, respectively. Figure 4 shows the compressive strength ratio of concrete mix proportions comparing to Type I cement according to concrete age. The ratio of 100% in Figure 5 means the compressive strength of Type I cement mix proportion at 91 days.

At this point, as FMC increased, compressive strength dropped, and there was a big difference of compressive strength in the initial age; however, the difference in compressive strength became smaller in the latter age. As a result, strength revelation grew at a wider span with the progress of age as FMC increased, which can be analyzed as a consequence of continuous hydration of unhydrated cement with the progress of age, whereas hydration happens slowly at initial age due to small surface area that can contact water because of the low fineness of high FMC.

For the estimation of compressive strength, the OriginPro 7.5 [14] program was used to set the cement fineness modulus and age as the independent variable and compressive strength as the dependent variable, to calculate parameters such as slope and intercept by conducting multiple regression analysis.

$$f_{cu} = 11.177 \times \log D - 11.364 \times FMC + 25.146, \quad (2)$$

TABLE 5: FMC based on (1).

| Name | Type | Residue (%) | | | | FMC |
		At 80 μm (R80)	At 40 μm (R40)	At 20 μm (R20)	At 10 μm (R10)	
FC	FC	0.6	4.0	14.5	30.6	0.50
FO1	FC + OPC	0.9	5.7	20.0	38.2	0.65
FO2	FC + OPC	1.2	7.5	25.5	45.8	0.80
FO3	FC + OPC	1.6	9.2	30.9	53.8	0.95
OPC	OPC	1.9	10.9	36.4	61.8	1.11
CO1	CC + OPC	3.0	15.8	41.8	64.9	1.25
CO2	CC + OPC	4.0	20.6	47.1	67.9	1.40
CO3	CC + OPC	5.1	25.1	52.8	71.4	1.54
CC	CC	6.1	29.6	58.4	74.8	1.69

TABLE 6: Fineness distribution of cement.

| FMC | Fineness distribution (%) | | | | Average particle diameter (μm) | Remains of 200 μm (%) |
	Pass of 1 μm	Pass of 24 μm	Pass of 64 μm	Remains of 64 μm		
FC	16.1	73.8	8.9	1.2	5.1	0.0
FC1	13.8	71.7	12.7	1.8	7.4	0.0
FC2	11.4	69.5	16.5	2.5	9.7	0.0
FC3	9.0	67.5	20.4	3.1	12.0	0.0
OPC	6.6	65.4	24.2	3.8	14.1	0.0
CO1	6.2	60.0	27.9	5.9	16.5	0.0
CO2	5.9	54.5	31.6	8.1	19.8	1.4
CO3	4.9	48.9	36.0	10.2	22.4	3.8
CC	4.0	43.3	40.4	12.3	24.9	6.1

TABLE 7: Mixture proportion.

| Name | S/a (%) | W (kg/m³) | SP/C (%) | AE/C (%) | Unit volume (kg/m³) | | | | |
					FC	C	CC	S	G
FC	48	183	0.71	0.017	366	—	—	315	341
FO1	47	182	0.66	0.018	91	274	—	309	348
FO2	47	181	0.61	0.018	180	180	—	310	349
FO3	46	180	0.53	0.018	91	271	—	304	357
OPC	46	178	0.51	0.018	—	356	—	305	359
CO1	46	178	0.51	0.020	—	268	88	305	359
CO2	46	180	0.49	0.019	—	180	180	304	357
CO3	46	185	0.48	0.019	—	91	277	300	352
CC	46	195	0.45	0.019	—	—	391	293	343

S/a, sand-coarse aggregate ratio; W, mixed water; S, sand; C, cement; G, aggregate.

where f_{cu} is compressive strength (MPa); FMC is the fineness modulus of cement; and D is the age based on log scale, day.

The result of (2) indicated that compressive strength declined as FMC increased and increased as age progressed. Compressive strength could be estimated and the estimation acquired through this study showed a good correlation with 0.942 of coefficient of determination. Figure 5 shows the comparison between the compressive strength estimated through (2) and the actually measured compressive strength by using the scatter plot. The correlation coefficient turned out to be 0.942, which can be interpreted that there is a decent level of estimation accuracy between the estimated and actually measured compressive strength.

4. Conclusion

This study reviewed the effects of the change of FMC on the mixing and strength revelation properties to analyze the manufacturing possibility of customized cement following the change of the FMC, which expresses particle-size distribution of cement as integers, and the results are summarized as follows:

(i) Regarding the mixing properties of fresh concrete, since the average particle diameter of cement grows as the FMC increases to maintain the same level of slump and air content which leads to lower viscosity, S/a and unit quantity had to be reduced until FMC 1.11, which then had to be increased afterwards and the usage amount of the SP agent had to be reduced to secure viscosity.

(ii) Setting time was delayed proportionally due to low fineness as FMC increased, which was a delay of about 0.42 hours per 0.1 increase of FMC.

(iii) The estimation formula for compressive strength using FMC and ages was derived through the

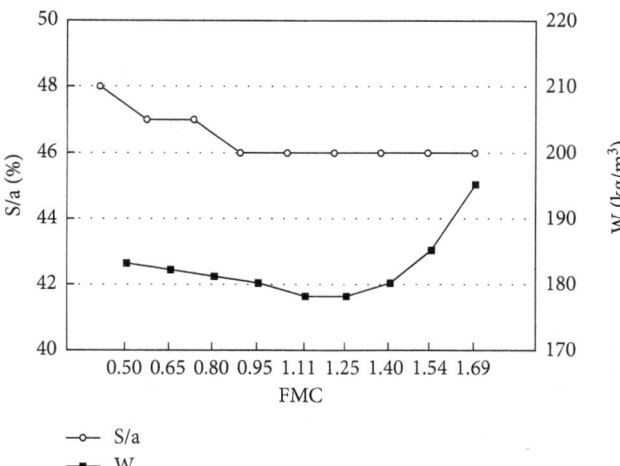

FIGURE 1: Fine aggregate ratio and water contents in accordance with FMC change.

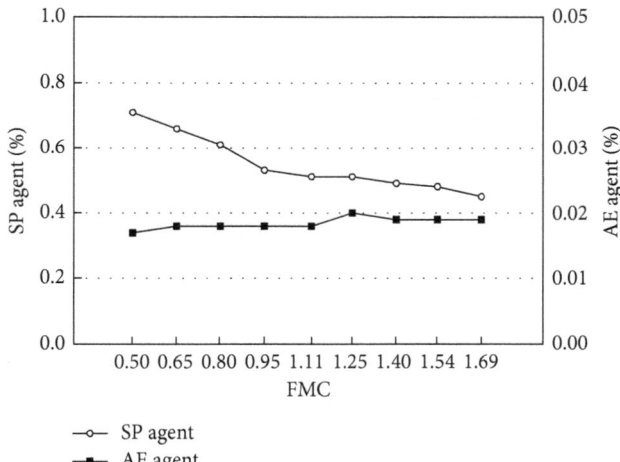

FIGURE 2: Amount of chemical admixture in accordance with FMC change.

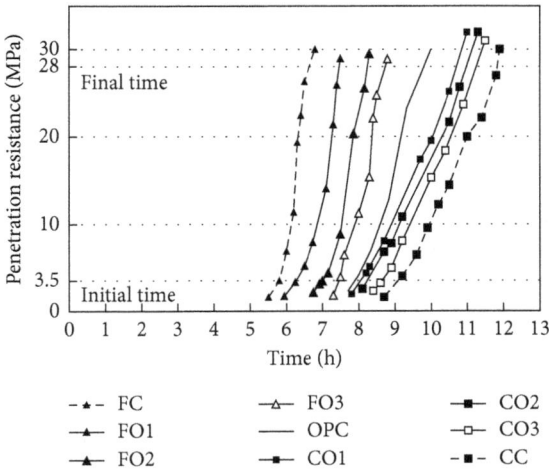

FIGURE 3: Penetration resistance.

TABLE 8: Results of fresh concrete test.

Name	Slump (mm)	Slump flow (mm)	Air content (%)	Setting time (hr) Initial	Final
FC	125	235	5.6	5.8	6.6
FO1	137	225	4.1	6.3	7.4
FO2	130	234	4.9	6.9	8.3
FO3	119	222	5.2	7.4	8.6
OPC	122	220	5.1	7.9	9.7
CO1	134	235	5.1	8.1	10.8
CO2	110	220	3.9	8.3	11.1
CO3	116	227	3.3	8.7	11.3
CC	130	231	4.4	8.9	11.8

TABLE 9: Result of the compression test (unit: MPa).

Name	1 day	3 days	7 days	28 days	91 days
FC	19.7	25.9	30.6	35.3	38.9
FO1	17.6	24.5	29.8	32.9	35.7
FO2	13.2	22.9	28.3	31.6	34.6
FO3	12.2	22.4	27.8	30.4	33.7
OPC	9.1	20.6	26.0	29.5	33.1
CO1	6.8	18.2	22.7	28.4	32.6
CO2	5.3	14.8	21.7	26.3	30.5
CO3	4.1	14.0	17.9	24.6	29.0
CC	3.0	11.0	17.0	22.7	28.4

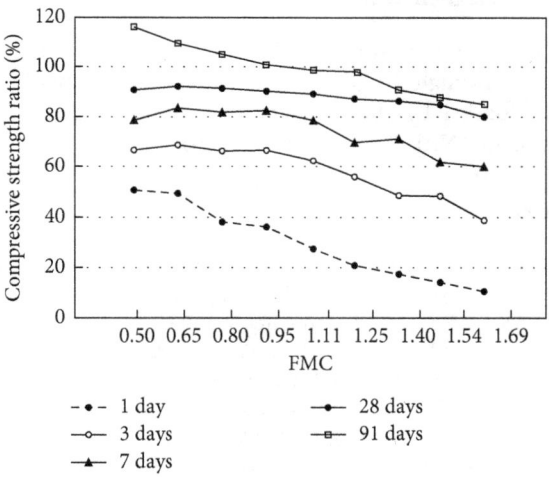

FIGURE 4: Compressive strength ratio against FMC values.

multiple regression analysis, and as a result of comparing the estimated and actually measured compressive strength, there was a high correlation with a correlation efficient of 0.942, demonstrating decent estimation accuracy.

(iv) It is concluded that customized cement or concrete that meet the diverse needs of the consumers can be manufactured when fine or assembled cement are by-passed, collected in the cement-grinding process, and then re-mixed with ASTM Type I cement according to FMC. Lower value of FMC has the effect of reducing the size of cement particles, which is

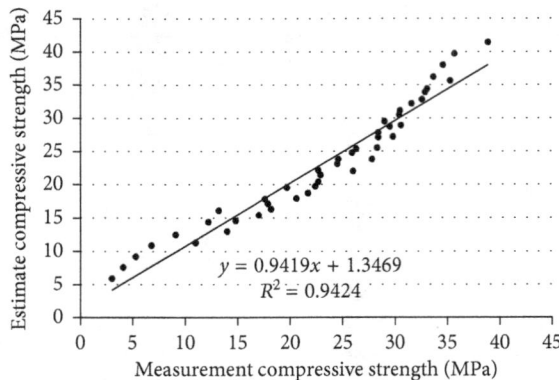

FIGURE 5: Comparison of estimated compressive strength by formula and measured compressive strength.

advantageous to the increase of initial strength. For construction in cold region, this customized cement by the change of FMC may contribute to improving quality of concrete due to early development of strength without costly action using ASTM Type III.

Conflicts of Interest

The authors declare that they have no conflicts of interest.

Acknowledgments

This research was supported by the Basic Science Research Program through the National Research Foundation of Korea (NRF) funded by the Ministry of Education, Science and Technology (NRF-2018R1C1B5045860).

References

[1] P. K. Mehta and P. J. Monteiro, *Concrete-Structure, Properties and Materials*, Prentice-Hall, Englewood Cliffs, NJ, USA, ISBN-0-13-175621-4, 2nd edition, 1993.

[2] M. S. Mamlouk and J. P. Zaniewski, *Materials for Civil and Construction Engineers*, Prentice-Hall, Upper Saddle River, NJ, USA, ISBN-13-978-0-13-611058-3, 3rd edition, 2011.

[3] D. P. Bentz, "Blending different fineness cements to engineer the properties of cement-based materials," *Magazine of Concrete Research*, vol. 62, no. 5, pp. 327–338, 2010.

[4] G. Frigione and S. Marra, "Relationship between particle size distribution and compressive strength in Portland cement," *Cement and Concrete Research*, vol. 6, no. 1, pp. 113–127, 1976.

[5] B. Osbaeck and V. Johansen, "Particle size distribution and rate of strength development of Portland cement," *Journal of the American Ceramic Society*, vol. 72, no. 2, pp. 197–201, 1989.

[6] C. G. Han, M. C. Han, and J. B. Kim, "Engineering properties of the particle classifying cement and the mortar using the particle classifying cement," *Journal of the Architectural Institute of Korea*, vol. 23, no. 7, pp. 111–118, 2007.

[7] J. Hu, Z. Ge, and K. Wang, "Influence of cement fineness and water-to-cement ratio on mortar early-age heat of hydration and set times," *Construction and Building Materials*, vol. 50, pp. 657–663, 2014.

[8] F. Sajedi and H. A. Razak, "Effects of curing regimes and cement fineness on the compressive strength of ordinary Portland cement mortars," *Construction and Building Materials*, vol. 25, no. 4, pp. 2036–2045, 2011.

[9] ASTM C143, *Standard Specification for Portland Cement Standard Test Method for Slump of Hydraulic-Cement Concrete*, ASTM International, West Conshohocken, PA, USA, 2015.

[10] ASTM C231, *Standard Test Method for Air Content of Freshly Mixed Concrete by the Pressure Method*, ASTM International, West Conshohocken, PA, USA, 2017.

[11] ASTM C403, *Standard Test Method for Time of Setting of Concrete Mixtures by Penetration Resistance*, ASTM International, West Conshohocken, PA, USA, 2016.

[12] ASTM C1688, *Standard Test Method for Density and Void Content of Freshly Mixed Pervious Concrete*, ASTM International, West Conshohocken, PA, USA, 2014.

[13] ASTM C39, *Standard Test Method for Compressive Strength of Cylindrical Concrete Specimens*, ASTM International, West Conshohocken, PA, USA, 2017.

[14] OriginLab, *OriginPro 7.5*, OriginLab Corporation, Northampton, MA, USA, 2003.

Bioimmobilized Limestone Powder for Autonomous Healing of Cementitious Systems

Nafeesa Shaheen,[1] Rao Arsalan Khushnood ⓘ,[2] and Siraj Ud din[1]

[1]NUST Institute of Civil Engineering (NICE), School of Civil and Environmental Engineering (SCEE),
 National University of Sciences and Technology (NUST), Sector H-12, Islamabad 44000, Pakistan
[2]Department of Structural, Geotechnical and Building Engineering (DISEG), Politecnico di Torino, Corso Duca degli Abruzzi 24,
 Torino 10129, Italy

Correspondence should be addressed to Rao Arsalan Khushnood; rao_nust@yahoo.com

Academic Editor: Wen Deng

For preserving concrete structures and hindering ingress of chemicals through cracks and fissures, repair is inevitable. Microbial calcite precipitation is an intrinsic approach for crack rectification and emulating way of sustainability for reducing anthropogenic greenhouse gases (GHGs) along with conserving the natural resources. In this study, *Bacillus subtilis* strain is applied for intrinsic repair of concrete's cracks because of its high pH endurance and capability of sporulation. For prolonged survival of microorganisms, immobilization technique was employed. *B. subtilis* was immobilized through limestone powder (LSP) before adding into cement matrix. Self-healing proficiency of *B. subtilis* was deliberated in terms of mechanical strength regain after cracking at 3, 7, 14, and 28 days. To examine the microstructure and characterization of healing precipitate, micrographical (field emission scanning electron microscopy), chemical (energy dispersive X-ray), and thermal (thermogravimetric analysis) analyses were performed after the healing period of 28 days. The results revealed evident signs of calcite precipitation in nano-/microcracks subsequent to microbial activity. Furthermore, immobilized LSP improved the compressive strength of the analyzed formulations.

1. Introduction

Concrete, an extensively used construction material, is a source of anthropogenic greenhouse gas (GHG) emissions, depleting the raw materials, consuming the fossil fuels, and intensifying the environmental concern [1]. Additionally, it is susceptible to crack under tensile stresses, ensuing in the dramatic increase in its porosity, consequences in declined strength, and durability [2]. Numerous advanced cementitious systems have been investigated and practiced for limiting the ingress of deleterious chemicals through pores for enhancing the durability of cementitious systems [3]. But they involved external interventions which are labor dependent and cost extensive [4]. Researchers are probing for sustainable solutions to reduce cost as well as the environmental impacts. One of the potential solutions is self-healing concrete, which can repair its cracks itself [5].

Self-healing mechanisms in concrete are categorized as autogenous and autonomous healing [6]. Autogenous self-healing can be achieved by adding anhydrous cement particles or some pozzolanic material. When cracks appear in humid environments, anhydrous particles endure secondary hydration that seals the internal microcracks and fissures. But it is a limited process with minimal recurring rate having ability of 0.1-0.2 mm trivial crack-healing widths [7, 8]. For accomplishing distinct crack-healing widths by persistent process of repairs, autonomous healing is recommended [9]. Concrete, autonomously, can be healed through addition of engineered cementitious crystalline admixtures, polymers, shape-memory alloys, and microbes [10]. Microbial calcite precipitation attained imperative position among all of them since calcite formation is a way of emulating sustainability in cementitious systems [11]. Microbes secrete calcite under favorable surroundings; the secreting process consumes CO_2 from the environment and acts as sink for CO_2 dumping [12]. But problems associated with such healing systems are survival of bacteria for relatively longer periods and reduction in mechanical properties of concrete [13]. Moreover, the healing rate depends upon the type of microbial strain, its endurance in high alkalinity, food source, and immobilization techniques [14]. Microbes having ability of sporulation can stay relatively long in dormant stage and are therefore preferred in the bio-influenced cementitious systems [15]. Microbes are either directly induced through mixing water or immobilized using different techniques [16]. These immobilization techniques include encapsulation, entrapment, and adsorption of microbes [17]. These immobilization techniques are reported efficient for enhancing the microbes' survival but deficient in attaining intended mechanical properties [18, 19].

Numerous researchers are probing for optimization of immobilization techniques and struggling for appropriate immobilizers. Wiktor and Jonker embedded microbe's *B. alkalinitrilicus* and calcium lactate into porous expanded clay (EC) particles for formulating healable cementitious mortar. Both components are adsorbed on the surface and entrapped into pores of EC. EC particles released microbes on ingress of water and achieved maximum of 0.46 mm crack-healing width [20]. In another study, *B. sphaericus* were adsorbed in diatomite earth (DE) and offered 0.15 to 0.17 mm crack-healing widths. When *B. sphaericus* immobilized through polyurethane sheet or silica gel, they contributed in lowering the permeability of the system [18, 21]. Then, microencapsulation technique was used for immobilization of microbes into melamine capsules using the condensation process. Maximum crack-healing width of 0.97 mm was attained, which was 40% to 80% higher as compared to purely melamine capsules, but it reduced the compressive strength by 15–34% [22]. Hydrogels were also employed for microbial immobilization, and microbes were attached to hydrogels through the cross-linking process. Hydrogels repaired cracks up to 0.5 mm in mortar [19]. Natural diatomite has been investigated for bacterial adsorption and entrapment into the pores with final insertion made in the form of pellets. Addition of pellets enhanced compressive strength but lowered the permeability of the resultant matrix [23]. Ceramsite (sand) was also used as a carrier, but microbes and nutrients were immobilized separately in that case, eventually providing an accomplished repair rate of 87.5% [24]. Expanded perlite (EP) was also tried for enhancing microbial survival, and the results were compared to direct induction and EC [25]. Furthermore, zeolite powder, graphite nanotubes (GNPs), and light weight aggregate (LWA) have also been explored for immobilization of microbes [15, 16]. The state of the art highlighted that each bioimmobilizer has some related pros and cons. Researchers are still struggling to explore viable alternatives to successfully immobilize microbes in cementitious systems for relatively longer periods.

This work is basically a contributing effort to explore potential usage of LSP as an immobilizer in self-healing cementitious systems. LSP has been employed in the construction industry since eras, and it is still divulging contemporary applications owing to its versatility. It is a by-product of aggregate formation from rocks and a prime source of calcium oxide (CaO) provision in the cement manufacturing industry [26]. Moreover, it has been investigated as secondary raw material (SRM) in the cementitious systems for dense microstructure and imparting durability as a cement and sand replacement for fabricating a sustainable concrete to shrink CO_2 footprints [26–28]. Furthermore, LSP is an inert media and acts as a filler, but it reacts with aluminates phases to form carboaluminates and influence the hydration rates of cementitious systems [29]. Initially, hemicarboaluminate component was formed, and then, it converted into monocarboaluminate (Al_2O_3-Fe_2O_3-mono) AFm group [30]. The newly formed AFm improved the microstructure of cementitious systems and enhanced its durability. In a recent study, LSP was used as a cement replacement to enhance calcite precipitation rate via CO_2 curing. Supplementary content replacement of LSP helped in more calcite formation owing to pore structure and nucleation site provision by LSP [31]. The promising results of that study motivated the researchers to investigate the feasibility of LSP as a microbial immobilizer in self-healing cementitious systems as calcite production is a key for microbial activity. Moreover, researchers supported the argument that LSP addition influenced the mechanical properties of cementitious systems positively such as melamine microencapsulation that reduced the compressive strength [22]. Furthermore, its chemical nature resembles instigated calcite that makes it compatible to provide sufficient sites for precipitation as GNPs serving as microbial immobilizers killed the microbes [16]. Additionally, LSP retains its shape and position by filling the empty voids that ensure its homogeneous distribution inside the host matrix for uniform healing.

2. Experimental Program

2.1. Materials

2.1.1. Cement and Sand. Ordinary Portland cement CEM-I (grade 53), conforming to ASTM C-150, was used for all mortar formulations. The average particle size of cement grains is 16.4 μm with density of 3.17 g/cm^3. The oxide composition as determined via X-ray fluorescence (XRF) test is presented in Table 1.

Locally available sand from Lawrencepur, Pakistan, was used in the study. The fineness modulus of sand was determined according to the ASTM C-136, and the calculated value was 2.018 with specific gravity of 2.65 having $D_{50} = 0.215$ mm. Absorption capacity of sand was 2.4% as determined according to ASTM C-128.

2.1.2. Microorganisms. Cementitious composite's environment is highly alkaline in nature having pH around 11–13 [32]. So, bacterial strains must be capable of enduring high pH and have ability of sporulation for ensuring survival within the harsh environment of cementitious composites [33]. Moreover, bacterial strains must be proficient in production of copious amount of calcite, which is

TABLE 1: Chemical composition of OPC and LSP (%).

Parameters	CaO	SiO$_2$	MgO	Al$_2$O$_3$	Fe$_2$O$_3$	TiO$_2$	Na$_2$O	K$_2$O	P$_2$O$_5$	MnO	LOI
OPC	65.00	19.19	2.23	4.97	3.27	0.29	0.58	0.51	0.08	0.04	3.84
LSP	52.67	3.00	0.67	0.69	0.27	0.04	0.30	0.10	—	0.01	42.24

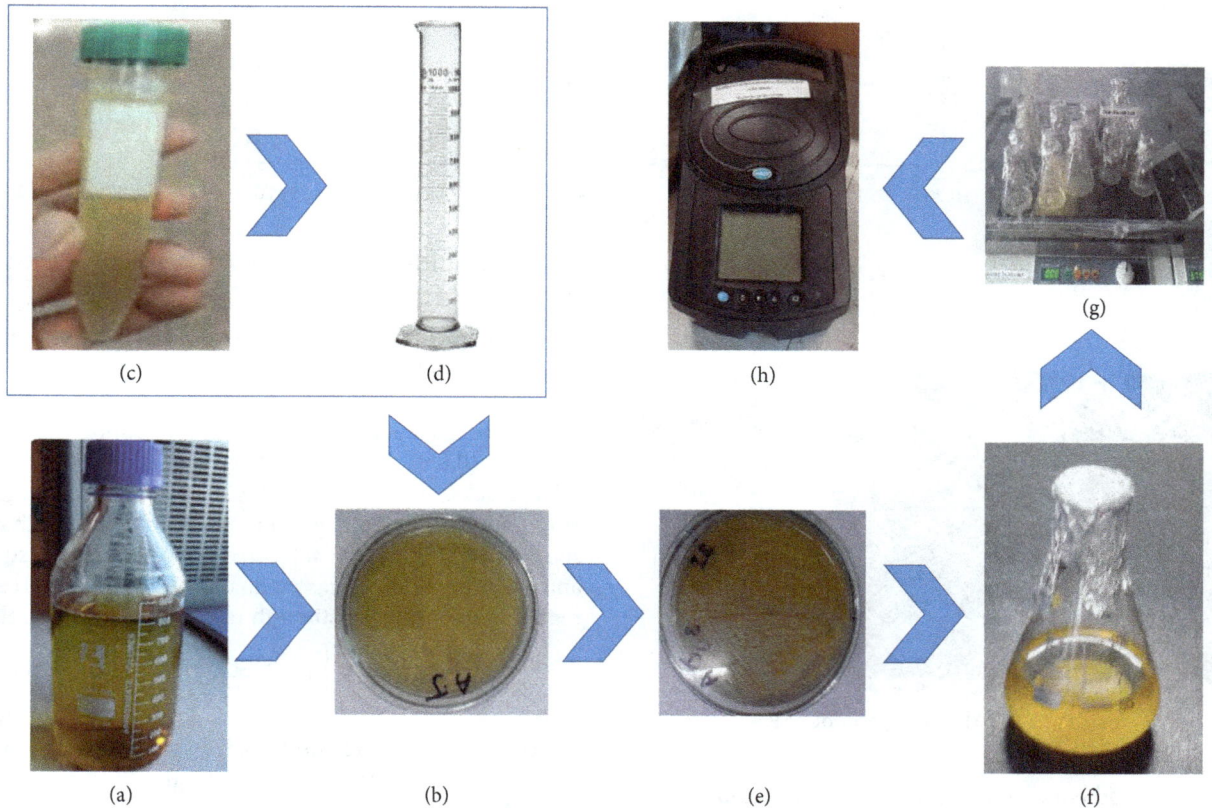

FIGURE 1: Microbial solution preparation. (a) LB media. (b) Agar plating for growth of microbial colonies. (c) *B. subtilis* spores. (d) Distilled saline. (e) Microbial colonies. (f) Inoculum in LB. (g) Incubation in shaker. (h) OD measurements.

responsible for crack healings [34]. *Bacillus subtilis*, soil-based bacteria, was selected for the purpose. *B. subtilis* is a gram-positive alkaliphilic bacterium having characteristics of sporulation and can survive for prolonged durations [35].

B. subtilis was revived from glycerol stock. 1 ml of frozen glycerol was mixed with 5 ml of 0.9% autoclaved saline solution. Then, this solution was spread on agar media by cotton swab and incubated for 24 hours. Nutrition broth used for *B. subtilis* was standard Luria broth (LB) consisting of 5 g tryptone, 5 g NaCl, and 2 g of LG broth in 1000 ml of distilled water. Incubation was done for 6 to 8 hours at 37°C while shaking at 200 rpm; after that, 15 ml of LB was inoculated on satirized colony of *B. subtilis*.

For sporulation, Difco sporulation medium (DSM) was used. DSM consisted of 2.5 g of peptone, 0.1 g of KCl, 1.5 g of meat extract, 0.5 ml of MgSO$_4$ (1 M), and 0.25 g of MnSO$_4$ in 500 ml of water and sterilized by autoclave. 0.25 ml of CaCl$_2$, 0.5 ml of FeSO$_4$, and 2.5 ml of LB was added into DSM shown in Figure 1(a). The entire solution was incubated for 4 days with shaking at 200 rpm at 37°C. Bacterial cells were pelletized at 9000 rpm for 20 minutes then washed 8 to 10 times. Spores were identified by light microscope. Bacterial solution is shown in Figure 1(c). Calcium lactate was used as food source for microbes.

The concentration of bacteria in the solution was calculated by HACH DR 2400 portable spectrophotometer shown in Figure 1(h). Spectrophotometer was calibrated by using 0.5 ml blank solution at 600 nm wavelength. Bacteria concentration in the solution was measured using the expression $Y = 8.59 \times 10^7 X^{1.3627}$ [36], where Y is bacterial concentration and X is wavelength value at OD$_{600}$. The cell concentration during the mixing of the mortar samples was kept constant at 6×10^6 cells/cm^3.

2.1.3. Limestone Powder. LSP used in the study was taken from local source and produced by milling of limestone rock. Average particle size of LSP was 22.3 μm with a specific surface area of 3048 m^2/kg and absorption capacity of 26% having density of 2.72 g/cm^2. Particle size distribution of LSP is given in Figure 2 and XRF analysis results are presented in Table 1.

The field emission scanning electron microscope (FESEM) image of LSP is shown in Figure 3. It is evident from SEM that the LSP particles are rhombohedral in nature with sharp indents having rough superficial textures and apparently seem porous. These characteristics contribute in

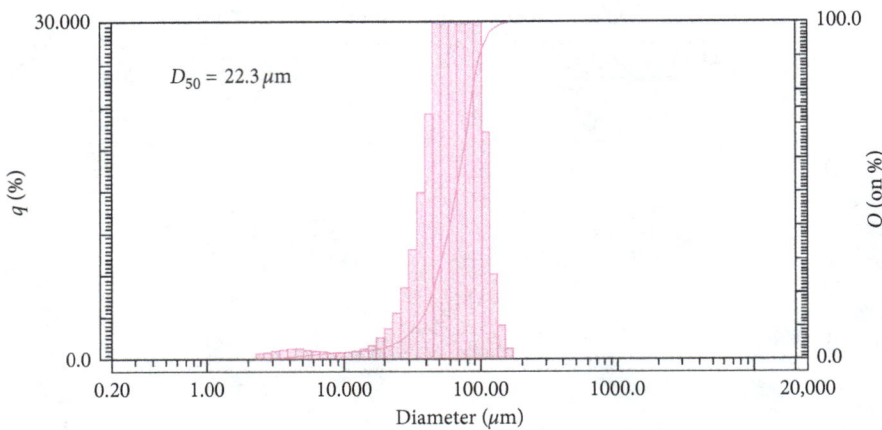

FIGURE 2: Particle size distribution of LSP.

FIGURE 3: FESEM micrograph of LSP.

developing good bond among grains of cementitious matrix [37, 38].

2.2. Mixing and Testing Regimes. Two types of formulations were investigated with their mix proportions mentioned in Table 2. Standard consistency was ensured in accordance with ASTM standard C191-11. Initial and final setting times of mortar formulations were monitored as per the standard set forth in ASTM C187-11. The analyzed formulations were designated as CM and LSP-B. CM corresponds to the controlled formulation without any addition of LSP, whereas LSP-B contains *B. subtilis* immobilized via 10% LSP in replacement to cement. Mix proportion of cement to sand was set as (1 : 1.4) with constant water to cement ratio (w/c) of 0.4. Quantities of cement and sand were $930 \, kg/m^3$ and $1400 \, kg/m^3$ with water $372 \, l/m^3$. Calcium lactate of $18.7 \, kg/m^3$ was used, which was 2% of the cement in both formulations, whereas bacterial solution of $7.6 \, Liter/m^3$ was only added in LSP-B formulation. Limestone powder particles were soaked in bacterial solution for 24 hours to ensure maximum adsorption and entrapment.

All ingredients were mixed in Hobart mixer of 5 L capacity in accordance with ASTM C-305. Mortar cube specimens of $50 \times 50 \times 50 \, mm^3$ dimensions were prepared for both the formulations. Specimens were demolded after 24 hours and moist cured till the age of testing in controlled conditions (25°C temperature, 100% relative humidity).

Compressive resistance of specimens was gauged at the age of 3, 7, 14, and 28 days in accordance with ASTM C-109.

For healing inspection, specimens were precracked at 3, 7, 14, and 28 up to 80% of their maximum compressive strength $(0.8 \times fc)$ of respective days to induce internal microcracks. The precracked specimens were immersion cured for the healing period of 28 days. After that, the specimens were again subjected to compressive strength analysis, and healing was estimated in terms of percentage regain in compressive strength using the following relation:

$$RCS\% = \left[1 - \frac{Cu_{28} - Cr}{Cu_{28}} \times 100 \right] \quad (1)$$

where Cu_{28} = ultimate compressive strength at 28 days and Cr = regained compressive strength after 28 days of curing.

In total, 48 samples were casted, and reported values are the average of three specimens. Field emission scanning electron microscope (FESEM), energy dispersive X-ray analysis (EDX), and thermogravimetric analysis (TGA) were employed to evidence the precipitation of $CaCO_3$.

3. Results and Discussions

For assessment of self-healing efficiency, two types of testing were conducted, and their results are discussed here. One was based on compressive strength analysis while the other was related to the affirmation of calcite precipitates in the induced cracks, through FESEM, EDX, and TGA.

3.1. Compressive Strength Analysis. Compressive strength of mortar specimens was measured using MCC8 compression-testing machine at a controlled loading rate of 0.2 MPa/sec conforming to standard ASTM C-109. The attained compressive resistance of the investigated formulations is given in Figure 4 at the age of 3, 7, 14, and 28 days. It can be seen that the addition of *B. subtilis* immobilized by LSP improved the strength of mortar specimens in compression during the entire hydration tenure.

The calcite precipitated by microbes within the cementitious matrix actually plugs the pores, and hence, contributes in improvement of compressive strength.

TABLE 2: Mix proportions of mortar formulations having mixing ratio of (1 : 1.4) with w/c of 0.4.

Specimens Units	Cement kg/m^3	Fine aggregates kg/m^3	Water cement ratio	Calcium lactate kg/m^3	Bacterial solution Liter/m^3	Immobilization media
CM	930	1400	0.4	18.7	—	None
LSP-B	840	1400	0.4	18.7	7.6	LSP (10% of C)

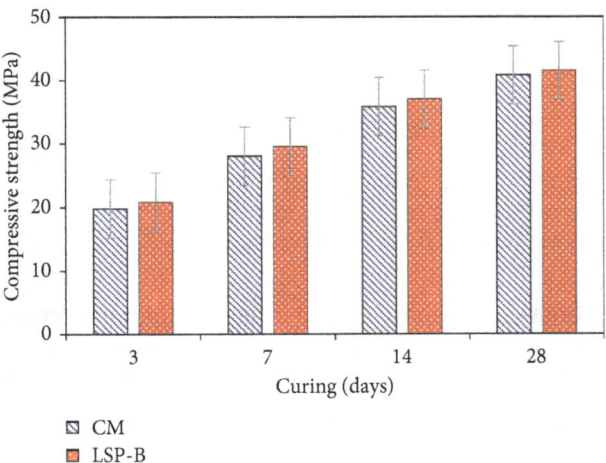

FIGURE 4: Compressive strength of mortar formulations with and without immobilized microbes at different curing ages.

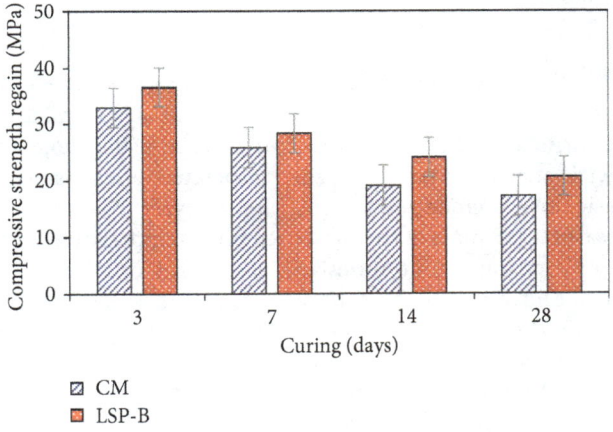

FIGURE 5: Compressive strength regain in mortar formulations with and without immobilized microbes at different precracking periods.

The reaction kinetics in the presence of microbes also endorses the formation of additional CSH gel as a result of CO_2 consumption as revealed by Li et al., which further adds in compressive strength [39]. The involved chemical reaction is as follows:

$$C_3S/C_2S + CO_2 + H_2O \longrightarrow C\text{-}S\text{-}H + CaCO_3 \qquad (2)$$

Generally, the addition of microorganisms through an immobilizer reduces the compressive strength of the cementitious matrix [19, 21, 22]. But on the contrary, immobilization via LSP increased the compressive strength of mortar formulations due to its filling ability that plugs the inner pores and homogenized distribution [40, 41]. At the age of 3 days, CM and LSP-B indicated 19.81 MPa and

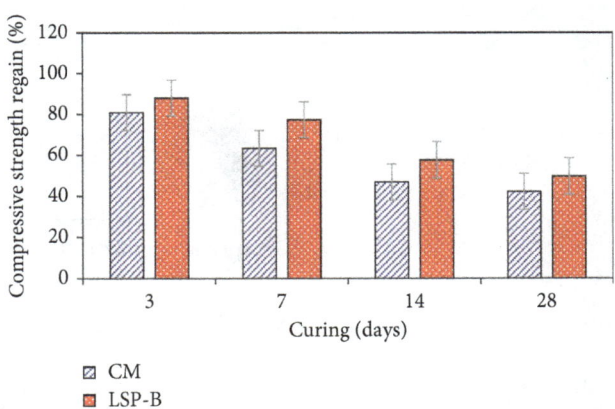

FIGURE 6: Percentage regain in compressive strength of mortar formulations with and without immobilized microbes at different precracked ages.

20.87 MPa values of compressive strength. LSP-B showed 5% more strength than CM formulation. Similar trend was observed at 7-day strength of both the formulations. At 14 days, CM and LSP-B showed 35.81 MPa and 37 MPa values of compressive strength, respectively, and LSP-B possessed 3.5% more strength than CM. At 28 days, both formulations attained maximum compressive strength showing values of 40.77 MPa and 41.98 MPa. The reported increase further enhanced to a maximum of 11.5% (instead of 5%) while focusing the isolated behavior of microbes with reference to 10% LSP diluted cement formulation. This reveals LSP as quite promising media that adds to the compressive strength of mortar specimens.

3.2. Compressive Strength Regain. Self-healing efficiency was evaluated via regain in compressive strength after precracking in mechanical mode. The compressive strength regain values are plotted in Figure 5 while Figure 6 shows the percentage regain in compressive strength using (1).

Both formulations showed regain in compressive strength. In CM formulation, compressive strength regain is attributed to autogenous healing due to secondary hydration of unhydrated cement grains and carbonation of calcium hydroxide into calcite crystals [42, 43]. While in LSP-B, microbial activity is responsible for regain in compressive strength of specimens. LSP-B exhibited more compressive strength regain than CM at each stage of curing as a result of calcite precipitation induced by microbial actions. The analyzed formulations showed maximum strength regain at 3 days which was around 81% for CM and 88% for LSP-B specimens. At 7 days of curing, compressive strength regain was reduced to 63% and 77% for CM and LSP-B, respectively. Then at 14 and 28 days of curing, similar trend was observed in regained strength with a noticeable

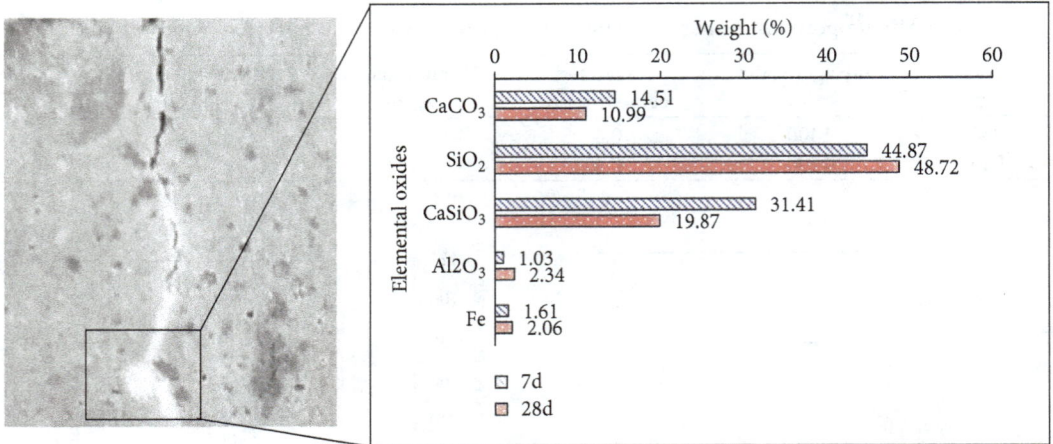

FIGURE 7: EDX spectrum of extracted powder out of healed crack in LSP-B formulation.

reduction to 42% and 50% in CM and LSP-B formulations, respectively.

The decrease in compressive strength regain with increase in the testing age of specimens may be attributed with reduction in microbial activity at later ages. Microbes' survival depends on food and carrier media. The involved phenomena that can justify the reduction in compressive strength regain are as follows: first, the production of microbial calcite ceased due to lack of food, and second, LSP did not sustain pressure at later stages subsequent to densified microstructure of mortar matrix in later ages ensuing in microbial crushing. However, the comparison of results of microbial formulation to referenced CM gives an isolated effect of bio-influenced self-healing process in the regain of mechanical strength. LSP showed higher preservation efficiency in the earlier phase of hydration that reduces in the later part.

3.3. Energy Dispersive X-Ray Analysis.
Energy dispersive X-ray (EDX) spectroscopy is a chemical mode of microstructural analysis. Precracked LSP-B specimens at 7 and 28 days were subjected to EDX spectroscopy after 28 days of healing period to inspect the instigation of possible calcite. The EDX spectrograph of LSP-B samples are given in Figure 7.

The results made the manifestion of calcite precipitation in both the formulations at the age of 7 and 28 days. But, at 7 days, calcite content was more as compared with 28 days. It is further concluded that microbial activity slowed down with the passage of time. However, it remained active till the age of 28 days that renders the investigated media efficient enough to effectively shield *B. subtilis* for relatively longer durations.

3.4. Field Emission Scanning Electron Microscopy.
To further explore the instigated calcite, field emission scanning electron microscopy (FESEM) was used. The FESEM micrograph was carefully sketched along the healed crack to observe any possible signs of calcite precipitation. FESEM micrograph of LSP-B formulations is shown in Figure 8 at the precracking age of 7 and 28 days.

In the presence of *B. subtilis*, calcium lactate converted into calcite, by consuming oxygen, and produced rhombohedra crystals. The reaction activity is shown in (3) [44].

$$CaC_6H_{10}O_6 + 6O_2 \longrightarrow CaCO_3 + 5CO_2 + 5H_2O \qquad (3)$$

Fabrication of calcite crystals can be clearly seen in both the micrographs of cryofractured cementitious formulations at 7 and 28 days of precracking consequent to microbial healing that ensures the persistence of microbes till the age of 28 days.

3.5. Thermogravimetric Analysis.
The thermogravimetric analysis (TGA) is a mode of thermal analysis which is performed in an instrument named as the thermogravimetric analyzer. Mass, time, and temperature are considered basic measurements of TGA. The analyzer continuously measures the mass of the substance while temperature of the sample changes continuously at a constant rate [45]. This change in mass of the substance assists in recognition of different chemical compounds categorized by their decomposition temperature.

TGA was conducted on the white powder scratched from the surface of healed specimens for confirmation of precipitated calcite after 28 days of healing. The decomposition temperature of $CaCO_3$ crystals ranges from 600 to 850°C [46]. The decomposition reaction of $CaCO_3$ is as given in (4), and the TGA result of healed powder is plotted in Figure 9.

$$CaCO_3(s) \longrightarrow CaO(s) + CO_2(g) \qquad (4)$$

It is clearly evident from the curve that major mass loss of 6.2% occurred in the predefined range of 600–850°C that endorses the presence of calcite ensuing in strong microbial action. Hence, it can be inferred that LSP effectively preserved the microbes "*B. subtilis*" up till the formation of cracks in the mortar formulations.

4. Concluding Remarks

LSP is locally available, cost-efficient alternative to immobilize *B. subtilis* in highly alkaline cementitious environment. The addition of microbes immobilized on LSP

(a)   (b)

FIGURE 8: (a) FESEM micrograph of LSP-B specimen. (b) FESEM micrograph of LSP-B specimen precracked at 7 and 28 days.

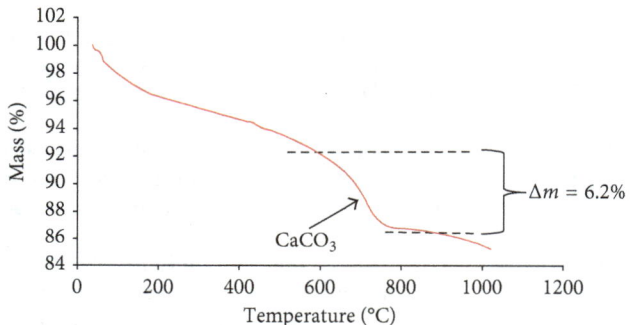

FIGURE 9: TGA curve of extracted powder out of healed crack in LSP-B formulation.

contributed in the enhancement of compressive strength at 3, 7, 14, and 28 days of curing with maximum increment of 5.75% at 14 days comparative to CM formulation. The regained compressive strength gauging the healing efficiency in mechanical mode endorsed the maximum regain at 3 days of precracking age, which declined later on with increased age of specimens. LSP-B formulation exhibited more regained compressive strength at each testing age. Moreover, micrographical analysis through FESEM, chemical analysis by EDX, and thermal analysis via TGA on the powdered specimens from the near vicinity of healed cracks further evidenced visible signs of calcite precipitation consequent to microbial activity. Hence, LSP can be claimed as a promising carrier media for *B. subtilis* ensuring its preservation for relatively longer durations in highly alkaline cementitious environment.

Abbreviations

FESEM: Field emission scanning electron microscopy
EDX: Energy dispersive X-ray
TGA: Thermogravimetry analysis
LSP: Limestone powder
XRF: X-ray florescence
CSH: Calcium-silicate-hydrate.

Conflicts of Interest

The authors declared no potential conflicts of interest with respect to the research, authorship, and/or publication of this article.

Acknowledgments

The authors would like to acknowledge the financial assistance from Higher Education Commission (HEC), Pakistan, and pay their gratitude to the lab staff of Structural Engineering Laboratory in National University of Sciences and Technology (NUST) for their assistance in the execution of work.

References

[1] V. Achal and A. Mukherjee, "A review of microbial precipitation for sustainable construction," *Construction and Building Materials*, vol. 93, pp. 1224–1235, 2015.

[2] H. Schlangen, H. Jonkers, S. Qian, and A. Garcia, "Recent advances on self healing of concrete," in *FraMCos-7: Proceedings of the 7th International Conference on Fracture Mechanics of Concrete and Concrete Structures*, pp. 23–28, Jeju Island, Replic of Korea, May 2010.

[3] K. Arbi, M. Nedeljković, Y. Zuo, and G. Ye, "A review on the durability of alkali-activated fly ash/slag systems: advances, issues, and perspectives," *Industrial & Engineering Chemistry Research*, vol. 55, no. 19, pp. 5439–5453, 2016.

[4] A. Vaysburd and P. Emmons, "How to make today's repairs durable for tomorrow—corrosion protection in concrete repair," *Construction and Building Materials*, vol. 14, no. 4, pp. 189–197, 2000.

[5] H. M. Jonkers, A. Thijssen, G. Muyzer, O. Copuroglu, and E. Schlangen, "Application of bacteria as self-healing agent for the development of sustainable concrete," *Ecological Engineering*, vol. 36, no. 2, pp. 230–235, 2010.

[6] W. Tang, O. Kardani, and H. Cui, "Robust evaluation of self-healing efficiency in cementitious materials–a review," *Construction and Building Materials*, vol. 81, pp. 233–247, 2015.

[7] D. Snoeck, P.-A. Smetryns, and N. De Belie, "Improved multiple cracking and autogenous healing in cementitious materials by means of chemically-treated natural fibres," *Biosystems Engineering*, vol. 139, pp. 87–99, 2015.

[8] M. G. Meharie, J. W. Kaluli, Z. Abiero-Gariy, and N. D. Kumar, "Factors affecting the self-healing efficiency of cracked concrete structures," *American Journal of Applied Scientific Research*, vol. 3, no. 6, p. 80, 2017.

[9] K. Van Tittelboom and N. De Belie, "Self-healing in cementitious materials—A review," *Materials*, vol. 6, no. 6, pp. 2182–2217, 2013.

[10] P. Minnebo, G. Thierens, G. De Valck et al., "A novel design of autonomously healed concrete: towards a vascular healing network," *Materials*, vol. 10, no. 1, p. 49, 2017.

[11] S. Sangadji, "Can self-healing mechanism helps concrete structures sustainable?," *Procedia Engineering*, vol. 171, pp. 238–249, 2017.

[12] R. Chang, S. Kim, S. Lee, S. Choi, M. Kim, and Y. Park, "Calcium carbonate precipitation for CO_2 storage and utilization: a review of the carbonate crystallization and polymorphism," *Frontiers in Energy Research*, vol. 5, p. 17, 2017.

[13] N. De Belie and W. De Muynck, "Crack repair in concrete using biodeposition," in *Proceedings of the International Conference on Concrete Repair, Rehabilitation and Retrofitting (ICCRRR)*, pp. 291-292, Cape Town, South Africa, 2008.

[14] M. Araujo, S. Van Vlierberghe, J. Feiteira et al., "Crosslinkable polyethers as healing/sealing agents for self-healing of cementitious materials," *Materials and Design*, vol. 98, pp. 215–222, 2016.

[15] S. Bhaskar, K. M. A. Hossain, M. Lachemi, G. Wolfaardt, and M. O. Kroukamp, "Effect of self-healing on strength and durability of zeolite-immobilized bacterial cementitious mortar composites," *Cement and Concrete Composites*, vol. 82, pp. 23–33, 2017.

[16] W. Khaliq and M. B. Ehsan, "Crack healing in concrete using various bio influenced self-healing techniques," *Construction and Building Materials*, vol. 102, pp. 349–357, 2016.

[17] M. Elakkiya, D. Prabhakaran, and M. Thirumarimurugan, "Methods of cell immobilization and its applications," *Methods*, vol. 5, no. 4, 2016.

[18] J. Wang, N. De Belie, and W. Verstraete, "Diatomaceous earth as a protective vehicle for bacteria applied for self-healing concrete," *Journal of Industrial Microbiology and Biotechnology*, vol. 39, no. 4, pp. 567–577, 2012.

[19] J. Wang, D. Snoeck, S. Van Vlierberghe, W. Verstraete, and N. De Belie, "Application of hydrogel encapsulated carbonate precipitating bacteria for approaching a realistic self-healing in concrete," *Construction and building materials*, vol. 68, pp. 110–119, 2014.

[20] V. Wiktor and H. M. Jonkers, "Quantification of crack-healing in novel bacteria-based self-healing concrete," *Cement and Concrete Composites*, vol. 33, no. 7, pp. 763–770, 2011.

[21] J. Wang, K. Van Tittelboom, N. De Belie, and W. Verstraete, "Use of silica gel or polyurethane immobilized bacteria for self-healing concrete," *Construction and building materials*, vol. 26, no. 1, pp. 532–540, 2012.

[22] J. Y. Wang, H. Soens, W. Verstraete, and N. De Belie, "Self-healing concrete by use of microencapsulated bacterial spores," *Cement and Concrete Research*, vol. 56, pp. 139–152, 2014.

[23] N. N. T. Huynh, N. M. Phuong, N. P. A. Toan, and N. K. Son, "*Bacillus subtilis* HU58 Immobilized in micropores of diatomite for using in self-healing concrete," *Procedia Engineering*, vol. 171, pp. 598–605, 2017.

[24] H. Chen, C. Qian, and H. Huang, "Self-healing cementitious materials based on bacteria and nutrients immobilized respectively," *Construction and Building Materials*, vol. 126, pp. 297–303, 2016.

[25] J. Zhang, Y. Liu, T. Feng et al., "Immobilizing bacteria in expanded perlite for the crack self-healing in concrete," *Construction and Building Materials*, vol. 148, pp. 610–617, 2017.

[26] O. M. Omar, G. D. A. Elhameed, M. A. Sherif, and H. A. Mohamadien, "Influence of limestone waste as partial replacement material for sand and marble powder in concrete properties," *HBRC Journal*, vol. 8, no. 3, pp. 193–203, 2012.

[27] D. P. Bentz, C. F. Ferraris, S. Z. Jones, D. Lootens, and F. Zunino, "Limestone and silica powder replacements for cement: early-age performance," *Cement and Concrete Composites*, vol. 78, pp. 43–56, 2017.

[28] S. A. Rizwan and T. A. Bier, "Blends of limestone powder and fly-ash enhance the response of self-compacting mortars," *Construction and Building Materials*, vol. 27, no. 1, pp. 398–403, 2012.

[29] M. Zajac, A. Rossberg, G. Le Saout, and B. Lothenbach, "Influence of limestone and anhydrite on the hydration of Portland cements," *Cement and Concrete Composites*, vol. 46, pp. 99–108, 2014.

[30] A. Ipavec, R. Gabrovšek, T. Vuk, V. Kaučič, J. Maček, and A. Meden, "Carboaluminate phases formation during the hydration of calcite-containing Portland cement," *Journal of the American Ceramic Society*, vol. 94, no. 4, pp. 1238–1242, 2011.

[31] Z. Tu, M.-Z. Guo, C. S. Poon, and C. Shi, "Effects of limestone powder on $CaCO_3$ precipitation in CO_2 cured cement pastes," *Cement and Concrete Composites*, vol. 72, pp. 9–16, 2016.

[32] O. Aviam, G. Bar-Nes, Y. Zeiri, and A. Sivan, "Accelerated biodegradation of cement by sulfur-oxidizing bacteria as a bioassay for evaluating immobilization of low-level radioactive waste," *Applied and environmental microbiology*, vol. 70, no. 10, pp. 6031–6036, 2004.

[33] H. Huang, G. Ye, C. Qian, and E. Schlangen, "Self-healing in cementitious materials: materials, methods and service conditions," *Materials and Design*, vol. 92, pp. 499–511, 2016.

[34] P. Anbu, C.-H. Kang, Y.-J. Shin, and J.-S. So, "Formations of calcium carbonate minerals by bacteria and its multiple applications," *SpringerPlus*, vol. 5, no. 1, p. 250, 2016.

[35] E. J. Hayhurst, L. Kailas, J. K. Hobbs, and S. J. Foster, "Cell wall peptidoglycan architecture in Bacillus subtilis," *Proceedings of the National Academy of Sciences*, vol. 105, no. 38, pp. 14603–14608, 2008.

[36] S. K. Ramachandran, V. Ramakrishnan, and S. S. Bang, "Remediation of concrete using micro-organisms," *ACI Materials Journal*, vol. 98, no. 1, pp. 3–9, 2001.

[37] B. Barra, L. Momm, Y. Guerrero, and L. Bernucci, "Characterization of granite and limestone powders for use as fillers in bituminous mastics dosage," *Anais da Academia Brasileira de Ciências*, vol. 86, no. 2, pp. 995–1002, 2014.

[38] S. A. Rizwan, *High-Performance Mortars and Concrete Using Secondary Raw Materials*, Ph.D. thesis, American Concrete Institute (ACI), Farmington Hills, MI, USA, 2006.

[39] M. Li, X. Zhu, A. Mukherjee, M. Huang, and V. Achal, "Biomineralization in metakaolin modified cement mortar to

improve its strength with lowered cement content," *Journal of Hazardous Materials*, vol. 329, pp. 178–184, 2017.

[40] Z. Tan, G. De Schutter, G. Ye, and Y. Gao, "The effect of limestone powder addition on strength of slag blended cement," in *Proceedings of Concrete under Severe Conditions: Environment and Loading (CONSEC-2013)*, pp. 1888–1898, RILEM Publications, Vancouver, BC, Canada, 2013.

[41] S. Türkel and Y. Altuntaş, "The effect of limestone powder, fly ash and silica fume on the properties of self-compacting repair mortars," *Sadhana*, vol. 34, no. 2, pp. 331–343, 2009.

[42] J. Reis, D. Moreira, L. Nunes, and L. Sphaier, "Evaluation of the fracture properties of polymer mortars reinforced with nanoparticles," *Composite Structures*, vol. 93, no. 11, pp. 3002–3005, 2011.

[43] H. M. Jonkers and E. Schlangen, "A two component bacteria-based self-healing concrete," in *Concrete Repair, Rehabilitation and Retrofitting II: 2nd International Conference on Concrete Repair, Rehabilitation and Retrofitting, ICCRRR-2*, pp. 24–26, CRC Press, Cape Town, South Africa, November 2008.

[44] O. Regnault, V. Lagneau, and H. Schneider, "Experimental measurement of portlandite carbonation kinetics with supercritical CO_2," *Chemical Geology*, vol. 265, no. 1-2, pp. 113–121, 2009.

[45] V. Kodur and M. Sultan, "Effect of temperature on thermal properties of high-strength concrete," *Journal of Materials in Civil Engineering*, vol. 15, no. 2, pp. 101–107, 2003.

[46] I. Halikia, L. Zoumpoulakis, E. Christodoulou, and D. Prattis, "Kinetic study of the thermal decomposition of calcium carbonate by isothermal methods of analysis," *European Journal of Mineral Processing and Environmental Protection*, vol. 1, no. 2, pp. 89–102, 2001.

3

The Correlation between Pore Structure and Macro Durability Performance of Road Concrete under Loading and Freeze-Thaw and Drying-Wetting Cycles

Sheng-bo Zhou,[1] Jun-lin Liang,[2] Wei-an Xuan,[1] and Ye Qiu[3]

[1]*Guangxi Key Laboratory of Road Structure and Materials, Guangxi Transportation Research Institute Co., Ltd., Nanning 530007, China*
[2]*School of Civil Engineering & Architecture, Guangxi University, Nanning 530004, China*
[3]*College of Civil and Transportation Engineering, Hohai University, Nanjing 210098, China*

Correspondence should be addressed to Jun-lin Liang; ljl_1217@126.com

Academic Editor: Xiao-Jian Gao

The grey correlation theory and multiple regression method are used to reveal macro performance degradation rules of road concrete under loading and freeze-thaw and drying-wetting cycles; then the correlation between mesoscopic pore structure and residual strength and antifreezing index of concrete is analyzed. Under the freeze-thaw and drying-wetting cycles with 50% loading level, the pore structure parameters that influence concrete strength show the following sequence: fractal dimension > most probable pore size > porosity > less harmful pore. The correlation between strength and pore parameters can be represented with multiple nonlinear equations. A negative correlation is shown between strength and fractal dimension and most probable pore size. Conversely, a positive correlation is shown between strength, porosity, and less harmful pore. Under the freeze-thaw and drying-wetting cycles with 80% loading level, the pore structure parameters that influence concrete strength show another sequence: fractal dimension > porosity > less harmful pore > most probable pore size. The correlation between antifreezing index and pore parameters should be described with multiple linear equations. The relative dynamic elastic modulus shows a positive correlation to most probable pore size, pore surface area, and porosity but a negative correlation to less harmful pore and pore spacing coefficient.

1. Introduction

Establishing the relationship between materials structure and performance has been a central content of materials science research. Professors Wu and Lian [1, 2] propose that the relation between different structure dimensions should be paid attention in research of concrete science and technology. Currently, scholars at home and abroad have carried out mechanical and durability research of concrete based on the mesostructure and microstructure level [3–8]. But the research achievement was rarely obtained about the relationship between pore structure and macro performance. So the fracture mechanics of concrete are urgent to be resolved. The destruction of road concrete is attributable to the interaction of loading and environment temperature-humidity. The performance degradation of concrete would be accelerated when these factors alternate frequently. The failure condition of concrete in seasonal frozen area is studied in this paper. The relationship between mesostructure and macro performance is established which provides theoretical basis for concrete application and performance evaluation of this area.

The mesopore structure of concrete has a significant influence on the macro strength and antifreezing performance [9–12]. The pore structure is composed of several parameters, such as porosity, pore surface area, area median aperture, average aperture, most probable aperture, and pore fractal dimension. The concrete performance is affected by these parameters in different aspects and degrees [13–17]. In order to determine influence degree of every parameter and the main parameter, the grey system theory [18, 19] is needed

TABLE 1: Fatigue test conditions under loading and freeze-thaw and drying-wetting cycles.

Climatic region	Loading	Factors	
		Temperature and humidity	
		Winter	Summer
Alternating cold and drying-wetting region	50%; 80%	Temperature: −18°C~+5°C; humidity: 20%~40%	Temperature: 25°C~35°C; humidity: 60%~80%

as traditional mathematical theory cannot be achieved. Grey correlation analysis is an effective system analysis method in grey system theory. It takes the uncertainty system with small sample and poor information as research object. After data processing, the main factors that influence research object would be found in random factors series. Therefore, it is especially suitable for the analysis of uncertain and complicated problems.

The basic steps of grey correlation analysis to deal with the problem are as follows.

(1) Determine reference sequence that reflects the behavior characteristics of system $X_i = \{x_i(k)|x_i(1), x_i(2), \ldots, x_i(n)|\}$, $i = 0, 1, 2, \ldots, m$, and comparative sequence composed of each factor which affects system behavior $Y_i = \{y_i(k)|y_i(1), y_i(2), \ldots, y_i(n)|\}$, $i = 0, 1, 2, \ldots, m$.

(2) In order to avoid the different data dimension as different physical meaning of each factor, make reference sequence and comparative sequence being dimensionless: $X_i' = \{x_i'(k)|x_i(1)/x_i(1), x_i(2)/x_i(1), \ldots, x_i(n)/x_i(1)|\}$, $i = 0, 1, 2, \ldots, m$; $Y_i' = \{y_i'(k)|y_i(1)/y_i(1), y_i(2)/y_i(1), \ldots, y_i(n)/y_i(1)|\}$, $i = 0, 1, 2, \ldots, m$. Calculate the difference sequence and range according to $\Delta_i(k) = |x_i(k) - y_i(k)|$. Then calculate grey correlation coefficient of reference sequence and comparative sequence based on $\gamma_i = (m + \xi M)/(\Delta_i(k) + \xi M)$, $(\xi = 0.5)$; Finally calculate the correlation degree by $\gamma_{0i} = (1/n)\sum_{k=1}^{n}\gamma_i(k)$.

(3) Sort the correlation degree so as to reflect influence degree of each factor.

2. Test Scheme and Method

2.1. Test Scheme Design. The test loading is 50% and ultimate flexural loading is 80%, which correspond to ordinary traffic and heavy traffic levels. Freeze-thaw and drying-wetting cycles' environment mainly occurs in the temperate monsoon climate region: −18°C~+5°C in winter and 25°C~35°C in summer. The relative humidity is 20%–40% in winter and 60%–80% in summer, which refers to the average humidity distribution given by Gao et al. [20] about five regions of Northeast China and North China, East China and Central China, Northwest China, South China, and Southwest China. The design conditions of concrete durability test under loading and freeze-thaw and drying-wetting cycles are shown in Table 1.

The test scheme is as follows.

First Stage. Firstly, the test beam is loaded 36 thousand times (50% stress level) or 12 thousand times (80% stress level) on the fatigue test machine. Then 50 freeze-thaw cycles (50% stress level) or 25 freeze-thaw cycles (80% stress level) at 4 hours intervals are conducted in environment box, with temperature range of −18°C~+5°C and humidity range of 20%~40% with uniform rate. Take out the samples for second same loading and then place them in the environment box and reset temperature-humidity program: temperature range of 25°C~35°C and humidity range of 60%~80% with control interval at 4 h and a period of a month. At the end, the samples are taken out to test the strength, antifreezing index, and mesopore structure.

Second Stage. On the basis of the first stage, the cumulative test is conducted in accordance with the same procedure as the first phase. At the end, the samples are taken for the tensile strength test and mesostructure test.

Third Stage. On the basis of the second stage, fatigue loading was 18 thousand times (50% stress level) or 6 thousand times (80% stress level). Then 25 freeze-thaw cycles at 4 hours intervals are conducted in environment box under winter conditions. Take out the samples for the second 18 thousand times (50% stress level) or 6 thousand times (80% stress level) loading and then place them in the environment box and set the summer temperature-humidity conditions, with control interval at 4 h and a period of a month. At the end, the samples are taken for the above characterization test.

Fourth Stage. The test is conducted in accordance with the same procedure as the third phase until fatigue destruction. Then the samples are taken for the above characterization test.

2.2. Materials and Test Methods. Cement is Qinling P.O 42.5R; mineral filler is S95 class; fly ash is class I; coarse aggregate with nominal maximum size is 19 mm; fineness modulus of river sand is 2.6; water-reducing rate of high performance water reducer is 26%; water is tap water. 28 d concrete tensile design strength is C1, 4.5 MPa, and C2, 5.0 MPa.

The fatigue test is carried out on a 10 t MTS-810 fatigue testing machine; loading frequency is 10 HZ, loading mode is three-point sine wave, and low and high stress ratio is 0.1. Mesopore structure parameters are measured by AutoPore IV 9510 mercury porosimetry. Strength and antifreezing index test is in accordance with the specification.

3. Results Analysis and Discussion

3.1. Relationship between Pore Structure Parameters and Residual Flexural Strength under Loading and Freeze-Thaw and Drying-Wetting Cycles. The analysis result of grey correlation between pore structure parameters and residual flexural strength under different loading and freeze-thaw and drying-wetting cycles is obtained, as is shown in Table 2. It is known that, in the case of 50% loading level and freeze-thaw and drying-wetting cycles, the sequence of the correlation between pore structure parameters and residual flexural

TABLE 2: The grey correlation degree between pore structure parameters and strength under different loading levels and freeze-thaw and drying-wetting cycles.

Strength	Parameters					
	Porosity/%	Total pore surface area/m²/g	Area median aperture/nm	Average aperture/nm	Most probable pore size/nm	Pore fractal dimension
50% loading						
C1	0.866	0.736	0.531	0.596	0.903	0.845
C2	0.889	0.749	0.549	0.538	0.893	0.850
80% loading						
C1	0.862	0.765	0.643	0.601	0.973	0.926
C2	0.886	0.728	0.670	0.575	0.909	0.910
Strength	Parameters					
	More harmful pore/nm	Harmful pore/nm	Less harmful pore/nm	Harmless pore/nm	Pore spacing coefficient	/
50% loading						
C1	0.777	0.627	0.858	0.815	0.716	/
C2	0.749	0.713	0.884	0.807	0.724	/
80% loading						
C1	0.856	0.771	0.915	0.823	0.843	/
C2	0.796	0.799	0.891	0.789	0.798	/

strength after three cycles is as follows: most probable pore size > porosity > less harmful pore > pore fractal dimension > harmless pore > more harmful pore > total pore surface area > pore spacing coefficient > harmful pore > area median aperture and average aperture. Results showed that the main pore structure parameters that influence the concrete strength are most probable pore size, porosity, less harmful pore, pore fractal dimension, and pore spacing coefficient. Similarly, in the case of 80% loading level and freeze-thaw and drying-wetting cycles, the sequence is as follows: most probable pore size > pore fractal dimension > less harmful pore > porosity > more harmful pore > pore spacing coefficient > harmless pore > harmful pore > total pore surface area > area median aperture and average aperture. The main pore structure parameters that influence the concrete strength are most probable pore size, pore fractal dimension, less harmful pore, porosity, and more harmful pore.

According to the results of grey correlation analysis, when establishing the relationship between strength and pore structure parameters under different loading and freeze-thaw and drying-wetting cycles, the four parameters of most probable pore size, porosity, less harmful pore, and pore fractal dimension can be chosen as the main influencing factors. Formulas (1)~(4) are obtained by multiple linear regression analysis and multiple nonlinear regression analysis. When the loading level is 50%, a higher accuracy can be displayed using multiple nonlinear equation with the correlation coefficient of 0.970. A negative correlation is shown between strength and fractal dimension and most probable pore size. Conversely, a positive correlation is shown between strength and porosity and less harmful pore. The sequence according to influence degree is as follows: fractal dimension > most probable pore size > porosity > less harmful pore. When the loading level is 80%, the 0.999 correlation coefficient can be displayed, using both multiple linear equations and multiple nonlinear equations. A negative correlation is shown between strength and less harmful pore. Conversely, a positive correlation is shown between strength and fractal dimension, most probable pore size, and porosity. The sequence according to influence degree is as follows: fractal dimension > porosity > less harmful pore > most probable pore size.

$$\frac{\sigma}{\sigma_0} = 8.310 + 0.647 \times \frac{P_{lh}}{P_{lh0}} - 7.432 \times \frac{F_p}{F_{p0}} - 1.633 \times \frac{R_p}{R_{p0}} + 1.099 \times \frac{P_g}{P_{g0}}$$

(1)

(multiple linear regression at 50% loading levels, $R = 0.931$),

$$\frac{\sigma}{\sigma_0} = 0.977 \times \left(\frac{P_{lh}}{P_{lh0}}\right)^{0.986} \times \left(\frac{F_p}{F_{p0}}\right)^{-10.457} \times \left(\frac{I_p}{R_{p0}}\right)^{-2.637} \times \left(\frac{P_g}{P_{g0}}\right)^{1.036}$$

(2)

(multiple nonlinear regression at 50% loading levels, $R = 0.970$),

TABLE 3: Grey correlation degree between pore structure parameters and antifreezing index under different loading levels and freeze-thaw and drying-wetting cycles.

Strength	Parameters					
	Porosity/%	Total pore surface area/m^2/g	Area median aperture/nm	Average aperture/nm	Most probable pore size/nm	Pore fractal dimension
50% loading						
C1	0.905	0.838	0.533	0.585	0.782	0.733
C2	0.930	0.885	0.587	0.569	0.787	0.785
80% loading						
C1	0.842	0.896	0.579	0.561	0.749	0.731
C2	0.819	0.943	0.624	0.563	0.737	0.737
Strength	Parameters					
	More harmful pore/nm	Harmful pore/nm	Less harmful pore/nm	Harmless pore/nm	Pore spacing coefficient	/
50% loading						
C1	0.692	0.600	0.835	0.931	0.655	/
C2	0.710	0.699	0.807	0.957	0.704	/
80% loading						
C1	0.721	0.644	0.762	0.873	0.695	/
C2	0.690	0.700	0.739	0.900	0.691	/

$$\frac{\sigma}{\sigma_0} = 0.382 - 0.279 \times \frac{P_{lh}}{P_{lh0}} + 0.440 \times \frac{F_p}{F_{p0}} + 0.025 \times \frac{R_p}{R_{p0}} + 0.433 \times \frac{P_g}{P_{g0}}$$

(3)

(multiple linear regression at 80% loading levels, $R = 0.999$),

$$\frac{\sigma}{\sigma_0} = \left(\frac{P_{lh}}{P_{lh0}}\right)^{-0.701} \times \left(\frac{F_p}{F_{p0}}\right)^{-0.547} \times \left(\frac{R_p}{R_{p0}}\right)^{0.564} \times \left(\frac{P_g}{P_{g0}}\right)^{0.411}$$

(4)

(multiple nonlinear regression at 80% loading levels, $R = 0.999$).

3.2. Relationship between Pore Structure Parameters and Antifreezing Index under Loading, Freeze-Thaw, and Drying-Wetting Cycles. The analysis result of grey correlation between pore structure parameters and antifreezing index under different loading and freeze-thaw and drying-wetting cycles is shown in Table 3. The sequence of grey correlation degree is as follows: harmless pore > porosity > total pore surface area > less harmful pore > most probable pore size > pore fractal dimension > more harmful pore > pore spacing coefficient > harmful pore > area median aperture and average aperture, in the case of 50% loading level and freeze-thaw and drying-wetting cycles. Similarly, in the case of 80% loading level and freeze-thaw and drying-wetting cycles, the sequence is as follows: total pore surface area > harmless pore > porosity > less harmful pore > most probable pore size > pore fractal dimension > more harmful pore > harmful pore and average aperture > area median aperture > pore spacing coefficient.

According to the results of grey correlation analysis and considering the importance of less harmful pore and pore spacing coefficient, when the relationship between antifreezing index and pore structure parameters is established under different loading and freeze-thaw and drying-wetting cycles, the five parameters, porosity, total pore surface area, less harmful pore, most probable pore size, and pore spacing coefficient, are chosen. Formulas (5)~(8) are obtained by regression analysis. Under these two loading levels and freeze-thaw and drying-wetting cycles, a higher accuracy can be displayed using multiple linear equation with the correlation coefficient of 0.999. The relative dynamic elastic modulus shows a positive correlation to most probable pore size, total pore surface area, and porosity but a negative correlation to less harmful pore and pore spacing coefficient. When the loading level is 50%, the sequence of pore structure parameters' influence on relative dynamic elastic modulus is as follows: pore spacing coefficient > total pore surface area > porosity > most probable pore size > less harmful pore. Meanwhile, when the loading level is 80%, the sequence is as follows: most probable pore size > porosity > pore spacing coefficient > less harmful pore > total pore surface area.

$$\frac{E}{E_0} = -0.336 + 0.291 \times \frac{R_p}{R_{p0}} + 0.336 \times \frac{S_g}{S_{g0}} - 0.048 \times \frac{P_{lh}}{P_{lh0}} + 0.296 \times \frac{P_g}{P_{g0}} - 0.365 \frac{L_p}{L_0} \tag{5}$$

(multiple linear regression at 50% loading levels, $R = 0.999$),

$$\frac{E}{E_0} = 0.998 \times \left(\frac{R_p}{R_{p0}}\right)^{0.146} \times \left(\frac{S_g}{S_{g0}}\right)^{0.267} \times \left(\frac{P_{lh}}{P_{lh0}}\right)^{-0.233} \times \left(\frac{P_g}{P_{g0}}\right)^{0.329} \times \left(\frac{L_p}{L_0}\right)^{-0.891} \tag{6}$$

(multiple nonlinear regression at 50% loading levels, $R = 0.985$),

$$\frac{E}{E_0} = -0.184 + 1.170 \times \frac{R_p}{R_{p0}} + 0.132 \times \frac{S_g}{S_{g0}} - 0.203 \times \frac{P_{lh}}{P_{lh0}} + 0.521 \times \frac{P_g}{P_{g0}} - 0.461 \frac{L_p}{L_0} \tag{7}$$

(multiple linear regression at 80% loading levels, $R = 0.999$),

$$\frac{E}{E_0} = \left(\frac{R_p}{R_{p0}}\right)^{4.612} \times \left(\frac{S_g}{S_{g0}}\right)^{0.058} \times \left(\frac{P_{lh}}{P_{lh0}}\right)^{-1.633} \times \left(\frac{P_g}{P_{g0}}\right)^{2.134} \times \left(\frac{L_p}{L_0}\right)^{-1.685} \tag{8}$$

(multiple nonlinear regression at 80% loading levels, $R = 0.998$).

4. Conclusion

(1) Under the ordinary traffic loading and freeze-thaw and drying-wetting cycles, the main pore structure parameters that influence concrete strength consist of most probable pore size, porosity, less harmful pore, pore fractal dimension, and pore spacing coefficient. Meanwhile, under the heavy traffic loading and freeze-thaw and drying-wetting cycles, the main pore structure parameters are most probable pore size, pore fractal dimension, less harmful pore, porosity, and more harmful pore. The reason is the stress concentration and stress diffusion process caused by superposition of multiple fields, along with the pore structure of the compression, and splitting effect occupied the leading position. The loading affects the pore size of the structure. As the loading level increased from 50% to 80%, the pore fractal dimension (pore complexity) has more influence on the stress diffusion than porosity. The amount of pore nucleation is only the inducing factor of the rapid development of concrete damage, but for the fracture of concrete under overloading, the effect of pore fractal dimension is more significant.

(2) Under the ordinary traffic loading and freeze-thaw and drying-wetting cycles, the correlation between strength and pore parameter should be represented with multiple nonlinear equation. A negative correlation is shown between strength and fractal dimension and most probable pore size. Conversely, a positive correlation is shown between strength and porosity and less harmful pore. The sequence according to influence degree is as follows: fractal dimension > most probable pore size > porosity > less harmful pore. Meanwhile, under the heavy traffic loading and freeze-thaw and drying-wetting cycles, a same high accuracy can be displayed using multiple linear equations and multiple nonlinear equations. A negative correlation is shown between strength and less harmful pore. Conversely, a positive correlation is shown between strength and fractal dimension, most probable pore

size, and porosity. The sequence according to influence degree is as follows: fractal dimension > porosity > less harmful pore > most probable pore size.

(3) Under freeze-thaw and drying-wetting cycles, under the ordinary traffic or heavy traffic loading, there is a higher correlation between antifreezing index and the five pore structure parameters: porosity, total pore surface area, less harmful pore, most probable pore size, and pore spacing coefficient.

(4) Under these two loading levels and freeze-thaw and drying-wetting cycles, the correlation between antifreezing index and pore parameters should be described with multiple linear equation. The relative dynamic elastic modulus shows a positive correlation to most probable pore size, total pore surface area, and porosity but a negative correlation to less harmful pore and pore spacing coefficient. Under the ordinary traffic, the sequence of pore structure parameters' influence on relative dynamic elastic modulus is as follows: pore spacing coefficient > total pore surface area > porosity > most probable pore size > less harmful pore. Meanwhile, under the heavy traffic, the sequence is as follows: most probable pore size > porosity > pore spacing coefficient > less harmful pore > total pore surface area.

Disclosure

Sheng-bo Zhou (1979-), Ph.D., senior engineer, mainly engaged in road cement concrete structure and material durability research; Jun-lin Liang, doctoral tutor, mainly engaged in highway engineering structural performance research.

Conflicts of Interest

The authors declare that they have no conflicts of interest.

Acknowledgments

This research was supported by Guangxi Natural Science Foundation (Grant no. GUIKE AC16380109), the National Natural Science Foundation of China (51278059), and Opening Fund Project of Guangxi Key Lab of Road Structure and Materials (Project no. 2016gxjgclkf-004).

References

[1] Z.-W. Wu, "Development prospects and problems of high performance concrete (HPC)," *Architecture Technology*, vol. 29, no. 1, pp. 8–13, 1998.

[2] Z.-W. Wu and H.-Z. Lian, *High Performance Concrete*, China Railway Publishing House, Beijing, China, 1999.

[3] P. K. Mehta and P. J. M. Monteiro, *Concrete Structure, Properties and Materials*, Prentice Hall, Upper Saddle River, NJ, USA, 1986.

[4] M. Y. Du, X. Y. Jin, H. L. Ye, N. G. Jin, and Y. Tian, "A coupled hygro-thermal model of early-age concrete based on micropore structure evolution," *Construction and Building Materials*, vol. 111, pp. 689–698, 2016.

[5] S. Goel, S. P. Singh, and P. Singh, "Flexural fatigue strength and failure probability of Self Compacting Fibre Reinforced Concrete beams," *Engineering Structures*, vol. 40, pp. 131–140, 2012.

[6] J. Wawrzeńczyk and W. Kozak, "Protected paste volume (PPV) as a parameter linking the air-pore structure in concrete with the frost resistance results," *Construction and Building Materials*, vol. 112, pp. 360–365, 2016.

[7] X.-C. Pu and Y.-W. Wang, "Study on pore structure and interface structure of super high strength and high performance concrete," *China Concrete and Cement Products*, vol. 3, pp. 9–13, 2004.

[8] W.-B. Xu, Z.-H. Shui, J.-T. Ma, and W. Chen, "Carbonation of Fly Ash Concrete Investigated with micro hardness analysis," *Bulletin of the Chinese Ceramic Society*, vol. 30, no. 1, pp. 7–12, 2011.

[9] A. G. Graeff, K. Pilakoutas, K. Neocleous, and M. V. N. N. Peres, "Fatigue resistance and cracking mechanism of concrete pavements reinforced with recycled steel fibres recovered from post-consumer tyres," *Engineering Structures*, vol. 45, no. 10, pp. 385–395, 2012.

[10] T. Matusinović, J. Šipušić, and N. Vrbos, "Porosity-strength relation in calcium aluminate cement pastes," *Cement and Concrete Research*, vol. 33, no. 11, pp. 1801–1806, 2003.

[11] A. U. Ozturk and B. Baradan, "A comparison study of porosity and compressive strength mathematical medals with image analysis," *Cement and Concrete Research*, vol. 43, no. 12, pp. 974–979, 2008.

[12] I. Odler and M. Rossler, "Investigations on the relationship between porosity, structure and strength of hydrated Portland cement pastes II. Effect of pore structure and of degree of hydration," *Cement and Concrete Research*, vol. 15, no. 3, pp. 401–410, 1985.

[13] R. Kumar and B. Bhattacharjee, "Porosity, pore size distribution and in situ strength of concrete," *Cement and Concrete Research*, vol. 33, no. 1, pp. 155–164, 2003.

[14] X.-L. Zhao, J. Wei, and H.-J. Ba, "Relationship between frost durability and pore structure of high performance concrete," *Industrial Construction*, vol. 33, no. 8, pp. 5–27, 2003.

[15] L. Chen, J.-H. He, and Y.-N. Zhao, "Relationship between pore structure of concrete and its frost durability degradation," *Communications Standardization*, vol. 2, pp. 70–75, 2009.

[16] S.-P. Zhang, M. Deng, J.-H. Wu, and M.-S. Tang, "Effect of pore structure on the frost resistance of concrete," *Journal of Wuhan University of Technology*, vol. 30, no. 6, pp. 56–59, 2008.

[17] Z. Giergiczny, M. A. Glinicki, M. Sokołowski, and M. Zielinski, "Air void system and frost-salt scaling of concrete containing slag-blended cement," *Construction and Building Materials*, vol. 23, no. 6, pp. 2451–2456, 2009.

[18] J.-L. Deng, *Gray Prediction and Decision*, HuaZhong University of Science and Technology Publishing House, Wuhan, China, 1989.

[19] M.-X. Jiang, *Grey Theory in the Study of Concrete Durability*, Nanjing University of Science & Technology, Nanjing, China, 2012.

[20] H. Gao, N.-Q. Weng, G. Sun et al., "Distribution feature of meteorology parameter of upper air of different areas in China," *Journal of Atmospheric and Environmental Optics*, vol. 7, no. 2, pp. 101–107, 2012.

4

Impact of Plastic Hinge Properties on Capacity Curve of Reinforced Concrete Bridges

Nasim Shatarat,[1] Mutasem Shehadeh,[2] and Mohammad Naser[3]

[1]*Civil Engineering Department, The University of Jordan, Amman 11942, Jordan*
[2]*Department of Mechanical Engineering, American University of Beirut, Beirut, Lebanon*
[3]*The Hashemite University of Jordan, Zarqa, Jordan*

Correspondence should be addressed to Nasim Shatarat; n.shatarat@ju.edu.jo

Academic Editor: Jun Liu

Pushover analysis is becoming recently the most practical tool for nonlinear analysis of regular and irregular highway bridges. The nonlinear behaviour of structural elements in this type of analysis can be modeled through automated-hinge or user-defined hinge models. The nonlinear properties of the user-defined hinge model for existing highway bridges can be determined in accordance with the recommendations of the Seismic Retrofit Manual by the Federal Highway Administration (FHWA-SRM). Finite element software such as the software SAP2000 offers a simpler and easier approach to determine the nonlinear hinge properties through the automated-hinge model which are determined automatically from the member material and cross section properties. However, the uncertainties in using the automated-hinge model in place of user-defined hinge model have never been addressed, especially for existing and widened bridges. In response to this need, pushover analysis was carried out for four old highway bridges, of which two were widened using the same superstructure but with more attention to seismic detailing requirements. The results of the analyses showed noticeable differences in the capacity curves obtained utilizing the user-defined and automated-hinge models. The study recommends that bridge design manuals clearly ask bridge designers to evaluate the deformation capacities of existing bridges and widened bridges using user-defined hinge model that is determined in accordance with the provisions of the FHWA-SRM.

1. Introduction

Several methods are available to capture the seismic behaviour of buildings and highway bridges. These methods range from simple equivalent static analysis to complex nonlinear dynamic analysis. Nonlinear time-history analysis constitutes the most reliable approach to estimate seismic behaviour because it can realistically predict the deformation demand on and capacity of structures, especially for irregular ones. However, complexities in the application of this method limit its use by practicing engineers [1]. The nonlinear static procedure often called "Pushover analysis" appears therefore as an interesting alternative approach due to its simplicity, yet ensuring reasonably accurate results [2].

Pushover analysis is not a recent development and its origin traces back to 1970s [3]. The validity and applicability of pushover analysis to seismic assessment of buildings

and highway bridges have been extensively investigated in literature [4–13]. Currently, pushover analysis is a very common method of analysis among the structural engineering profession and researchers and is recommended by most guidelines and codes, such as in FEMA 273 [14], ATC-40 [15], FEMA 356 [16], Eurocode 8 [17], FEMA-440 (ATC-55) [18], and ASCE/SEI 41-06 Standard [19].

In pushover analysis, the results depend on the approach used to define the plastic hinges, whether it is lumped or distributed plasticity model [2]. Concentrated plasticity is the most commonly used approach for estimating the deformation capacity in the seismic codes, manuals, and structural analysis software [2, 20]. However, a proper definition of the concentrated plastic hinge model depends on many factors such as mechanical properties of longitudinal and transverse reinforcement, reinforcement details, reinforcement ratio, concrete compressive strength, cross-sectional shape, axial force level, level of confinement, plastic hinge length, and possible local failure mechanisms within the plastic hinge zone [21–23]. Practically, most often the nonlinear properties recommended by FEMA-356 [16], ATC-40 [15], and Caltrans [24] documents are used to describe the deformation capacity of the plastic hinge without any consideration to possible local failure mechanisms due to convenience and simplicity [20].

Many commercial software programs are available to perform pushover analysis. A survey on engineering firms showed that the software SAP2000 [25] is the most used software by bridge engineers for nonlinear static analysis of highway bridges [2]. In SAP2000, the nonlinear behaviour of structural elements for pushover analysis can be modeled through automated-hinge or user-defined hinge models. The nonlinear properties of the user-defined hinge model for existing highway bridges can be determined in accordance with the recommendations of the FHWA-SRM [26], which provides the most current state-of-practice for evaluating the seismic vulnerability of old and existing highway bridges. The document demonstrates detailed procedures for calculating the curvature capacity of the structural elements based on potential local failure mechanisms within the plastic hinge. On the other hand, the nonlinear properties of the automated-hinge model may be based on FEMA-356, Caltrans specifications, or can be determined automatically from the member material and cross section properties.

The literature review showed that very limited research is available on the possible differences in the capacity curve when using user-defined versus automated-hinge models. Inel and Ozmen [20] investigated the influence of different plastic hinge properties on the capacity curve of four- and seven-storey reinforced concrete buildings. The results of their study showed that the improper use of default-hinge model could lead to displacement capacities that might be unreasonable. The authors concluded that the user-defined hinge model is better than the default-hinge model as it reflects nonlinear behaviour compatible with the element properties. It is worth noting that this study was explicitly oriented to reinforced concrete buildings and that default-hinge properties were defined based on FEMA-356. Furthermore, user-defined hinge properties were obtained from moment-curvature analysis based on extreme compression fiber reaching the ultimate concrete compressive strain, as determined by Priestley et al. [1], or the longitudinal steel reaching a tensile strain of 50% of the ultimate strain capacity.

Shatarat [27] studied the effect of plastic hinge properties on the capacity curve of highway bridges. A bridge that was built in the 1940s was considered for the purpose of the study. The curvature capacity of the plastic hinge zone was controlled by buckling of longitudinal rebars. Pushover analysis was carried out for the bridge in one direction only using user-defined and automated plastic hinge models. The results of the study showed a difference in the capacity curve when using the two different models. However, the study was limited to one bridge with one type of local failure mechanism of the plastic hinge zone.

Due to the growing use of pushover analysis in seismic assessment of old highway bridges, there is a need to assess the suitability of utilizing automated plastic hinge properties, which is based on material and section properties, as an alternative to user-defined plastic hinge properties that are determined based on FHWA-SRM for obtaining bridge capacity curves utilizing the software SAP2000. This paper addresses that need by examining the capacity curves of four highway bridges that were built in the 1960s, of which two were widened in 1980s, using two alternatives in defining the properties of the plastic hinge zone. The four bridges were selected to represent old bridges and widened bridges and to cover all types of possible local failure mechanisms within the plastic hinge zone.

2. Plastic Hinge Local Failure Mechanisms

According to the FHWA-SRM [26], the plastic curvature capacity of a reinforced concrete element should be based on the controlling limit state of the potential plastic hinge zone, shear strength of the member, and strength of the joint. Shear strength of the members and the joints is not addressed in this study. A complete discussion of these limit states is covered in the FHWA-SRM [26]. However, a summary of the equations corresponding to each limit state is given in Table 1.

3. Pushover Analysis

3.1. User-Defined Plastic Hinge Properties. For user-defined hinge properties, moment-curvature relationship and the axial-moment interaction surface of the potential plastic hinge are determined by the user and input into the software SAP2000. A minimum of five axial load levels are required to define the interaction surface. An elastic perfectly plastic moment-curvature relationship is established for each axial load level based on the controlling plastic curvature limit state that is determined in accordance with the FHWA-SRM [26]. For axial load levels other than the five values used to create the interaction surface, the software SAP2000 uses linear interpolation to determine the plastic hinge curvature. Plastic hinge length and location are determined in accordance with the FHWA-SRM [26] and are input by the user. MathCAD sheets [28] were developed to handle the large amount of calculations under this part.

3.2. Automated Plastic Hinge Properties. For this model, the moment-curvature relationship and the axial-moment interaction surface of the potential plastic hinge are determined by the software SAP2000. The parameters required for generating moment-curvature relationship and interaction surface compromise the definition of the column cross section geometry, longitudinal and transverse reinforcement details, and unconfined concrete and steel stress-strain curve parameters. Unconfined stress-strain parameters are set based on the expected concrete strength and strain levels as required in the AASHTO guide specifications for LRFD Seismic Bridge Design [29]. Longitudinal and transverse stress-strain parameters are set based on the expected strengths following the models suggested by Caltrans specifications. Mander et al. [30] confined concrete stress-strain curve of the column

TABLE 1: Curvature limit states.

Failure criteria	Plastic curvature (ϕ_p)
Compression failure of unconfined concrete	$\phi_p = \dfrac{\xi_{cu}}{c} - \phi_y$
Compression failure of confined concrete	$\phi_p = \dfrac{\varepsilon_{cu}}{(c - d'')} - \phi_y$
Buckling of longitudinal bars	$\phi_p = \dfrac{2 f_y}{E_s (c - d')} - \phi_y \quad 6 d_b < s < 30 d_b$
Fracture of the longitudinal reinforcement	$\phi_p = \dfrac{\varepsilon_{s\,max}}{(d - c)} - \phi_y$
Low cycle fatigue of longitudinal reinforcement	$\phi_p = \dfrac{2\varepsilon_{ap}}{D'}$
Failure in the lap-splice zone	$\phi_p = (\mu_{lap\phi} + 7)\phi_y$

ξ_{cu} is the ultimate concrete compression strain for unconfined concrete and is limited to 0.005; c is the distance from the extreme compression fiber to the neutral axis; ϕ_y is the yield curvature given by the following equation: $\phi_y = 2\varepsilon_y / D'$; ε_y is equal to the expected yield strength of the longitudinal reinforcement f_{ye} divided by the modulus of elasticity; D' is the length between the center lines of the transverse reinforcement; d'' is distance measured from centerline of the perimeter stirrup to the extreme compression fiber of the cover concrete; ε_{cu} is the ultimate strain of the confined concrete and is determined by $\varepsilon_{cu} = 0.005 + 1.4\rho_s f_{yh}\varepsilon_{su} / f'_{cc}$; ε_{su} is strain corresponding to the maximum stress of the transverse rebars; f_{yh} is yield stress of the transverse steel; ρ_s is volumetric ratio of transverse steel; f'_{cc} is confined concrete strength; d' is the distance from the extreme compression fiber to the centroid of the nearest compression rebars; ε_{smax} is limited to a value of 0.10 or less and d is the effective depth of the cross section; ε_{ap} is plastic strain amplitude, as given by $\varepsilon_{ap} = 0.08(2N_f)^{-0.5}$; N_f is given by $N_f = 3.5(T_n)^{-1/3}$ provided that: $2 \le N_f \le 10$; T_n is natural period of vibration of the bridge; l_{lap} is the length of the lap-splice that is provided; l_s is the lap-splice length $l_s = 0.4(f_{ye}/\sqrt{f'_{ce}})d_b$; L_p is plastic hinge length. $L_p = [0.08L + 4400\varepsilon_y d_b]$ and $L_p = l_{lap}$ if l_{lap} is smaller than l_s; f'_{ce} is the expected strength of concrete surrounding the lap-splice zone; L is the shear length of the member; $\mu_{lap\phi}$ is the curvature ductility at the initial breakdown of bond in the lap-splice zone.

core is consequently generated based on the aforementioned parameters. The software divides the cross section into fiber elements to generate the moment-curvature relationship. The plastic hinge length and its relative location are taken identical to user-defined hinge properties.

4. Description of Selected Bridges

A preliminary evaluation of plastic hinge properties was performed on a group of old bridges that were built in early 1960s with little attention to seismic forces and reinforcement details. This step helped in selecting four candidates that would have different local failure mechanisms within the potential plastic hinge zones. Two of the selected bridges were widened with same superstructure; however the substructure received more attention to seismic detailing requirements. The following is a description of the selected bridges.

4.1. Bridge #1-W Description. Bridge #1-W was originally built in 1961 and consists of four simply supported spans supported on three intermediate piers. The original superstructure consisted of precast concrete girders with span lengths of 60 ft, 98 ft, 98 ft, and 80 ft (3.28 ft = 1 m). The original roadway width was 27 ft with a 6 ft wide sidewalk. The intermediate piers are supported on three 3-ft circular columns that are connected through a beam of 4.5 ft height. The intermediate piers are constructed on isolated footings, while the bridge abutments are seat type abutments supported on concrete piles. In 1983, the roadway was widened by 45 ft to an overall width of 72 ft and the sidewalk was expanded by 6 ft, to an overall width of 12 ft. At each intermediate pier, the new precast concrete girders are supported on four columns, each having a diameter of 3 ft. Unlike the original columns,

longitudinal reinforcement lap splices were removed from the plastic hinge zone and the transverse reinforcement was changed to spiral reinforcement with a larger bar size. Table 2 shows columns' geometric properties, concrete strength, and reinforcement details.

4.2. Bridge #2-E Description. Bridge #2-E consists of three simply supported spans totaling 213 ft in length, with a 22-ft roadway width. The superstructure consists of precast concrete girders supported by 45-degrees skewed intermediate piers that are comprised of two circular 3-ft diameter columns constructed on footings with timber piles. The bridge abutments are seat type abutments founded on precast concrete piles. Poor seismic detailing of the column ends showed the existence of a lap splice of the longitudinal reinforcement within the potential plastic hinge zone with a length of 40 in. Also, the transverse reinforcement consisted of #3 hoop rebars that are spaced at 12.0 in. Table 3 shows columns' geometric properties, concrete strength, and reinforcement details.

4.3. Bridge #3-E Description. Bridge #3-E consists of a 4.5-ft deep multicell cast-in-situ reinforced concrete box girder with spans of 47, 73, 73, and 47 feet. The bridge roadway width is 49 ft with two 6-ft wide sidewalks. The intermediate Piers 2 and 4 are comprised of four square 2.5 ft columns, while intermediate Pier 3 is comprised of four rectangular 2.0 ft by 2.5 ft columns. The bridge abutments are spill-through type abutments with four square 2.5 ft columns hinged at the top and bottom in the bridge longitudinal direction. The columns of the abutments and the intermediate Piers are founded on square spread footings. Column longitudinal reinforcement was spliced in the potential plastic hinge zone with poor

TABLE 2: Bridge #1-W columns' geometric properties and reinforcement details*.

	Pier #2		Pier #3		Pier #4	
	Old	Widened	Old	Widened	Old	Widened
Column						
Clear length (ft)	28.6	25.8	32.2	29.2	26.8	24.2
Shape	Circular	Circular	Circular	Circular	Circular	Circular
Dimensions (in.)	36	36	36	36	36	36
Compressive strength (ksi)	4.0	4.0	4.0	4.0	4.0	4.0
Longitudinal reinforcement						
Size and number	12#9	12#9	12#9	12#9	12#9	12#9
Yield strength (ksi)	40	60	40	60	40	60
Transverse reinforcement						
Type	Hoop	Spiral	Hoop	Spiral	Hoop	Spiral
Size and number	#3	#4	#3	#4	#3	#4
Spacing (in.)	12.0	3.3	12.0	3.3	12.0	3.3
Lap splice						
Within plastic hinge	Yes	No	Yes	No	Yes	No
Length (in.)	50.0	—	50.0	—	50.0	—

*(1 in. = 25.4 mm; 1 ksi = 6.89 MPa).

TABLE 3: Bridge #2-E columns' geometric properties and reinforcement details.

	Pier #2	Pier #3
Column		
Clear length (ft)	20.75	19.50
Shape	Circular	Circular
Dimensions (in.)	36	36
Compressive strength (ksi)	4.0	4.0
Longitudinal reinforcement		
Size and number	12#9	12#9
Yield strength (ksi)	40	40
Transverse reinforcement		
Type	Hoop	Hoop
Size and number	#3	#3
Spacing (in.)	12.0	12.0
Lap splice		
Within plastic hinge	Yes	Yes
Length (in.)	40.0	40.0

transverse confinement that is comprised of a single #3 hoop spaced at 12 in on centre. Table 4 shows columns' geometric properties, concrete strength, and reinforcement details.

4.4. Bridge #4-W Description. Bridge #4-W consists of three simply supported spans, totaling 260 ft in length. The roadway width was 44 ft before widening. The superstructure is comprised of precast concrete girders, with span lengths of 78 ft, 93 ft, and 89 ft. Each intermediate pier is comprised of five 3-ft diameter circular columns supported by single footings. The bridge abutments are seat type abutments

founded on spread footings. The roadway was widened by 23 ft to an overall width of 67 ft. This bridge widening included the addition of precast concrete girders and two 3-ft diameter columns founded on 4.5-ft diameter shafts at each intermediate bent. The original and new column longitudinal reinforcements were spliced within the potential plastic hinge zone. However, column transverse confinement was improved through the use of #4 spiral reinforcement spaced at 3.0 in. on centre. Table 5 shows columns' geometric properties, concrete strength, and reinforcement details.

5. Modeling of the Selected Bridges

A three-dimensional spine-type model was created for each bridge, as shown in Figure 1 for Bridge #2-E. The superstructure and pier elements are represented by frame elements that pass through their centroid. The columns are split into three frame elements as required by Section 5.4.3 of the AASHTO Guide Specifications for LRFD Seismic Bridge Design. Effective moments of inertia were used to reflect concrete cracking and reinforcement yielding, as provided in Table 7-1 of the FHWA-SRM [26]. Gross section properties were used for the elements representing the super structure, the cross beam, and the foundation as they are assumed to behave elastically. Figure 2 shows a typical bridge pier, the frame elements, and their associated stiffness. Spread footings were represented by spring elements which were determined utilizing the method for spread footings outlined in the FHWA-SRM [26]. L-pile software [31] was used to generate equivalent linear-elastic springs for the piles.

A modal analysis was performed to identify fundamental natural periods and mode shapes. Pushover analysis was then carried out for each pier individually and independently in

TABLE 4: Bridge #3-E columns' geometric properties and reinforcement details.

	Pier #2	Pier #3	Pier #4
Column			
Clear length (ft)	22.5	22.0	21.5
Shape	Square	Rectangular	Square
Dimensions (in.)	30	Width = 30 Depth = 24	30
Compressive strength (ksi)	4.0	4.0	4.0
Longitudinal reinforcement			
Size and number	8#10	8#9	8#10
Yield strength (ksi)	40	40	40
Transverse reinforcement			
Type	Hoop	Hoop	Hoop
Size and number	1#3	1#3	1#3
Spacing (in.)	12.0	12.0	12.0
Lap splice			
Within plastic hinge	Yes	Yes	Yes
Length (in.)	50.0	44.0	50.0

TABLE 5: Bridge #4-W columns' geometric properties and reinforcement details.

	Pier #2 and Pier #3	
	Old	Widened
Column		
Clear length (ft)	26.0	26.0
Shape	Circular	Circular
Dimensions (in.)	36	36
Compressive strength (ksi)	4.0	4.0
Longitudinal reinforcement		
Size and number	12#9	13#10
Yield strength (ksi)	40	60
Transverse reinforcement		
Type	Hoop	Spiral
Size and number	#3	#4
Spacing (in.)	12.0	3.0
Lap splice		
Within plastic hinge	Yes	Yes
Length (in.)	40	40

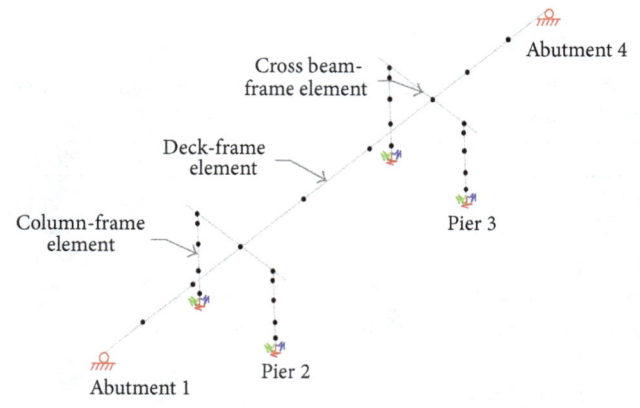

FIGURE 1: Three-dimensional model of Bridge #2-E.

both the longitudinal and the transverse direction, as per the requirements of the FHWA-SRM [26].

6. Approach for Studying the Effect of Plastic Hinge Properties

Accurate determination of the nonlinear capacity curve of a highway bridge depends on the properties of the plastic hinge zone. Typically, design documents and seismic manuals proposed plastic hinge properties based on mechanical properties of longitudinal and transverse reinforcement, reinforcement details, reinforcement ratio, concrete compressive strength, cross-sectional shape, axial force level, and level of confinement. The FHWA-SRM [26] is the only document to include the effect of possible local failure mechanisms in the hinge zone on the properties of the hinge model. The following summarizes the approach used in this study to determine the capacity curve based on user-defined and automated plastic hinge properties:

(1) A three-dimensional model was created to determine the natural periods of the selected bridges.

(2) For user-defined plastic hinge properties, the curvature capacity of each limit state for each pier column at three axial load levels: balanced point P_b, dead and seismic loads P_s, and pure flexure P_f, in the transverse and longitudinal directions, is determined using the in-house developed MathCAD sheets.

(3) The ordinates of the moment-curvature relationships at the associated axial load levels, axial-moment interaction surface, and plastic hinge length were

TABLE 6: Plastic curvature capacity of Pier #2 old columns, Bridge #1-W[*].

Axial load level	Axial force[**] (kips)	Direction	Compression failure Unconfined	Buckling of long. rebars	Fracture of long. rebars	Low-cycle fatigue of long. rebars	Failure in the lap-splice zone
				rad/in.			
P_b	−1543.0	T & L	0.0002197	**0.0001173**	0.0056480	0.0020726	0.0011435
P_s	−584.0	T	0.0004549	**0.0003543**	0.0040599	0.0020726	0.0014724
P_s	−302.0	L	0.0006400	**0.0006224**	0.0037028	0.0020726	0.0017314
P_f	0.0	T & L	**0.0011457**	0.0024662	0.0033434	0.0020726	0.0024387

[*]1 kip = 4.45 kN; 1 k·ft = 0.00135 kN·m; [**]P_o = −3868.6 kips and P_t = 782.6 kips.

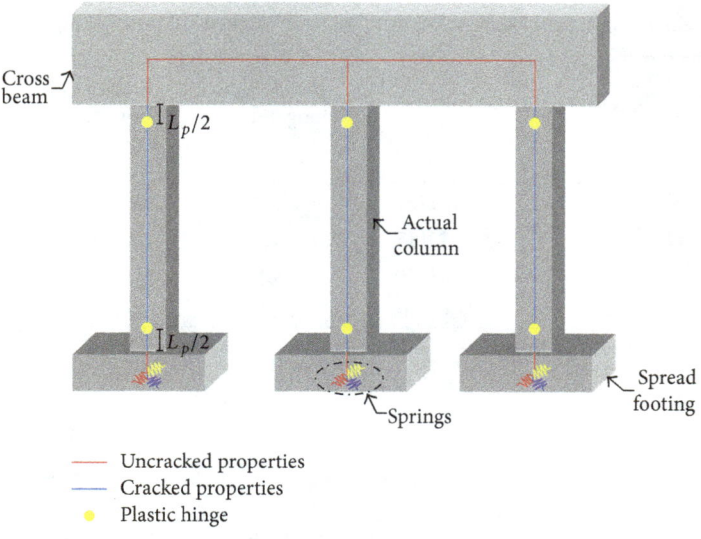

FIGURE 2: Bridge pier elements and associated stiffness.

7. Analysis Results and Discussions

input into the software SAP2000 under user-defined plastic hinge properties.

(4) The capacity curve using user-defined hinge properties was obtained by pushing each pier individually and independently in both the longitudinal and the transverse direction, as per the requirements of the FHWA-SRM.

(5) For automated plastic hinge properties, the moment-curvature relationship and the axial-moment interaction surface of the potential plastic hinge are determined by the software SAP2000 based on the column cross section geometry, longitudinal and transverse reinforcement details, and unconfined concrete and steel stress-strain curve parameters.

(6) The capacity curve using automated plastic hinge properties was obtained by pushing each pier individually and independently in both the longitudinal and the transverse direction.

(7) Capacity curves using user-defined and automated plastic hinge properties were then compared and discussed.

For user-defined plastic hinge properties, the value of the controlling limit state is shown in bold type font. Plastic curvature and plastic rotation for pure compression P_o and pure Tension P_t are not included because they are equal to zero. The ordinates of the moment-curvature relationships at the associated axial load levels, axial-moment interaction surface, and plastic hinge length to be input into the software SAP2000 are shown in Tables 6, 7, 8, and 9.

7.1. Bridge #1-W. Pier #2, Pier #3, and Pier #4 columns showed similar behaviour in regard to curvature capacity limit states. Therefore, for brevity, results for Pier #2 are reported. It is clear from Table 6 that buckling of longitudinal rebars is the controlling limit state of the columns of the old piers at P_b and P_s axial load levels in both directions. This behaviour is attributed to the poor confinement details, #3 hoops at 12.0 in. spacing. As the axial load decreases and reaches zero, P_f, compression failure of the unconfined concrete becomes the controlling limit state. The unconfined concrete in here refers to the cover concrete and the core concrete. The change in the type of the controlling limit state supports the effect of the axial load level on the deformation capacity of the plastic

TABLE 7: User-defined plastic hinge properties of Pier #2 old columns, Bridge #1-W.

Direction	Axial load level	Overstrength Axial (kips)	Moment (kip·ft)	Plastic hinge length (L_p) (in.)	Yield curvature rad/in$*10^{-6}$	Yield rotation Rad	Plastic curvature rad/in$*10^{-6}$	Plastic rotation Rad
T	P_b	−1543.0	1554.4				**0.0001173**	0.00248
	P_s	−584.0	1290.1	21.18	99.6	0.00211	**0.0003543**	0.00750
	P_f	0.0	870.2				**0.0011457**	0.02427
L	P_b	−1543.0	1554.4				**0.0001173**	0.00409
	P_s	−302.0	1111.8	34.91	99.6	0.00348	**0.0006224**	0.02173
	P_f	0.0	870.2				**0.0011457**	0.04000

TABLE 8: Plastic curvature capacity of Pier #2 widened columns, Bridge #1-W.

Axial load level	Axial Force* (kips)	Direction	Compression failure confined	Buckling of long. rebars	Fracture of long. rebars	Low-cycle fatigue of long. rebars	Failure in the lap-splice zone
					rad/in.		
P_b	−1705.7	T & L	**0.0011781**	NA**	0.0055760	0.0020899	NA
P_s	−584.0	T	0.0025015	NA	0.0040056	**0.0020899**	NA
P_s	−302.0	L	0.0035630	NA	0.0037024	**0.0020899**	NA
P_f	0.0	T & L	0.0068851	NA	0.0033970	**0.0020899**	NA

*$P_o = -4585.3$ kips and $P_t = 1174.0$ kips; **NA = Not applicable.

TABLE 9: User-defined plastic hinge properties of Pier #2 widened columns, Bridge #1-W.

Direction	Axial load level	Overstrength Axial (kips)	Moment (kip·ft)	Plastic hinge length (L_p) (in.)	Yield curvature rad/in$*10^{-6}$	Yield rotation Rad	Plastic curvature rad/in$*10^{-6}$	Plastic rotation rad
T	P_b	−1705.7	1958.51				**0.0011781**	0.02776
	P_s	−584.0	1661.35	23.57	150.6	0.00355	**0.0020899**	0.04925
	P_f	0.0	1271.37				**0.0020899**	0.04925
L	P_b	−1705.7	1958.51				**0.0011781**	0.04235
	P_s	−302.0	1493.15	35.95	150.6	0.00541	**0.0020899**	0.07513
	P_f	0.0	1271.37				**0.0020899**	0.07513

hinge. Also, the values of the controlling curvature capacities shown in Table 6 identify the effect of the axial force on the deformation capacity of the plastic hinge. For example, the curvature capacity was equal to 0.0001173 rad/in at axial force of 1543.0 kips and 0.0011457 rad/in. at zero axial force level. Table 7 shows the moment-curvature ordinates and the moment-axial force interaction surface.

The columns of the new piers had confined concrete details where transverse reinforcement consisted of #4 spirals at 3.3 in. and column longitudinal reinforcement was not spliced within the plastic hinge zone. Therefore, buckling of longitudinal rebars and failure in the lap splice zone are not applicable, as shown in Table 8. Accordingly, low cycle fatigue of longitudinal rebars was the controlling limit state at P_s and P_f axial load levels, while compression failure of the confined concrete controlled the curvature capacity at P_b axial load level. Failure limit state pertaining

to unconfined concrete is not considered for the widened columns because the core concrete has the capability to withstand higher plastic curvature demands after spalling off the cover concrete. Comparison of curvature capacities between the old and the new columns shows the effect of the confinement on the deformation capacity of the plastic hinge. For example, the plastic curvature capacity of the old column was 0.0011457 rad/in, while the curvature capacity of the new column was 0.0020899 rad/in. at the zero axial load level.

Tables 7 and 9 show that the plastic hinge length in the transverse direction is less than the plastic hinge length in the longitudinal direction which is basically attributed to the difference in the shear span in both directions. Subsequently, higher deformation capacities are expected in the longitudinal direction. It should be noted that the plastic hinge length was determined in accordance with the equations in the FHWA-SRM which depends on the

TABLE 10: Plastic curvature capacity of Pier #2 columns, Bridge #2-E.

Axial load level	Axial Force* (kips)	Direction	Compression failure unconfined	Buckling of long. rebars	Fracture of long. rebars	Low-cycle fatigue of long. rebars	Failure in the lap-splice zone
				rad/in.			
P_b	−1543.0	T & L	0.0002197	**0.0001173**	0.0056480	0.0020010	0.0011435
P_s	−405.0	T	0.0005576	**0.0004913**	0.0038295	0.0020010	0.0016161
P_s	−302.0	L	0.0006404	**0.0006231**	0.0037024	0.0020010	0.0017319
P_f	0.0	T & L	**0.0011457**	0.0024662	0.0033434	0.0020010	0.0024387

*P_o = −3868.6 kips and P_t = 782.6 kips.

shear span of the member, and the yield strain and the diameter of the longitudinal bars. However, other researchers [21, 23] identified that the plastic hinge length depends on the axial load level, aspect ratio of the member, type of reinforcement, longitudinal and transverse reinforcement ratios, and concrete strength.

Figures 3 and 4 show the capacity curves of Bridge #1-W in the longitudinal and transverse directions, respectively. The figures also show the capacity curves based on user-defined and automated plastic hinge properties. It is clear from Figure 3 that the base shear capacity of Pier #2, Pier #3, and Pier #4 in the longitudinal direction using the automated-hinge model is less than base shear capacity using user-defined hinge model by 17.7%, 15.2%, and 15.4%, respectively. On the other hand, the displacement capacity of Pier #2, Pier #3, and Pier #4 in the longitudinal direction using the automated-hinge model is higher than the displacement capacity using user-defined hinge model by 9.7%, 18.9%, and 9.7%, respectively. Similarly, Figure 4 shows that base shear capacity of Pier #2, Pier #3, and Pier #4 in the transverse direction using the automated-hinge model is less than base shear capacity using user-defined hinge model by 15.9%, 15.7%, and 16.0%, respectively. However, the displacement capacity of Pier #2, Pier #3, and Pier #4 in the transverse direction using the automated-hinge model is higher than the displacement capacity using user-defined hinge model by 10.9%, 20.2%, and 15.0%, respectively.

The differences in the base shear and displacement capacities of the bridge piers using the two hinge models are retained to the differences in the associated moment-curvature relationship. At an axial load level of −584.0 kips, P_s, the ultimate curvature capacity is equal to 6.33×10^{-4} rad/in. and the associated idealized moment capacity is equal to 1155.4 k·ft using the automated-hinge model in SAP2000. The corresponding values using the user-defined model are equal to 3.54×10^{-4} rad/in. and 1290.1 k·ft, respectively.

7.2. Bridge #2-E. Analysis results showed that the behaviours of Pier #2 and Pier #3 are similar; thus only the deformation capacities for Pier #2 are discussed. Table 10 shows that buckling of longitudinal rebars is the controlling limit state of Pier #2 columns at P_b and P_s axial load levels in the longitudinal and transverse directions. This behaviour is attributed to the poor confinement details, #3 hoops at 12.0 in.

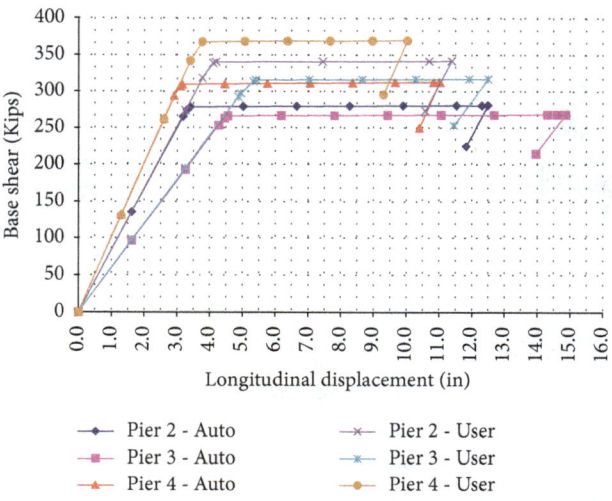

FIGURE 3: Longitudinal pushover capacity curve of Bridge #1-W.

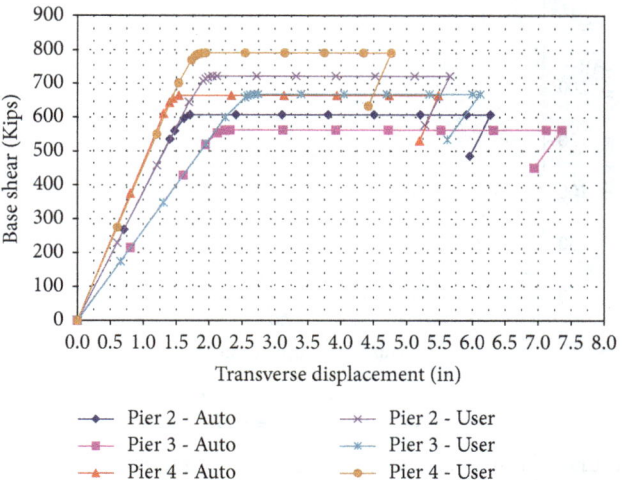

FIGURE 4: Transverse pushover capacity curve of Bridge #1-W.

spacing. As the axial load level decreases, the value of the plastic curvature capacity increases until the axial load level reaches zero, P_f, whereat compression of the unconfined concrete is the controlling limit state. Table 11 shows the moment-curvature ordinates and the plastic hinge lengths of Pier #2 columns. Similar to the findings for Bridge #1, the

TABLE 11: User-defined plastic hinge properties of Pier #2 columns, Bridge #2-E.

Direction	Axial load level	Overstrength		Plastic hinge length (L_p) (in.)	Yield curvature rad/in$*10^{-6}$	Yield rotation Rad	Plastic curvature rad/in$*10^{-6}$	Plastic rotation Rad
		Axial (kips)	Moment (kip·ft)					
T	P_b	−1543.0	1554.4				**0.0001173**	0.00204
	P_s	−405.0	1182.2	17.41	99.6	0.00173	**0.0004913**	0.00856
	P_f	0.0	870.2				**0.0011457**	0.01995
L	P_b	−1543.0	1554.4				**0.0001173**	0.00321
	P_s	−302.0	1111.8	27.37	99.6	0.00273	**0.0006231**	0.01706
	P_f	0.0	870.2				**0.0011457**	0.03136

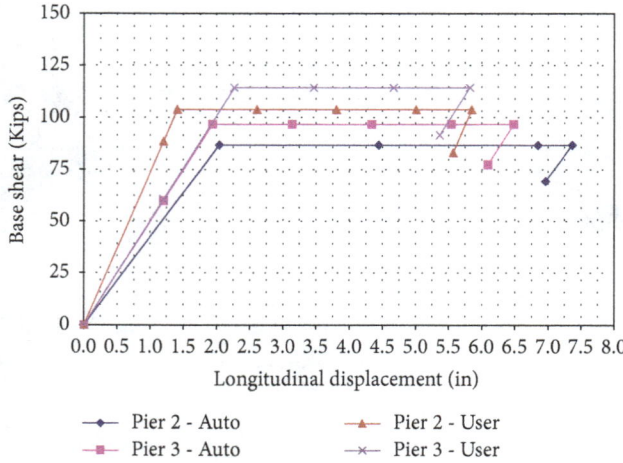

FIGURE 5: Longitudinal pushover capacity curve of Bridge #2-E.

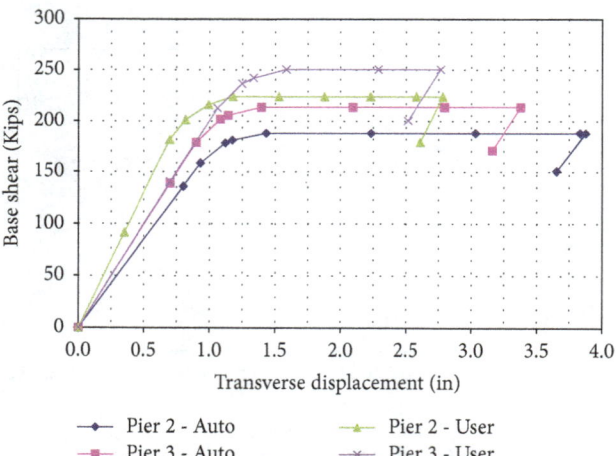

FIGURE 6: Transverse pushover capacity curve of Bridge #2-E and Bridge #3-E.

change in the type of the controlling limit state supports the effect of the axial load level on the deformation capacity of the plastic hinge. For example, the plastic curvature capacity was equal to 0.0001173 rad/in at axial force of 1543.0 kips and 0.0011457 rad/in. at zero axial force level.

Figures 5 and 6 show the capacity curves of Bridge #2-E in the longitudinal and transverse directions, respectively.

Figure 5 shows that the base shear capacity of Pier #2 and Pier #3 in the longitudinal direction using the automated-hinge model is less than base shear capacity using user-defined hinge model by 16.4% and 15.4%, respectively. This difference is basically due to the fact that the idealized flexural capacity in the FHWA-SRM is an overstrength moment which includes an overstrength factor in addition to the factor associated with the expected material properties. The displacement capacity of Pier #2 and Pier #3 in the longitudinal direction using the automated-hinge model is higher than the displacement capacity using user-defined hinge model by 25.9% and 11.3%, respectively. Similarly, Figure 6 shows that base shear capacity of Pier #2 and Pier #3 in the transverse direction using the automated-hinge model is less than base shear capacity using user-defined hinge model by 16.0% and 14.6%, respectively. However, the displacement capacity of Pier #2 and Pier #3 in the transverse direction using the automated-hinge model is higher than the displacement capacity using user-defined hinge model by 39.2% and 22.0%, respectively. A close look at the capacity curves shows higher differences in the displacement capacity in the transverse direction than in the longitudinal direction between the two different hinge models. This difference is due to the fact that the expected seismic and gravity axial load level in the transverse direction is higher than the axial load level in the longitudinal direction which in turn shifts the controlling capacity limit state to buckling of the longitudinal bars. The higher this shift is, the higher the difference between the curvature values obtained from buckling of longitudinal rebars limit state and the failure compression of the unconfined concrete limit state is. The latter is one of the two controlling limit states for the automated-hinge model.

7.3. *Bridge #3-E.* The pier columns of this bridge are rectangular columns with poor confinement details, #3 hoops at 12.0 in. spacing. Table 12 supports the notion that buckling of longitudinal rebars is the controlling limit state at P_b axial load level as expected from the behaviour of the previous bridges. Compression failure of the unconfined concrete limited the curvature capacity of the columns at P_s axial load level, while low cycle fatigue of longitudinal rebars was the controlling limit state at P_f axial load level. Table 13 shows the user-defined hinge properties of Pier #2 columns, wherein the plastic hinge length has the same value in the longitudinal

TABLE 12: Plastic curvature capacity of Pier #2 columns, Bridge #3-E.

Axial load level	Axial Force* (kips)	Direction	Compression failure unconfined	Buckling of long. rebars	Fracture of long. rebars	Low-cycle fatigue of long. rebars	Failure in the lap-splice zone
					Rad/in.		
P_b	−1300.1	T & L	0.0003102	**0.0002003**	0.0063260	0.0021910	0.0015229
P_s	−432.0	T	**0.0008436**	0.0011569	0.0044562	0.0021910	0.0023165
P_s	−303.0	L	**0.0010593**	0.0021517	0.0042670	0.0021910	0.0026375
P_f	0.0	T & L	0.0023579	Tension	0.0038788	**0.0021910**	0.0045696

*$P_o = -3262.9$ kips and $P_t = 662.4$ kips.

TABLE 13: User-defined plastic hinge properties of Pier #2 columns, Bridge #3-E.

Direction	Axial load level	Overstrength Axial (kips)	Moment (kip·ft)	Plastic hinge length (L_p) (in.)	Yield curvature rad/in$*10^{-6}$	Yield rotation Rad	Plastic curvature rad/in$*10^{-6}$	Plastic rotation Rad
	P_b	−1300.1	1309.2				**0.0002003**	0.00384
T & L	P_s	−432.0	1053.0	19.18	125.0	0.0024	**0.0008436**	0.01618
	P_f	0.0	734.7				**0.0021910**	0.04202
L	P_s	−303.0	971.3	19.18	125.0	0.0024	**0.0010593**	0.02032

and transverse directions. This is because the shear span is identical in both directions. A comparison between the ultimate curvature capacities of the columns of this bridge and the columns of Bridge #2 shows the effect of the column cross section dimensions on the value of the curvature capacity. For the compression failure of the unconfined concrete limit state at zero axial load level, the plastic curvature capacity was equal to 0.0001173 rad/in for Bridge #2 while the plastic curvature capacity was 0.0023579 rad/in. for Bridge #3. This finding is related to the smaller depth of the column section in Bridge #3 which is equal to 30 in. compared to 36 in. for the columns of Bridge #2. This difference in the section size will support the capability of the smaller column section to undergo higher curvature levels for the same level of concrete strain.

Figures 7 and 8 show the capacity curves of the bridge in the transverse and longitudinal directions, respectively. The capacity curve in the longitudinal direction was obtained for the whole bridge because the soil behind the bridge abutments was part of the seismic load resisting system. According to Figure 7, the displacement capacity of Pier #2, Pier #3, and Pier #4 determined using the automated-hinge model is higher than the displacement capacity using the user-defined model by 0.3%, 0.5%, and 1.6%, respectively. However, the corresponding base shear capacity was lower by 19.2%, 23.4%, and 20.6%, respectively. A similar trend was experienced in the longitudinal direction for the whole bridge with 26.8% and 4.4% difference in the displacement and base shear capacities, respectively. The differences in the capacity curves between the two hinge models are attributed to differences in the moment-curvature values. At an axial load level of −432.0 kips, the ultimate curvature capacity is equal to 1.09×10^{-3} rad/in. and the associated idealized

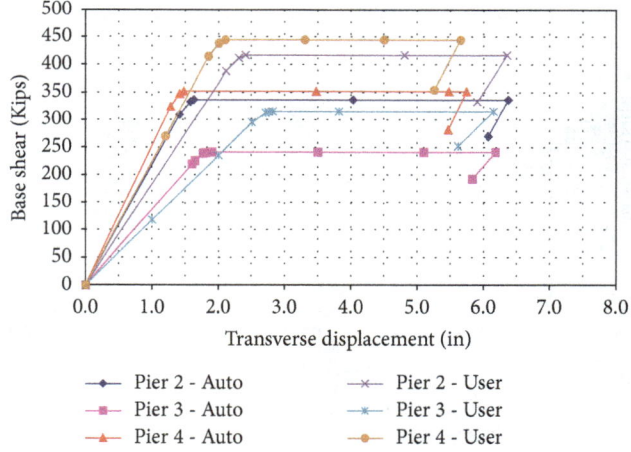

FIGURE 7: Transverse pushover capacity curve of Bridge #3-E.

moment capacity is equal to 901.3 k·ft using the automated-hinge model in SAP2000. The corresponding values using the user-defined model are equal to 0.84×10^{-3} rad/in. and 1053.0 k·ft, respectively.

7.4. Bridge #4-W. For brevity, the results of Pier #2 widened columns will be discussed. The columns in the widened part of Bridge #4-W were provided by #4 transverse reinforcement at 3.0 in. on centre. However, the longitudinal rebars were spliced within the plastic hinge zone with a length of 40.0 in. Table 14 demonstrates that failure of the confined concrete is the controlling limit state at P_b axial load level, whereas failure in the lap splice zone is the controlling limit state at all other axial load levels.

TABLE 14: Plastic curvature capacity of new columns, Pier #2 & #3 of Bridge #4-W.

Axial load level	Axial Force[*] (kips)	Direction	Compression failure confined	Buckling of long. rebars	Fracture of long. rebars	Low-cycle fatigue of long. rebars	Failure in the lap-splice zone
				Rad/in.			
P_b	−1824.8	T & L	**0.0011169**	NC[**]	0.0056628	0.0019116	0.0011484
P_s	−268.0	T & L	0.0030921	NC	0.0037002	0.0019116	**0.0012960**
P_f	0.0	T & L	0.0046158	NC	0.0034514	0.0019116	**0.0013836**

[*]$P_o = −5264.7$ kips and $P_t = 1615.2$ kips; [**]NC = Not controlling.

TABLE 15: User-defined plastic hinge properties of new columns, Pier #2 Bridge #4-W.

Direction	Axial load level	Overstrength Axial (kips)	Overstrength Moment (kip·ft)	Plastic hinge length (L_p) (in.)	Yield curvature rad/in $*10^{-6}$	Yield rotation Rad	Plastic curvature rad/in $*10^{-6}$	Plastic rotation Rad
	P_b	−1824.8	2418.5	25.05			**0.0011169**	0.02798
T	P_s	−268.0	1923.1		146.4	0.00367	**0.0012960**	0.02997
	P_f	0.0	1737.9	23.12			**0.0013836**	0.03199
	P_b	−1824.8	2418.5	37.53			**0.0011169**	0.04191
L	P_s	−268.0	1923.1		146.4	0.00549	**0.0012960**	0.02997
	P_f	0.0	1737.9	23.12			**0.0013836**	0.03199

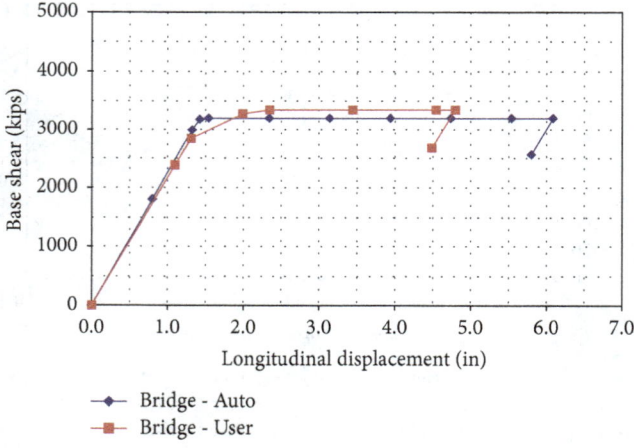

FIGURE 8: Longitudinal pushover capacity curve of Bridge #3-E.

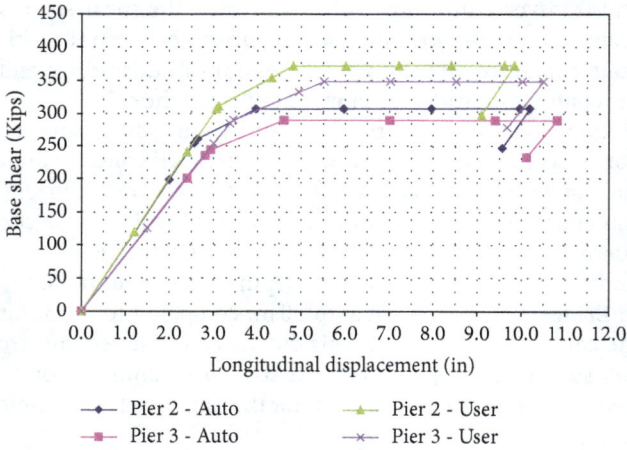

FIGURE 9: Longitudinal pushover capacity curve of Bridge #4-W.

A comparison between the ultimate curvature capacities of the columns of this bridge and the columns of Bridge #2 shows the effect of the confinement on the value of the curvature capacity. For the compression failure of the unconfined concrete limit state at zero axial load level, the plastic curvature capacity was equal to 0.0001173 rad/in for Bridge #2 while the plastic curvature capacity was 0.0046158 rad/in. for Bridge #4. This finding is related to the huge difference in the details of the transverse reinforcement of the column section in Bridge #2 which has #3 hoops at 12 in. and the transverse reinforcement of the column section in Bridge #4 which has #4 spirals at 3.0 in. This difference in the confinement details will allow the section to undergo higher plastic deformations.

Table 15 includes the nonlinear properties of the user-defined hinge model. It is for the first time that two different plastic hinge lengths are reported for the column in the same direction, as shown in Table 15. This result is pertained to change in the controlling deformation limit state from failure of the confined concrete at P_b axial load level to low cycle fatigue of the longitudinal bars at P_s and P_f axial load levels. Based on the FHWA-SRM, when at the deformation capacity is controlled by low cycle fatigue of longitudinal bars, the effective plastic hinge length is reduced in length and the effective plastic hinge is concentrated at the beginning of the lap. This finding brings attention to the limited capabilities of the lumped plasticity model in dealing with conditions where the plastic hinge length changes with the axial load level.

Figures 9 and 10 show the capacity curves of the bridge in the longitudinal and transverse directions, respectively.

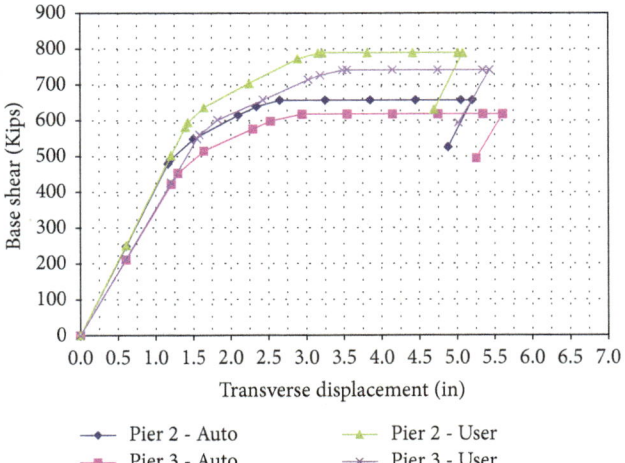

FIGURE 10: Transverse pushover capacity curve of Bridge #4-W.

According to Figure 9, the displacement capacity of Pier #2 and Pier #3 determined using the automated-hinge model is higher than the displacement capacity using the user-defined model by 3.5% and 3.0%, respectively. However, the corresponding base shear capacity was lower by 17.2% and 16.7%, respectively. A similar trend was experienced in the transverse direction with 2.5% and 3.2% difference in the displacement capacity, and 16.7% and 16.6% difference in the base shear capacity for Pier #2 and Pier #3, respectively.

8. Conclusions

The conclusions of the study may be summarized as follows:

(1) The results of the analyses clearly showed that the capacity curve depends on the plastic hinge properties. The base shear capacity of models with automated hinges is less than the shear capacity of models with user-defined hinge properties. Also, the displacement capacity of models with automated hinges is higher than the shear capacity of models with user-defined hinges.

(2) User-defined hinge model determined using the provisions of the FHWA-SRM is more reliable than the automated-hinge model since it deals with the effect of possible local failure mechanisms within the plastic hinge zone.

(3) Capacity curves of existing and widened highway bridges are impacted approximately in the same manner when using different hinge models.

(4) It is recommended that bridge design manuals clearly ask bridge designers to evaluate the deformation capacities of existing bridges and widened bridges, which have members that do not meet current seismic detailing standards, using the provisions of the FHWA-SRM.

(5) It is recommended that the software industry, like the software SAP2000, includes in its future versions the capability to model plastic hinges that do not meet current seismic detailing standards.

The scope of the present study was limited to flexure dominated conventionally reinforced concrete bridge piers with the plastic hinge length defined in accordance with the FHWA-SRM. Further research needs to be carried out considering shear dominant piers, other types of reinforcement such as shape memory alloys, and other models for plastic hinge lengths.

Conflicts of Interest

The authors declare that there are no conflicts of interest regarding the publication of this paper.

References

[1] M. J. N. Priestley, F. Seible, and G. M. Calvi, *Seismic Design and Retrofit of Bridges*, John Wiley Sons, New York, NY, USA, 1996.

[2] N. K. Shattarat, M. D. Symans, D. I. McLean, and W. F. Cofer, "Evaluation of nonlinear static analysis methods and software tools for seismic analysis of highway bridges," *Engineering Structures*, vol. 30, no. 5, pp. 1335–1345, 2008.

[3] N. Panandikar Hede and K. S. B. Narayan, "Sensitivity of pushover curve to material and geometric modelling—an analytical investigation," *Structures*, vol. 2, pp. 91–97, 2015.

[4] R. Pinho, C. Casarotti, and S. Antoniou, "A comparison of single-run pushover analysis techniques for seismic assessment of bridges," *Earthquake Engineering and Structural Dynamics*, vol. 36, no. 10, pp. 1347–1362, 2007.

[5] R. Pinho, R. Monteiro, C. Casarotti, and R. Delgado, "Assessment of continuous span bridges through nonlinear static procedures," *Earthquake Spectra*, vol. 25, no. 1, pp. 143–159, 2009.

[6] T. Isaković, M. P. Nino Lazaro, and M. Fischinger, "Applicability of pushover methods for the seismic analysis of single-column bent viaducts," *Earthquake Engineering and Structural Dynamics*, vol. 37, no. 8, pp. 1185–1202, 2008.

[7] T. S. Paraskeva, A. J. Kappos, and A. G. Sextos, "Extension of modal pushover analysis to seismic assessment of bridges," *Earthquake Engineering and Structural Dynamics*, vol. 35, no. 10, pp. 1269–1293, 2006.

[8] S. Antoniou and R. Pinho, "Advantages and limitations of adaptive and non-adaptive force-based pushover procedures," *Journal of Earthquake Engineering*, vol. 8, no. 4, pp. 497–522, 2004.

[9] A. S. Elnashai, "Advanced inelastic static (pushover) analysis for earthquake applications," *Structural Engineering and Mechanics*, vol. 12, no. 1, pp. 51–69, 2001.

[10] F. Naeim, *Ten Commandments on Pushover Analysis*, John A. Martin and Associates, Los Angeles, Calif, USA, 1999.

[11] A. K. Chopra and R. K. Goel, *A Modal Pushover Analysis Procedure to estimate Seismic Demands for Buildings: Theory and Preliminary Evaluation*, Pacific Earthquake Engineering Research Center, College of Engineering University of Berkeley, Berkeley, Calif, USA, 2001.

[12] J. Mao, C. Zhai, and L. Xie, "An improved modal pushover analysis procedure for estimating seismic demands of structures," *Earthquake Engineering and Engineering Vibration*, vol. 7, no. 1, pp. 25–31, 2008.

[13] H. Krawinkler and G. D. P. K. Seneviratna, "Pros and cons of a pushover analysis of seismic performance evaluation," *Engineering Structures*, vol. 20, no. 4-6, pp. 452–464, 1998.

[14] FEMA, "NEHRP guidelines for the seismic rehabilitation of buildings," Report FEMA-273 (Guidelines) and Report FEMA-274 (Commentary), Washington, DC, USA, 1997.

[15] ATC, "Seismic evaluation and retrofit of concrete buildings," Tech. Rep. ATC-40, Applied Technology Council, Redwood, Calif, USA, 1997.

[16] FEMA, "Prestandard and commentary for the seismic rehabilitation of buildings," Tech. Rep. FEMA 356, American Society of Civil Engineers for the Federal Emergency Management Agency, Washington, DC, USA, 2000.

[17] CEN, "Eurocode 8: Design of structures for earthquake resistance, Part 1: general rules, seismic actions and rules for buildings," Tech. Rep. EN1998-1, European Committee for Standardization, 2004.

[18] FEMA, "Improvement of nonlinear static seismic analysis procedures," Applied Technology Council (ATC-55 Project) FEMA 440, Federal Emergency Management Agency, Washington, DC, USA, 2005.

[19] ASCE, "Seismic rehabilitation of existing buildings," Tech. Rep. ASCE/SEI 7- 05, American Society of Civil Engineers, 2005.

[20] M. Inel and H. B. Ozmen, "Effects of plastic hinge properties in nonlinear analysis of reinforced concrete buildings," *Engineering Structures*, vol. 28, no. 11, pp. 1494–1502, 2006.

[21] A. H. M. M. Billah and M. Shahria Alam, "Plastic hinge length of shape memory alloy (SMA) reinforced concrete bridge pier," *Engineering Structures*, vol. 117, pp. 321–331, 2016.

[22] K. C. Shrestha, M. S. Saiidi, and C. A. Cruz, "Advanced materials for control of post-earthquake damage in bridges," *Smart Materials and Structures*, vol. 24, no. 2, Article ID 025035, 2015.

[23] A. H. M. Muntasir Billah and M. Shahria Alam, "Performance-based seismic design of shape memory alloy-reinforced concrete bridge piers. I: development of performance-based damage states," *Journal of Structural Engineering*, vol. 142, no. 12, Article ID 4016140, 2016.

[24] CALTRANS, "Caltrans seismic design criteria," Tech. Rep. Version 1.6, California Department of Transportation, 2010.

[25] CSI, "SAP2000: integrated software for structural analysis and design," Tech. Rep. Version 18.2, Computers and Structures, 2016.

[26] FHWA, "Seismic retrofitting manual for highway structures: part 1-bridges," Publication No FHWA-HRT-06-032, Federal Highway Administration, US Department of Transportation, Washington, DC, USA, 2006.

[27] N. Shatarat, "Effect of plastic hinge properties in nonlinear analysis of highway bridges," *Jordan Journal of Civil Engineering*, vol. 6, no. 4, pp. 501–510, 2012.

[28] PTC MathCAD, "Engineering math software," Tech. Rep. Version 15, PTC, Needham, Mass, USA, 2010.

[29] AASHTO, *Guide Specifications for LRFD Seismic Bridge Design*, American Association of State Highway and Transportation Officials (AASHTO), Washington, DC, USA, 2nd edition, 2011, with 2012, 2014, and 2015 Interim Revisions.

[30] J. B. Mander, M. J. N. Priestley, and R. Park, "Theoretical stress-strain model for confined concrete," *Journal of Structural Engineering*, vol. 114, no. 8, pp. 1804–1826, 1988.

[31] Lpile, *A Program for the Analysis of Piles and Drilled Shafts under Later Load*, Version 2014, Ensoft, Austin, Tex, USA, 2014.

5

Effects of Reinforcing Fiber and Microsilica on the Mechanical and Chloride Ion Penetration Properties of Latex-Modified Fiber-Reinforced Rapid-Set Cement Concrete for Pavement Repair

Woong Kim,[1] Jong-Chan Jeon,[2] Byung-Hwan An,[2] Joo-Ha Lee,[3] Hae-Do Kim,[4] and Chan-Gi Park (ID)[2]

[1]Department of Bio-Industry Mechanical Engineering, Kongju National University, Yesan 32439, Republic of Korea
[2]Department of Rural Construction Engineering, Kongju National University, Yesan 32439, Republic of Korea
[3]Department of Civil Engineering, University of Suwon, Hwaseong 18323, Republic of Korea
[4]Rural Research Institute, Korea Rural Community Corporation, Ansan 15634, Republic of Korea

Correspondence should be addressed to Chan-Gi Park; cgpark@kongju.ac.kr

Academic Editor: Young Hoon Kim

This study evaluated the influence of reinforcement fiber type and microsilica content on the performance of latex-modified fiber-reinforced roller-compacted rapid-hardening cement concrete (LMFRCRSC) for a concrete pavement emergency repair. Experimental variables were the microsilica substitution ratio (1, 2, 3, and 4%), and the reinforcement fiber (jute versus macrosynthetic fiber). In the tests, compressive, flexural, and splitting tensile strength; chloride ion penetration resistance; and abrasion resistance were assessed. From the compressive and flexural strength tests with microsilica substitution, the 4-hour curing strength decreased as the microsilica substitution ratio increased. From the chloride ion penetration test, as the microsilica substitution ratio increased, chloride ion penetration decreased. The abrasion resistances increased with the substitution ratio of microsilica increase. Based on these test results, microsilica at a substitution ratio of 3% or less and macrosynthetic fiber as the reinforcement improved the performance of LMFRCRSC for a concrete pavement emergency repair and satisfied all of the target strength requirements.

1. Introduction

The repair of deteriorated concrete pavements, using roller-compacted latex-modified rapid-hardening cement concrete, has been studied recently [1]. A roller-compacted concrete (RCC) is a concrete with low fluidity [2–6]. RCC does not require a consistency for compacting, and it must be compacted with an external vibrator, such as a vibration roller or a vibration pressure tamper [2]. Roller-compacted concrete pavement (RCCP) offers a fast, successive construction and use of a wide range of construction equipment; thus, it has higher economic feasibility than other concrete pavements [2–6]. Compared with a general concrete

pavement, RCCP has a lower water-to-cement (W/C) ratio [2]. A lower W/C ratio has the effect of increasing the strength of concrete, and it may reduce the risk of contraction cracks due to moisture evaporation [2–6]. Thus, RCCP also has the advantage of increased durability of pavement over the long term [2–6]. When RCC and latex are used with rapid-hardening cement for a concrete pavement, it is possible to achieve both easy construction and durability [1]. Currently, roller-compacted rapid-hardening cement concrete pavement uses up to 15% latex [1]. The use of latex up to 15% may delay the concrete's initial strength development. As such, it may be difficult to secure sufficient initial target strength [7, 8]. Economically, the overall concrete pavement repair is more costly due to the high latex amounts. The latex-modified rapid-hardening cement offers superior workability, crack resistance, and durability compared to the general rapid-hardening cement concrete [8–15], but it has the problems of initial strength development delay and lower economic feasibility. Also, the increase in initial fluidity makes roller compaction difficult because latex-modified fiber-reinforced roller-compacted rapid-hardening cement concrete (LMFRCRSC) uses a mix that has low slump, due to the characteristics of RCC.

In the previous study [16], the performance of LMFRCRSC according to the type of fiber reinforcement was evaluated. The possibility of the roller compaction method was evaluated by measuring the slump value according to the type of fiber reinforcement. Mechanical properties such as compressive strength and flexural strength were evaluated, and durability such as permeability and abrasion resistance were evaluated. However, the using amount of rapid-hardening cement is not reduced, so there is a limit to solving the problems caused by the increase in hydration heat in the early age. Therefore, it is necessary to reduce the using amount of rapid-hardening cement. The study was conducted to apply microsilica as a substitute material for rapid-hardening cement in order to reduce the using amount of rapid-hardening cement. The addition of microsilica brings about the pozzolanic reaction and a fine pore-filling effect, providing increased strength and improved water tightness [16, 17]. Also, this study added macrosynthetic fibers and natural jute fibers which showed good results in the previous study. A fiber reinforcement minimizes crack formation and propagation in concrete through the fiber's fracture, pullout, debonding, and bridging effects [13–15]. As a result, the addition of the reinforcement fiber improves the tensile strength of concrete [13–15]. The influence of reinforcement fiber type and microsilica content on the performance of LMFRCRSC was evaluated for a concrete pavement emergency repair.

2. Materials

2.1. Materials. A rapid-hardening cement used was the product manufactured by Jungang Polytech, Korea. Physical and chemical characteristics of the rapid-hardening cement are shown in Table 1. A microsilica, a product from Micro Chemical, Korea, was used for this study, and its chemical compositions are shown in Table 2. Also, the microsilica consists of spherical particles with an average particle size of $0.15\,\mu m$ and a specific surface area of $20\,m^3/g$. The coarse aggregate was the crushed aggregate with a maximum diameter of 13 mm. The fine aggregate was river sand, with a specific gravity of 2.58. The physical characteristics of the aggregates are listed in Table 3. A styrene butadiene latex (SB latex) from Jungang Polytech, Korea, was used for this study, and its characteristics are given in Table 4. The macrosynthetic fiber and natural jute fiber were purchased from Nycontech, Korea, and the fiber characteristics are shown in Table 5. The shapes of the fibers are shown in Figure 1 [16].

TABLE 1: Chemical compositions of rapid-hardening cement.

SiO_2 (%)	Al_2O_3 (%)	Fe_2O_3 (%)	CaO (%)	MgO (%)	K_2O (%)	SO_3 (%)
13 ± 3	17.5 ± 3	$3>$	50 ± 3	$2.5>$	0.21	14 ± 3

TABLE 2: Chemical compositions of microsilica.

SiO_2 (%)	Al_2O_3 (%)	Fe_2O_3 (%)	CaO (%)	Others (%)
90–98	0.4–0.9	1–2	0.2–0.7	2–3

TABLE 3: Physical properties of coarse aggregates.

Properties	Density (g/mm^3)	Absorption (%)	Fineness modulus
Value	2.61	0.35	6.92

2.2. Mix Proportions. In case of concrete pavement repaired using rapid-hardening cement, the traffic open time is specified as a 4-hour minimum curing time by the American Association of State Highway and Transportation Officials (AASHTO) [18], the road traffic administrations of each state in the United States, and the Korea Expressway Corporation [19]. The traffic open standard is a compressive strength of at least 21 MPa and a flexural strength of at least 3.5 MPa. After curing for 28 days, the compressive strength is required to be at least 35 MPa, with a flexural strength of 4.5 MPa and a splitting tensile strength of 4.2 MPa. The abovementioned strength criteria were used as target mix strengths for this study. In addition, permeability has the biggest influence on concrete pavement life cycle degradation. In terms of durability, a chloride ion penetration test result of ≤2000 Coulombs (C) after 28 days of curing was set as a target for permeability, based on the Korea Expressway Corporation's ASTM C1202 test method. Also, in this study, to ensure initial permeability, the target chloride ion penetration at 4 hours of curing was set at 4000 C or less.

The W/C ratio was set at 0.28, and latex at 5% (solid-based) of the binder (cement + microsilica) weight was used. In the preliminary study [16], it was decided to apply about 5%, considering the range of latex that can be roller compaction methods on the type of fiber reinforcement. The reinforcement fibers, macrosynthetic fiber and jute fiber, were added at a volume ratio of 0.10%. The addition of reinforcement fibers facilitates slump reduction for roller compaction and is effective for controlling crack formation/growth and reducing water penetration. Microsilica was substituted for cement at weights of 0, 1, 2, 3, and 4% to evaluate the influence of the addition of microsilica. The study mix ratios are shown in Table 6.

2.3. Manufacturing of Test Specimens. For test specimens of LMFRCRSC, this study manufactured specimens using a pressure tamper, which mimicked the roller compaction process. In the first step of specimen fabrication, one-third of the mixed latex-modified fiber-reinforced rapid-hardening

TABLE 4: Properties of styrene butadiene latex.

Solid content (%)	Styrene content (%)	Butadiene content (%)	pH	Specific gravity	Surface tension (dyne/cm)	Particle size (A)	Viscosity (cps)
49	34 ± 1.5	66 ± 1.5	11.0	1.02	30.57	1700	42

TABLE 5: Properties of fibers.

Properties	Macrosynthetic fiber	Natural jute fiber
Elastic modulus (GPa)	10	61
Density (g/mm^3)	0.91	1.26
Fiber length (mm)	30	3
Fiber diameter (mm)	1	0.015
Tensile strength (MPa)	550	510

cement concrete was poured for specimens, and the vibration and pressure tamper was used to apply vibration pressure for 30 s. Similarly, the second third was poured, and 30 s of pressure vibration compaction was carried out. The final third was poured, pressure vibration compaction was carried out, and the surface was finished. The specimens were then evaluated in terms of performance. Figure 2 shows the vibration and pressure tamper used and the specimens manufactured using the vibration and pressure tamper [16].

3. Test Methods

3.1. Compressive Strength Tests. Compressive strength tests were performed in accordance with the ASTM C 39 standard [20]. Tests were performed after 4 hours and 28 days of curing. Each variable was investigated using six specimens.

3.2. Splitting Tensile Tests. Splitting tensile tests were conducted in accordance with the ASTM C 496 standard [21]. Tests were performed after 4 hours and 28 days of curing. Specimens (Ø100 × 200 mm) were cured in water at 23 ± 2°C. Each variable was investigated using six specimens.

3.3. Flexural Tests. Flexural tests were conducted in accordance with the ASTM C 496 standard [22]. Tests were performed after 4 hours and 28 days of curing. Specimens (100 mm × 100 mm × 400 mm) were cured in water at 23 ± 2°C. Each variable was investigated using six specimens.

3.4. Chloride Ion Penetration Tests. Chloride ion penetration tests were conducted in accordance with the ASTM C 1202-94 standard [23]. Specimens, 150 mm × 50 mm in size, were tested after 28 days of curing. Each variable was investigated using six specimens. The test apparatus for the chloride ion penetration test is shown in Figure 3 [16].

3.5. Abrasion Tests. Abrasion tests were conducted in accordance with the ASTM C 944 standard [24]. Specimens,

150 mm × 60 mm in size, were tested after 7 days of curing. Each variable was investigated using six specimens. The test apparatus for the abrasion test is shown in Figure 4 [16].

4. Results and Discussion

4.1. Compressive Strength. Figure 5 shows the compressive strength test results of LMFRCRSC for a concrete pavement emergency repair according to the reinforcement fiber type and microsilica substitution ratio. Generally, when pressure vibration compaction is used, the resulting concrete is more dense and may show increased compressive strength. On mixing with jute fibers, the 4-hour curing target compressive strength of at least 21 MPa was satisfied up to a microsilica substitution ratio of 2%. Also, on mixing with macrosynthetic fibers, the target compressive strength was satisfied up to a microsilica substitution ratio of 3%. As the microsilica substitution ratio increased, the compressive strength decreased. Rapid-hardening cement promotes hydration, with an active reaction of $3CaO \cdot SiO_2$, generating acicular crystals of ettringite by the reaction of calcium silicate hydrate (CSH) gel and calcium sulfoaluminate (CSA); this increases the initial strength. With microsilica substituting for rapid-hardening cement, the $3CaO \cdot SiO_2$ ingredient is present in a smaller proportion, causing a delay in compressive strength development in the early stages. For the 4-hour compressive strength, in cases using jute fibers, as the microsilica substitution ratio increased from 0 to 1, 2, 3, and 4%, the compressive strengths were 22.1, 21.8, 21.3, 16.2, and 15.4 MPa, respectively. Using macrosynthetic fibers, as the microsilica substitution ratio increased from 0 to 1, 2, 3, and 4%, the compressive strengths were 27.5, 24.5, 23.0, 20.8, and 15.8 MPa, respectively. The cases using macrosynthetic fibers showed higher compressive strengths than those using jute fibers. Fiber-reinforcing materials can be classified into structural fibers to improve structural performance, such as strength, and nonstructural fibers to be used for crack control and durability improvement [15, 16]. In this study, macrosynthetic fibers, which can replace steel fibers as structural fibers, were applied, while jute fibers were incorporated as nonstructural fibers [15, 16]. Also, jute fiber is a natural fiber; thus, it is more difficult to maintain the quality. The jute fiber's main ingredient contains cellulose and a small amount of lignin. Lignin has the effect of delaying the concrete's compressive strength development.

The 28-day curing compressive strength results showed that the target of 35 MPa was satisfied by the results with all mixes. As the microsilica substitution ratio increased, compressive strength increased. The microsilica additive improved the performance of the concrete via the pozzolanic reaction and the fine pore-filling effect. These tendencies

(a)   (b)

FIGURE 1: Geometry of fibers [16]. (a) Jute fiber. (b) Macrosynthetic fiber.

TABLE 6: Mix proportions of LMFRCRSC for a pavement repair.

| Mix no. | W^*/B^{**} (%) | S/A (%) | W | C | Macrosilica | Unit weight (kg/m³) | | | | | |
						S	G	Latex (solid)	Water in latex	Macrosynthetic fiber	Jute fiber
Macrosynthetic-0				400	0						
Macrosynthetic-1				396	4						
Macrosynthetic-2				392	8					0.91	—
Macrosynthetic-3				388	12						
Macrosynthetic-4	28	55	91	384	16	1015	831	20	21		
Jute-0				400	0						
Jute-1				396	4						
Jute-2				392	8					—	1.26
Jute-3				388	12						
Jute-4				384	16						

*W + water in latex. **Cement + microsilica.

were also seen in the mixes with added jute and macrosynthetic fibers. The compressive strength was higher with macrosynthetic fibers than with jute fibers.

4.2. Flexural Strength.

The flexural strength test results of LMFRCRSC for a concrete pavement emergency repair by reinforcement fiber types and the microsilica substitution ratio are shown in Figure 6. On mixing with jute fibers, the results showed that the 4-hour curing target flexural strength of 3.5 MPa was satisfied up to a microsilica substitution ratio of 2%. When the macrosynthetic fiber was included, the target flexural strength was satisfied up to a microsilica substitution ratio of 3%. As the microsilica substitution ratio increased, the flexural strength decreased. Rapid-hardening cement promotes hydration, with an active reaction of $3CaO \cdot SiO_2$, generating acicular crystals of ettringite by the reaction of CSH gel and CSA. With microsilica substituting for rapid-hardening cement, the $3CaO \cdot SiO_2$ ingredient is present in a relatively lower proportion, causing a delay in strength development in the early stages. Thus, the strength is lower. The cases using macrosynthetic fibers showed higher flexural strengths than those using jute fibers. The

28-day curing flexural strength results showed that the target of 4.5 MPa was satisfied by all mixes. Generally, latex and reinforcement fibers have more influence on flexural strength than on compressive strength. Thus, in this study, latex and reinforcement fibers were used. The mix with no microsilica added had the highest flexural strength. Moreover, pressure vibration compaction was used to densify the concrete structure; it was effective in increasing the concrete's strength. Thus, after curing for 28 days, the flexural strength satisfied the target strength of ≥ 4.5 MPa. As the microsilica substitution ratio increased, the flexural strength increased slightly. Microsilica improves the performance of the concrete, due to the pozzolanic reaction and the fine pore-filling effect. Thus, in the case of long-term strength, the flexural strength increased with the substitution ratio. This tendency was also seen in the mixes with added jute and macrosynthetic fibers. The flexural strength was slightly higher with macrosynthetic fibers than with jute fibers.

4.3. Splitting Tensile Strength.

The splitting tensile strength results of LMFRCRSC for a concrete pavement emergency

(a)

(b)

FIGURE 2: Manufacturing of test specimens [16]. (a) Manufacturing of compressive strength specimens. (b) Manufacturing of flexural strength specimens.

FIGURE 3: Chloride ion penetration test setup [16].

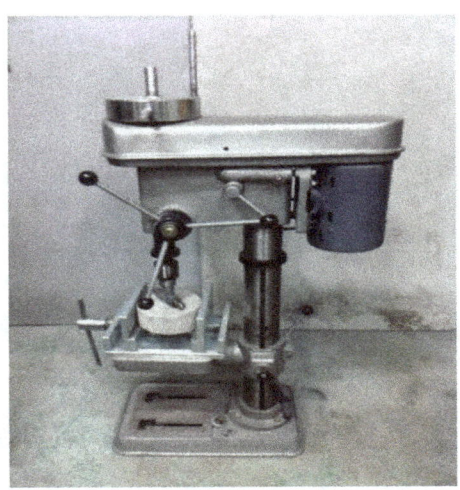

FIGURE 4: Abrasion test setup [16].

repair by the reinforcement fiber type and the microsilica substitution ratio are shown in Figure 7. The results showed that the mixes satisfied the 28-day splitting tensile strength target of ≥4.2 MPa. In this study, pressure vibration compaction was used to densify the concrete structure and was effective in increasing the concrete's strength. Thus, after curing for 28 days, the splitting tensile strength satisfied the target of ≥4.2 MPa. The addition of latex and reinforcement fibers has more influence on tensile strength than on compressive strength. Thus, when latex and reinforcement fibers were added, splitting tensile strength also increased. All mixes satisfied the target of ≥4.2 MPa splitting tensile strength. Splitting tensile strength increased slightly as the microsilica substitution ratio increased; this was attributed to the pozzolanic reaction and filling of fine pores by the microsilica additive. The pozzolanic reaction has the effect of increasing the long-term strength. Moreover, while latex delays initial strength development, it increases the long-term strength. The splitting tensile strength was higher with macrosynthetic fibers than with jute fibers. This is because the macrosynthetic fiber is a structural fiber, which is used to replace the steel fiber and improves the structural performance of the concrete [16, 25, 26]. However, jute fibers are crack-control fibers, which are used to improve durability rather than structural performance [16, 27].

4.4. Chloride Ion Penetration. The chloride ion penetration resistance results of LMFRCRSC for a concrete pavement emergency repair by the reinforcement fiber type and the microsilica substitution ratio are shown in Figure 8. In the chloride ion penetration tests conducted after 28 days of curing, all mixes showed values lower than the target 2000 C. In this study, pressure vibration compaction was used to densify the concrete structure. Thus, with 28 days of curing, all mixes satisfied the target chloride ion penetration amount of ≤2000 C. Also, the target value of 4000 C or less at 4 h of curing was satisfied for all mix ratios. In the case of rapid-hardening cement concretes, cracks occur inside the cement due to high hydration heat in the early stages [1–3]. The addition of reinforcement fibers appeared to offset hydration heat-induced crack formation [5, 12]. Because the reinforcing fibers inhibit the internal cracks due to the generation of hydration heat before sufficient strength is developed, the chlorine ion penetration amount decreases. For the 4-hour curing stage, all mixes showed moderate water penetration properties. The reinforcing fibers inhibit internal crack formation due to the generation of hydration heat before sufficient strength develops. Therefore, it can be concluded that the addition of reinforcing fibers is effective in reducing chloride ion penetration at the initial curing period in the repair concrete. After curing for 28 days, all

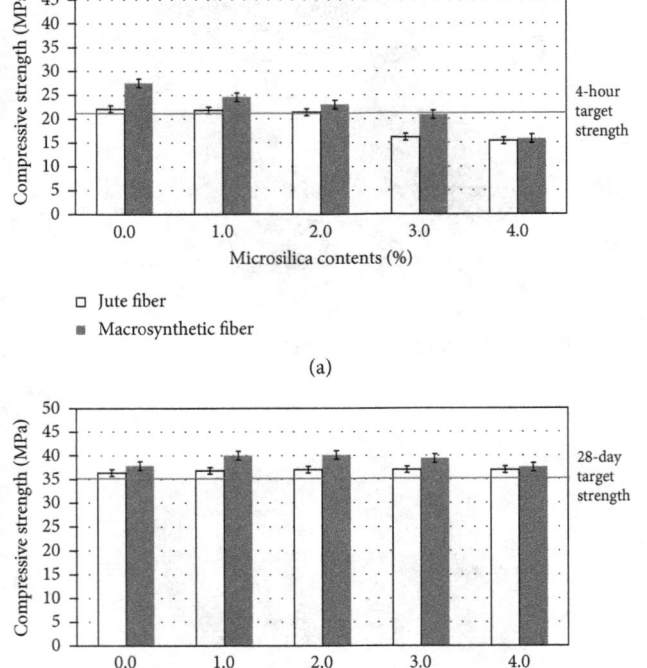

FIGURE 5: Compressive strength of LMFRCRSC. (a) 4-hour curing. (b) 28-day curing.

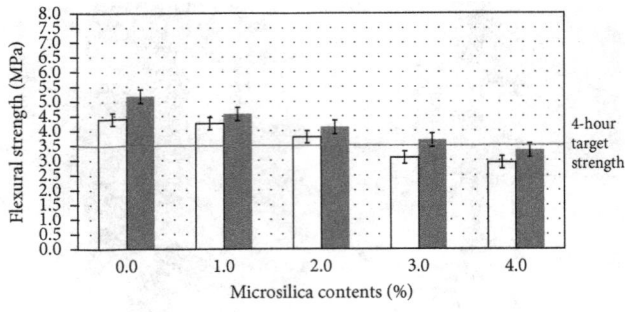

(a)

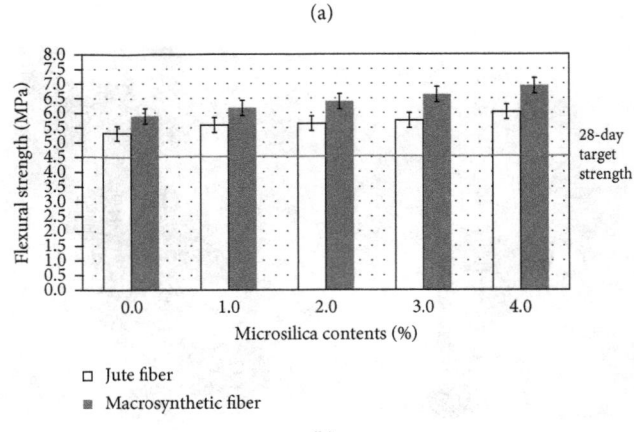

(b)

FIGURE 6: Flexural strength of LMFRCRSC. (a) 4-hour curing. (b) 28-day curing.

mixes showed low-level penetration at up to 2% microsilica content. When the microsilica content exceeded 3%, low-level permeability was observed. Also, the results showed decreased chloride ion penetration in cases using macrosynthetic fibers versus those using jute fibers. The main ingredient of the jute fiber is cellulose, plus a small amount of lignin. Lignin has the effect of delaying the concrete's strength development [16, 27]. Therefore, the hardening was insufficient, and the chloride ion permeability was enhanced.

4.5. Abrasion Resistance. The abrasion test results of LMFRCRSC for a concrete pavement emergency repair by the reinforcement fiber type and the microsilica substitution ratio are shown in Figure 9. As the microsilica substitution ratio increased, the results showed decreased abrasion on the concrete surface. This is because when microsilica is added to the concrete, the pozzolanic reaction and the fine pore-filling effect create a more dense concrete structure with improved abrasion resistance. By the fiber reinforcement type, the results showed increased abrasion resistance in the cases using macrosynthetic fibers versus those using jute fibers. However, the difference in the results was not significant.

5. Conclusions

This study evaluated the influence of reinforcement fiber types and microsilica on the performance of LMFRCRSC for

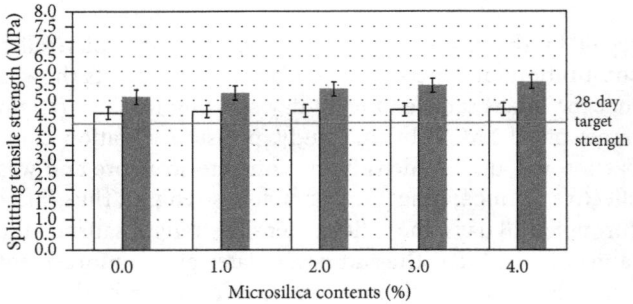

FIGURE 7: Splitting tensile strength of LMFRCRSC.

a concrete pavement emergency repair. A summary of the test results is given below:

(i) In the compressive strength and flexural strength tests with curing for 4 h, the strength decreased as the microsilica substitution ratio increased. For the jute fiber-reinforced concrete mix with 4-hour curing, a target compressive strength of ≥21 MPa and target flexural strength of ≥3.5 MPa were achieved up to a microsilica substitution ratio of 2%. When the macrosynthetic fiber was mixed, the 4-hour curing target strength was satisfied up to a microsilica substitution ratio of 3%.

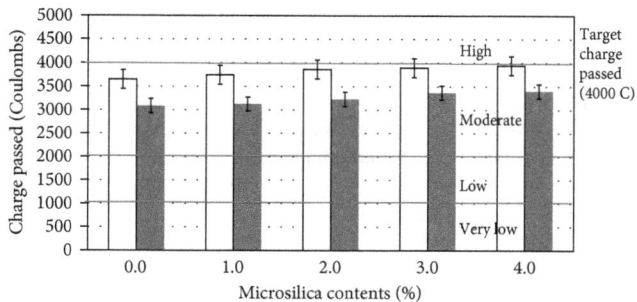

(a)

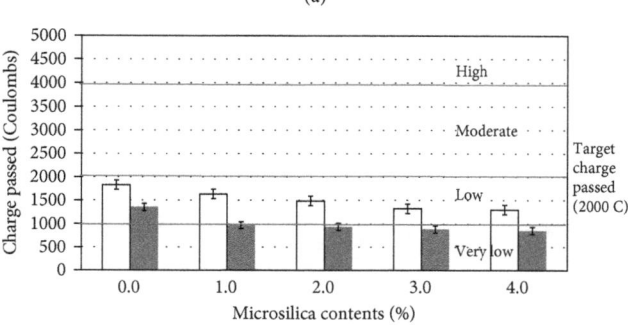

(b)

Figure 8: Chloride ion penetration of LMFRCRSC. (a) 4-hour curing. (b) 28-day curing.

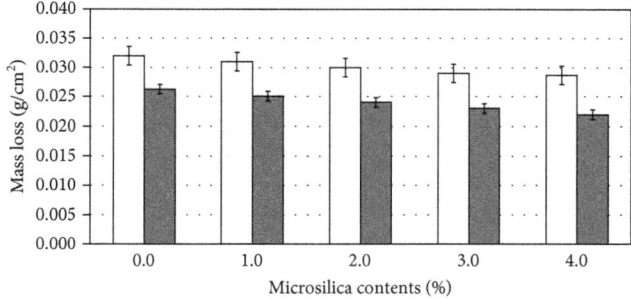

Figure 9: Abrasion resistance of LMFRCRSC.

(ii) With curing for 28 days, the compressive strength, flexural strength, and splitting tensile strength increased as the microsilica substitution ratio increased. Also, with curing for 28 days, the target compressive strength ($\geq$35 MPa), flexural strength ($\geq$4.5 MPa), and splitting tensile strength ($\geq$4.2 MPa) were all satisfied.

(iii) In terms of compressive, flexural, and splitting tensile strength test results, cases with macrosynthetic fibers showed higher values than cases with jute fibers.

(iv) From the chloride ion penetration tests, as the microsilica substitution ratio increased, the chloride

ion penetration decreased. All mixes satisfied the target chloride ion penetration of $\leq$2000 C after curing for 28 days. Regarding the reinforcement fiber, chloride ion penetration decreased more with macrosynthetic fibers than with jute fibers, but the difference was minor.

(v) From the abrasion, as the microsilica substitution ratio increased, the abrasion resistance properties increased. With respect to the fiber type, the macrosynthetic fibers showed slightly better results than jute fibers.

(vi) For mixes that satisfied both the target strength goals and the target chloride ion penetration amount, with improved abrasion resistance, the results showed that when the microsilica substitution ratio was 3% or less and macrosynthetic fiber was used as the reinforcement fiber, LMFRCRSC performance improved for a concrete pavement emergency repair and satisfied the target values. Also, the microsilica would likely be used because of the high cost of microsilica and the marginal increase in benefits with increasing amounts.

Conflicts of Interest

The authors declare that they have no conflicts of interest.

Acknowledgments

This work was supported by the research grant of the Kongju National University in 2017.

References

[1] J. P. Won, J. M. Kim, S. J. Lee, S. W. Lee, and S. K. Park, "Mix proportion of high-strength, roller-compacted, latex-modified rapid-set concrete for rapid road repair," *Construction and Building Materials*, vol. 25, no. 4, pp. 1796–1800, 2011.

[2] ACI, "State-of-the-art report on roller compacted concrete pavements," 2 Tech. Rep.32510R-95, American Concrete Institute, Farmington Hills, MI, USA, 1995.

[3] J. La Hucik, S. Dahal, J. Roesler, and A. N. Amirkhanian, "Mechanical properties of roller-compacted concrete with macro-fibers," *Construction and Building Materials*, vol. 135, pp. 440–446, 2017.

[4] S. Krishna Rao, P. Sravana, and T. Chandrasekhar Rao, "Abrasion resistance and mechanical properties of roller compacted concrete with GGBS," *Construction and Building Materials*, vol. 114, pp. 925–933, 2016.

[5] S. Krishna Rao, P. Sravana, and T. Chandrasekhar Rao, "Investigating the effect of M-sand on abrasion resistance of fly ash roller compacted concrete (FRCC)," *Construction and Building Materials*, vol. 118, pp. 352–363, 2016.

[6] J. N. Karadelis and Y. Lin, "Flexural strengths and fibre efficiency of steel-fibre-reinforced, roller-compacted, polymer modified concrete," *Construction and Building Materials*, vol. 93, pp. 498–505, 2015.

[7] J.-W. Han, J.-H. Jeon, and C.-G. Park, "Mechanical and permeability characteristics of latex-modified pre-packed

pavement repair concrete as a function of the rapid-set binder content," *Materials*, vol. 8, no. 10, pp. 6728–6737, 2015.

[8] J. P. Won, J. H. Kim, C. G. Park, and J. W. Kang, "Shrinkage cracking of styrene butadiene polymeric emulsion-modified concrete using rapid-hardening cement," *Journal of Applied Polymer Science*, vol. 112, no. 4, pp. 2229–2234, 2009.

[9] M. Sprinkel, "Latex-modified concrete overlay containing type K cement," Tech. Rep. FHWA/VTRC 05–R26, Virginia Transportation Research Council, Charlottesville, VA, USA, 2005.

[10] M. Sprinkel, "High early strength latex modified concrete overlay," Tech. Rep. VTRC 88–R12, Virginia Transportation Research Council, Charlottesville, VA, USA, 1998.

[11] M. Sprinkel, "Very-early-strength latex-modified concrete over-lays," Tech. Rep. TAT99–TAR3, Virginia Transportation Research Council, Charlottesville, VA, USA, 1998.

[12] Z. J. Ricardo, "Development of rapid, cement-based repair materials for transportation structures," M.S. thesis, The University of Texas at Austin, Austin, TX, USA, 2013.

[13] D. H. Kim and C. G. Park, "Permeability, abrasion, and impact resistance of latex-modified fibre reinforced concrete for precast concrete pavement application," *Progress in Rubber, Plastics & Recycling Technology*, vol. 29, pp. 239–254, 2013.

[14] R. O. Oh, D. H. Kim, and C. G. Park, "Durability performance of latex modified nylon fiber reinforced concrete for precast concrete pavement applications," *Indian Journal of Engineering and Materials Sciences*, vol. 21, pp. 49–56, 2014.

[15] D. H. Kim and C. G. Park, "Strength, permeability and durability of hybrid fiber-reinforced concrete containing styrene butadiene latex," *Journal of Applied Polymer Science*, vol. 129, pp. 1499–1505, 2013.

[16] S. K. Lee, M. J. Jeon, S. S. Cha, and C. G. Park, "Mechanical and permeability characteristics of latex-modified fiber-reinforced roller-compacted rapid-hardening-cement concrete for pavement repair," *Applied Sciences*, vol. 7, no. 7, p. 694, 2017.

[17] A. Dousti, M. Shekarchi, R. Alizadeh, and A. Taheri-Motlagh, "Binding of externally supplied chlorides in micro silica concrete under field exposure conditions," *Cement and Concrete Composites*, vol. 33, no. 10, pp. 1071–1079, 2011.

[18] AASHTO, *Standard Specification of Rigid Pavement*, American Association of State Highway and Transportation Officials, Washington, DC, USA, 1998.

[19] Korea Expressway Corporation, *Construction Material Quality and Standard for Highway Construction*, Korea Expressway Corporation, Seongnam, Republic of Korea, 2005.

[20] ASTM C39, *Standard Test Method for Compressive Strength of Cylindrical Concrete Specimens*, American Society for Testing Materials, Philadelphia, PA, USA, 2015.

[21] ASTM C496/C496M, *Standard Test Method for Splitting Tensile Strength of Cylindrical Concrete Specimens*, American Society for Testing Materials, Philadelphia, PA, USA, 2011.

[22] ASTM C 78/C78M, *Standard Test Method for Flexural Strength of Concrete (Using Simple Beam with Third-Point Loading)*, American Society for Testing Materials, Philadelphia, PA, USA, 2015.

[23] ASTM C1202, *Standard Test Method for Electrical Indication of Concrete's Ability to Resist Chloride Ion Penetration*, American Society for Testing Materials, Philadelphia, PA, USA, 2012.

[24] ASTM C944, *Standard Test Method for Abrasion Resistance of Concrete or Mortar Surfaces by the Rotating-Cutter Method*, American Society for Testing Materials, Philadelphia, PA, USA, 2012.

[25] J. P. Won, D. H. Lim, C. G. Park, and H. G. Park, "Bond behavior and flexural performance of structural synthetic fiber reinforced concrete," *Magazine of Concrete Research*, vol. 28, no. 6, pp. 401–410, 2003.

[26] J. P. Won, C. G. Park, S. W. Lee, C. I. Jang, and H. Y. Kim, "Performance of synthetic macrofibers in reinforced concrete for tunnel linings," *Magazine of Concrete Research*, vol. 61, no. 3, pp. 165–172, 2009.

[27] A. Bentur and S. Mindess, *Fiber Reinforced Cementitious Composites*, Elsevier Applied Science, London, UK, 1992.

6

Performance Evaluation of Cement-Stabilized Oil Shale Semicoke as Base or Subbase Course Construction Material

Guojin Tan[ID],[1] Zhiqing Zhu,[1] Yafeng Gong[ID],[1] Chenglin Shi,[2] and Ziyu Liu[1]

[1]*College of Transportation, Jilin University, Changchun, Jilin 130025, China*
[2]*Jilin Provincial Transport Scientific Research Institute, Changchun, Jilin 130012, China*

Correspondence should be addressed to Yafeng Gong; gongyf@jlu.edu.cn

Academic Editor: Jun Liu

Oil shale semicoke is a hazardous byproduct in oil shale retorting process. In Jilin province, China, abundant oil shale resources are mainly used for retorting shale oil accompanying with a large number of oil shale semicoke slags, which will generally cause environmental pollution and disposal problems. If oil shale semicoke can be utilized as a road base or subbase course construction material, it will be a great help to the disposal of a long-term accumulated oil shale semicoke in landfill sites. Moreover, the resource treatment of oil shale semicoke in road project construction could obtain positive social and economic benefits. Subsequently, we conducted a series of laboratory tests to study the practicability of cement-stabilized oil shale semicoke as a road base or subbase course construction material, including the raw material test, modified compaction test, unconfined compressive strength (UCS) test, splitting tensile strength (STS) test, compressive resilient modulus (CRM) test, and freezing-thawing test. In this paper, test results were compared with the technical requirements of Chinese road base construction specification, preliminarily confirming that cement-stabilized oil shale semicoke can be used as a subbase course material of a highway or a base course material of a low-grade highway.

1. Introduction

The worldwide petroleum industry is now making a historic leap from conventional to unconventional oil and gas. Oil shale is one of the most important sources of unconventional oil and gas, and the shale oil extracted from oil shale is an ideal alternative energy source and has broad prospects [1]. China is one of the largest producers for unconventional oil extracted from oil shale. The oil shale resources in China are mainly distributed in Jilin, Liaoning, and Guangdong provinces. Oil shale can yield shale oil, retorting gas, and semicoke by retorting technology [2, 3]. Natural oil shale has high mineral composition, so the retorting solid byproduct, that is, oil shale semicoke, is usually in good integrity of rock. Owing to the absence of efficient disposal techniques, landfill is always adopted to dispose of oil shale semicoke. However, this method not only occupies much land resource but also causes great hazards to environment [4–7].

A good solution for disposing of semicoke is partly recycling to burn it as a low-grade fuel, which needs to cocombusted with some high calorific fuels at a certain rate in the circulating fluidized bed combustor. Another common utilization of semicoke is to reprocess it as a raw material of fertilizers and construction products [8–16]. All these processes are complex, and the treatment costs are relatively high. However, the abovementioned methods to dispose of oil shale semicoke are not mature enough to be widely applied currently, and the consumption of semicoke is very limited. Thus, the remainder is still a waste requiring large disposal areas, causing land and environmental problems. The main chemical compositions of oil shale and oil shale semicoke are SiO_2, Al_2O_3, CaO, and Fe_2O_3, which can be successfully utilized in the pozzolanic reaction [17, 18]. Oil shale ash is a byproduct after oil shale or oil shale semicoke combustion. To use the pozzolanic characteristics of oil shale, some scholars studied on the performance of oil shale ash as binder and filler materials of road construction. For instance, Raado et al. [19] used oil shale ash as the main binder for low strength concrete. The laboratory results showed that the water resistance was improved and expansion diminished with proper oil shale ash content in the concrete. Al-Massaid et al. [20] studied the use of oil shale ash as a partial substitute for the asphalt binder in bituminous paving mixtures under normal conditions as well as under freezing and thawing conditions. The test results indicated that the substitution of ash up to 10 percent by volume of asphalt would improve the performance of mixtures under both conditions. It follows that the chemical composition of oil shale ash is partly active. However, no attempt has been made to evaluate the practicability of oil shale semicoke as a road construction aggregate. Through the preliminary investigation, we found that oil shale semicoke of Jilin province is in good integrity of rock and has a certain strength. Meanwhile, oil shale semicoke in the landfill site has a relatively continuous gradation. These characteristics have certain similarity to graded crushed gravel. On the other hand, the active ingredient in semicoke may be beneficial for the pozzolanic reaction with cement. These allow semicoke to have potential applications as aggregate of cement-stabilized road base or subbase course construction, and oil shale semicoke would be utilized in large quantities. Therefore, the problems caused by accumulation of oil shale semicoke can be solved significantly. Furthermore, the environmentally friendly target, saving natural gravel resource, and resource utilization of oil shale semicoke could be achieved. In this connection, researchers have investigated the practicability of cement-stabilized oil shale semicoke as a road base or subbase course construction material.

In recent decades, many researches have been done on using different kinds of industrial waste residue instead of sand or gravel to prepare the pavement base materials, which provide valuable preferences for the research on performance of oil shale semicoke acting as a base course material in this paper. In the Netherlands, blast furnace slags are used as a self-cementing road base material on a wide scale for many decades [21]. Singh et al. [22] carried out investigations on cement-stabilized fly ash-granulated blast furnace slag mixes in order to evaluate its performance for road embankments and for base and subbase courses of highway pavements. They confirmed its suitability for use in base and subbase courses in highway pavements with proper combinations of raw materials. Sharma and Sivapullaiah [23] investigated the effect of the joint activation of fly ash- and ground-granulated blast furnace slag on the unconfined compressive strength of mixtures of these two materials. This study suggested that properly designed combinations of fly ash-slag-lime can be used as construction materials for infrastructure projects such as structural fills or subgrade and subbase courses in pavements without requiring large quantities of lime. Based on experiment study in gangue base course mixture materials for pavement, Wang et al. [24] concluded that coal gangue, acting as the solid skeleton of pavement base materials, can efficiently reduce the drying shrinkage. Additionally, similar to fly ash in mixture material, gangue can react with lime (so called pozzolanic reaction), resulting in the formation of a certain strength, good water stability, and frost resistance, which is good overall for the pavement base. Similarly, additional strength

may be provided, utilizing oil shale semicoke as the solid skeleton of base course materials, because of the pozzolanic characteristics in oil shale semicoke. Zhang et al. [25] conducted test research on temperature shrinkage performance of cement-stabilized coal gangue base course materials. Analysis of test results shows that cement cinderstabilized coal gangue base course materials have less temperature shrinkage coefficient, which are propitious to pavement base course in cold areas. Cheng and Zhang [26] studied anticracking performance of cement-stabilized coal gangue. To have both road performance and anticracking performance, researchers recommended that reasonable proportion of cement in cement-stabilized coal gangue is five percent to six percent. Pasetto et al. [27, 28] carried out experimental evaluation of high-performance base course and road base asphalt concrete with electric arc furnace steel (EAF) slags. The mixtures with EAF slags presented better mechanical characteristics than those of corresponding asphalts with natural aggregate, and the test result of rheological characterization indicated that the environmental sustainability of modified asphalt concrete with EAF slags can be further enhanced. Androjić and Dimter [29] conducted compressive strength testing on stabilisation mixes with steel furnace slag to check whether slag can be used as aggregate in pavement base courses and confirmed that slag can be used in stabilized base courses of pavement structures. Zhang and Ren [30] studied strength and thermal conductive characteristics of cement-stabilized cinder base and concluded that cinder content should be in 35%~40% to have both strength and thermal conductivity of fillings. Liu et al. [31] investigated the pavement performance of cementstabilized municipal solid waste incineration bottom ash aggregate (BAA) and crushed stones. Based on the performance analysis, it is recommended that the dosage of BAA is 20% to 30%. Qin et al. [32] evaluated the corrosion resistance of manganese slag cement-stabilized macadam base. It was observed that corrosion resistance and scour resistance performances were significantly improved with the incorporation of manganese.

In this paper, the properties of raw materials were first studied. Then, the performances of cement-stabilized oil shale semicoke were tested, including compaction characteristic, unconfined compressive strength, splitting strength, compressive modulus of resilience, and frost resistance. Two types of aggregate gradation structures, suspend-dense structure (SDS) and framework-dense structure (FDS), were used throughout the tests to study and compare the performances. Finally, the suitability of oil shale semicoke as a road base or subbase course construction material was discussed.

2. Background

In China, abandoned oil shale semicoke landfill occupies a large number of forest land and farmland owing to the absence of efficient disposal techniques, as shown in Figure 1. In order to alleviate the environmental and land occupation pressure of oil shale semicoke landfill, oil shale retorting plants need to afford a great deal of extra cost every

FIGURE 1: A landfill site of abandoned oil shale semicoke in Jilin province of China.

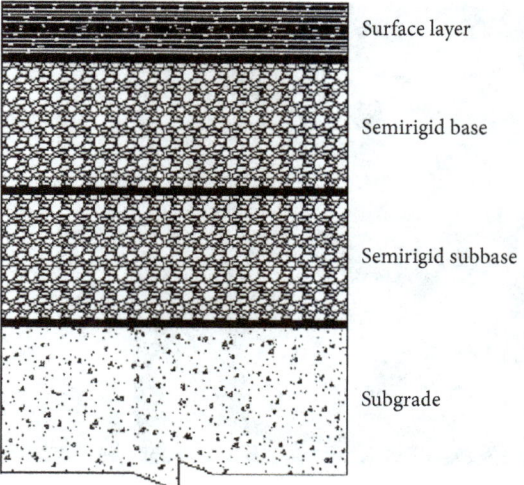

FIGURE 2: Typical semirigid asphalt pavement structure in China.

year. Therefore, a method of disposing large amount of oil shale semicoke is of great significance to environmental protection and ecological sustainability.

Semirigid base has been widely used in high-grade asphalt pavement in China for decades. Semirigid pavement is composed of asphalt pavement, semirigid base, semirigid subbase, and soil base, whose main bearing layer is semirigid base and auxiliary bearing layer is semirigid subbase. Semirigid base or semirigid subbase is usually composed of cement or lime, fly ash, and other inorganic binders mixed with gravel or graded industrial waste aggregate, whose rigidity is between flexible base and rigid base. In the early curing period, semirigid base shows a mechanical property approximated to the flexible base course, while its strength and stiffness increase greatly in the later curing period. Therefore, the mechanical property of semirigid base especially caters to performance requirements of asphalt pavement. Figure 2 gives the typical semirigid pavement structure in China.

Preliminary investigation results show that oil shale semicoke in Jilin province is in good integrity of rock and has a certain strength. Combining with the pozzolanic characteristics of oil shale, there is a potential possibility that graded oil shale semicoke can replace gravel as a road base or subbase course construction material. In terms of above analysis, the suitability of oil shale semicoke acting as the solid skeleton of cement-stabilized pavement base materials was studied. A series of laboratory tests were conducted to investigate the performance of cement-stabilized oil shale semicoke mixture in this paper.

3. Raw Materials

The oil shale semicoke used in this investigation was collected from processing plant in Wangqing County of Jilin province; its appearance is shown in Figure 3. The oil shale semicoke was sieved to different grain size ranges, including coarse aggregates ranging from 5 mm to 31.5 mm and fine aggregates smaller than 5 mm, to prepare the mixture.

A scanning electron microscope (SEM) was used to investigate the microsurface characteristics of oil shale semicoke. Then, X-ray photoelectron spectroscopy (XPS) and X-ray fluorescence spectrometry (XRF) were used to test and analyze the elemental and chemical compositions of oil shale semicoke. The result of SEM is shown in Figure 4.

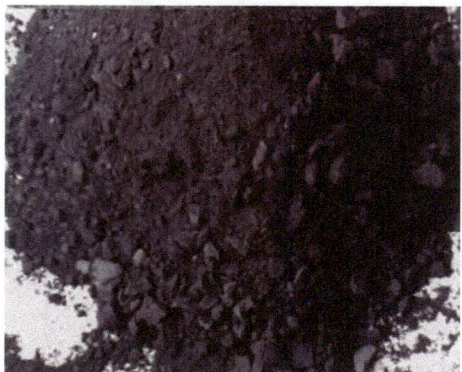

FIGURE 3: Appearance of oil shale semicoke used in this paper.

According to Figure 4, it can be seen that the surface of oil shale semicoke is rough, and it is densely embedded on the surface by irregular granular and flaky structure. The oil shale semicoke is featured in dense structures, and there are only some irregular pores.

Elemental composition of oil shale semicoke was studied by XPS. The XPS is illustrated in Figure 5, and the result of elemental composition analysis is shown in Table 1.

The chemical composition of oil shale semicoke was determined by XRF. Table 2 gives the test result.

As shown in Table 2, the main chemical compositions of oil shale semicoke are SiO_2, Al_2O_3, CaO, and Fe_2O_3, whose active parts probably can be utilized in the pozzolanic reaction, forming an extracertain strength.

The physical properties of oil shale semicoke, including apparent density, gross volume density, water absorption, crushing value, and free expansion rate, were tested as per the Chinese specification of JTG E42-2005 [33]. All oil shale semicoke materials needed for tests should be selected from the stack by using the sample quartering method, and the indoor temperature should be controlled around 25°C. These physical properties of oil shale semicoke were tested at least twice, and Table 3 shows the averages of the test results.

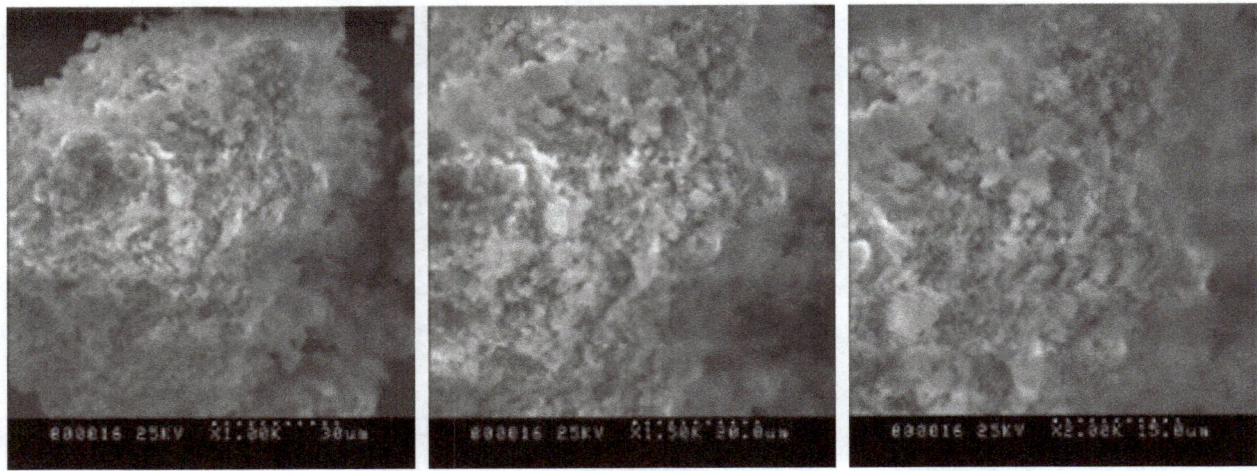

FIGURE 4: The result of SEM.

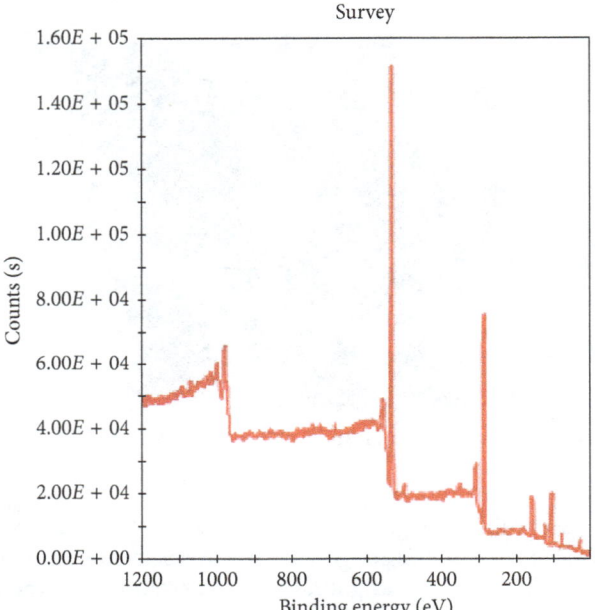

FIGURE 5: XPS of oil shale semicoke.

As shown in Table 3, oil shale semicoke has smaller density, higher porosity, higher water absorption, and lower strength than natural gravel, which means lower strength and stiffness than natural gravel. Therefore, it is not recommended to use oil shale semicoke as a load-bearing aggregate of heavy traffic road construction material.

The ordinary Portland cement of grade 32.5 was procured from Jilin Yatai Dinglu Cement Ltd., and its performance was tested using the method of Chinese specification GB 175 [34]. All testing instruments should be checked to comply with the specification. Laboratory temperature is $20 \pm 2°C$ and the relative humidity should not be less than 50%, while the controlled chamber temperature is $20 \pm 1°C$ and the relative humidity should not be less than 90%. In compressive strength and flexural strength test, the water-binder ratio should be 0.50 and the fluidity of cement mortar is not less than 180 mm. The results are shown in Table 4.

4. Preparation of Specimens

Suspend-dense structure (SDS) and framework-dense structure (FDS) were selected as the gradation structure types of oil shale semicoke aggregate in this paper. SDS and FDS are the two most commonly used gradation structure types of semirigid base or semirigid subbase in China. As shown in Figure 6, FDS contains more coarse aggregates than SDS. That is, the structure model of FDS is fine aggregates filling the skeleton formed by the extrusion of the coarse aggregates, while the structure model of SDS is coarse aggregates distributing in very dense fine aggregates. The two gradation structure types were designed as per Chinese specification JTG D50 [35], which gives a detailed introduction to design standards of Chinese highway asphalt pavement. The gradations and gradation curves of SDS and FDS in this paper are shown in Table 5 and Figure 7, respectively.

To simulate the engineering practice of highway base course construction in China, considering that the density of oil shale semicoke is less than natural gravel, cement contents of SDS and FDS mixes of cement-stabilized oil shale semicoke are both 7% in this paper. The Chinese specification JTG E51 [36] introduces road performance test methods of cement stabilized materials, which is the main reference of this paper.

The moisture contents versus dry density relationships for SDS and FDS mixes of cement-stabilized oil shale semicoke were determined by using the modified compaction test as per Chinese specification JTG E51 [36]. Model LD-140 electric compaction device was used in this paper as shown in Figure 8. According to the dry density and moisture content relationship, as shown in Figure 9, the optimum moisture content (OMC) and maximum dry density (MDD) were calculated. The OMC and MDD are summarized in Table 6.

According to Chinese specification JTG E51 [36], cylinder specimens of SDS and FDS were prepared at their respective OMC and MDD by applying static compressive force in a constant volume sampler. The compacted specimens were 150 mm in diameter and 150 mm in height. All of the samples were cured in a temperature and humidity

TABLE 1: Elemental composition of oil shale semicoke.

Elementals	C	N	Mg	Al	S	Si	Na	Ca	Fe	O	K
Contents (%)	46.51	0.72	2.46	3.52	0.13	10.29	0.40	0.65	0.52	34.59	0.22

TABLE 2: The chemical composition of oil shale semicoke.

Chemical composition	SiO_2	Al_2O_3	CaO	Fe_2O_3	Na_2O	K_2O	MgO	TiO_2	P_2O_5	MnO
Contents (%)	54.57	14.63	6.13	5.95	3.65	2.98	2.67	0.70	0.49	0.10

TABLE 3: The physical properties of oil shale semicoke.

Apparent density (g/cm^3)	Gross volume density (g/cm^3)	Water absorption (%)	Crushing value (%)	Free expansion rate (%)
2.567	1.614	20.5	48.7	3.75

TABLE 4: Experimental results of cement.

Normal consistency (%)	Setting time (min)		Surface area ratio (Blaine method) (m^2/kg)	Compressive strength (MPa)		Flexural strength (MPa)	
	Initial setting time	Final setting time		3 d	28 d	3 d	28 d
25.0	255	328	358	12.9	37.4	3.7	7.0

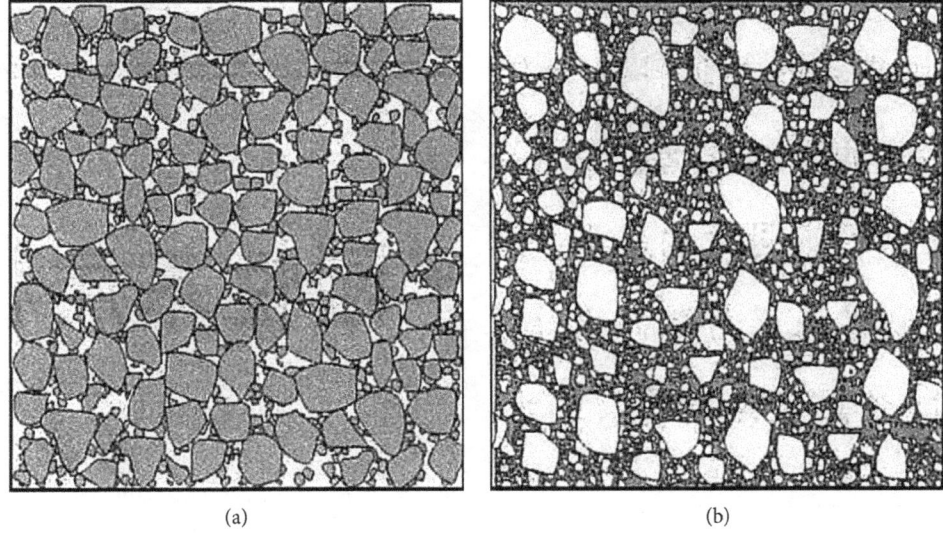

(a)   (b)

FIGURE 6: Gradation structure models: (a) structure model of FDS and (b) structure model of SDS.

TABLE 5: The gradations of SDS and FDS.

Diameter (mm)	37.5	31.5	26.5	19	9.5	4.75	2.36	0.6	0.075
Passing ratio of SDS (%)	100	100	97	90	63	40	25	12	3
Passing ratio of FDS (%)	100	100	97.5	75	45	30	22	10	1.5

controlled chamber (curing temperature = 20°C ± 2°C; curing relative humidity >95%) for 7, 28, and 90 days. The cylinder specimens were used for the tests on the UCS, STS, CRM, and frost resistance performance.

5. Methodology

5.1. Experiments. In this paper, the engineering properties of cement-stabilized oil shale semicoke were evaluated.

Laboratory tests such as unconfined compressive strength (UCS) test, splitting tensile strength (STS) test, compressive resilient modulus (CRM) test, and freezing-thawing test were conducted. The cured specimens of SDS and FDS mixes were prepared as per the specific test requirements. The experiments were divided into 12 test groups, as shown in Table 7. In order to make the test results more representative, each test group consisted of 13 repeated samples, and the test results were processed with statistical methods.

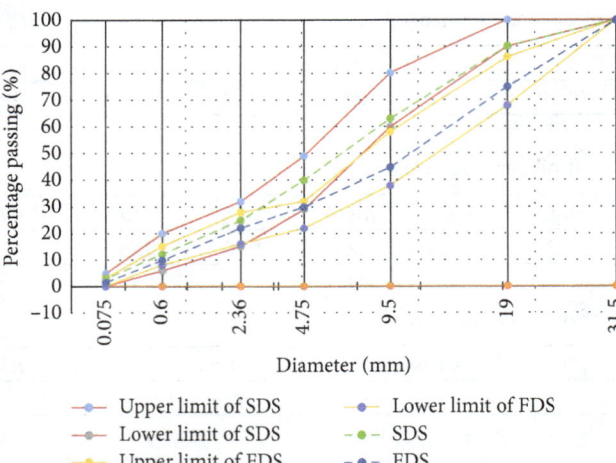

FIGURE 7: Gradation curves of SDS and FDS.

TABLE 7: The experimental programme.

Number	Cement (%)	Aggregate gradation structure	Test
1	7	FDS	UCS (7 d)
2	7	SDS	UCS (7 d)
3	7	FDS	UCS (28 d)
4	7	SDS	UCS (28 d)
5	7	FDS	UCS (90 d)
6	7	SDS	UCS (90 d)
7	7	FDS	STS (90 d)
8	7	SDS	STS (90 d)
9	7	FDS	CRM (90 d)
10	7	SDS	CRM (90 d)
11	7	FDS	Freezing-thawing
12	7	SDS	Freezing-thawing

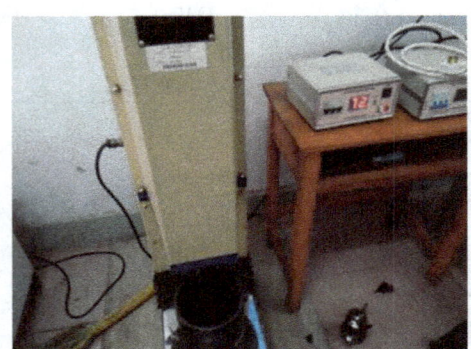

FIGURE 8: Compaction test.

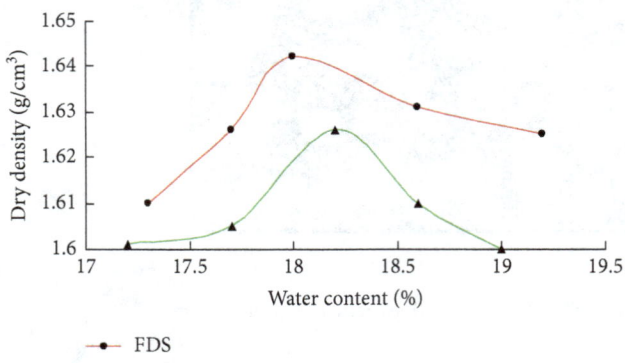

FIGURE 9: Moisture-density curves of SDS and FDS.

TABLE 6: The OMC and MDD of SDS and FDS.

	OMC (%)	MDD (g/cm^3)
FDS	18.0	1.642
SDS	18.2	1.626

5.1.1. Unconfined Compressive Strength Test. UCS is the basic technical index of a base course material. It reflects the ability of the base structure to bear load, which is determined by the coarse aggregate extrusion structure, effect of fine aggregate filling, and bond strength of binder. UCS of SDS

FIGURE 10: Unconfined compressive strength test.

and FDS mixes of cement-stabilized oil shale semicoke was determined as per Chinese specification JTG E51 [36]. As shown in Figure 10, the mechanical testing and simulation (MTS) device was adopted to test the UCS of cured samples, using the strain-control condition at an axial strain rate of 1 mm/min. UCS of samples is calculated using

$$R_c = \frac{P}{A}, \tag{1}$$

where R_c is the UCS of one sample (MPa), P is the maximum loading leading to sample failure (N), and A is the sectional area of one sample (mm^2). Then, the UCS values at curing times of 7, 28, and 90 days were determined.

5.1.2. Splitting Tensile Strength Test. In the STS test, the sample is subjected to the tangential tensile stress and radial compressive stress at the same time, and its stress state is closer to the stress state of actual pavement structure than the direct uniaxial tension. Therefore, STS can reflect the

FIGURE 11: Splitting tensile strength test.

FIGURE 12: Compressive resilient modulus test.

tensile capacity of the base course material, which is mainly determined by the bond strength of the binder. The STS tests were conducted according to Chinese specification JTG E51-T0855-2009 [36]. The samples were cured for 90 days. As shown in Figure 11, the tests were conducted using an aligning jig to apply a force along the longitudinal side of the samples, with an axial strain rate of 1 mm/min. The STS is calculated out by

$$R_i = \frac{0.004178P}{H}, \qquad (2)$$

where R_i is the STS of one sample (MPa), P is the maximum loading leading to sample failure (N), and H is the height of the sample (mm).

5.1.3. Compressive Resilient Modulus Test. The CRM test is used to study the stiffness performance of the base course structure, reflecting the compressive strength and the ability to resist vertical deformation of the base course material. The CRM tests were conducted in accordance with Chinese specification JTG E51 [36]. The samples were cured for 90 days. As shown in Figure 12, the mechanical testing and simulation (MTS) device was used to apply a pressure to samples at a loading rate of 1 mm/min. The maximum applied loading is about 70% of UCS and divided into six progressively loading. The loading gradually increased from a low load to a high load; the deformations of each loading level were recorded. The compressive resilient modulus is calculated using

$$E_c = \frac{PH}{L} \qquad (3)$$

where E_c is the compressive resilient modulus (MPa), P is the progressively pressure (MPa), H is the height of the sample (mm), and L is the resilient deformation (mm).

5.1.4. Freezing-Thawing Test. The freezing-thawing test was conducted to evaluate the ability of frost resistance and fatigue resistance of the road base course structure. The frost resistance was tested according to Chinese specification JTG E51 [36]. The frost resistance of samples after 28 days of curing was evaluated by the loss ratio of UCS after 5 freeze-thaw cycles of samples. The samples of cement-stabilized oil shale semicoke mixes were exposed to 5 freeze-thaw cycles consisting of freezing for 16 h in a temperature controlled chamber of $-18 \pm 1°C$ and thawing for 8 h in water of $+20 \pm 1°C$. Figure 13 gives the samples' appearance change of two gradation structure types, that is, SDS and FDS, during the freezing-thawing test. The frost resistance index is finally given by

$$\text{BDR} = \frac{100R_{DC}}{R_c}, \qquad (4)$$

where BDR, the frost resistance index, is the loss ratio of UCS of the sample after 5 freeze-thaw cycles (%), R_{DC} is the UCS after 5 freeze-thaw cycles (MPa), and R_c is the UCS of the reference sample not subjected to any freeze-thaw cycles (MPa).

6. Results and Discussion

6.1. Unconfined Compressive Strength. The UCS test results of samples with different gradation structure types and curing times are shown in Table 8. According to Chinese specification JTG E51 [36], the strength representative values in Table 8 can be used in the strength performance evaluation of cement-stabilized oil shale semicoke. The UCS at 7 days of cement-stabilized materials should comply with the requirements of Chinese specification JTG/T F20 [37], which rules the UCS standard at 7 days for cement-stabilized materials, as shown in Table 9. It is observed that the UCSs at

FIGURE 13: Freezing-thawing test: (a) appearance of SDS before freeze-thaw cycle, (b) appearance of FDS before freeze-thaw cycle, (c) appearance of SDS after 5 freeze-thaw cycles, and (d) appearance of FDS after 5 freeze-thaw cycles.

TABLE 8: UCS test results.

Curing times (days)	Gradation structure types	Minimum strength value (MPa)	Maximum strength value (MPa)	Mean value (MPa)	Standard deviation	Coefficient variation (%)	Strength representative value (MPa)
7	FDS	3.98	5.01	4.51	0.291	6.45	4.03
	SDS	4.01	4.85	4.39	0.306	6.97	3.89
28	FDS	5.01	6.24	5.36	0.385	7.17	4.73
	SDS	6.19	7.71	6.75	0.507	7.51	5.92
90	FDS	5.82	6.31	6.11	0.472	7.21	5.33
	SDS	7.18	7.88	7.49	0.503	7.60	6.66

TABLE 9: UCS standard at 7 days for cement-stabilized materials.

Structure layers	Highway grade	Very heavy traffic (MPa)	Heavy traffic (MPa)	Medium and light traffic (MPa)
Base course	Freeway, class 1 highway	5.0–7.0	4.0–6.0	3.0–5.0
	Class 2 and below highway	4.0–6.0	3.0–5.0	2.0–4.0
Subbase course	Freeway, class 1 highway	2.5–4.5	2.5–4.5	2.0–4.0
	Class 2 and below highway	2.0–4.0	2.0–4.0	1.0–3.0

7 days of cement-stabilized oil shale semicoke with two gradation structure types, SDS and FDS, are both greater than 3.5 MPa, which meets the subbase requirements of all highway grades and traffic grades and the base requirements of highway with lower grade or lighter traffic.

UCS of cement-stabilized oil shale semicoke grows with the increase of curing time. Figure 14 gives the trends of strength growth for two gradation structure types with curing time.

From Figure 14, UCS of cement-stabilized oil shale semicoke increases with increasing curing time. The UCS

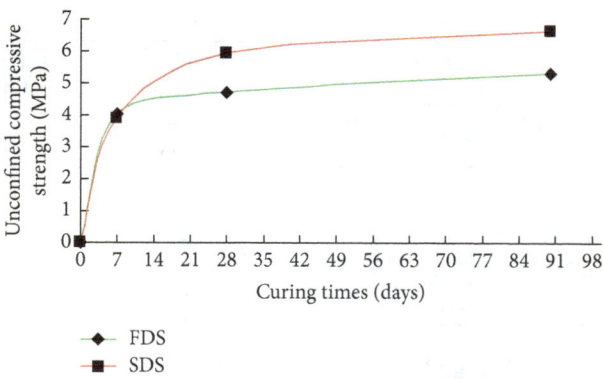

FIGURE 14: UCS of FDS and SDS growing with curing time.

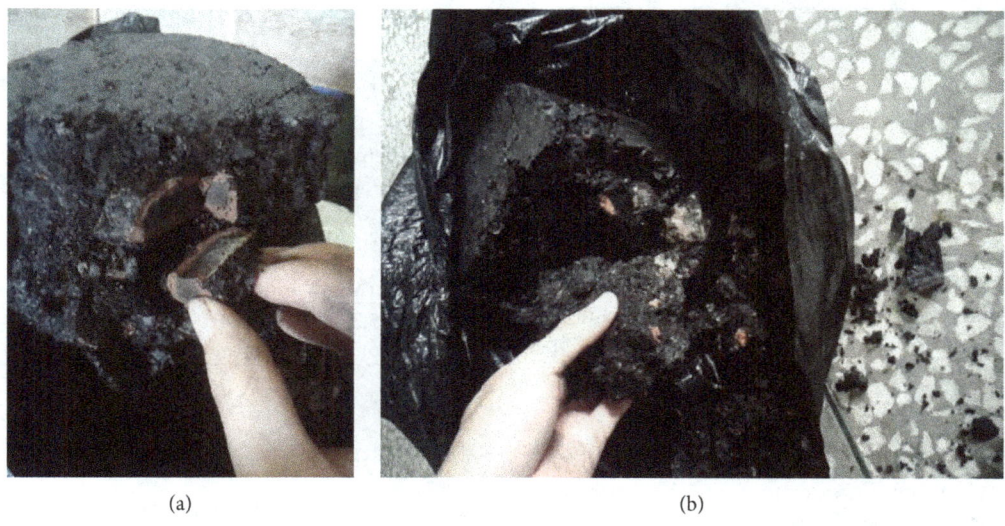

(a)   (b)

FIGURE 15: The internal cracking surfaces of samples: (a) FDS and (b) SDS.

results at 7 days of FDS and SDS are similar, whereas the strength growth of SDS at 28 days is significantly higher than that of FDS. From then on, the strength growth trends of the two gradation structure types are similar until the 90 days. It's obvious that the strength growth difference of the two gradation structure types mainly occurs during the period from 7 days to 28 days. The mixture structure is composed of aggregate, fine material filling, and cement bonding, whose strength is mainly determined by the strength of aggregate itself and the bond strength of hydration of cement. During the 7 days of curing time, the bond strength of cement is initially formed, which is smaller than the strength of aggregate itself and the UCS is mainly determined by the bond strength of cement. Therefore, the UCS results at 7 days of FDS and SDS are similar because of the same cement content. During the period from 7 days to 28 days, the bond strength of cement increases significantly. Considering that the crushing value of oil shale semicoke aggregate is larger than natural gravel, the strength of the sample is largely determined by the crushing of the aggregate. FDS mixture contains more coarse aggregate than SDS mixture, which means that the FDS sample is easier to be crushed than SDS, showing smaller UCS than SDS.

The crushed samples at 28 days of two gradation structure types were artificially cracked, as shown in Figure 15. Comparing the internal cracking surfaces of the two samples, it can be seen that FDS appeared to be an obvious phenomenon of locally crushed coarse aggregate leading to failure. Analyzing the gradation difference of FDS and SDS, SDS has more 2.36 mm–9.5 mm aggregate content, less 19 mm–31.5 mm aggregate content than FDS, which means less influence from framework aggregate crushing of oil shale semicoke on the sample failure than FDS. Secondly, gradation of SDS makes the mix graded better than FDS, thus increasing the compacted density and hence the mechanical strength of the compacted mixture. There is also an appreciable gain in strength with addition of fine oil shale semicoke at constant cement content, because fine oil shale semicoke contains reactive SiO_2, Al_2O_3, CaO, and Fe_2O_3, which can be effectively utilized in the pozzolanic reaction with cement.

6.2. *Splitting Tensile Strength.* Table 10 shows the STS test results of SDS and FDS at curing time of 90 days.

As shown in Table 10, the STS of FDS is 0.46 MPa, while the STS of SDS is 0.65 MPa, indicating that the aggregate

TABLE 10: The test results of STS.

	Minimum strength value (MPa)	Maximum strength value (MPa)	Mean value (MPa)	Standard deviation	Coefficient variation (%)	Strength representative value (MPa)
FDS	0.48	0.68	0.59	0.080	13.57	0.46
SDS	0.64	0.75	0.71	0.035	4.99	0.65

TABLE 11: The test results of CRM.

	Progressively load (kN)	Mean value of the resilient deformation (mm)
FDS	10	0.324
	20	0.401
	30	0.460
	40	0.527
	50	0.574
	60	0.660
SDS	10	0.320
	20	0.406
	30	0.475
	40	0.525
	50	0.589
	60	0.658

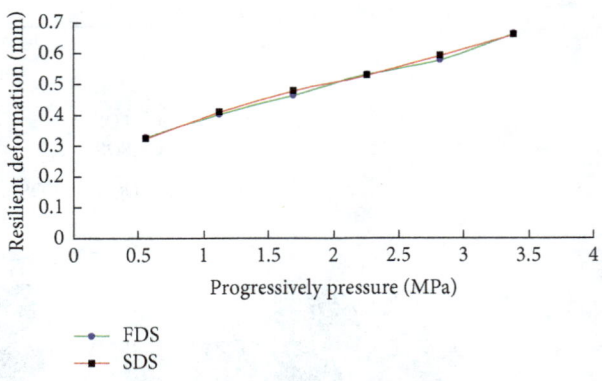

FIGURE 16: The relation curve between progressively pressure and resilient deformation.

TABLE 12: The results of freezing-thawing tests.

	R_c (MPa)	R_{DC} (MPa)	BDR (%)	Mass loss rate (%)
FDS	5.36	4.41	82.28	1.7
SDS	6.51	5.93	91.10	0.7

structure has obvious influence on the tensile strength of mixtures, and SDS exhibits better tensile properties. In addition, the coefficient variation of FDS is 13.57%, while the coefficient variation of SDS is 4.99%, indicating that SDS has better performance stability and smaller discreteness.

6.3. Compressive Resilient Modulus.

The CRM of SDS and FDS was tested in parallel. In order to make the test results representative, a set number of repeated tests were conducted by using three times the mean square deviation method to eliminate outliers. The mean values of the resilient deformation at 6 loading levels are listed in Table 11, and the relation curve between progressively pressure and resilient deformation is shown in Figure 16.

It can be seen from Figure 16 that the CRM results of SDS and FDS are similar. Modifying the false deformation of curve in Figure 16 according to Chinese specification JTG E51 [36], the CRM was calculated using (3). The calculation results are as follows: The CRM values of SDS and FDS were 1010 MPa and 1130 MPa, respectively. During the curing period of 90 days, there is a protracted and complex physical and chemical reaction between cement and the reactive SiO_2, Al_2O_3, CaO, and Fe_2O_3 in oil shale semicoke, resulting in additional filling and bonding effect, which contributes to the increase of the stiffness performance of samples.

6.4. Frost Resistance.

The results of freezing-thawing tests with 5 freeze-thaw cycles are shown in Table 12. It is observed that the BDR and mass loss rate of FDS are 82.28% and 1.7%, respectively, while the BDR and mass loss rate of SDS are 91.1% and 0.7%, respectively. Therefore, it can be said that cement-stabilized oil shale semicoke mixture shows good frost resistance, and SDS has better frost resistance than FDS.

7. Conclusions

In this paper, the suitability of cement-stabilized oil shale semicoke as a road base or subbase course construction material was investigated, and the following conclusions are reached:

(1) The UCS at 7 days of cement-stabilized oil shale semicoke could satisfy Chinese subbase requirements of all highway grades and traffic grades and base requirements of highway with lower grade or lighter traffic.

(2) The CRM values of SDS and FDS at 90-day curing time could reach 1010 MPa and 1130 MPa, respectively. The STS of SDS and FDS at 90-day curing time are 0.65 MPa and 0.46 MPa, respectively. The results of freezing-thawing tests indicate that cement-stabilized oil shale semicoke shows good frost resistance. Oil shale semicoke contains pozzolanic reactive SiO_2, Al_2O_3, CaO, and Fe_2O_3, which is responsible for the positive performance of cement-stabilized oil shale semicoke mixture.

(3) The SDS mixtures presented better mechanical characteristics than those of FDS throughout the laboratory tests. It is because the crushing value of oil shale semicoke is larger than natural gravel, while FDS contains more coarse aggregate of oil shale semicoke than SDS. When the bond strength of cement increases rapidly with the curing time, locally

crushed coarse aggregate has decisive influence on the sample failure.

(4) An appropriate mix of cement-stabilized oil shale semicoke with the gradation of SDS can be prepared for construction of various subbase course or base course with lower grade or lighter traffic, enabling the use of these waste materials in highway construction. Additionally, it is of great significance in broadening the pavement material source, eliminating environmental hazards, and recycling oil shale semicoke wastes.

Further investigation on cement-stabilized oil shale semicoke-gravel mixes as a highway construction material is in progress.

Conflicts of Interest

The authors declare that there are no conflicts of interest regarding the publication of this paper.

Acknowledgments

The authors express their appreciation for financial supports of National Natural Science Foundation of China (Grant no. 51478203); Transportation Science and Technology Project of Jilin Province, which research on strengthening technology of small and medium span beams based on the transverse tension prestressed method; and Training Program for Outstanding Young Teachers of Jilin University.

References

[1] C. Zou, Z. Yang, J. Cui et al., "Formation mechanism, geological characteristics and development strategy of nonmarine shale oil in China," *Petroleum Exploration and Development*, vol. 40, no. 1, pp. 15–27, 2013.

[2] C. Chen, "The new model of mining oil shale in Jilin province," *China Mining Magazine*, vol. 16, no. 5, p. 55, 2007.

[3] J. Bai, Z. Bai, Q. Wang, and S. Li, "Process simulation of oil shale comprehensive utilization system based on Huadian-type retorting technique," *Oil Shale*, vol. 32, no. 1, pp. 66–81, 2015.

[4] H. Qin, Y. Yue, H. Liu, and Q. Wang, "Current status and prospect of oil shale retorting technologies in China," *Chemical Industry and Engineering Progress*, vol. 34, no. 5, pp. 1191–1198, 2015.

[5] L. Pllumaa, A. Maloveryan, M. Trapido, H. Sillak, and A. Kahru, "Study of the environmental hazard caused by the oil shale industry solid waste," *Alternatives to Laboratory Animals*, vol. 29, no. 3, pp. 259–267, 2001.

[6] H. Raave, S. Kapak, and K. Orupld, "Phytotoxicity of oil shale semi-coke and its aqueous extracts: a study by seed germination bioassay," *Oil Shale*, vol. 24, no. 1, pp. 59–71, 2007.

[7] L. Vallnera, O. Gavrilovab, and R. Vilub, "Environmental risks and problems of the optimal management of an oil shale semi-coke and ash landfill in Kohtla-Järve, Estonia," *Science of the Total Environment*, vol. 524-525, pp. 400–415, 2015.

[8] K. Brendow, "Global oil shale issues and perspectives," *Oil Shale*, vol. 20, no. 1, pp. 81–92, 2003.

[9] H. Liu, W. Liang, H. Qin, and Q. Wang, "Synergy in co-combustion of oil shale semi-coke with torrefied cornstalk," *Applied Thermal Engineering*, vol. 109, pp. 653–662, 2016.

[10] Y. Yang, X. Lu, and Q. Wang, "Investigation on the co-combustion of low calorific oil shale and its semi-coke by using thermogravimetric analysis," *Energy Conversion and Management*, vol. 136, pp. 99–107, 2017.

[11] Q. Wang, Y. Xuan, H. Liu, X. Li, and M. Chi, "Gaseous emission and thermal analysis during co-combustion of oil shale semi-coke and sawdust using TG-FTIR," *Oil Shale*, vol. 32, no. 4, pp. 356–372, 2015.

[12] H. Fadaei, M. Sennoune, S. Salvador, A. Lapene, and G. Debenest, "Modelling of non-consolidated oil shale semi-coke forward combustion: Influence of carbon and calcium carbonate contents," *Fuel*, vol. 95, pp. 197–205, 2012.

[13] B. An, W. Wang, G. Ji et al., "Preparation of nano-sized α-Al$_2$O$_3$ from oil shale ash," *Energy*, vol. 35, no. 1, pp. 45–49, 2010.

[14] L. Zhang, X. Zhang, S. Li, and Q. Wang, "Comprehensive utilization of oil shale and prospect analysis," *Energy Procedia*, vol. 17, pp. 39–43, 2012.

[15] S. H. Aljbour, "Production of ceramics from waste glass and Jordanian oil shale ash," *Oil Shale*, vol. 33, no. 3, pp. 260–271, 2016.

[16] M. M. Smadi and R. H. Haddad, "The use of oil shale ash in Portland concrete," *Cement and Concrete Composites*, vol. 25, no. 1, pp. 43–50, 2003.

[17] H. Liu, X. Sun, C. Jia, H. Qin, and Q. Wang, "Analysis of mineral composition of oil shale semicoke and its co-combustion performance with cornstalks," *Transactions of the Chinese Society of Agricultural Engineering*, vol. 32, no. 10, pp. 226–231, 2016.

[18] L. M. Raado, R. Kuusik, T. Hain, M. Uibu, and P. Somelar, "Oil shale ash based stone formation-hydration, hardening dynamics and phase transformations," *Oil Shale*, vol. 31, no. 1, pp. 91–101, 2014.

[19] L. M. Raado, T. Hain, E. Liisma, and R. Kuusik, "Composition and properties of oil shale ash concrete," *Oil Shale*, vol. 31, no. 2, pp. 147–160, 2014.

[20] H. Al-Massaid, T. Khedaywi, and M. Smadi, "Properties of asphalt-oil shale ash bituminous mixtures under normal and freeze-thaw conditions," *Journal of the Transportation Research Board*, vol. 1228, no. 1228, pp. 54–62, 1989.

[21] A. A. A. Molenaar, S. Akbarnejad, and L. J. M. Houben, "Performance of pavements with blast furnace base courses," in *Proceedings of Geoshanghai International Conference*, vol. 2010, pp. 476–483377, Shanghai, China, June 2010.

[22] S. P. Singh, D. P. Tripathy, and P. G. Ranjith, "Performance evaluation of cement stabilized fly ash–GBFS mixes as a highway construction material," *Waste Management*, vol. 28, no. 8, pp. 1331–1337, 2008.

[23] A. K. Sharma and P. V. Sivapullaiah, "Strength development in fly ash and slag mixtures with lime," in *Proceedings of the Institution of Civil Engineers: Ground Improvement*, vol. 169, pp. 194–205 no. 3, London, UK, August 2016.

[24] F. J. Wang, J. Luo, and H. R. Zhu, "Experimental study in gangue base-course mixture materials for pavement," *Advanced Materials Research*, vol. 860-863, pp. 1270–1273, 2013.

[25] H. Z. Zhang, P. F. Cheng, H. J. Shao, and L. Wang, "Test research on temperature shrinkage performance of cement and cinder stabilized coal gangue base course materials," *Journal of Highway and Transportation Research and Development*, vol. 24, no. 11, pp. 29–32, 2007.

[26] P. F. Cheng and H. Z. Zhang, "A study on anti-cracking performance of cement stabilized coal gangue," *Highway*, vol. 52, no. 10, pp. 85–88, 2007.

[27] M. Pasetto and N. Baldo, "Experimental evaluation of high performance base course and road base asphalt concrete with electric arc furnace steel slags," *Journal of Hazardous Materials*, vol. 181, no. 1–3, pp. 938–948, 2010.

[28] M. Pasetto, A. Baliello, G. Giacomello, and E. Pasquini, "Rheological characterization of warm-modified asphalt mastics containing electric arc furnace steel slags," *Advances in Materials Science and Engineering*, vol. 2016, Article ID 9535940, 11 pages, 2016.

[29] I. Androjić and S. Dimter, "Compressive strength of steel slag stabilized mixes," *Gradevinar*, vol. 64, no. 1, pp. 15–21, 2012.

[30] X. Zhang and K. Ren, "Experimental study on strength and thermal conductive characteristics of cement stabilized cinder base," *Non-Metallic Mines*, vol. 40, no. 4, pp. 47–49, 2017.

[31] D. Liu, L. Li, and H. Cui, "Pavement performance of cement stabilized municipal solid waste incineration bottom ash aggregate and crushed stones," *Journal of Tongji University (Natural Science)*, vol. 43, no. 3, pp. 405–409, 2015.

[32] F. Qin, Z. B. He, and Q. N. Huang, "Research on the corrosion resistance of manganese slag cement stabilized macadam base," *Advanced Materials Research*, vol. 280, pp. 13–18, 2011.

[33] Ministry of Communications of the People's Republic of China, *Test Methods of Aggregate for Highway Engineering*, JTG E42, Ministry of Communications, Beijing, China, 2005.

[34] CIQ China Inspection and Quarantine, Standardization Administration of China, *Common Portland Cement*, GB175, CIQ China, Beijing, China, 2007.

[35] Ministry of Communications of the People's Republic of China, *Specifications for Design of Highway Asphalt Pavement*, JTG D50, Ministry of Communications, Beijing, China, 2006.

[36] Ministry of Communications of the People's Republic of China, *Test Methods of Materials Stabilized with Inorganic Binders for Highway Engineering*, JTG E51, Ministry of Communications, Beijing, China, 2009.

[37] Ministry of Communications of the People's Republic of China, *Technical Guidelines for Construction of Highway Roadbases*, JTG/T F20, Ministry of Communications, Beijing, China, 2015.

7

Evaluation of Selected Physicomechanical Properties of SFRC according to Different Standards

Tereza Komárková ⓘ**, Jaromír Láník, and Ondřej Anton**

Faculty of Civil Engineering, Brno University of Technology, Veveří 95, Brno, Czech Republic

Correspondence should be addressed to Tereza Komárková; tereza.komarkova@vutbr.cz

Academic Editor: Rishi Gupta

Steel fiber reinforced concretes are currently very popular, especially in the construction of industrial floors of warehouses and other halls with relatively large floor areas. However, it is important to mention that despite the rapid development of steel fiber reinforced concretes, the standards and regulations for their designing and testing have not been unified yet. This paper presents findings about the physicomechanical parameters of the steel fiber reinforced concretes manufactured by adding steel fibers into the truck mixer on the building site. The experimentally obtained results from the performed tests of tensile strength in bending according to various procedures are compared, and the suitability of the methods used is assessed according to these procedures.

1. Introduction

In the current building practice, steel fiber reinforced concrete is still mostly considered a new material. However, the idea of adding hard fibers to lower the undesirable natural fragility of the common building material is relatively old. The beginning of the 1960s marked a new period of a rising interest in reinforcing fragile building materials with fibers. First steel fiber reinforced concretes and glass fiber cement appeared. The various fiber technologies underwent rapid development, with ever-improving technical parameters, such as tensile strength and elasticity modulus (carbon fibers, aramid fibers, etc.) [1]. Various fibers were being experimentally added to concrete. The first three types to be employed in the building practice were steel wires, glass fibers, and polypropylene fibers. Their purpose was to improve the basic physicomechanical property of the composite, that is, to increase the tensile strength in bending at which the first crack caused by the dispersed fiber reinforcement appears, compared to the bending strength of common concrete. The dispersed fibers, however, can substantially increase the final strength, that is, the maximum load that the test specimen can sustain. However, this is preceded by a significant deformation, which in most

practical applications exceeds the acceptable values, accompanied by an appearance of a number of very narrow cracks. The problems with strength lead to the promotion of steel fiber reinforced concretes focusing on their uniquely high toughness [2]. Steel fiber reinforced concretes usually display high-transformation work related to the ability to keep the useful load capacity even in large deformations. The toughness and the related properties are based on the fracture process of the composite, during which fibers are drawn and broken simultaneously with the increase in the number and width of cracks in the structure of the manufactured composite. Toughness, that is, the ability of steel fiber reinforced concrete to transfer load even after a relatively high deformation (compared to common concrete), is a useful property which is increasingly demanded but rarely appropriately specified. The even dispersion of fibers leads to the effect of spatial action of fibers in the structure of a composite and to stiffening of its whole structure [3]. As such, the fibers cannot be regarded as a reinforcement in the traditional sense of the word but rather as a stiffening element of the relatively fragile structure of plain concrete [4, 5]. The initial optimistic predictions of the wide usage of steel fiber reinforced concretes assumed full-fledged interaction between the fibers and the structure of the original hardened common concrete. The interaction depends on cohesion between the fibers and the cement paste of the composite. There are many technologies for steel fiber manufacturing, which causes that they have various physical mechanical properties. Dimensions are also very important, such as length, thickness, or shape. These parameters are very important for required cohesion of fibers and cement matrix, which improves mechanical properties of steel-reinforced concrete. A vast majority of fibers with a practical use in steel fiber reinforced concretes show very low natural cohesion with the hardened concrete [4]. The efficiency of dispersed fibers does not depend on the cohesion only [5]. In practice, all steel fiber reinforced concretes face the inevitable problems of the correct amount and concentration of fibers and their even dispersion and orientation [5, 6]. Fiber-reinforced concrete with steel fibers gradually became commonplace in applications where it was required for the structural element to keep its loading capacity even after a considerable deformation, not to exhibit wide cracks, and, in the case of damage, not to fall apart for as long as possible. Its higher strength against impact load and other dynamic impacts proved to be also quite useful in such cases. This is reflected by a common use of steel fiber reinforced concretes for large industrial floors with heavy duty traffic or high load.

The insufficient homogeneity can be partially due to the manufacturing process of the fresh mixture or its processing. During the manufacturing process, the steel fibers, which determine the characteristics of the resulting steel fiber reinforced concrete as its principal component, must be dosed in the order required by the type of aggregate used. The diversity of fiber types results in the need to respect the recommended main principles determined on the basis of the long-term experience obtained from testing the manufacture of steel fiber reinforced concretes. All types of steel fibers must be added as the last ingredient of the mixture. Considering the variety of the existing machinery used for the manufacture of common concrete, which should also be used for the manufacture of common steel fiber reinforced concrete, it is always necessary to determine in advance that the machinery and given procedure of dosing fibers in the manufacture of fresh steel fiber reinforced concrete lead to achieving the necessary homogeneity of the mixture. A reliable path to manufacturing homogenous steel fiber reinforced concrete leads through the use of the machinery used in concrete plants, that is, compulsory mixers. The standard manufacture of concrete, done so far mostly in truck mixers, is not an optimal solution. Although the main principle for the use of steel fibers—to add steel fibers as the last component—is fulfilled, achieving homogeneity is difficult and at a cost of prolonging the time of mixing because the performance of truck mixers is considerably lower than that of compulsory mixers. In practice, where only the minimum amount of steel fibers per unit volume of fresh concrete is applied (as in the case of SFRC floors of the industrial halls), achieving the homogeneity of steel fiber reinforced concrete remains a pipe dream. The same can be

said about its characteristics when applied, considering the influence of fibers on the hardening of the structure of cement composite. The relatively extensive usage of steel fiber reinforced concrete in floors leads to the emergence of new steel fiber manufacturers, new types of fibers, and thus a wide range of fibers which allows for choosing a suitable type of fiber for the required application.

Achieving broader knowledge of the properties of steel fiber reinforced concrete similar to that of other materials is only possible through a set of tests and experience obtained from real applications over a sufficiently long period of time. Only by connecting the material testing and the practice, we can achieve the defined objective. At the moment, it is possible to contribute to accelerating the aforementioned process by achieving a higher number of applications with a practical use of steel fiber reinforced concrete in a short time. One of the possibilities is a gradual unification of the testing procedures, which will lead to a unified assessment of the main characteristics of steel fiber reinforced concrete and to outputs allowing for a common procedure during preparations and implementation. For certain structures, the properties of steel fiber reinforced concretes can not only bring economic benefits but also be crucial for the implementation of the structures themselves. This applies even in the cases where steel fiber reinforced concrete is to be used in structures with traditional reinforcing-steel reinforcements or in prestressed structures. The method of testing the properties of steel fiber reinforced concretes, which slowly became a common part of structure designs, is currently not unified in terms of size of the test specimens, test arrangement, and their assessment. Fiber reinforced concrete with steel fibers must be included into building materials. The prerequisite for its use in concrete applications is having a good knowledge of the properties of the resulting composite obtained from mechanical tests with an appropriate assessment [7–14].

In this article, there are compared designed mixtures of steel fiber reinforced concrete, and they are tested according to standards common in Middle Europe. As it was mentioned before, resulting physicomechanical properties of the composite are affected by many factors. Authors focused on the test evaluation options according to described legislative standards with purpose of confirming of uniting methodology of determination bending strength, which is crucial for the steel fiber reinforced concrete classification.

2. Experiment

The experiment target was, according to the findings, to compare commonly used mixtures of the concrete of specified strength classes with concrete, where various amount of steel fibers were added, according to the flexural strength after cracks formations. The next target of the experiment was to compare the results evaluated according to various recommendations and standards for designing the steel fiber reinforced concrete constructions. Tests results were evaluated by three chosen recommendations for

TABLE 1: Design of construction concrete with individual amounts of steel fibers [7].

Concrete	Amount of fibers (kg per m³)
C 16/20	—
	20
	25
	30
C 20/25	—
	20
	25
	30
C 25/30	—
	20
	25
	30
C 30/37	—
	20
	25
	30

TABLE 2: Properties of the steel fibers.

Fibers	Length (mm)	Diameter (mm)	Material	Tensile strength
DE 50/1,0 N	50 ± 10%	1.0 ± 10%	C7D	1100 ± 15%

construction designing and steel fiber reinforced concrete testing:

(1) Technické podmínky 1: Vláknobeton–Část 1 Zkoušení vláknobetonu: Vyhodnocení destruktivních zkoušek a stanovení charakteristického pracovního diagramu vláknobetonu pro navrhování vláknobetonových konstrukcí. hereinafter referred as TP FC 1-1 [15].

(2) According to the Directive Österreichische Vereinigung für Beton- und Bautechnik, hereinafter referred as ÖVBB [16].

(3) Performance classes of steel fiber reinforced concrete [17].

The concrete mixture was prepared in a laboratory. From each batch prepared, three test prisms with the dimensions of $150 \times 150 \times 700$ mm were made for the tests of tensile strength in bending and for the determination of the residual strength after the appearance of cracks. 4 mixtures of fresh concrete with 4 variations of steel fibers amount, totally 48 test specimens, were made considering the required minimum amount of specimens for statistic evaluation and scale of experiment (Table 1).

Every fresh mixture was tested in consistency by the method of the slump test and flow test, and the air content was also tested. The specimens were cured and kept according to ČSN EN 12390-2 Testing hardened concrete—Part 2: Making and curing specimens for strength tests [18].

The steel fibers, from the German manufacturer Krampe-Harex s.r.o., marked as DE50/0,1 N, were used in every mixture. The parameters of these fibers taken from the technical list are described in Table 2. Due to the requirement for pumpability of fresh concrete, there was chosen a procedure of dosing fibers in steps by 20 kg–25 kg–30 kg per m³ of fresh concrete.

For experimental testing, reference mixtures were used, which contained not only cement and aggregate but also fine slag (JMS), plasticizer, and power fly ash (Tables 3 and 4).

During the experiment, the manufactured test prisms with the dimensions of $150 \times 150 \times 700$ mm were subjected to bending tests with the testing span of supports set to 600 mm. The bending tests were carried out on a testing device which allowed for loading by a continuous controlled deformation up to the point of damage. The bending tests of prisms were always performed in four-point bending setup according to regulation TP FC 1-1 [15]. The four-point test setup was chosen on purpose so that the specimen is broken by bending in the critical cross section dependent on the distribution of steel fibers in the test specimen (Figure 1). The results of the test of steel fiber reinforced concrete prisms in four-point bending without a notch are mostly on the safe side. That was also the main reason why this type of arrangement of the test had been chosen. The bending tests were carried out on a mechanical testing press with a set range of 0–40 kN. The values of the strength parameters were read by a calibrated strain gauge. The rate of the controlled deformation was set to 0.04 mm/min¹, and the strain rate was constant during the test. The deflection of the test prism was read above the support with the use of a rigid steel measuring frame located on the upper surface of the test specimen, and the values were doubled afterwards. The reading was performed by a calibrated inductance trajectory sensor HBM with a measuring range of up to 100 mm and a sensitivity of 0.001 mm. The output values were recorded with the data storage rate of 5 Hz. The duration of the executed bending tests of standard steel fiber reinforced concrete prisms ranged from 45 to 60 minutes. In the case of reference concretes without steel fibers, the tests took from half to one third of the time. The values of deflection were considered as half of the value at the point of reading, and the values of monitored parameters (loadings and deflections) were rounded to 4 decimal places.

On the basis of the data obtained from the bending tests executed on the series of SFRC specimens, tensile strengths were calculated according to the individual regulations. The results are shown in Table 5 and Figures 2–11.

The values of the tensile strength in bending of the reference concrete mixtures tested according to the individual regulations do not show significant differences. When the individual values are compared, it is not possible to determine unambiguously which assessment approach is the most suitable because no marked differences were recorded during the evaluation, nor is it possible to observe any dependence between the individual approaches to the calculation of the tensile strength in bending.

By adding 20 kg of steel fibers into the reference concrete mixtures of the individual strength classes, no significant increase in the tensile strength in bending was achieved. The increase of the tensile strength depended only on the class of the concrete.

TABLE 3: Composition of the referential concrete mixture in kg per m^3 [19].

Concrete	Cement CEM I 42,5 R	JMS	Fly ash	Aggregates 0–4	Aggregates 8–16	Aggregates 11–22	Amount of additive	Water
C 16/20	220	50	30	901	536	425	2.3	176
C 20/25	240	50	30	880	536	425	2.5	177
C 25/30	260	50	30	860	536	425	2.6	178
C 30/37	280	50	30	855	546	435	2.1	170

TABLE 4: Composition of the steel fiber reinforced concrete mixtures in kg per m^3.

Concrete	Amount of fibers	Cement CEM I 42,5 R	JMS	Fly ash	Aggregates 0–4	Aggregates 8–16	Aggregates 11–22	Amount of additive	Water
	20						425		180
C 16/20	25	220	50	30	890	530	420	2.30	181
	30						415		182
	20						425		181
C 20/25	25	240	50	30	870	530	420	2.47	182
	30						415		183
	20						425		182
C 25/30	25	260	50	30	850	530	420	2.64	183
	30						415		184
	20						435		174
C 30/37	25	280	50	30	845	540	430	2.23	175
	30						425		176

FIGURE 1: Process of the bending test of the SFRC prisms.

By adding 25 kg of steel fibers into the reference concrete mixtures of the individual strength classes, no significant change in the values of the tensile strength in bending was achieved. The values increase only in relation to the strength class of the concrete.

Admixture of 30 kg of steel fibers into the reference concrete mixtures caused a decrease in the tensile strength in bending in the strength class C 25/30. This fact could have been caused, for example, by exhaustion of the cement paste. Adding steel fibers caused an increase of the specific surface area which had to be bound to the cement paste, probably causing the exhaustion of the cement paste and the decrease of values.

Then the laboratory tests for residual bending strengths after crack formation were executed. The summary results of average values in MPa are stated in Table 6.

Calculation of the residual strength for reference concretes without dispersed steel reinforcement is rather misleading. The calculation results are strongly influenced by friction of aggregate in the emerging macrocrack. During the test prism loading controlled by deformation at low rates (0.2 mm/min^1), the friction of aggregate grains in the emerging microcrack is relatively high and "artificially" influences the values of residual strengths at larger deflections. In case the prism was loaded with a controlled force (hydraulic testing press), the values of residual forces would not be high, and the prism damage would suddenly occur at lower deflections. The phenomenon of grain friction in the macrocrack would be almost negligible, as is the case with common tests of tensile strength in bending executed according to the procedure described in ČSN EN 12390-5 [20].

A dose of 20 kg of steel fibers per cubic meter of fresh concrete mixture is regarded as a minimum in most steel fiber reinforced concrete formulas in terms of the resulting composite keeping its ability to transfer load after the appearance of a microcrack. It is apparent from the results of the experiments executed that the dose of 20 kg of steel fibers is a borderline amount for ensuring sufficient residual strengths after the appearance of a crack in all the monitored strength classes. The individual assessment procedures showed different results depending on the method of recommended calculation. The lowest results were obtained when calculated according to the procedure described in the regulation TP FC 1-1 [15], where the minimum recommended residual strength after the emergence of a macrocrack is 0.4 N/mm^2. The best results were achieved when calculated according to the recommendation of the Performance classes of SFRC [17]. These differences were caused by different approaches of individual formulas to the calculation of residual strength.

By increasing the dose of steel fibers to 25 kg/m^3, a higher residual strength was achieved in all monitored strength classes. However, when evaluated according to TP FC 1-1, the resulting values of residual strength were on the boundary of applicability, especially for strength class C

TABLE 5: Assessment of the tensile strength in bending according to the Austrian standard (ÖVB) [20], Performance classes of SFRC [6], and TP FC 1-1 [10] in MPa.

Concrete mixtures		ÖVBB	Performance classes of SFRC	TP FC 1-1
Referential	C 16/20	3.8	4.5	4.1
	C 20/25	4.9	5.0	5.1
	C 25/30	4.2	4.7	4.4
	C 30/37	5.0	4.8	5. 0
20 kg	C 16/20	3.8	4.2	4.0
	C 20/25	3.7	4.2	3.9
	C 25/30	4.6	5.1	4.8
	C 30/37	4.9	4.8	5.0
25 kg	C 16/20	3.9	4.5	4.2
	C 20/25	3.9	3.9	4.0
	C 25/30	4.3	4.2	4.5
	C 30/37	5.1	5.5	5.4
30 kg	C 16/20	3.8	3.7	4.0
	C 20/25	4.9	5.2	5.1
	C 25/30	3.8	3.9	3.8
	C 30/37	5.7	5.6	5.8

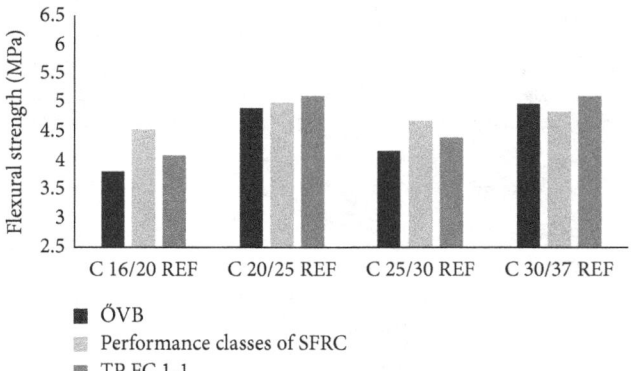

■ ÖVB
■ Performance classes of SFRC
■ TP FC 1-1

FIGURE 2: Bending strength of reference concrete mixtures.

20/25. In the case of strength class C 16/20, a relatively significant increase in residual strength was observed when calculated according to the Performance classes of SFRC, where the performance calculation procedure involves a comparison between the resulting SFRC composite and the reference mixture.

Adding a dose of 30 kg of steel fibers into the individual reference mixtures caused an increase in the residual tensile strength in bending of the resulting steel fiber reinforced concrete. The most significant increase was observed in strength classes C 16/20 and C 20/25. The assessment of residual strengths according to the individual procedures showed that the highest benefit of the dose of steel fibers for the residual strength was found within strength class C 20/25. The calculation results according to the procedure in TP FC 1-1 are again on the boundary of applicability in terms of the required residual strength after the emergence of a macrocrack. However, in strength class C 20/25, a significant increase in this strength was observed even when calculated according to the procedure in TP. The best results

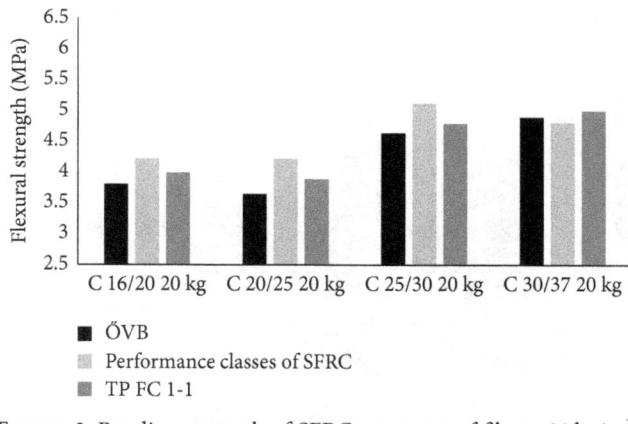

■ ÖVB
■ Performance classes of SFRC
■ TP FC 1-1

FIGURE 3: Bending strength of SFRC—amount of fibers: 20 kg/m³.

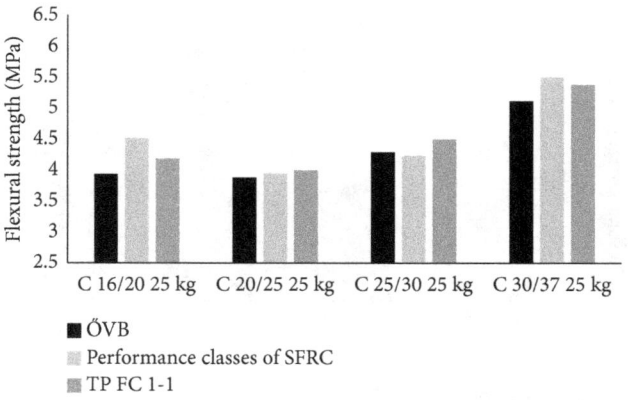

■ ÖVB
■ Performance classes of SFRC
■ TP FC 1-1

FIGURE 4: Bending strength of SFRC—amount of fibers: 25 kg/m³.

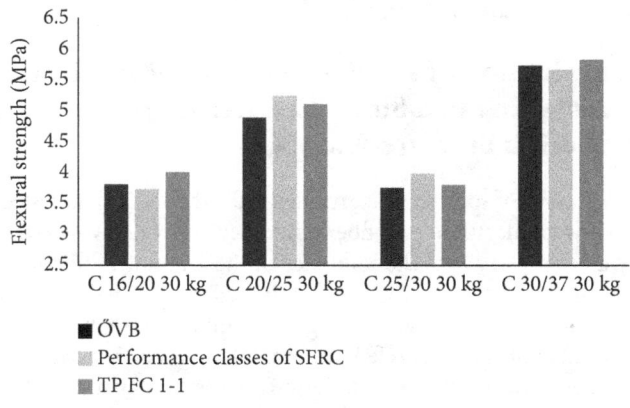

■ ÖVB
■ Performance classes of SFRC
■ TP FC 1-1

FIGURE 5: Bending strength of SFRC—amount of fibers: 30 kg/m³.

of the residual tensile strength in bending were, again, achieved when the procedure described in the Performance classes of SFRC was used.

Table 7 shows the results of the individual calculations according to the selected procedures and recommendations for designing and assessment of manufactured SFRC composites on the basis of tests of tensile strength in bending performed on the manufactured standard test specimens after 28 days of maturing in a controlled environment.

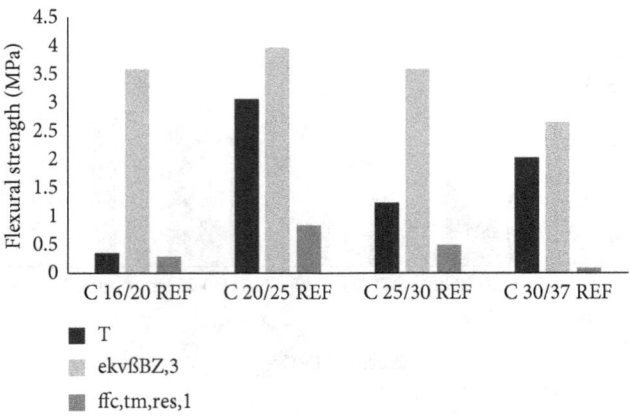

FIGURE 6: Residual tensile strength in bending at the moment of achieving a deflection of 3.5 mm, reference mixture of common concrete.

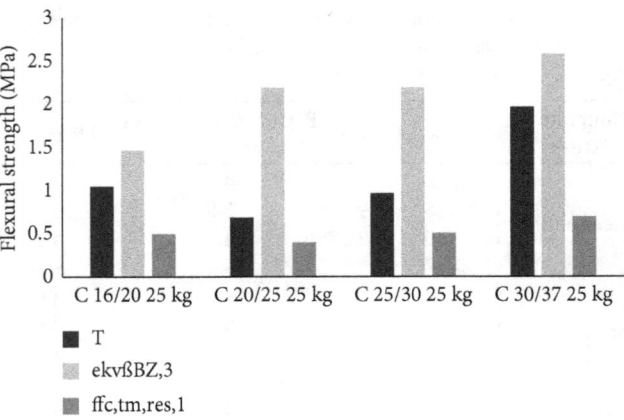

FIGURE 8: Residual tensile strength in bending at the moment of achieving a deflection of 3.5 mm, steel fiber reinforced concrete with a dose weight of 25 kg/m³.

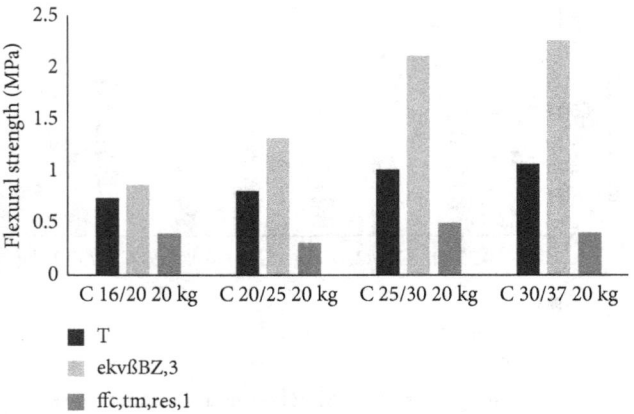

FIGURE 7: Residual tensile strengths in bending at the moment of achieving a deflection of 3.5 mm, steel fiber reinforced concrete with a mass dose of 20 kg/m³.

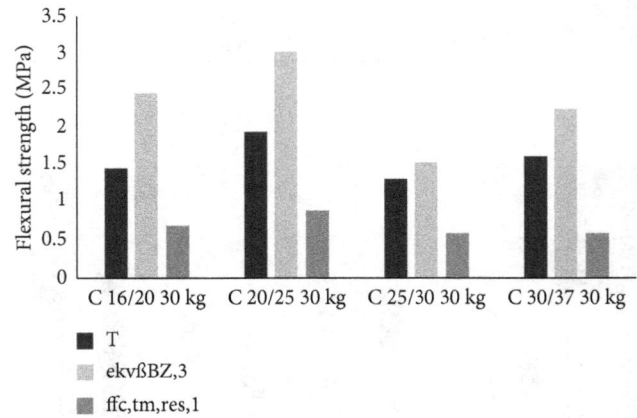

FIGURE 9: Residual tensile strengths in bending at the moment of achieving a deflection of 3.5 mm, steel fiber reinforced concrete with a dose weight of 30 kg/m³.

3. Evaluation of the Performed Bending Tests according to a Simplified Calculation of Specific Fracture Energy

Calculation of specific fracture energy for the individual tested sets of manufactured steel fiber reinforced concretes was driven by an effort to assess the influence of the amount of fibers on the energy necessary for the bending damage of the tested specimen to occur and the degree of influence of the given amount of fibers in the reference cement matrix on the bending damage. The results of calculations are summarized in the following table and included an evaluation of a standard deviation and a coefficient of variation for the individual sets. The values are only approximate because the specific fracture energy is calculated by a highly simplified procedure, and the ambient and other factors are disregarded in this case. The calculations are designed to compare the individual sets of test specimens with each other and cannot be used in more complex applications, such as MKP models (finite element method). If they were to be used for a modelling of the given matter by means of MKP methods, it would be necessary to use more complex procedures for determining the specific fracture energy and to approach experts on the given field (Table 8).

On the basis of the graphic representation of the results of simplified calculations of specific fracture energy, it is possible to conclude that the increasing content of steel fibers in concrete leads to an increase of fracture work, or fracture energy, compared to the reference mixture of common concrete. In some cases, it was possible to observe an opposite phenomenon where fibers decreased the value of the specific fracture energy compared to the reference concrete mixture. This phenomenon could have been caused, for example, by incorrect distribution of fibers in the reference concrete or by the loosening of the concrete due to admixture of steel fibers into a fresh mixture, potentially increasing the content of air pores in the hardened concrete. A higher content of air pores in the hardened concrete, acting as perfectly compressible nonhomogeneities in the structure of the composite, can influence the ability of the concrete composite to transmit.

Calculations of the standard deviation and the coefficient of variation when determining the specific fracture energy showed that increasing the amount of steel fibers in concrete causes an increase in the dispersion of measured values (tensile strength in bending). When higher doses of

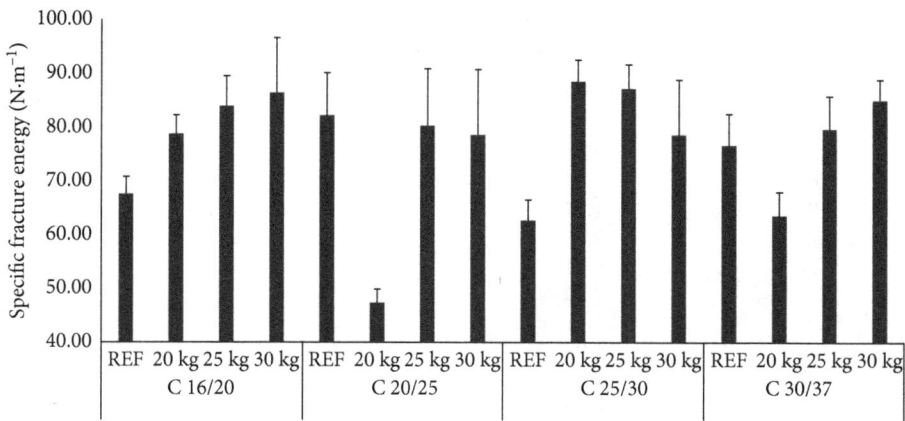

FIGURE 10: Results of calculations of specific fracture energy.

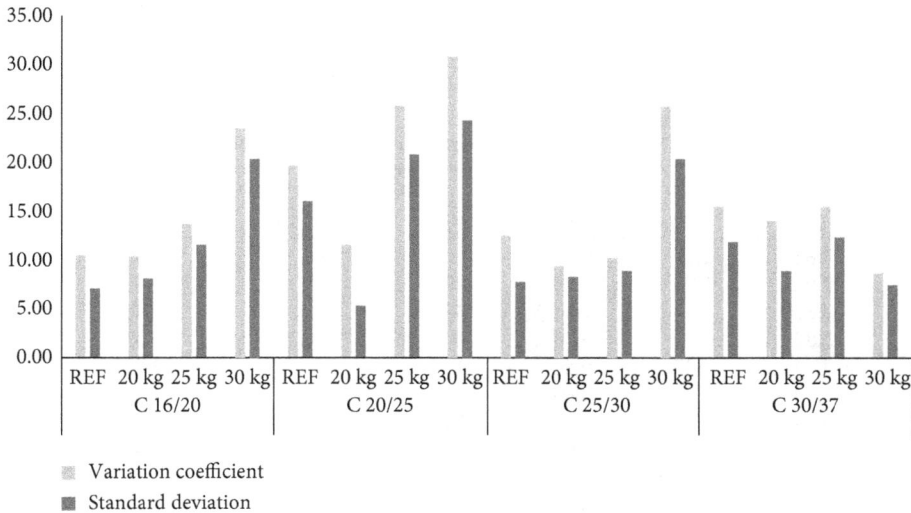

- Variation coefficient
- Standard deviation

FIGURE 11: Representation of the coefficient of variation and standard deviation in the calculations of specific fracture energy from the working diagrams of the performed tests.

dispersed steel reinforcement (30 kg/m³) are used, the distribution of fibers in the critical cross section is more random than in the case of lower doses. The dispersion of steel fibers when using higher doses is so significant that it considerably influences the dispersion of values in the bending tests where the critical cross section of the damage is not determined in advance, such as in the case of tests performed according to ČSN EN 14651 + A1 [21], where a three-point bending with a central notch is considered. When performing the bending tests of SFRC according to ČSN EN 14651 + A1 [21], the dispersion of values is lower even with higher doses of steel fibers. However, the question is which the method of damage testing is closer to reality and to the real influence of steel fibers in the manufactured cross section of, for example, floor panels in relation to the dosing of fibers to the truck mixer on the construction site.

4. Conclusion

During the bending tests of the manufactured steel fiber reinforced concretes, it was observed that the values of

bending strength increase with the increasing amount of fibers. However, the increase of values was not as significant as expected. In relation to the increase in bending strengths, an increase in the residual strengths was observed. Based on the results of the performed tests, it is possible to say that the steel fiber reinforced concrete made from the basic mixture of the C 16/20 strength class with a 30 kg dose of steel fibers showed very similar properties in the residual tensile strength in bending as the SFRC one made from the basic mixture C 30/37 with a 30 kg dose of steel fibers. The assessment of the executed bending tests of steel fiber reinforced concrete according to the three selected procedures (TP FC 1-1, Performance classes of SFRC, and Austrian standard for SFRC) showed that it is very important to choose a correct procedure according to which the tests of SFRC will be executed. The methods for determination of the tensile strength and residual tensile strength are rather different in the individual procedures and the calculation results vary significantly. The best results for residual strengths were always obtained using the procedure given by the "Performance classes of SFRC," while the residual strength

TABLE 6: Assessment of the residual tensile strength in bending at a deflection of 3.5 mm according to the Austrian Standard [16], Performance classes of SFRC [17], and TP FC 1-1 [15] in MPa.

Concrete mixture		ÖVBB T	Performance classes of SFRC ekvß$_{BZ,3}$	TP FC 1-1 f$_{fc,tm,res,1}$
REF	C 16/20	0.38	3.59	0.30
	C 20/25	3.09	3.96	0.90
	C 25/30	1.27	3.61	0.50
	C 30/37	2.06	2.68	0.10
20 kg	C 16/20	0.74	0.86	0.40
	C 20/25	0.80	1.31	0.30
	C 25/30	1.01	2.10	0.50
	C 30/37	1.07	2.26	0.40
25 kg	C 16/20	1.06	1.48	0.50
	C 20/25	0.69	2.21	0.40
	C 25/30	0.97	2.21	0.50
	C 30/37	1.98	2.60	0.70
30 kg	C 16/20	1.48	2.48	0.70
	C 20/25	1.95	3.03	0.90
	C 25/30	1.33	1.54	0.60
	C 30/37	1.63	2.26	0.60

TABLE 7: Summary table with the results of assessment of the performed bending tests in MPa.

Identification		ÖVB			Performance classes of SFRC			TP FC 1-1	
		BZ	G	T	ß$_{BZ}$	ekvß$_{BZ,2}$	ekvß$_{BZ,3}$	f$_{fc,tm,fl}$	f$_{fc,tm,res,1}$
C 16/20	20 kg	3.82	1.73	0.74	4.22	1.73	0.86	4.00	0.40
	25 kg	3.94	1.98	1.06	4.52s	1.76	1.48	4.20	0.50
	30 kg	3.81	2.47	1.48	3.74	2.05	2.48	4.00	0.70
	REF	3.82	1.12	0.38	4.53	3.27	3.59	4.10	0.30
C 20/25	20 kg	3.66	1.47	0.80	4.22	1.74	1.31	3.90	0.30
	25 kg	3.90	2.55	0.69	3.95	2.21	2.21	4.00	0.40
	30 kg	4.88	2.84	1.95	5.23	2.36	3.03	5.10	0.90
	REF	4.91	4.81	3.09	5.00	3.65	3.96	5.10	0.90
C 25/30	20 kg	4.63	2.22	1.01	5.13	2.41	2.10	4.80	0.50
	25 kg	4.31	2.05	0.97	4.24	2.36	2.21	4.50	0.50
	30 kg	3.75	2.41	1.33	3.98	1.64	1.54	3.80	0.60
	REF	4.18	3.77	1.27	4.70	3.32	3.61	4.40	0.50
C 30/37	20 kg	4.90	2.94	1.07	4.81	2.25	2.26	5.00	0.40
	25 kg	5.12	3.05	1.98	5.51	254	2.60	5.40	0.70
	30 kg	5.71	3.15	1.63	5.64	2.48	2.26	5.80	0.60
	REF	4.99	4.72	2.06	4.84	3.34	2.68	5.10	0.10

TABLE 8: Results of calculations of specific fracture energy from the working diagrams in N/m.

Identification		Specimen A	Specimen B	Specimen C	Average value	Standard deviation	Coefficient of variation
C 16/20	REF	74.79	57.46	69.66	67.30	7.27	10.80
	20 kg	86.95	80.82	67.15	78.31	8.28	10.57
	25 kg	100.19	79.1 2	72.79	84.03	11.71	13.94
	30 kg	64.69	113.95	81.11	86.582	20.48	23.65
C 20/25	REF	83.00	61.93	101.94	82.29	16.34	19.86
	20 kg	55.23	42.96	43.96	47.38	5.56	11.74
	25 kg	108.99	58.81	74.25	80.68	20.98	26.01
	30 kg	66.65	56.75	113.04	78.81	24.54	31.13
C 25/30	REF	59.90	73.84	54.92	62.89	8.01	12.73
	20 kg	79.79	100.24	85.84	88.62	8.58	9.68
	25 kg	78.03	99.94	84.55	87.51	9.19	10.50
	30 kg	73.11	106.36	56.96	78.81	20.57	26.10
C 30/37	REF	80.04	60.69	89.89	76.87	12.13	15.78
	20 kg	53.54	62.11	75.59	63.75	9.08	14.24
	25 kg	88.52	62.12	89.04	79.89	12.57	15.73
	30 kg	79.60	96.21	80.41	85.41	7.65	8.95

calculations according to the recommendations of the Austrian standard for SFRC also lead to relatively good results. The results of calculations of residual strengths of SFRC according to the Czech Technical Regulation TP FC 1-1 were relatively low, often on the boundary of usability of the manufactured composite in the given application. These differences were caused by different approaches to the determination of residual strengths based on the bending tests. The foreign procedures perform calculations either on the basis of performance, where the benefit of steel fibers is assessed by comparing a reference mixture of common concrete and the same mixture with the steel fibers added, or by means of an "energetic" approach in which the area below the curve of the working diagram of the SFRC bending test up to the conventional deformation is evaluated. It is thus very important to know which approach to the assessment of residual strength of SFRC to choose so that the resulting composite has the required properties. If someone specifies, or guarantees, the properties or steel fiber reinforced concrete, it is necessary to know according to which regulation and procedure these properties, especially residual strength, were calculated. From the results achieved with the manufactured steel fiber reinforced concretes in the experimental part, it is apparent that the individual procedures and recommendations for SFRC should not be combined because the results deemed satisfactory according to one procedure but can be nonsatisfactory when calculated according to another. Based on the results, the authors recommend the procedure according to the TP FC 1-1.

Conflicts of Interest

The authors declare that they have no conflicts of interest.

Acknowledgments

This paper has been worked out under the Project no. LO1408 "AdMaS UP-Advanced Materials, Structures and Technologies," supported by Ministry of Education, Youth and Sports under the "National Sustainability Programme I."

References

[1] T. F. Flory, J. Hearle, H. Mckenna, and M. Parsey, "About 75 years of synthetic fiber rope history," in *Proceedings of the OCEANS MTS/IEEE Conference*, Washington, DC, USA, October 2015.

[2] T. Simões, H. Costa, D. Dias-da-Costa, and E. Júlio, "Influence of fibres on the mechanical behaviour of fibre reinforced concrete matrixes," *Construction and Building Materials*, vol. 137, no. 1, pp. 548–556, 2017.

[3] G. Appa Rao and A. Sreenivasa Rao, "Toughness indices of steel fiber reinforced concrete under mode II loading," *Materials and Structures*, vol. 42, no. 9, pp. 1173–1184, 2009.

[4] V. Veselý, J. Vodička, J. Vašková, and J. Krátký, "Tests of structural fibre concrete," in *Beton: Technologie, Konstrukce, Sanace*, pp. 43–47, Czech Concrete Society, Praha, Czech Republic, 2010.

[5] R. Gettu, D. R. Gardner, H. Saldívar, and B. E. Barragán, "Study of the distribution and orientation of fibers in SFRC specimens," *Materials and Structures*, vol. 38, no. 1, pp. 31–37, 2005.

[6] P. J. M. Bartoš, "Fibre reinforced concrete," in *Beton: Technologie, Konstrukce, Sanace*, pp. 69–77, Czech Concrete Society, Praha, Czech Republic, 2012.

[7] J. Vodička, V. Veselý, K. Kolář, and J. Krátký, "Fibre concrete in practice," in *Beton: Technologie, Konstrukce, Sanace*, pp. 53–55, Czech Concrete Society, Praha, Czech Republic, 2010.

[8] A. Blanco, P. Pujadas, A. de la Fuente, S. H. P. Cavalaro, and A. Aguado, "Assessment of the fibre orientation factor in SFRC slabs," *Composites Part B: Engineering*, vol. 68, no. 1, pp. 343–354, 2015.

[9] P. Stähli, R. Custer, and J. G. M van Mier, "On flow properties, fibre distribution, fibre orientation and flexural behaviour of FRC," *Materials and Structures*, vol. 41, no. 1, pp. 189–196, 2008.

[10] P. Pytlík, *Technologie Betonu*, VUTIUM, Brno, Czech Republic, 2000.

[11] J. Vodička, V. Veselý, and J. Krátký, "Specifics of fibre concrete technology," in *Beton: Technologie, Konstrukce, Sanace*, pp. 38–42, Czech Concrete Society, Praha, Czech Republic, 2010.

[12] J. Krátký, J. Vodička, and J. Vašková, "Design of fibre concrete structural elements," in *Beton: Technologie, Konstrukce, Sanace*, pp. 87–89, Czech Concrete Society, Praha, Czech Republic, 2010.

[13] J. Vašková, "Experimental verifying of fibre concrete elements behaviour," in *Beton: Technologie, Konstrukce, Sanace*, pp. 74–78, Czech Concrete Society, Praha, Czech Republic, 2010.

[14] J. Štoller and E. Zezulová, "The application of fibre reinforced concrete for protective shelter from auxiliary material," *Key Engineering Materials*, vol. 755, pp. 374–381, 2017.

[15] TP FC 1-1, *Technické podmínky 1: Vláknobeton – Část 1 Zkoušení vláknobetonu: Vyhodnocení Destruktivních Zkoušek a Stanovení Charakteristického Pracovního Diagramu Vláknobetonu pro Navrhování Vláknobetonových Konstrukcí*, Českomoravský Beton, a.s., Praha, Czech Republic, 2007.

[16] ÖVBB-Richtlinie Faserbeton, *Österreichische Vereinigung für Beton und Bautechnik: Faserbeton*, Austrian Structural Engineering Association, Vienna, Austria, 2008.

[17] H. Falkner, M. Teutsch, and H. Klinkert, *Massivbau und Brandschutz: Leistungsklassen von Stahlfaserbeton*, IBMB, Braunschweig, Germany, 1999.

[18] ČSN EN 12390-2, *Testing Hardened Concrete: Part 2: Making and Curing Specimens for Strength Tests*, Czech Office for Standards, Metrology and Testing, Praha, Czech Republic, 2009.

[19] T. Komárková, J. Láník, and P. Dvořák, "Influence of dispersed reinforcement on the physico-mechanical properties of the SFRC," *Key Engineering Materials*, vol. 755, pp. 75–81, 2017.

[20] ČSN EN 12390-5, *Testing Hardened Concrete–Part 5: Flexural Strength of Test Specimens*, Czech Office for Standards, Metrology and Testing, Praha, Czech Republic, 2009.

[21] ČSN EN 14651 + A1, *Test Method for Metallic Fibered Concrete: Measuring the Flexural Tensile Strength (Limit of Proportionality (LOP), Residual)*, Czech Office for Standards, Metrology and Testing, Praha, Czech Republic, 2008.

8

Experimental Study on a Self-Centering Earthquake-Resistant Masonry Pier with a Structural Concrete Column

Lijun Niu and Wenfang Zhang

College of Architecture and Civil Engineering, Taiyuan University of Technology, Taiyuan 030024, China

Correspondence should be addressed to Wenfang Zhang; zhangwenfang@tyut.edu.cn

Academic Editor: Carlo Santulli

This paper proposes a slotting construction strategy to avoid shear behavior of multistory masonry buildings. The aspect ratio of masonry piers increases via slotting between spandrels and piers, so that the limit state of piers under an earthquake may be altered from shear to rocking. Rocking piers with a structural concrete column (SCC) form a self-centering earthquake-resistant system. The in-plane lateral rocking behavior of masonry piers subjected to an axial force is predicted, and an experimental study is conducted on two full-scale masonry piers with an SCC, which consist of a slotting pier and an original pier. Meanwhile, a comparison of the rocking modes of masonry piers with an SCC and without an SCC was conducted in the paper. Experimental verification indicates that the slotting strategy achieves a change of failure modes from shear to rocking, and this resistant system with an SCC incorporates the self-centering and high energy dissipation properties. For the slotting pier, a lateral story drift ratio of 2.5% and a high displacement ductility of approximately 9.7 are obtained in the test, although the lateral strength decreased by 22.3% after slotting. The predicted lateral strength of the rocking pier with an SCC has a margin of error of 5.3%.

1. Introduction

Multistory unreinforced masonry (URM) buildings have been used extensively in the last several decades, and a large number of these buildings are still being erected at moderate cost during the process of urbanization in China. These structures have many advantages, but their performances during earthquakes are not satisfactory, such as in the Tangshan earthquake of 1976 [1]. After this earthquake, which led to the collapse of a vast number of URM buildings, an important innovation for the seismic protection of multistory URM structures is the addition of SCC. For a masonry structure with SCC, the masonry walls are the dominating lateral force-resisting elements, and, unlike confined masonry construction, the strength of masonry confined by SCC would not increase when used in seismic calculation. SCC is not involved in the seismic calculation of walls in most cases, which only belong to details of seismic design in existing Chinese code for seismic design of buildings. The location of SCC should be determined according to seismic fortification intensity and number of floors. Figure 1 provides several photographs of construction of SCC and masonry walls containing SCC and ring beam. The concrete of SCC would be poured after masonry of one story completed. SCC connect with masonry part by horizontal reinforcements, as shown in Figure 1(b). The vertical interval of horizontal reinforcements is usually 500 mm, and more details about horizontal reinforcements might be observed from design details of experimental specimens, which are shown in Figure 7. The longitudinal reinforcement ratio of SCC is low, and the largest ratio is just over 1.0% [2]. In general, SCC is a primary structural configuration in seismic design of masonry buildings in China, and masonry structures with SCC are still considered as URM systems owing to the low reinforcement ratio and small section of SCC [3].

Observation of seismic damage to URM piers, as well as laboratory experimental tests, showed that masonry piers subjected to in-plane loading may exhibit two typical types of behavior: flexural and shear deformations, and corresponding possible failure modes appear such as rocking, diagonal tension, diagonal stepped cracking, toe compression, and bed-joint sliding [4–6]. Previous research has revealed that

(a)

(b)

(c)

(d)

FIGURE 1: Photographs of construction of SCC and actual masonry walls containing SCC and ring beam (■: SCC; ●: ring beam; ♦: merger of ring and lintel beam).

the occurrence of different failure modes depends on several parameters: the geometry of the pier; the boundary conditions; the acting axial load; and the mechanical characteristics of the masonry constituents (mortar, blocks, and interfaces). The height-to-length aspect ratio of the geometry has a significant influence on the failure mode of masonry walls under seismic action in the above respects. The paper proposes a slotting construction strategy which aims to increase the aspect ratio of masonry walls. For walls with window openings, they might be slotted between spandrels and piers, so that the limit state piers under an earthquake may be altered from brittle shear to ductile rocking. More details about the slotting strategy are shown in Section 2.

The rocking mode is an expected mode owing to its self-centering behavior in a lateral loading. The notion of self-centering stemmed from the rocking idea was applied in the design of the rail bridge at first [7]. In order to improve the lateral capability of a rocking unreinforced wall under in-plane seismic loading, unbonded prestressed technology has been incorporated in masonry walls to provide self-centering behavior [8, 9]. Unbonded prestressed tendons were applied in the center of a concrete wall or a masonry wall in some tests, such as tests conducted by Restrepo et al. [10–12]. Several schematic diagrams and some simple information of these tests are shown in Figures 2(a)–2(c). Rocking walls in these tests showed a high strength capability, and an S-shaped global hysteresis loop was generally obtained. Stanton proposed a rocking construction method of a long concrete wall with prestressed technology, as shown in Figure 2(d) [13].

If rocking walls contain special energy dissipators, such as those shown in Figures 2(e) and 2(f), then a full hysteresis loop would be obtained [10].

In recent years, several tests on strengthening URM buildings using external prestressed technique were also conducted in China [14–16]. Figure 3 shows several experimental pictures, and these experimental piers generally showed a shear behavior owing to the low aspect ratio and the high compressive stress of piers.

For masonry piers containing an SCC, the SCC is generally located at the center of section, and the location is similar to that of prestressed tendons in Figure 2. Unlike the above existing researches about rocking wall, for rocking masonry piers with an SCC formed by slotting, it is the conventional reinforcements instead of prestressed tendons in the center of walls. The self-centering earthquake-resistant system is expected to recenter subjected to the relatively, moderate earthquakes and show plastic behavior subjected to the strong earthquakes owing to the low strength of longitudinal reinforcements. The shear mode of URM piers is the only failure mode, and the rocking mode of piers, especially piers with an SCC, is not presented in existing Chinese code for seismic design of buildings [3]. The primary aim of the paper is to conduct seismic performance evaluation of rocking masonry piers with an SCC.

The first part of this paper presents briefly a slotting strategy for Chinese multistory masonry buildings to avoid shear failure. The rocking strength equations of masonry piers with an SCC are presented in the second part. Meanwhile, in order

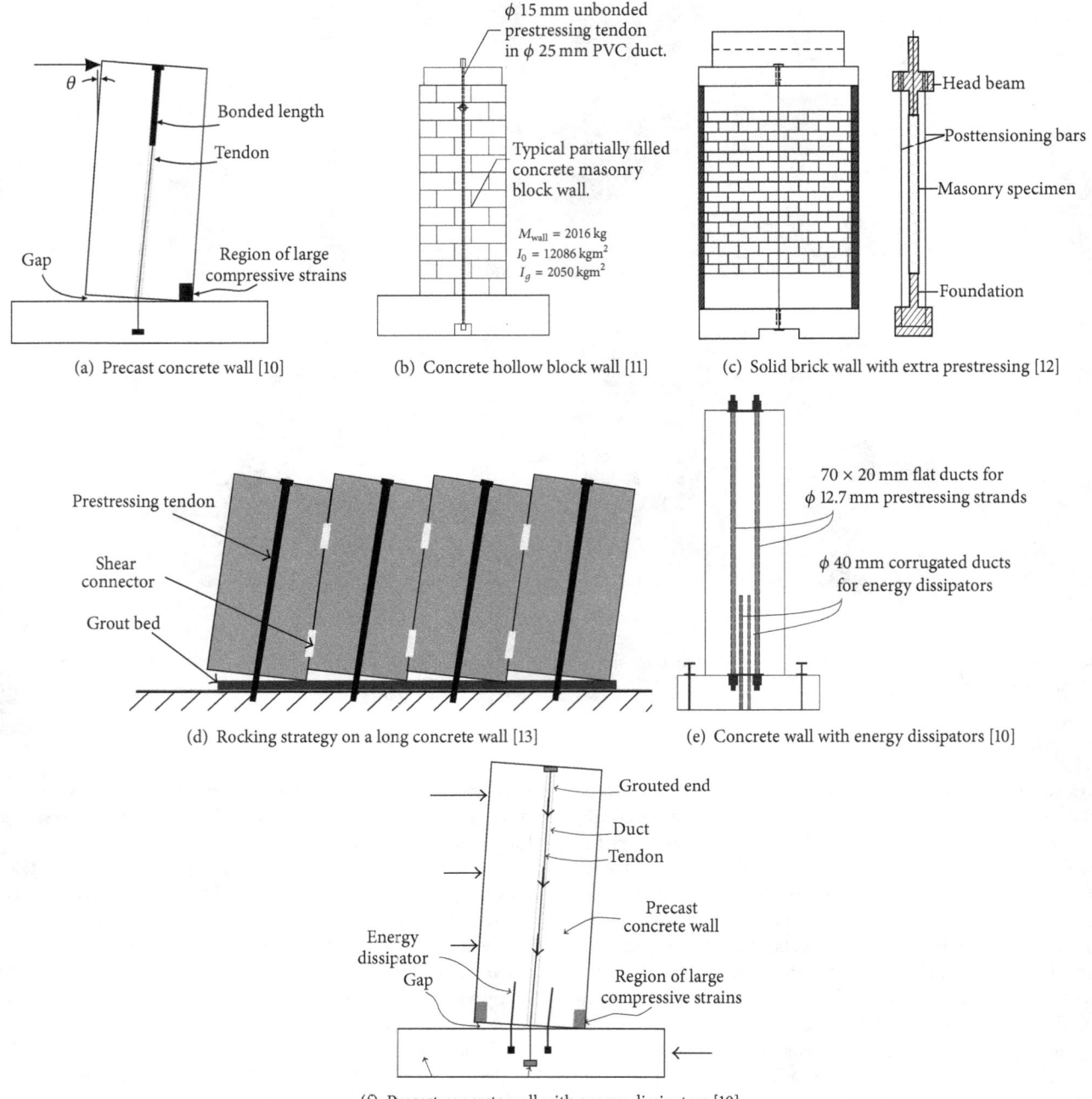

(a) Precast concrete wall [10]

(b) Concrete hollow block wall [11]

(c) Solid brick wall with extra prestressing [12]

(d) Rocking strategy on a long concrete wall [13]

(e) Concrete wall with energy dissipators [10]

(f) Precast concrete wall with energy dissipators [10]

FIGURE 2: Existing tests and strategy of rocking walls with prestressed tendon.

to predict the displacement ductility of rocking masonry piers, existing equations are also given in this part. In the last part, pseudo-static tests of two full-scale masonry piers, with or without slotting, and with an SCC, were conducted, and the equations given in the second part would be validated with the test results.

2. Slotting Strategy of Rocking Masonry Buildings

Figure 4 illustrates a slotting strategy to avoid shear behavior of multistory masonry buildings. The strategy presented

might be applied in construction of new buildings. A masonry building could be divided into two seismic zone at the 1/3 height of buildings, the stories above the 1/3 height belong to ordinary seismic zone, and the other stories belong to key seismic zone. Slots used to increase the aspect ratio of walls are applied in key seismic zone. For walls with window openings, slots might be set between piers and spandrels, and soft materials are filled inside the slots. Above window openings, adopting either slots above the lintel beam or merger of ring and lintel beam might be determined by the height of the lintel wall. After slotting, stability of cantilever spandrels constrained by circumjacent reinforced concrete belts can

(a) Wall with opening [14]

(b) Two-story building [15]

(c) Four-story building [16]

FIGURE 3: Photographs of masonry structures reinforced with external prestressed tendons in China.

be ensured. The stories which belong to ordinary seismic zone might be constructed as ordinary treatments, and those belonging to key seismic zone would be constructed by using strong details of seismic design.

3. In-Plane Rocking Behaviors of Masonry Walls

3.1. Rocking Strength of URM Piers. Two possible types of in-plane damage can occur with rocking mode: rocking and rocking/toe crushing. For the rocking mode of URM piers, if the tensile strength of masonry is neglected, then the lateral strength V_r may be determined using (1) based on FEMA 306 (1999) and FEMA 273 (1997) [17, 18].

$$V_r = 0.9\alpha P_{\mathrm{CE}}\left(\frac{L}{h_{\mathrm{eff}}}\right), \tag{1}$$

where α is factor equal to 0.5 for a fixed-free cantilever wall or equal to 1.0 for a fixed-fixed pier, P_{CE} is expected axial compressive force per load combination in FEMA 273, L is length of the wall, and h_{eff} is effective height of the pier. The parameter h_{eff} may be varied to reflect the observed crack patterns.

Flexural compressive stress in the compressive toe is assumed to be distributed across an equivalent rectangular stress block in rocking/toe crushing mode [19]. If the clamping force created at the corners is defined as a parameter P_u, then

$$P_u = \kappa t l_c f_{\mathrm{me}}, \tag{2}$$

where κ is compressive strength reduction factor (a common assumption is an equivalent rectangular stress block with $\kappa = 0.85$), t is thickness of masonry walls, l_c is depth from the neutral axis to the fiber of maximum compressive strain, and f_{me} is expected compressive strength of the masonry.

Equation (3) gives the lateral strength V_{tc} for rocking/toe crushing mode presented in FEMA 273, and a similar expression is also presented by Magenes and Calvi [20].

$$V_{\mathrm{tc}} = \alpha P_{\mathrm{CE}}\left(\frac{L}{h_{\mathrm{eff}}}\right)\left(1 - \frac{f_{\mathrm{ae}}}{0.7 f_{\mathrm{me}}}\right), \tag{3}$$

where f_{ae} is expected vertical axial compressive stress as defined in FEMA 273.

3.2. Rocking/Toe Crushing Strength for Masonry Piers with an SCC. Rocking behavior in piers with an SCC would be more likely to result in toe crushing upon repeated cycles, because higher drift levels are expected. Ignoring the tension capacity of masonry and concrete, based on compression theory with large eccentricity, the mechanical model of rocking masonry piers with an SCC is shown in Figure 5, and equations of the peak lateral strength are expressed as

$$V_{\mathrm{tcc}} = \frac{M_u}{h_{\mathrm{eff}}} \tag{4}$$

$$M_u = M_e + P_{\mathrm{CE}} \cdot \left(l_{\mathrm{tm}} + \frac{b}{2}\right) \tag{5}$$

$$M_e = f_y A_{\mathrm{as}}\left(l_{\mathrm{tm}} + \frac{b}{2}\right) + l_{\mathrm{cm}} P_u\left(1 - 0.5\kappa\right) \tag{6}$$

$$l_{\mathrm{cm}} = \frac{\left(P_{\mathrm{CE}} + f_y A_{\mathrm{as}}\right)}{\kappa t f_{\mathrm{me}}}, \tag{7}$$

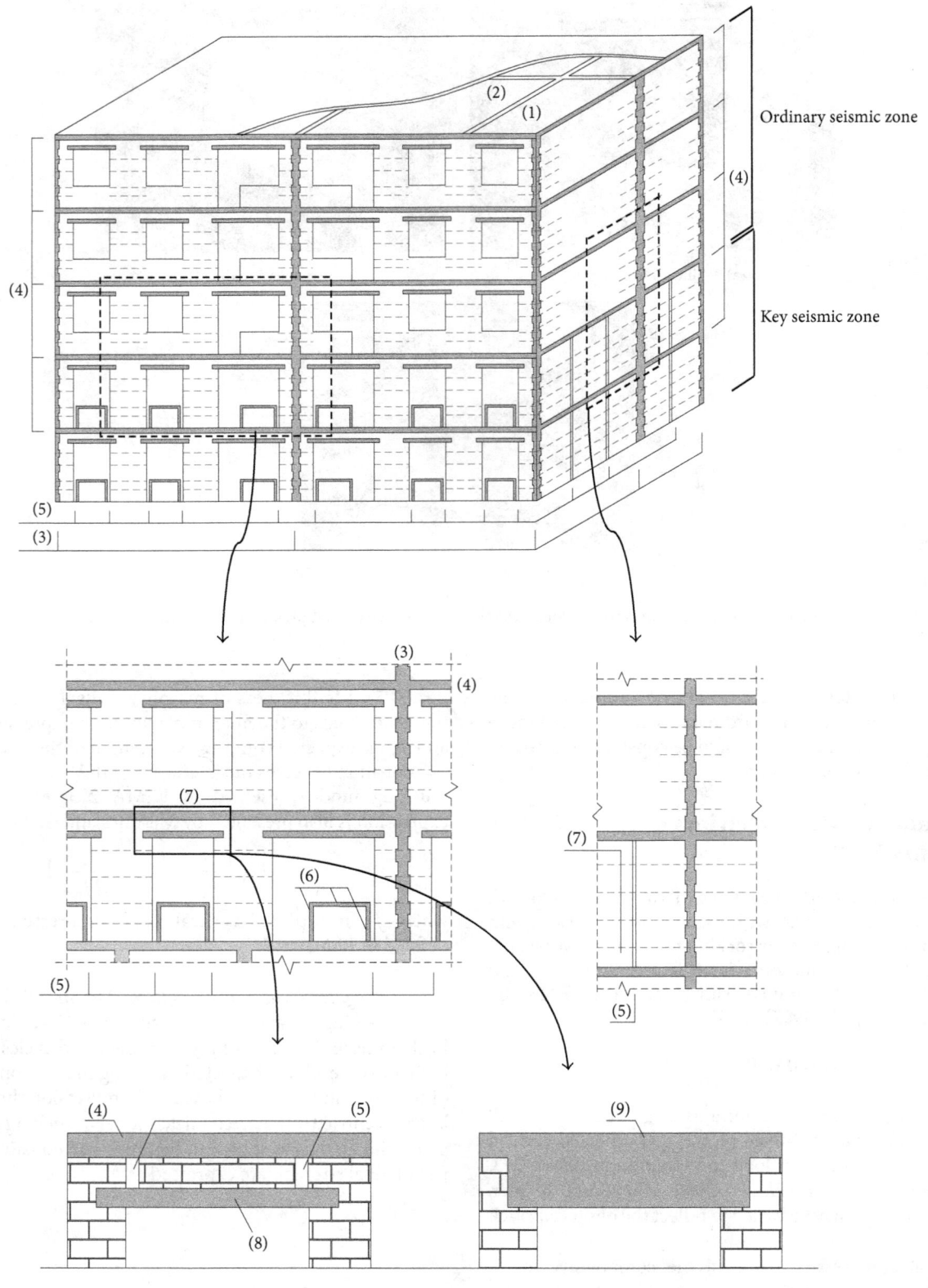

FIGURE 4: Construction strategy of rocking multistory masonry buildings.

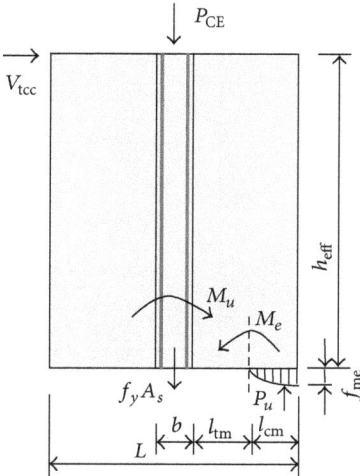

FIGURE 5: Mechanical model of rocking masonry walls with an SCC.

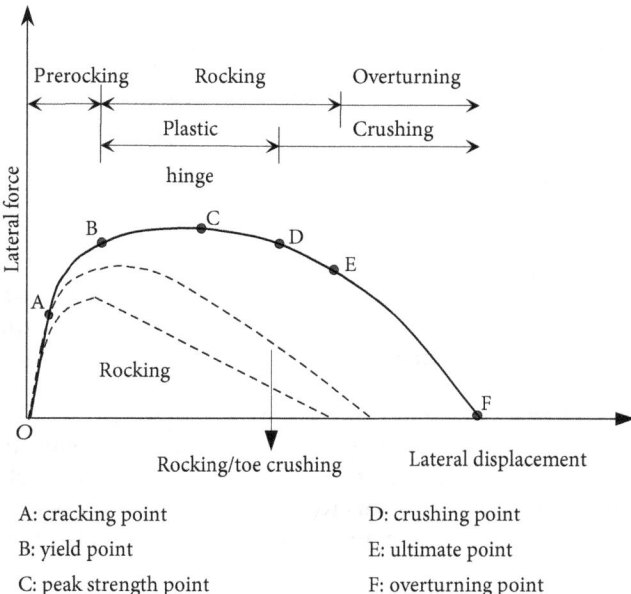

FIGURE 6: Excepted lateral force-displacement of rocking masonry piers.

where V_{tcc} is expected rocking strength of piers with an SCC, M_u is bending strength based on compression theory with large eccentricity, M_e is is defined about the "stress axis," l_{cm} is depth of compression zone at the peak strength, l_{tm} is depth from the neutral axis to the side of SCC at the peak strength, b is width of the SCC, f_y is expected yield strength of reinforcements, and A_{as} is area of all longitudinal reinforcements.

3.3. Expected Force/Displacement Relationships for Rocking Behavior.

Several models have been proposed to predict the force-displacement response of URM walls with a dominant rocking mode [21, 22]. Figure 6 indicates expected lateral force-displacement relationships of rocking masonry walls, and the solid curve represents that of piers with an SCC; the

dashed curves represent that of piers without an SCC. Several expected critical points in the solid curve are also indicated in Figure 6. The yield point means that the strain on longitudinal reinforcements at cracks just reaches the yield strain, and the ultimate point represents that the residuary lateral strength is equal to 85% of the peak strength. It is expected that significant strength degradation would not occur during a large deformation from the yield point to toe crushing point.

For rocking piers with an SCC, displacements may be predicated through a concentrated plastic hinge model, and a plastic hinge/rigid plane model is considered as the calculation diagram. The plastic hinge in masonry is not a point per se; it is a zone exceeding the yield curvature at large displacements. For a vertical cantilever masonry wall with an SCC, the equivalent plastic hinge length may be estimated from the following equation [18]:

$$l_p = 0.2L + 0.04h_e, \qquad (8)$$

where l_p is equivalent plastic hinge length and h_e is height to the resultant of the lateral force.

Paulay and Priestley provide a simple model for calculating the yield displacement [23], which is

$$\Delta_y = \frac{\phi_y H^2}{3}, \qquad (9)$$

where ϕ_y is the yield curvature of a masonry section and H is the shear height of masonry walls.

The yield curvature ϕ_y may be calculated as

$$\phi_y = \frac{\varepsilon_y}{\left(L/2 - l_{cy}\right)}, \qquad (10)$$

where $\varepsilon_y = f_y/E_s$ and l_{cy} is the neutral-axis depth at yield curvature and E_s is the elastic modulus of steel.

If a rocking pier with an SCC is idealized as a cantilever element with a zone of concentrated plastic rotation at the base, then the ultimate displacement capacity at the top of walls is

$$\Delta_p = \left(\phi_m - \phi_y\right) h_{eff}, \qquad (11)$$

where ϕ_m is the maximum curvature of a masonry section and $h_{eff} = H - 0.5l_p$.

The maximum curvature of a section is controlled by the maximum compression strain at the extreme fiber, since steel strain ductility capacity is typically high. The maximum curvature ϕ_m may be expressed as

$$\phi_m = \frac{\varepsilon_{cm}}{l_{cu}}, \qquad (12)$$

where ε_{cm} is the maximum compression strain of masonry material and l_{cu} is the neutral-axis depth at ultimate curvature.

The displacement ductility can be defined as

$$u_\Delta = \frac{\Delta_p}{\Delta_y}. \qquad (13)$$

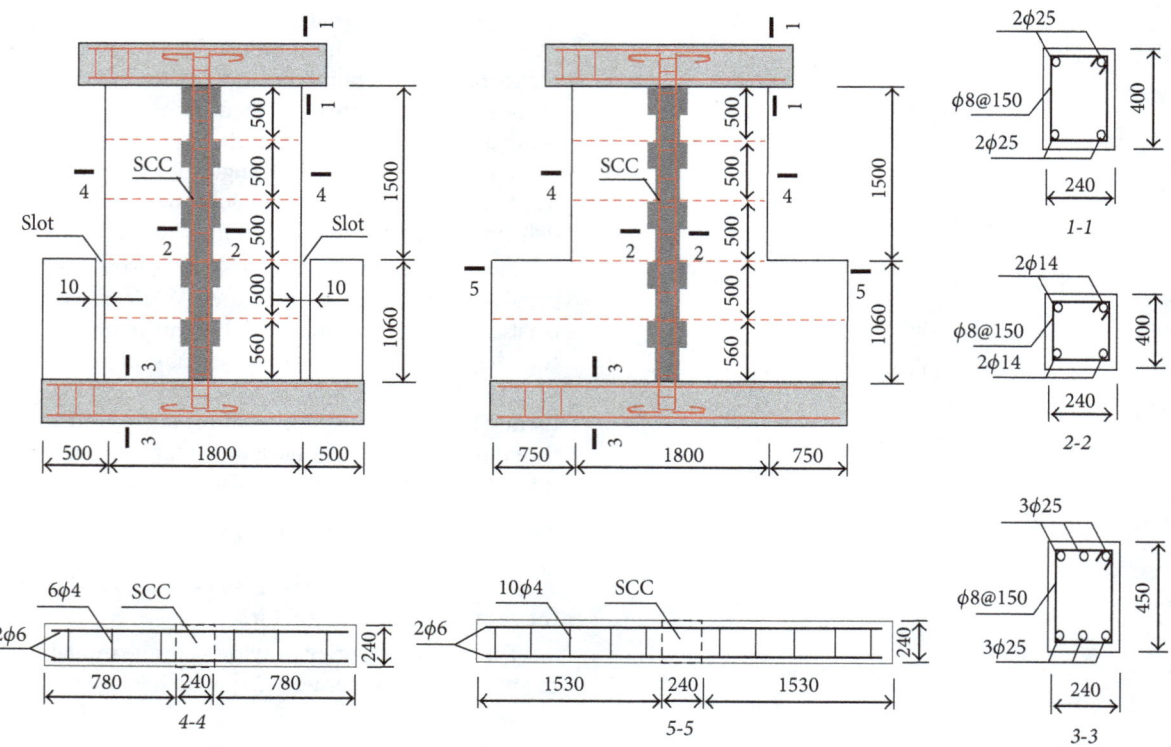

FIGURE 7: Details of specimens (mm).

The minimum demand of the width of slots should be determined based on the predicted ultimate displacement, and it is

$$d_s = \frac{h_s - 0.5 l_p}{h_{\text{eff}}} u_\Delta,$$ (14)

where d_s is the minimum demand of the width of slots and h_s is the height of spandrels.

4. Experimental Test of Masonry Walls with an SCC

4.1. Design of Specimens. Design details of both specimens are shown in Figure 7. Two full-scale masonry walls denoted as 1TB2 and 1TS2, respectively, had the same configuration; only one distinction between these two walls was that the specimen 1TB2 containing slots was expected to show a rocking failure, while the specimen 1TS2 without slots was expected to show a prevalent shear failure. Both specimens consisted of a reinforced concrete footing, a top beam, and a brick wall with an SCC. The geometry of walls was adopted in accordance with common constructions in China. Piers were 2560 mm high (clear height), 1800 mm wide, and 240 mm thick, and spandrels were 1000 mm high. The aspect ratio of specimen 1TB2 was 1.73 and that of the specimen 1TS2's pier was 0.87. It is particularly worth pointing out that the slotting width of specimen 1TB2 was 10 mm and hollow carton boards were applied to fill slots. Masonry walls were constructed from solid clay-bricks (240 × 115 × 53 mm) and mortar. The thickness of joints was 10 mm. Longitudinal reinforcements

inside the SCC were 4ϕ14, and the reinforcement ratio was 0.14%. The other details of walls were designed in accordance with the corresponding Chinese codes.

4.2. Material Properties. The designed strength grades of all materials such as brick, mortar, concrete, and reinforcements were selected in accordance with the prevailing values in fact. Samples of plain round bars were subjected to tensile testing and the original gage length of ϕ14 bars, and L_o calculated by (15) was 70 mm.

$$L_o = 5.65 \sqrt{A_s},$$ (15)

where A_s is the area of one longitudinal reinforcement. The average yield strengths for the ϕ6 and ϕ14 reinforcements were 342.1 MPa and 293.1 MPa, respectively. Because the strength yield ratio of ϕ14 bars was 1.49 and the percentage elongation after fracture was 38.5%, the bars have a high ductility capacity. Also, the average compressive strengths measured for the brick unit and the concrete of the SCC were 15.1 MPa and 35.3 MPa, respectively. The mixing ratio of the mortar after trial batches was 0.5 : 2 : 9 (cement : lime : sand), and the average compressive strength of the mortar measured from nine cubic samples was 17.1 MPa. The compressive strength of masonry estimated according to the strengths of brick and mortar was 6.76 MPa.

4.3. Test Setup. The test setup, which is shown in Figure 8, primarily consisted of a reacting-force wall providing horizontal supports and a reaction frame providing vertical supports. Constant vertical pressure was applied using a 500 kN jack.

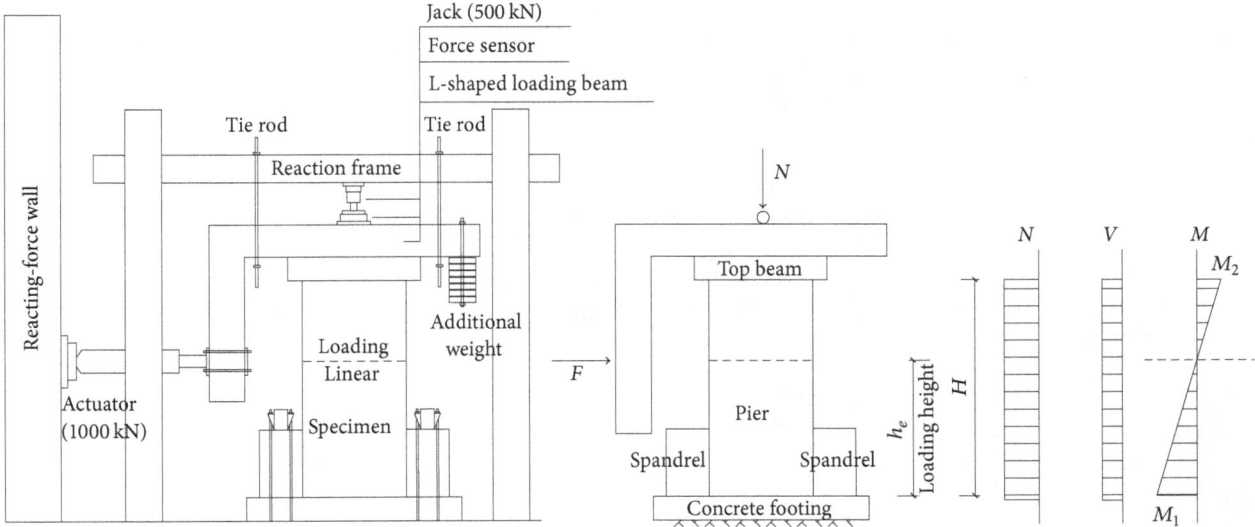

FIGURE 8: Equipment figure and the internal force diagram of specimens.

The vertical loading, which included additional weight, was 132 kN, and the corresponding average compressive stress was 0.3 MPa. Pseudo-static tests were performed by a horizontal hydraulic actuator (1000 kN) providing a lateral force to a welded-steel L-shaped loading beam with an I-shaped cross section, and a contraflexure point would be formed on the wall by using the L-shaped loading beam. Under this loading pattern, the interior force of specimens in elastic stage is also shown in Figure 8. The loading beam was safeguarded by two vertical tie rods hinged to the reaction frame. A vertical height from the top of concrete footing to the lateral loading resultant was defined as the loading height. The loading height was the same height as the contraflexure point on the wall and it was equal to 1830 mm in the test, which was located at approximately two-thirds of one-story height.

4.4. Instrumentation. Figures 9 and 10 schematically show the test instrumentation. Observations of walls mainly included in-plane deformations; strains of steel bars, which included longitudinal reinforcements inside the SCC and horizontal reinforcements inside masonry; the pattern of cracks; and measurement of loading forces. Distances of strain gages are marked in Figure 10, and the vertical direction means the data at longitudinal reinforcements and the horizontal direction means the data at horizontal reinforcements. One face of walls was used to monitor the creation and progression of cracks, whereas the other face was used to install external instrumentation. Vertical loading was verified by a load cell positioned in series with the vertical jack, and verification of the lateral force was achieved by using a strain gage at the corner of the L-shaped loading beam, because the strain is approximately proportional to the lateral force.

4.5. Loading History. The test specimens were subjected to cyclic lateral loading under constant vertical axial load, and the loading histories are shown in Figure 11, which consisted of force-controlled stages and displacement-controlled stages. Solid curves represent force-controlled mode and

dashed curves represent displacement-controlled stages in Figure 11. Note that the displacements in Figure 11 represent data produced by the hydraulic actuator, not the displacements on the top of walls read by equipment. Therefore, the actual magnitudes of displacement are estimated.

4.6. Experimental Results

4.6.1. Cracking and Failure Procedure

1TS2. Flexural cracks first appeared at the corner of the window opening when the lateral force was equal to 70 kN, and the propagating direction was diagonally downward. These flexural cracks only followed the joints. The first flexural cracks were observed to open and close subjected to the cyclic loading. When the lateral loading is equal to 220 kN, which is near the peak lateral loading, the diagonal shear cracks initiated near the center of the pier and propagate outward. Appearing of shear cracks indicated the transfer from the bending behavior to the shear behavior. Diagonal shear cracking caused by the principal tensile stresses occurred in the units as well as the joints. Test terminated after the diagonal tension cracks significantly widened and longitudinal reinforcements were engaged. The specimen 1TS2 exhibited a typical flexural cracking/diagonal tension failure mode, as shown in Figure 12, and may be considered as the shear failure wall.

1TB2. After applying an axial force, no visible cracks were observed in the pier. The first flexural cracks appeared below the first row of horizontal reinforcements when the lateral force was equal to 60 kN. Then, these cracks propagated in level and inclined directions after an increase in the lateral loading, and slight rocking was observed during the loading cycle. Note that flexural tension cracking does not cause any damage to the entire wall. It was observed that the width of slots with filler material was insufficient for deformation of the pier. If this did not widen the slots, transfer from bending

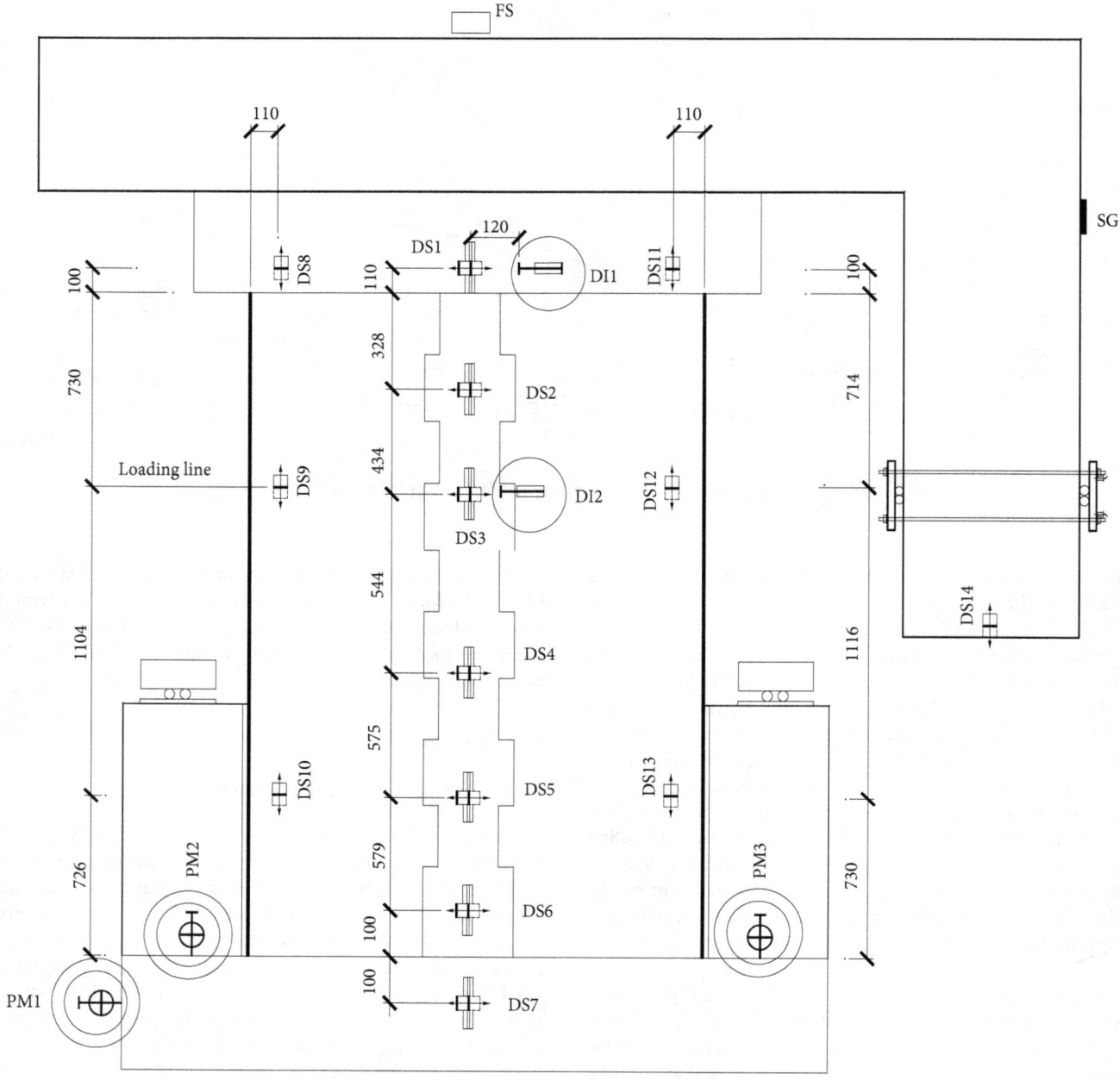

DI: linear variable differential transformer (LVDT)

DS: strain-type displacement transducer

FS: force sensors

PM: dial indicators

SG: strain gage at the corner of the L-shaped loading beam

FIGURE 9: External instrumentation.

behavior to shear behavior would occur as that appearing on the specimen 1TS2. Therefore, the slots were widened by taking down the part of the spandrels. With an increase in the lateral loading, flexural cracks also initiated at the base of the pier and propagated through the cross-section of the SCC. The phenomenon in this phase observed was that the upper four-fifths of the pier, as a rigid plane, exhibited rocking motion around the bottom of the SCC in the lateral loading cycle, and flexural cracks were observed to open and close. A stable rocking mechanism with a large deformation was established. The lateral loading arrived at the peak value before the toes cracked in the vertical direction, and then the inclined compressive cracks were also observed above the first row of horizontal reinforcements. The distribution

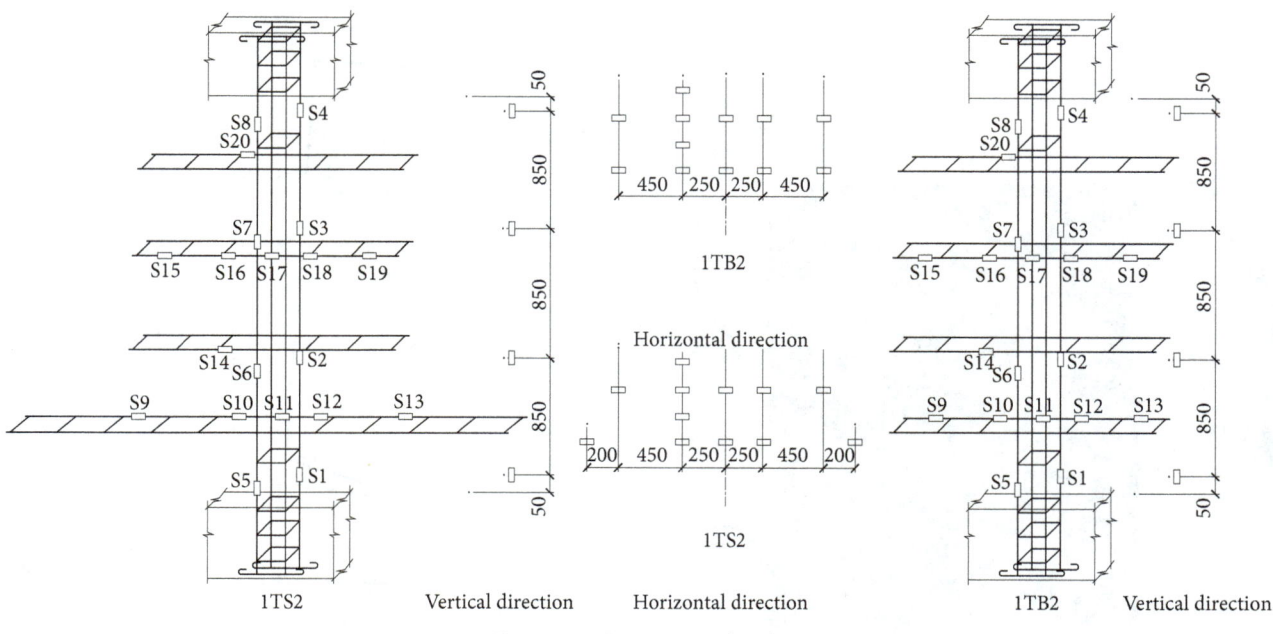

FIGURE 10: Strain gages (mm).

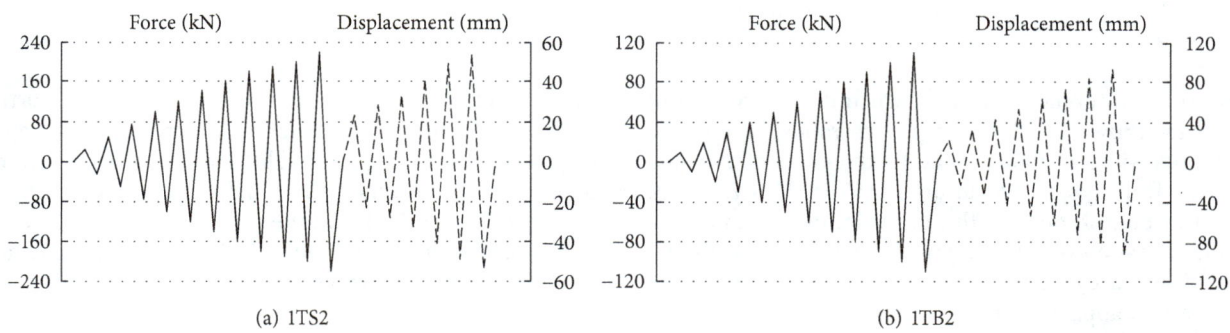

(a) 1TS2

(b) 1TB2

FIGURE 11: Loading history of test specimens.

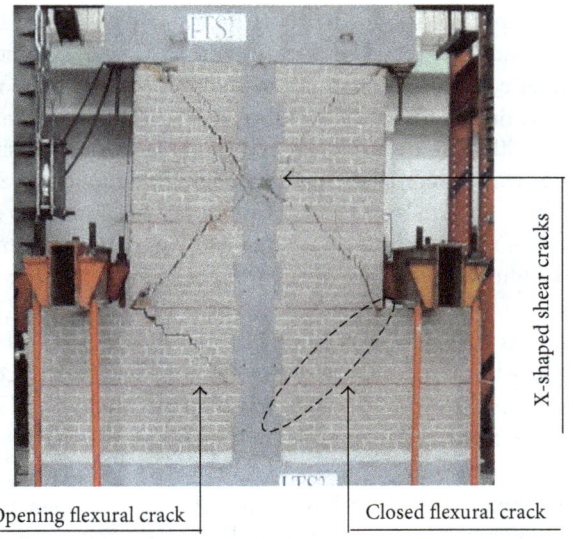

FIGURE 12: Failure photographs of the specimen 1TS2.

of cracks indicates that the horizontal reinforcements inside masonry have a direct influence on the propagation of compressive cracks. After the peak strength, the toes were crushed gradually and some bricks were crushed to powder. The failure photograph of specimen 1TB2 is shown in Figure 13(a), and Figure 13(b) shows the pattern of cracks and illustrates the behavior observed during the test. The specimen 1TB2 exhibited a typical rocking/toe crushing failure mode.

In tests, both specimens firstly showed the flexural cracks; however, different failure modes ultimately occurred owing to various aspect ratios. Experimental research confirms that the slotting strategy which is used to develop a rocking failure mode is feasible. Particularly note that the sufficient width of slots and enough soft filler materials are critical factors for the transfer; if not, first flexural cracking also might transfer to the shear behavior.

4.6.2. Self-Centering Behavior and Energy Dissipation. A comparison was made by employing the benchmark test

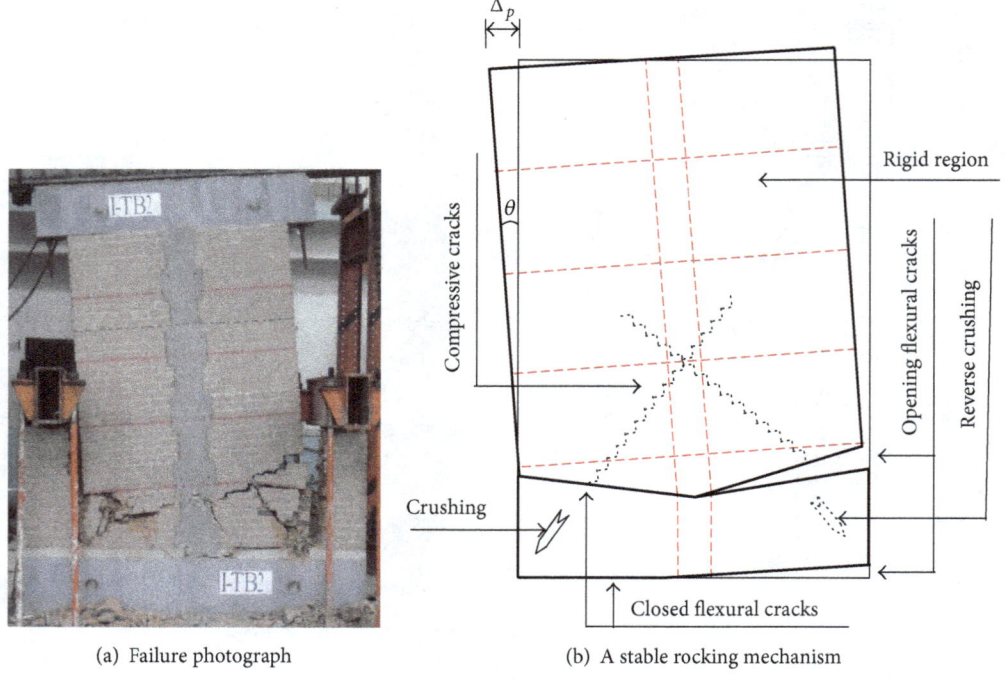

(a) Failure photograph

(b) A stable rocking mechanism

FIGURE 13: Failure photograph and the rocking mechanism of specimen 1TB2.

developed by Anthoine et al. [24], which utilizes two URM piers characterized by different aspect ratios. These two walls, with a width of 1000 mm and a thickness of 250 mm, were made from typical Italian bricks (55 × 120 × 250 mm) arranged in English bond with mortar joints of a thickness of 10 mm. Rotation of the horizontal edges of the walls was prevented and a cyclic displacement loading of increasing amplitude was applied to the upper side, while a constant vertical load of 150 kN was applied in order to produce a mean compressive stress of 0.6 MPa. In this paper, a wall with an aspect ratio equal to 2.0 was used for comparison, and the corresponding experimental S-shaped global hysteresis loop is shown in Figure 14(a). Similar S-shaped curves for rocking behavior of URM piers could be provided by other tests such as those conducted by Calvi et al. [25, 26]. Besides, the entire hysteresis curve obtained in the test is shown in Figure 14(c); Figure 14(b) also shows the early hysteresis curve in order to better understand the hysteretic behavior of specimen 1TB2. The typical S-shape with the self-centering and almost nonlinear elastic property, which can be observed in Figure 14(b), is similar to that observed in Figure 14(a), owing to their typical rocking response with a little residual displacement. For the case in Figure 14(a), if no other failure mechanism occurs, the displacements can be limited only by a decrease in strength owing to second-order effects associated with overturning, whereas the dissipating energy of the URM wall would not significantly increase.

Unlike URM piers without an SCC, the latter hysteresis curve of specimen 1TB2 exhibited a significant dissipated energy capability. Negligible degradation in strength and a little slip at the wall base which is limited by a clamping force along the wall base can be observed in subsequent cycles.

The amount of dissipated energy is essentially proportional to the area of the hysteresis loops shown when the imposed displacements are large enough to induce a significant nonlinear response. A schematic of a dissipative coefficient, E, is shown in Figure 15(a), and the many sequences of specimen 1TB2 are also shown in Figure 15. The dissipative coefficient is defined by the following equation:

$$E = \frac{S_{(ABC+CDA)}}{S_{(OBE+ODF)}},\tag{16}$$

where $S_{(ABC+CDA)}$ are the areas of the hysteresis loop and $S_{(OBE+ODF)}$ are the areas of the trilateral.

For specimen 1TB2, high dissipative coefficients in later stages indicate that the rocking mode has a significant hysteretic energy absorption capability. A perfect dissipative effect is due to the cyclic yielding of longitudinal reinforcements with a high ductility capacity and the gradual crushing of toe. The hysteretic strain curve of the gauging point S1 of specimen 1TB2 in large lateral displacements, which is 50 mm away from the pier bottom, is shown in Figure 16. If assuming the yielding microstrain of reinforcements is 1400, then the grey area is a plastic region in Figure 16. It is noticed the hysteresis curve shape is similar to that of a metal bucking-restrained brace, which is symmetrical in tensile and compressive behavior after yield. For reinforcements in plastic zone, the compressive yield rather than buckling behavior can provide a large energy dissipation capacity.

Compared with the hysteresis curve of specimen 1TS2, which is shown in Figure 14(d) and which exhibited a typical "bow" shape with a sliding effect, the rocking hysteresis curve

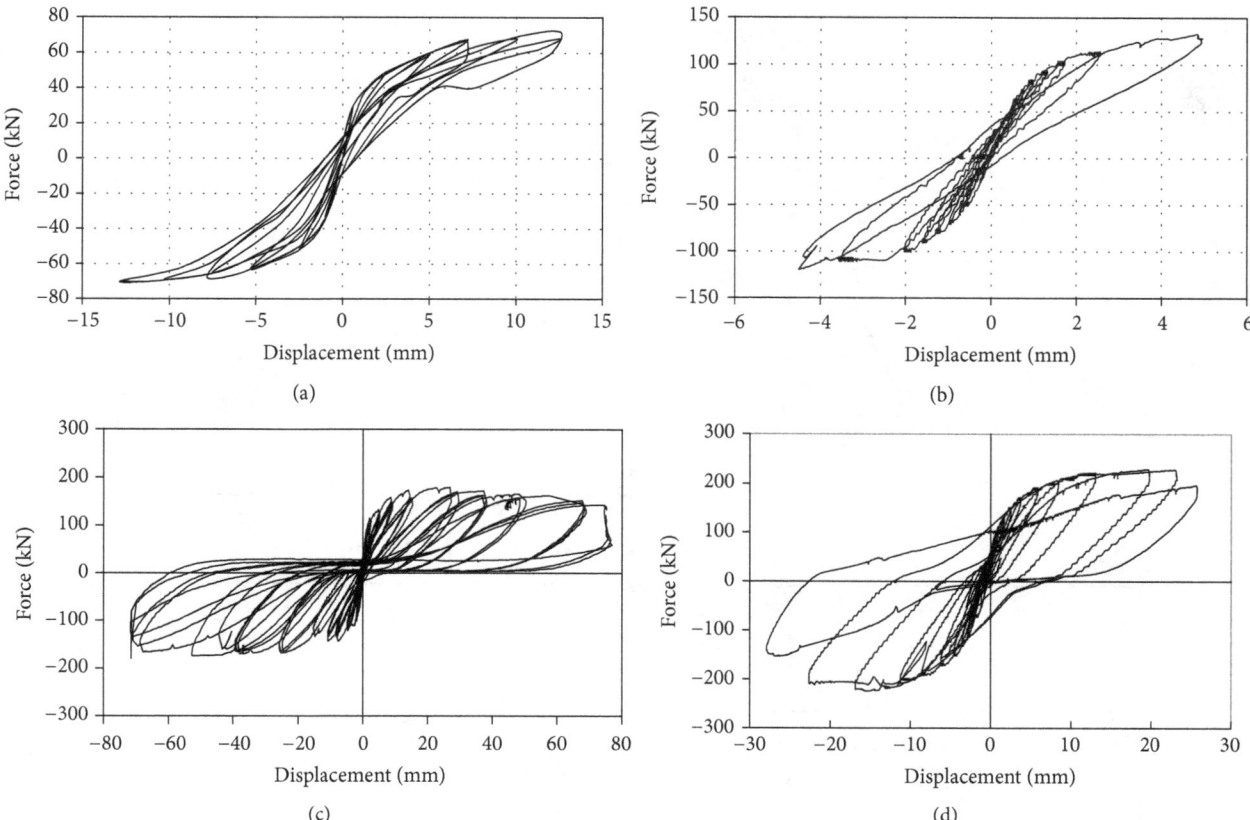

FIGURE 14: Hysteresis loop according to Anthoine et al. (a) and those of the specimens ((b) is that found in the earlier stage, (c) is the whole curve of specimen 1TB2, and (d) is the curve of specimen 1TS2).

of specimen 1TB2 exhibits a better seismic performance consisting of the self-centering property before steel yield and the high dissipating energy after steel yield. Rocking walls with prestressed technique and with no energy dissipation devices in Figures 2(a)–2(c) display a nonlinear elastic response with essentially no energy dissipation capacity, whereas specimen 1TB2 displays a plastic behavior with self-centering property.

4.6.3. Plastic Hinge Length. Predicated plastic hinge length of the specimen 1TB2, l_p, is 0.43 m based on (8). The plastic hinge length l_p in (8) is an equivalent length which is used to solve the wall's flexural deflection and plastic rotation capacity. The test value of physical plastic hinge length should be determined by the distribution of strain and cracks, and it was an approximate data in the test. Figure 17 describes the distribution of the maximum strain of longitudinal reinforcements at ultimate curvature, and it indicates that the mechanics assumption of a plastic hinge/rigid plane is appropriate. The maximum length of plastic zone should not exceed 0.67 m according to a yield strain of 0.14%. If the distribution of cracks was considered, a spacing value of horizontal reinforcements, 0.56 m, may be an appropriate experimental result.

4.6.4. Displacement Performance and Ductility. The backbone curves derived from the envelope curve of hysteresis loops are shown in Figure 18. Backbone curves of specimen

1TB2 and 1TS2 are plotted using a solid curve and a dashed curve in Figure 18, respectively. The corresponding points on the solid curve define specific damage states in test. Note that the story drifts in Figure 18 were calculated by assuming that the story height was equal to 2.8 m.

For specimen 1TB2, rocking behavior governs the maximum strength achieved in the test, and the lateral strength was maintained at high levels of cyclic deformation. As a typical deformation-controlled failure mode, the section of the solid curve from point b to point d is close to a level straight line, the rotation capacity of a plastic hinge is associated with its length, and the corresponding bending moment is the value borne by the plastic hinge. Beyond point e, which is considered as the ultimate displacement point, and when the corresponding story drift is about 2.5%, as shown by the dashed lines in Figure 18, this procedure was not conducted in the test for safety reasons, because the wall would exhibit overturning behavior. Curvatures of a section may be calculated according to the flat section assumption in rocking mode. The estimated yield curvature, ϕ_y, is 0.0029/mm. Assuming that the maximum compressive strain of masonry material equals 0.005, the result of maximum curvature ϕ_m is equal to 0.022/mm. Based on (9), the predicted yield displacement at the level of top, Δ_y, is 6.3 mm, which has an error of 11.2% with the test result equal to 7.1 mm. The maximum displacement Δ_p calculated according to (11) is equal to 50.8 mm, which is much small than the test result

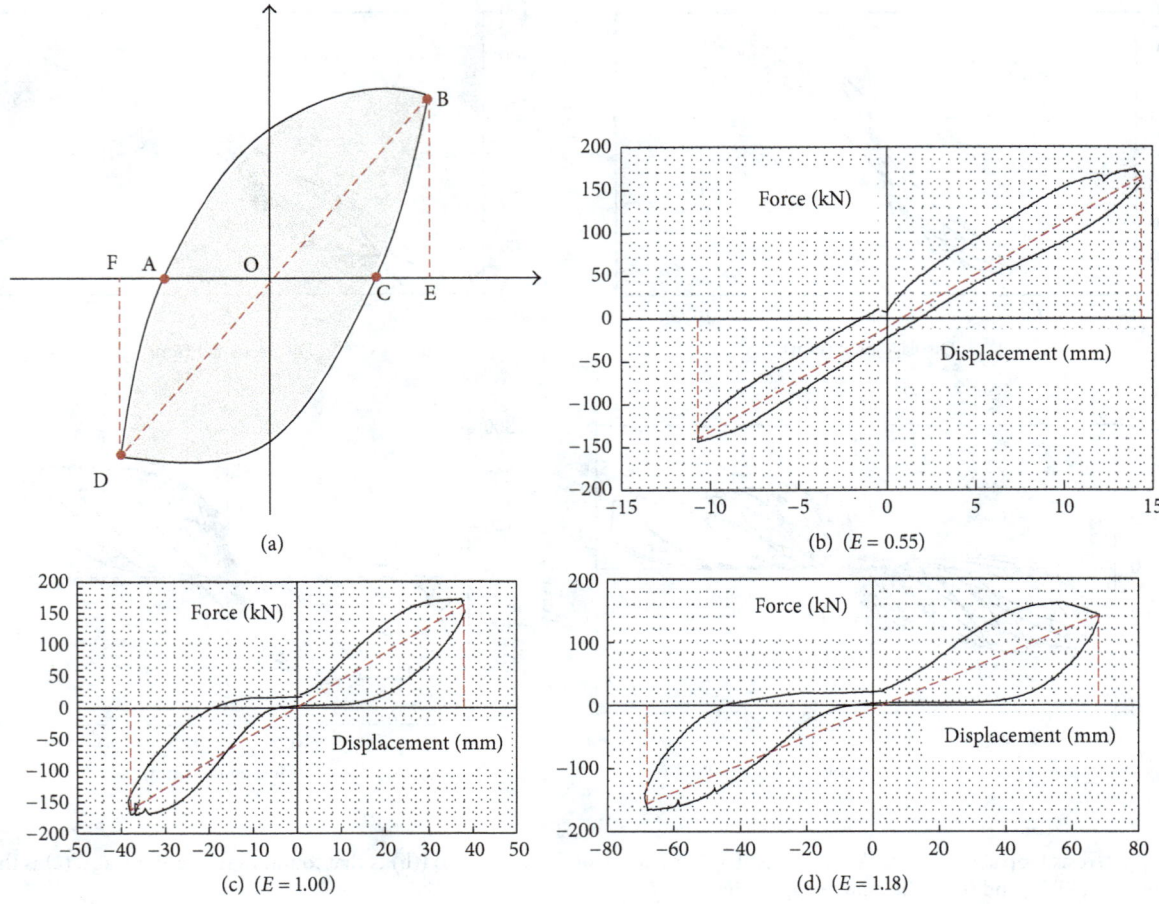

FIGURE 15: Schematic of dissipative coefficient (a) and the corresponding results in cycle of specimen 1TB2.

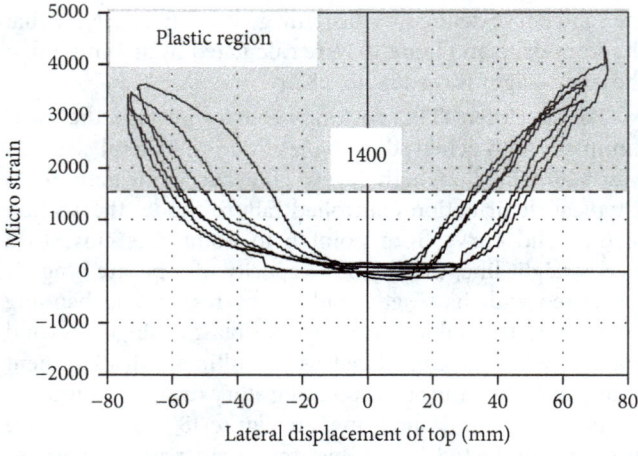

FIGURE 16: Strain hysteresis loop at point S1 in larger lateral displacements.

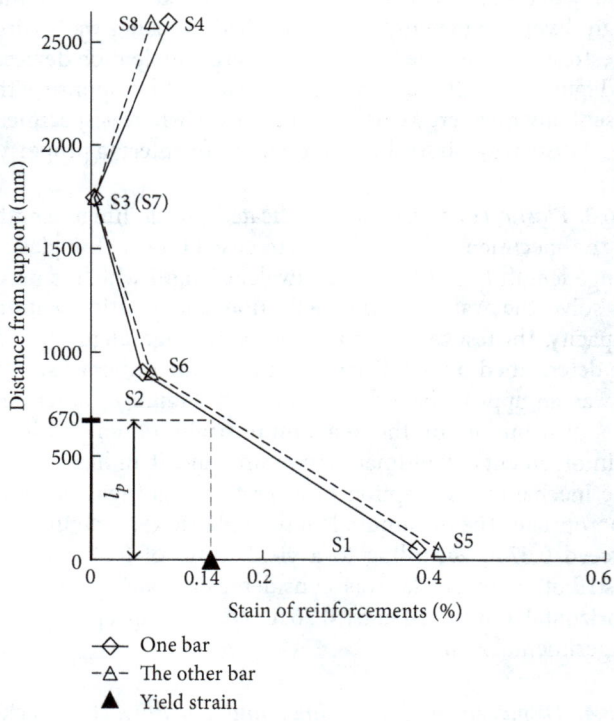

FIGURE 17: Strain distribution of longitudinal reinforcements of specimen 1TB2 at maximum curvature.

equal to 69.1 mm. The error in Δ_p is owing to the fact that the experimental ultimate state was determined by 85% of the peak strength, while the calculating result is based on the maximum compressive strain of masonry, which is the displacement at the crushing point in fact. Small distance between the experimental crushing point and the calculating

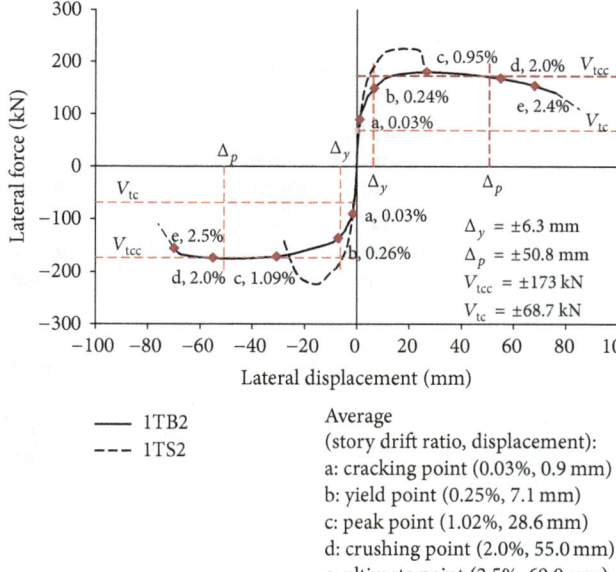

FIGURE 18: Backbone curves of specimens.

ultimate point can be observed in Figure 18. The predicated displacement ductility based on (13) equals 8.0, and the test value is equal to 9.7. Because the displacement ductility is greater than 4, specimen 1TB2 shall be determined as a high ductility component. The predicated width of slots is about 16.5 mm, and the demand in test is about 18.8 mm. Considering the compression capacity of filler material, the width of slots applied in buildings shall be determined by the predicted value multiplied by an amplification factor.

For the backbone curve of specimen 1TS2, a typical brittle mode can be observed; with loading increase, diagonal tension cracks developed, and lateral capacity was rapidly lost. As expected, specimen 1TB2 compared to 1TS2 exhibits a better displacement ductility capacity.

4.6.5. Comparison and Verification of Strengths. The experimental lateral peak strength of specimen 1TB2 and 1TS2 was 182.7 kN and 235.0 kN, respectively. After slotting, the lateral peak strength of the specimen decreased by 22.3% in the test.

As mentioned in Section 4.6.3, for the specimen ITB2, the predicted value of plastic hinge length is considered equal to 0.43 m, and then the parameter, h_{eff}, which is used to predict the lateral strength of the pier should be equal to 1.62 m. The predicted lateral strength V_{tcc} of specimen 1TB2, which is calculated based on (4), is equal to 173.0 kN. V_{tcc} has a margin of error of 5.3% with the experimental result. Assuming that specimen 1TB2 did not consist of an SCC, then the predicted lateral strength V_{tc} is equal to 68.7 kN based on (3). Both of the predicted strengths V_{tcc} and V_{tc} are also shown in Figure 18. Since an SCC is at the center of piers, the first flexural cracking strength of piers under rocking behavior should be close to that of piers without an SCC. V_{tc} is only 14.5% higher than the experimental cracking strength. After adding an SCC, the predicted strength V_{tcc} is 151.8% higher than V_{tc}; this significant increased strength is mainly provided

by steel, despite the fact that longitudinal reinforcement ratio is low in the SCC.

5. Conclusions

Although both specimens firstly showed the flexural cracks in test, different failure modes ultimately occurred owing to the various aspect ratios. In most cases, boundary limits of slotting piers might be sufficient to avoid shear failure. This slotting strategy could be expected to apply in seismic design of masonry buildings. The following points summarize the conclusions:

(1) A construction strategy of multistory masonry buildings with rocking mode is proposed in the paper. A masonry building is divided to ordinary seismic zone and key seismic zone in the construction. Slots are applied in key seismic zone.

(2) Experimental verification shows that slotting masonry piers with an SCC achieves a change of failure modes from shear to rocking, and the rocking specimen exhibits a stable mechanism with a large deformation.

(3) Rocking earthquake-resistant system for a masonry pier with an SCC consists of self-centering property, high energy dissipation, and high ductility capacity. Although the lateral strength decreased by 22.3% after slotting, a lateral story drift ratio of approximately 2.5% and a high displacement ductility of approximately 9.7 are achieved in test.

(4) For the specimen 1TB2, the full hysteretic curve was observed and the high dissipative coefficient was obtained in the later stages. Compared with a rocking URM pier, or a rocking wall with prestressed technique and with no energy dissipation devices, the hysteretic energy absorption capability of a rocking pier with an SCC increases significantly under a large lateral displacement.

(5) Equations proposed in the paper are successful in predicting rocking strengths of piers with an SCC. The test result of rocking displacements has a good agreement with the predicted value according to existing equations.

(6) Test procedure indicated that the initial slotting width of 10 mm is insufficient for rocking behavior with a large deformation level. The predicated width of slots is about 16.5 mm, which is close to the demanded value of the deformation of the rocking pier in test. The width of slots applied in buildings shall exceeds the predicted value, which helps to reduce the uncertainty in the change of failure mode, and also note that the filler material should be sufficiently soft.

(7) Distribution of horizontal reinforcements has an influence on the propagation of cracks and the plastic hinge length in the rocking failure mode.

Conflicts of Interest

The authors declare that they have no conflicts of interest.

Acknowledgments

This research is supported by the National Science Foundation of China under Grant no. 50978177.

References

[1] Y. C. Yang, L. Yang, Y. X. Gao et al., "Relation between damage to multi-storey brick buildings and their strength during the TangShan earthquake," *Earthquake Engineering and Engineering Vibration*, vol. 1, no. 1, pp. 21–35, 1981.

[2] GB 50203-2011, *Code for Acceptance of Constructional Quality of Masonry Structures*, China Architecture & Building Press, Beijing, China, 2011.

[3] GB 50011-2010, *Code for Seismic Design of Buildings*, China Architecture & Building Press, Beijing, China, 2010.

[4] N. Augenti and F. Parisi, "Learning from construction failures due to the 2009 L'Aquila, Italy, earthquake," *Journal of Performance of Constructed Facilities*, vol. 24, no. 6, pp. 536–555, 2010.

[5] T. Yi, F. L. Moon, R. T. Leon, and L. F. Kahn, "Lateral load tests on a two-story unreinforced masonry building," *Journal of Structural Engineering*, vol. 132, no. 5, pp. 643–652, 2006.

[6] J. K. Bothara, R. P. Dhakal, and J. B. Mander, "Seismic performance of an unreinforced masonry building: an experimental investigation," *Earthquake Engineering and Structural Dynamics*, vol. 39, no. 1, pp. 45–68, 2010.

[7] L. G. Cormack, "The design and construction of the major bridges on the mangaweka rail deviation," *Transactions of the Institute of Professional Engineers of New Zealand*, vol. 15, no. 1, pp. 6–23, 1998.

[8] D. Ryu, A. C. Wijeyewickrema, M. A. Elgawady, and M. A. K. M. Madurapperuma, "Effects of tendon spacing on in-plane behavior of posttensioned masonry walls," *Journal of Structural Engineering*, vol. 140, no. 4, Article ID 4013096, 2014.

[9] R. Hassanli, M. A. ElGawady, and J. E. Mills, "Strength and seismic performance factors of posttensioned masonry walls," *Journal of Structural Engineering*, vol. 141, no. 11, Article ID 04015038, 2015.

[10] J. I. Restrepo and A. Rahman, "Seismic performance of self-centering structural walls incorporating energy dissipators," *Journal of Structural Engineering*, vol. 133, no. 11, pp. 1560–1570, 2007.

[11] G. D. Wight, M. J. Kowalsky, and J. M. Ingham, "Shake table testing of posttensioned concrete masonry walls with openings," *Journal of Structural Engineering*, vol. 133, no. 11, pp. 1551–1559, 2007.

[12] M. A. ElGawady, P. Lestuzzi, and M. Badoux, "Performance of URM walls under in-plane seismic loading," *The Masonry Society Journal*, no. 23, pp. 85–104, 2005.

[13] J. F. Stanton and S. D. Nakaki, "Design Guidelines for Precast Concrete Structural Systems," PRESSS Report 01/03-09, Dept. of Civil Engineering, University of Washington, 2002.

[14] H. Liu, M. J. Han, C. G. Lan, and X. Z. Yang, "Pseudo-dynamic test and quasi-static test of full-scale model of a two-story brick building retrofitted with prestressed tendons," *China Civil Engineering Journal*, vol. 49, no. 3, pp. 43–55, 2016.

[15] H. Liu, L. R. Ban, C. G. Lan, T. Wang, and Y. J. Tian, "Quasi-static test of unreinforced brick walls retrofitted with post-tensioning tendons," *China Journal of Building Structures*, vol. 36, no. 8, pp. 142–149, 2015.

[16] R. Ma, L. Jiang, M. He, C. Fang, and F. Liang, "Experimental investigations on masonry structures using external prestressing techniques for improving seismic performance," *Engineering Structures*, vol. 42, pp. 297–307, 2012.

[17] Council BSS, *NEHRP Guidelines for the Seismic Rehabilitation of Buildings*, Federal Emergency Management Agency, Washington, DC, USA, FEMA-273 edition, 1997.

[18] Council BSS, *Evaluation of earthquake damaged concrete and masonry wall buildings*, Federal Emergency Management Agency, Washington, DC, USA, FEMA 306 edition, 1999.

[19] D. C. Rai, "Hysterretic behavior of unreinforced masonry piers strengthened with steel elements," in *Proceedings of the Eleventh World Conference on Earthquake Engineering*, pp. 501–509, Japan, 1996.

[20] G. Magenes and G. M. Calvi, "In-plane seismic response of brick masonry walls," *Earthquake Engineering & Structural Dynamics*, vol. 26, no. 11, pp. 1091–1112, 1997.

[21] A. H. Salmanpour, N. Mojsilovic, and J. Schwartz, "Deformation capacity of unreinforced masonry walls subjected to in-plane loading: a state-of-the-art review," *Advanced Structural Engineering*, vol. 5, no. 1, article no. 22, 2013.

[22] S. Petry and K. Beyer, "Force-displacement response of in-plane-loaded URM walls with a dominating flexural mode," *Earthquake Engineering and Structural Dynamics*, vol. 44, no. 14, pp. 2551–2573, 2015.

[23] T. Paulay and M. J. N. Priestley, *Seismic Design of Reinforced Concrete and Masonry Buildings*, John Wiley & Sons, New York, NY, USA, 1992.

[24] A. Anthoine, G. Magonette, and G. Magenes, "Shear-compression testing and analysis of brick masonry walls," in *Proceedings of the 10th European Conference on Earthquake Engineering*, pp. 1657–1662, 1995.

[25] G. M. Calvi, G. R. Kingsley, and G. Magenes, "Testing of masonry structures for seismic assessment," *Earthquake Spectra*, vol. 12, no. 1, pp. 145–162, 1996.

[26] F. Parisi, N. Augenti, A. Balsamo, A. Prota, and G. Manfredi, "Lateral loading tests on a masonry system with and without external reinforcement," in *Proceedings of the 14th European Conference on Earthquake Engineering*, Ohrid, Macedonia, 2010.

9

Water Permeability Characteristics of Normal Strength Concrete Made from Crushed Clay Bricks as Coarse Aggregate

Syed Ishtiaq Ahmad and Mohammad Anwar Hossain

Department of Civil Engineering, Bangladesh University of Engineering and Technology, Dhaka, Bangladesh

Correspondence should be addressed to Syed Ishtiaq Ahmad; siahmad@ce.buet.ac.bd

Academic Editor: Kaveh Edalati

Water permeability characteristic of concrete made from crushed clay bricks as coarse aggregate is investigated and compared with concrete made from natural stone aggregate. For this, six different brick and five different natural stone samples were selected. Crushing strength of brick samples and water absorption of aggregate produced from them were also measured. Concrete samples of three different compressive strengths were prepared as per ACI mix design method from each of these aggregate samples. Compressive strength of concrete that could be achieved with brick aggregate varied between 19 and 28 MPa, whereas, for stone aggregate, compressive strength varied between 24 and 46 MPa. These samples were then tested for water permeability using the AT 315 machine as per EN 12390-8: "Depth of Penetration of Water under Pressure." Experimental results and subsequent analysis indicate that water permeability of brick aggregate concrete is 225% to 550% higher than that of concrete made from natural stone aggregate of identical compressive strength. Water permeability was found to be directly related to compressive strength, water absorption, and porosity of hardened concrete. It was also observed that water permeability of concrete is influenced by water absorption of brick aggregate and crushing strength of brick.

1. Introduction

Concrete is the main ingredient in ever growing construction industry of Bangladesh. As natural stone is in short supply and hence expensive, crushed burnt clay bricks are extensively used as an economic alternative coarse aggregate in preparation of concrete in Bangladesh for construction of midrise residential and factory buildings, rigid pavements, and small and medium span bridges and culverts [1]. Properties of brick aggregate vary appreciably from natural stone aggregate in the context of strength, toughness, and other related indices [2]. Since coarse aggregate occupies a large share of concrete volume, therefore, it is presumable that the properties of concrete made from brick aggregate will vary appreciably from that made of stone aggregate. Though compressive strength of concrete in the normal range can be achieved satisfactorily, however, it is the durability properties like water permeability, creep, and shrinkage that has always been a concern for brick aggregate concrete. Water permeability is an important issue for concrete made from crushed clay bricks because brick aggregate is far more porous and hence permeable than granite and other natural stone aggregates [3]. There are a number of works that have been reported till today on properties of concrete made from brick aggregate [1–8]. However, none of these have extensively investigated water permeability of brick aggregate concrete except Debeib who has shown that it is possible to manufacture concrete containing crushed bricks (coarse and fine) with water permeability characteristics similar to those of natural aggregate concrete provided that the percentage of brick aggregate is limited to 25% and 50% for coarse and fine aggregates, respectively [5]. However, authors of this work feel that a systematic and comparative study of both natural stone and crushed clay brick aggregate concrete will assist in understanding the salient features of water permeability characteristics of brick aggregate concrete. This will also help designers and engineers using brick aggregate concrete in predicting the expected water permeability behavior. Outcome of this research would be a significant addition to the existing knowledge in this field because

TABLE 1: Properties of brick and brick aggregate.

Brick type	Crushing strength (MPa)	Specific gravity (SSD)	Density (Kg/m^3)	Water absorption (%)	LA abrasion value (%)
1	28.25	2.16	1450	9.8	39.25
2	27.65	2.12	1422	9.95	40.15
3	18.30	2.10	1400	13.40	44.10
4	17.25	2.06	1390	14.05	44.80
5	14.85	2.02	1380	14.60	46.10
6	13.95	1.97	1350	17.90	49.50

permeability is one of the main parameter responsible for concrete deterioration. Information regarding expected value of water permeability coefficient of brick aggregate concrete would help practicing engineers to design more durable and sustainable structures using brick aggregate concrete. This may also help to modify existing building codes on brick aggregate concrete in areas like clear cover over reinforcing bars of concrete as well as provisions for construction of water retaining structures using brick aggregate concrete. For this, an experimental program was conducted at Bangladesh University of Engineering and Technology, Dhaka, Bangladesh, to study the water permeability behavior of concrete made from crushed clay bricks. For this, six different brick and five different natural stone aggregates were selected. In the experimental program, concrete samples having three different compressive strengths were prepared from each of the natural stone and crushed clay brick aggregate. These samples were then subjected to water permeability testing using European standard AT 315 machine as per BS EN 12390-8: "Depth of Penetration of Water under Pressure" [9]. Test results were analyzed to examine the increase in water permeability associated with brick aggregate concrete compared to corresponding concrete made from natural stone aggregate. Several brick, brick aggregate, and concrete properties that include crushing strength of brick, water absorption of brick aggregate, water absorption, and porosity in hardened concrete were also measured. Influence of these properties on water permeability behavior of corresponding concrete was also investigated.

2. Materials Used

2.1. Cement.
Ordinary Portland cement (Type 1) having 28-day compressive strength of 46 MPa as per ASTM C 150 [10] was used for preparation of all concrete samples. By using one type of cement the effect of varying the types of coarse aggregate in concrete was investigated.

2.2. Fine Aggregate.
One type of natural coarse sand was used throughout the experimental work so as to keep the fine aggregate parameter constant. Sieve analysis was carried out in accordance with ASTM C136 [11]. The results of this analysis showed that the sand used fitted within the limits set out in ASTM C33 [12]. Unit weight of aggregates was also determined in accordance with ASTM C29/C29M [13] whereas water absorption and specific gravity of fine aggregate were found in accordance with ASTM C128 [14]. From these testing procedures, fineness modulus, unit weight,

water absorption, and specific gravity of fine aggregate were found as 2.70, 1630 Kg/m^3, 1.26%, and 2.66, respectively.

2.3. Clay Bricks.
In this work, six different types of brick samples, numbered as 1 to 6, were collected from different brick manufacturing factories. These factories use two types of widely used kiln in Bangladesh, that is, Bulls Trench Kiln and Fixed Chimney Kiln. Before these bricks were crushed down to aggregate, compressive (crushing) strength test was conducted according to ASTM C 67 [15]. Test results are reported in Table 1 which show that crushing strength of brick varied between 14 and 29 MPa. A large variation in crushing strength of brick was selected so that its effect on water permeability of concrete could be observed.

2.4. Brick and Stone Aggregate.
Brick aggregate was produced by breaking down whole new bricks on a solid concrete surface using a hammer. Natural crushed sandstone boulders were used as stone aggregate. In this work, five different types of natural stone boulders were collected from diverse sources. For comparison purpose, bricks and stone boulders were crushed in a way so that they possess similar gradation and approximately same fineness modulus to negate the effect, if any, of size and shape on water permeability behavior of concrete. Additionally, it was also ensured that grading limits set out in ASTM C33 [12] is strictly maintained. Before preparing concrete, different properties of both brick and stone aggregates were measured. This includes water absorption and specific gravity as per ASTM C127 [16] and Los Angeles (LA) abrasion test as per ASTM C131 [17]. Test results are presented in Tables 1 and 2. Observation of these results show that stronger bricks have higher density and lower LA abrasion value. Additionally, all brick aggregates have lower density than that of crushed stone aggregate. Therefore, concrete of lower density may be achieved through use of brick aggregate. Water absorption of brick aggregate, on the other hand, was found to be several times higher than that of stone aggregate.

3. Testing Scheme

3.1. Mix Design and Mixing Method.
The procedure for design of concrete mixes with normal aggregate can be used to design mixes using crushed brick aggregate [3]. In this work, mix design ratios for both stone and brick aggregate concrete with target compressive strength of 20.0, 30.0, and 40.0 MPa were evaluated from ACI method [18] with a water

Water Permeability Characteristics of Normal Strength Concrete Made from Crushed Clay Bricks as Coarse...

81

TABLE 2: Properties of natural stone aggregate.

Sl. number	Specific gravity (SSD)	Density (Kg/m^3)	Water absorption (%)	LA abrasion value (%)
1	2.63	1580	1.62	28.70
2	2.20	1550	1.93	30.85
3	2.69	1615	0.82	25.20
4	2.67	1605	1.22	26.90
5	2.64	1590	1.36	27.70

cement ratio (w/c) of 0.4, 0.5, and 0.6, respectively, considering slump value in the range of 25 to 50 mm. Required quantities for cement, water, coarse, and fine aggregate for all brick and stone aggregate are presented in Table 3. Since water absorption of brick aggregate is much higher, it is recommended to soak the brick aggregates in water prior to adding to the concrete mixture [2, 3, 6]. Otherwise, a large part of water from mix design calculation will be soaked by the aggregate and will not be available to react with cement changing the water cement ratio. Hence, both stone and brick aggregate were soaked in water for 48 h and added to the mixture in a saturated surface dry condition. Water absorbed in the aggregate is in addition to the water requirement from mix design calculation as shown in Table 1. Fine aggregate was dried for 48 h in oven at 110°C and was allowed to cool to room temperature before adding to the mixture. Water requirement from mix design process was adjusted to consider effect of water absorption of fine aggregates. The aggregates, cement, and water were then combined and mixed in a machine mixer as per ASTM C 192 [19]. Slump tests were also conducted on fresh concrete as per ASTM C143 [20] and corresponding values are reported in Table 3. Concrete mixes for which slump value exceeded the design range of 25 to 50 mm were discarded and remixed.

For every set of concrete with particular target compressive strength, a total of three 300 mm × 150 mm cylinder specimens and six 150 mm × 150 mm cubes were cast. Cylinder specimens were subjected to compressive strength test as per ASTM C39 [21] maintaining a loading rate between 0.25 and 0.30 MPa/sec. Three cube specimens were subjected to water permeability testing. The remaining three cubes were used to determine density, water absorption, and porosity in hardened concrete as per ASTM C642 [22].

3.2. *Water Permeability Testing.* European standard AT 315 apparatus was used to determine the water permeability of concrete according to EN 12390-8 [9]. The apparatus was connected to a normal air compressor capable of ensuring at least 5 bar compressed air continuously and equipped with dehumidifier and oil filter. Connection then was made to the laboratory water supply and to a drainage system. A specimen was subjected to test when its age was at least 28 days. For testing, the specimen was placed on the apparatus in such a manner that the water pressure act on the test area which actually is a 75 mm diameter area at the center of the bottom surface of the 150 mm by 150 mm cube. Water pressure of (500 ± 50) kPa for (72 ± 2) hours was applied on this surface.

FIGURE 1: Water penetration area and penetration front in a brick aggregate concrete sample.

After the pressure had been applied for the specified time, the specimen was removed from the apparatus. The face on which the water pressure was applied was wiped to remove excess water. The specimen was then split in half, perpendicularly to the face on which the water pressure was applied. As soon as the split face has dried to such an extent that the water penetration front can be clearly seen, maximum depth of penetration under the test area was recorded and measured to the nearest millimeter. Figure 1 shows example of such penetration area and marked penetration front in a brick aggregate concrete sample.

The depth of water penetration inside the specimen can be converted to its equivalent coefficient of water permeability using Valenta's equation [23]:

$$k = \frac{e^2 v}{2ht} \text{ m/sec}, \tag{1}$$

where e is depth of penetration of concrete in meters, h is hydraulic head in meters, t is time under pressure in seconds, and v is the fraction of the volume of concrete occupied by pores.

The value of v represents discrete pores, such as air bubbles, which do not become filed with water except under pressure and can be calculated from the increase in the mass of concrete during the test.

4. Results and Discussion

4.1. *Strength of Concrete, Water Absorption, and Porosity.* Table 4 shows compressive strength test results on concrete samples prepared. As can be seen, compressive strength

TABLE 3: Concrete mix proportions.

Aggr. types	w/c	Cement Kg/m^3	Fine aggregate Kg/m^3	Coarse aggregate Kg/m^3	Water Kg/m^3	Slump (mm)
				Mix design		
Brick-1	0.4	440	578	1000	176	30
	0.5	360	634	1000	180	35
	0.6	300	684	1000	180	45
Brick-2	0.4	440	582	982	176	25
	0.5	360	642	982	180	40
	0.6	300	690	982	180	40
Brick-3	0.4	440	592	966	176	30
	0.5	360	648	966	180	45
	0.6	300	698	966	180	45
Brick-4	0.4	440	571	959	176	30
	0.5	360	627	959	180	40
	0.6	300	678	959	180	45
Brick-5	0.4	440	549	952	176	35
	0.5	360	606	952	180	45
	0.6	300	656	952	180	50
Brick-6	0.4	440	545	931	176	35
	0.5	360	601	931	180	50
	0.6	300	652	931	180	50
Stone-1	0.4	440	722	1075	176	30
	0.5	360	779	1075	180	35
	0.6	300	829	1075	180	40
Stone-2	0.4	440	743	1050	176	25
	0.5	360	799	1050	180	35
	0.6	300	849	1050	180	45
Stone-3	0.4	440	698	1115	176	30
	0.5	360	755	1115	180	35
	0.6	300	805	1115	180	40
Stone-4	0.4	440	705	1100	176	25
	0.5	360	762	1100	180	35
	0.6	300	812	1100	180	50
Stone-5	0.4	440	715	1085	176	30
	0.5	360	772	1085	180	30
	0.6	300	820	1085	180	45

achieved using stone aggregate concrete is fairly close and lies within 15% of the target compressive strength. On the other hand, achieved compressive strength using brick aggregate was far less than target strength. For example, for concrete of 40 MPa target strength, the achieved compressive strength varied between 21 and 27.9 MPa, that is, about 30 to 47.5% lower than the target strength. For 20 MPa concrete, the difference is, however, smaller. As can be seen in Table 4, clear difference is found in water absorption and porosity between brick and stone aggregate concrete. In the current testing scheme, water absorption and porosity for stone aggregate concrete were found to vary between 1.5 to 4% and 3.8 to 8.9%, respectively, whereas, for brick aggregate concrete, water absorption and porosity varied between 5.9 to 9.9% and 7.6 to 15.8%, respectively. That is, for equivalent compressive

strength, water absorption and porosity in brick aggregate concrete were 60% to 80% higher. Similar trend was observed for brick aggregate concrete by other researchers [5, 8].

4.2. Compressive Strength of Concrete and Water Permeability. Water permeability of concrete made from natural stone and brick aggregate with respect to its compressive strength are presented in Figures 2 and 3, respectively. Coefficient of water permeability in this testing scheme for natural stone aggregate concrete was found to vary between 0.02×10^{-11} and 1.2×10^{-11} m/s. These values are in agreement with available results on natural stone aggregate concrete [24–26]. For brick aggregate concrete, coefficient of water permeability varied between 2.2×10^{-11} and 6×10^{-11} m/s for the range of compressive strength of concrete tested (16.7

TABLE 4: Compressive strength, water absorption, and porosity of hardened concrete.

Aggr. types	Compressive strength (MPa)			Water absorption after immersion (%)			Porosity (%)		
	$w/c = 0.4$	$w/c = 0.5$	$w/c = 0.6$	$w/c = 0.4$	$w/c = 0.5$	$w/c = 0.6$	$w/c = 0.4$	$w/c = 0.5$	$w/c = 0.6$
Brick-1	27.9	23.8	19.4	5.9	6.9	8.0	11.9	12.9	14.0
Brick-2	25.8	21.9	19.1	6.5	7.5	8.6	12.5	13.5	14.5
Brick-3	24.8	21.5	18.3	7.1	8.1	9.0	12.9	14.1	14.9
Brick-4	22.5	20.8	17.7	7.3	8.3	9.1	13.3	14.1	15.2
Brick-5	21.7	20.6	17.4	7.5	8.5	9.2	13.5	14.4	15.1
Brick-6	21.0	19.8	16.7	7.9	8.8	9.9	13.8	14.7	15.8
Stone-1	39.3	30.0	23.5	2.2	3.1	3.8	5.2	7.1	8.4
Stone-2	33.3	27.9	21.3	2.8	3.3	4.0	6.4	7.5	8.9
Stone-3	46.2	34.5	26.2	1.5	2.6	3.5	3.8	6.2	7.9
Stone-4	43.7	32.7	25.3	1.7	2.8	3.6	4.3	6.5	8.1
Stone-5	41.2	30.9	24.1	2.0	3.0	3.7	4.8	6.9	8.3

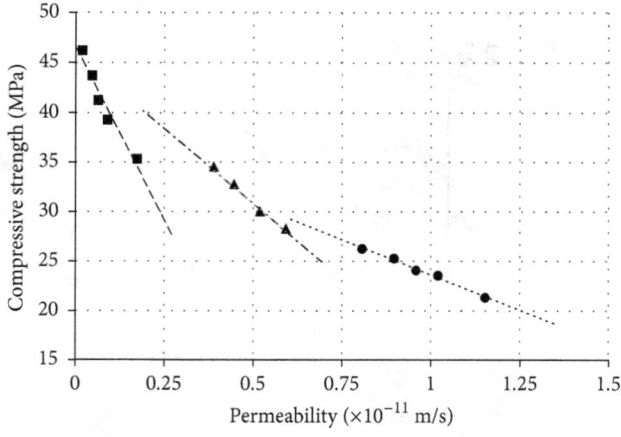

FIGURE 2: Water permeability coefficient with respect to compressive strength of stone aggregate concrete.

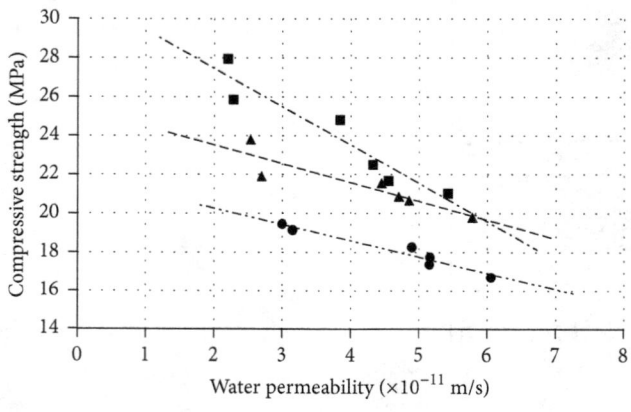

FIGURE 3: Water permeability coefficient with respect to compressive strength of brick aggregate concrete.

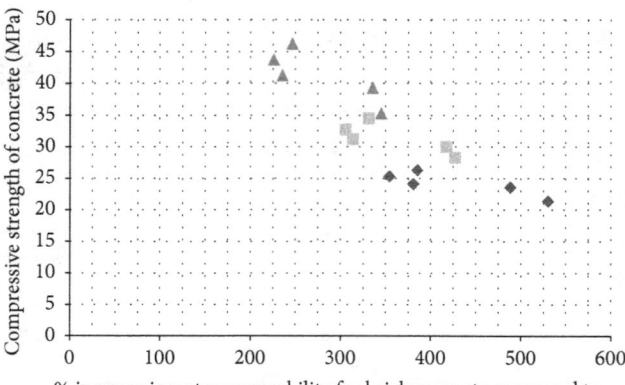

FIGURE 4: Comparative water permeability of brick and stone aggregate concrete.

to 27.9 MPa). For a particular w/c ratio, a well fitted straight line relationship was observed between coefficient of water permeability and compressive strength of stone aggregate concrete (Figure 2). Although not as fitting, an approximate straight line relationship between compressive strength and coefficient of water permeability for a specific w/c ratio can still be identified for brick aggregate concrete (Figure 3). Observation of Figures 2 and 3 shows that, for both stone and brick aggregate concrete, increased w/c ratio in the concrete mix results in corresponding increase in water permeability coefficient. Also, steeper curves suggest that percent increase in water permeability coefficient with respect to increase in compressive strength is more significant in stone aggregate concrete.

Next, coefficient of water permeability of brick aggregate concrete was compared to that of stone aggregate concrete of identical compressive strength. Linear variation

was observed between coefficient of water permeability and compressive strength for both brick and stone aggregate concrete. Accordingly, coefficient of water permeability was evaluated by linear extrapolation of obtained test data in the range of compressive strength for which test data was not available. Comparison shows that water permeability coefficient of brick aggregate is several times higher than corresponding water permeability coefficient of stone aggregate concrete of identical compressive strength. Figure 4 shows such comparison where percent increase in water permeability for brick aggregate compared to that of stone aggregate concrete of identical compressive strength is plotted. Three data sets are plotted in this figure, one for each w/c ratio, that is, 0.4, 0.5, and 0.6. For w/c ratio of 0.4 and for concrete of equivalent compressive strength, coefficient of water permeability of brick aggregate concrete is 225 to 350% times higher. For w/c ratio of 0.5, increase in coefficient of water permeability of brick aggregate concrete ranges from 300 to 425%, whereas, for w/c ratio of 0.6, coefficient of water permeability of brick aggregate concrete is 350 to 550% higher than that of stone aggregate concrete of identical compressive strength. Therefore, depending on compressive strength and w/c ratio, water permeability of concrete with crushed clay brick as coarse aggregate is 225 to 550% higher than corresponding concrete with natural stone as coarse aggregate.

4.3. Water Permeability Related to Water Absorption and Porosity in Hardened Concrete. Both porosity and water absorption of brick aggregate concrete were found to be 60% to 80% higher (Table 4) making it much more pervious than natural stone aggregate concrete of equivalent strength. Porosity and water absorptions are indication of pores or voids in concrete through which water permeates. Therefore, increase in these parameters results in corresponding increase in water permeability [23]. Water permeability coefficient with respect to water absorption and porosity in hardened brick aggregate concrete are shown in Figures 5

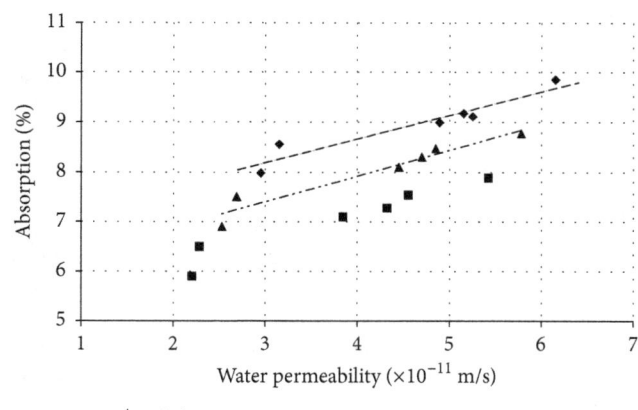

FIGURE 5: Water permeability coefficient with respect to water absorption of hardened concrete.

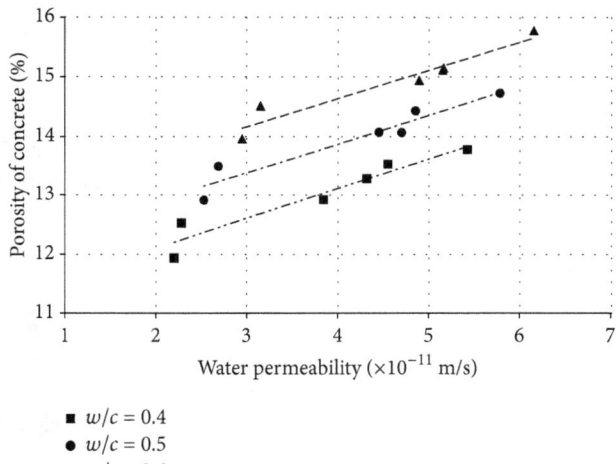

FIGURE 6: Water permeability coefficient with respect to porosity of hardened concrete.

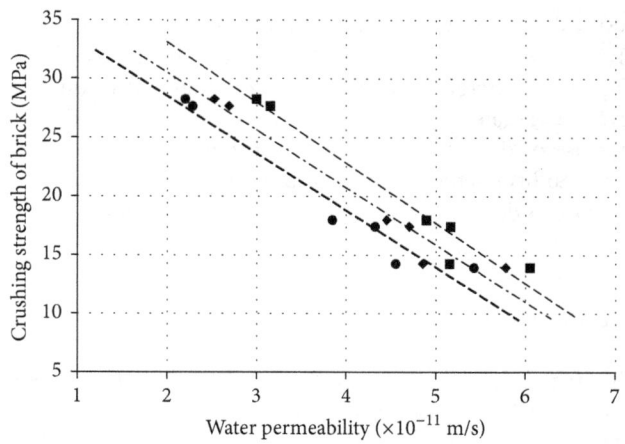

FIGURE 7: Water permeability with respect to crushing strength of brick.

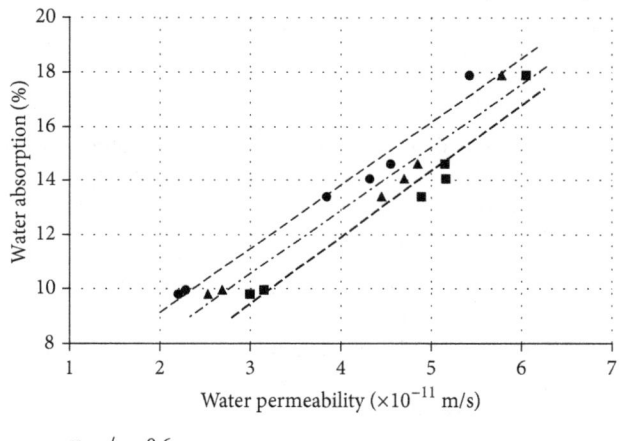

FIGURE 8: Water permeability with respect to water absorption of brick aggregate.

and 6, respectively. Observation of these figures shows that a linear relationship exists between these parameters and water permeability coefficient for a specific w/c ratio. Also, as can be seen, relatively flat slope of these straight lines indicates that water permeability coefficient of brick aggregate concrete is very much sensitive to these parameters. A similar linear behavior between porosity and water permeability was also observed for pervious concrete [26].

4.4. Water Permeability Related to Brick and Brick Aggregate Properties.
Two important indicative properties of brick and brick aggregate are crushing strength of brick and water absorption of brick aggregate. Observation of Table 2 shows that bricks with higher compressive strength produce aggregates that have higher specific gravity and density as well as lower water absorption and LA abrasion values. Consequently, concrete produced from bricks with higher

compressive strength has lower porosity (Table 4) and therefore it is less permeable. This is reflected in Figures 7 and 8, where crushing strength of brick and water absorption of brick aggregate is correlated with water permeability coefficient of concrete made from those aggregates, respectively. It may be seen from these figures that, for a specific w/c ratio, an approximately linear relationship exists between water permeability of brick aggregate concrete and these two parameters. Understandably, increase in crushing strength of brick is associated with decrease in water permeability. On the other hand, increase in water absorption of brick aggregate results in corresponding increase in water permeability of concrete. As an example, if crushing strength of brick is increased from 15 to 25 MPa, water permeability of concrete is reduced from 4.75×10^{-11} m/s to 2.75×10^{-11} m/s; that is, 66% increase in crushing strength of brick reduces water permeability of corresponding concrete by about 42%. 1×10^{-11} m/s increase in coefficient of water permeability of brick aggregate concrete is observed when water absorption of brick aggregate is increased by 2 to 3%, whereas, for 2% increase or decrease in water absorption of brick aggregate, permeability coefficient of corresponding concrete is found to increase or decrease, respectively, in a range of 0.8×10^{-11} m/s to 1.1×10^{-11} m/s.

5. Conclusion

This paper examined the water permeability properties of crushed clay brick aggregate concrete and compared it with concrete made from natural stone aggregate. Based on experimental results found in this study, it can be concluded that coefficient of water permeability of brick aggregate concrete is always higher than natural stone aggregate concrete of equivalent strength. Depending on compressive strength and w/c ratio, coefficient of water permeability of brick aggregate concrete may be 225% to 550% higher than that of stone aggregate concrete.

Additionally, following conclusions can be made based on the observations and experimental results in this investigation:

(i) For equivalent compressive strength, water absorption and porosity in hardened brick aggregate concrete are 60% to 80% higher compared to those in stone aggregate concrete.

(ii) For a specific w/c ratio, compressive strength and water permeability of brick aggregate concrete are linearly related. Increase in strength shows a corresponding decrease in water permeability coefficient and vice versa.

(iii) For a specific w/c ratio, linear relationship exists between water absorption and porosity in hardened brick aggregate concrete and its water permeability coefficient. Also, water permeability coefficient is very much sensitive to these parameters.

(iv) Coefficient of water permeability of brick aggregate concrete is linearly related to crushing strength of brick. 66% increase in crushing strength of brick reduces water permeability of corresponding concrete by about 42%.

(v) Coefficient of water permeability in brick aggregate concrete is linearly related to water absorption of brick aggregate. For 2% increase or decrease in water absorption of brick aggregate, permeability coefficient of corresponding concrete is found to increase or decrease, respectively, in a range of 0.8×10^{-11} m/s to 1.1×10^{-11} m/s.

Conflicts of Interest

The authors declare that they have no conflicts of interest.

References

[1] M. A. Mansur, T. H. Wee, and L. S. Cheran, "Crushed bricks as coarse aggregate for concrete," *ACI Materials Journal*, vol. 96, no. 4, pp. 478–484, 1999.

[2] F. M. Khalaf and A. S. DeVenny, "Properties of new and recycled clay brick aggregates for use in concrete," *Journal of Materials in Civil Engineering*, vol. 17, no. 4, article 456, 2005.

[3] F. M. Khalaf, "Using crushed clay brick as coarse aggregate in concrete," *Journal of Materials in Civil Engineering*, vol. 18, no. 4, pp. 518–526, 2006.

[4] A. A. Akhataruzzaman and A. Hasnat, "Properties of concrete using crushed brick as aggregate," *ACI Concrete International*, vol. 5, pp. 58–63, 1983.

[5] F. Debieb and S. Kenai, "The use of coarse and fine crushed bricks as aggregate in concrete," *Construction and Building Materials*, vol. 22, no. 5, pp. 886–893, 2008.

[6] P. B. Cachim, "Mechanical properties of brick aggregate concrete," *Construction and Building Materials*, vol. 23, no. 3, pp. 1292–1297, 2009.

[7] S. I. Ahmad and S. Roy, "Creep behavior and its prediction for normal strength concrete made from crushed clay bricks as coarse aggregate," *Journal of Materials in Civil Engineering*, vol. 24, no. 3, pp. 308–314, 2012.

[8] M. Adamson, A. Razmjoo, and A. Poursaee, "Durability of concrete incorporating crushed brick as coarse aggregate," *Construction and Building Materials*, vol. 94, pp. 426–432, 2015.

[9] BSI, "Hardened concrete—part 8: depth of penetration of water under pressure," BS EN 12390-8, BSI, London, Uk, 2009.

[10] ASTM, "Specification for Portland cement," ASTM C 150, ASTM, West Conshohocken, Pa, USA, 2007.

[11] ASTM, "Standard test method for sieve analysis of fine aggregate," ASTM C 136, ASTM, West Conshohocken, Pa, USA, 2014.

[12] ASTM, "Standard specification for concrete aggregates," ASTM C 33, ASTM, West Conshohocken, Pa, USa, 2011.

[13] ASTM, "Standard test method for bulk density ("unit weight") and voids in aggregate," ASTM C29/C29M, ASTM, West Conshohocken, Pa, USA, 2009.

[14] ASTM, "Standard test method for density, relative density (specific gravity), and absorption of fine aggregate," ASTM C 128, ASTM, West Conshohocken, Pa, USA, 2015.

[15] ASTM, "Standard test methods for sampling and testing brick and structural clay tile," ASTM C67, ASTM, West Conshohocken, Pa, USA, 2014.

[16] ASTM, "Standard test method for density, relative density (specific gravity), and absorption of coarse aggregate," ASTM C 127, ASTM, West Conshohocken, Pa, USA, 2015.

[17] ASTM, "Standard test method for resistance to degradation of small-size coarse aggregate by abrasion and impact in the los angeles machine," ASTM C 131, ASTM, West Conshohocken, Pa, USA, 2014.

[18] American Concrete Institute (ACI), "Standard practice for selecting proportions for normal, heavyweight and mass concrete," ACI 211.1-91, American Concrete Institute, Famington Hills, Mich, USA, 2002.

[19] ASTM, "Standard practice for making and curing concrete test specimen in the laboratory," ASTM C 192/C192M, ASTM, West Conshohocken, Pa, USA, 2007.

[20] ASTM, "Standard test method for compressive strength of cylindrical concrete specimens," ASTM C 39, ASTM, West Conshohocken, Pa, USA, 2015.

[21] ASTM, "Standard test method for slump of hydraulic cement concrete," ASTM C 143, ASTM, West Conshohocken, Pa, USA, 2015.

[22] ASTM, "Standard test method for density, absorption and voids in hardened concrete," ASTM C 642, ASTM, West Conshohocken, Pa, USA, 2013.

[23] O. Valenta, "The Permeability of Concrete in Aggressive Condition," in *Proceedings of the 10th International Congress on Large Dams*, pp. 103–117, 1970.

[24] A. M. Neville, *Properties of Concrete*, Longman Scientific and Technical, England, UK, 3rd edition, 1995.

[25] S. E. Hedegaard and T. C. Hansen, "Water permeability of fly ash concretes," *Materials and Structures*, vol. 25, no. 7, pp. 381–387, 1992.

[26] P. B. Bamforth, "The water permeability of concrete and its relationship with strength," *Magazine of Concrete Research*, vol. 43, no. 157, pp. 233–241, 1991.

10

Experiment Analysis of Concrete's Mechanical Property Deterioration Suffered Sulfate Attack and Drying-Wetting Cycles

Wei Tian and Nv Han

School of Civil Engineering, Chang'an University, Xi'an 710061, China

Correspondence should be addressed to Wei Tian; tianwei_816@163.com

Academic Editor: Andres Sotelo

The mechanism of concrete deterioration in sodium sulfate solution is investigated. The macroperformance was characterized via its apparent properties, mass loss, and compressive strength. Changes in ions in the solution at different sulfate attack periods were tested by inductively coupled plasma (ICP). The damage evolution law, as well as analysis of the concrete's meso- and microstructure, was revealed by scanning electron microscope (SEM) and computed tomography (CT) scanning equipment. The results show that the characteristics of concrete differed at each sulfate attack period; the drying-wetting cycles generally accelerated the deterioration process of concrete. In the early sulfate attack period, the pore structure of the concrete was filled with sulfate attack products (e.g., ettringite and gypsum), and its mass and strength increased. The pore size and porosity decreased while the CT number increased. As deterioration progressed, the swelling/expansion force of products and the salt crystallization pressure of sulfate crystals acted on the inner wall of the concrete to accumulate damage and accelerate deterioration. The mass and strength of concrete sharply decreased. The number and volume of pores increased, and the pore grew more quickly resulting in initiation and expansion of microcracks while the CT number decreased.

1. Introduction

Concrete is an important construction material that has been extensively researched and developed over many decades. The durability of concrete structures depends on environmental conditions. Sulfate attacks are a primary chemical attack of concrete, and thus analyzing their impact can help to elucidate the durability of concrete structures and predict their service life as they are subject to damage and deterioration. The main hydration products of cement and sulfate react to generate expansive products resulting in loss of strength [1–4]. The drying-wetting cycle can accelerate this process. The causes and mechanisms of sulfate attack deterioration have been researched extensively from the perspective of sulfate attack alone [5–12] as well as the coupling of drying-wetting cycles and sulfate attack [13–15].

Concrete is deteriorated under sulfate attack because the sulfate ions pass through the pores and the material components undergo a chemical reaction. This leads to the initiation and expansion of pores/cracks and loss of strength [16]. The macromechanics at work during the damage process can be determined by measuring the internal pore structure changes under different deterioration conditions. Many researchers have used the mercury intrusion porosimetry (MIP), nitrogen adsorption, scanning electron microscopy (SEM), and other methods to study this process; the distribution of pores' measurements is relatively accurate over a wide range of measurements. However, traditional microresearch methods have notable drawbacks. The specimen preparation process typically results in grinding damage because the initial state distribution of pores in the same concrete specimen cannot be repeated. X-ray CT scanning technology is an effective and nondestructive technique for observing material microstructures. Yuan et al. [15] studied sulfate attack and drying-wetting cycles to observe the concrete's mesodamage process using CT. Qian et al. [17] used nanoindentation and micro-CT to research concrete pore and mechanical properties under sulfate attack. Naik et al. [18] studied the effects of cement type and water-to-cement ratio on concrete sulfate attack by micro-CT and XRD. El-Hachem et al. [19] identified sulfate attack products and cracks at different points in the attack process via X-ray microtomography.

Despite these and other valuable contributions to the literature, however, these current studies are generally limited to two-dimensional pore structure evolution characteristics under the sulfate attack environment. Accordingly, further research is necessary concerning the three-dimensional pore structure evolutions, as well as the relationship between the mechanical properties of concrete material and its microstructure. In this paper, we investigated deterioration in concrete specimens subjected to sulfate attack per the macroscopic physical mechanical properties of the material. The mesostructure of the concrete was characterized using the CT technique, and then the three-dimensional (3D) pore structure of the concrete was reconstructed with digital image processing (DIP) technology. The variation in porosity and pore distribution characteristics was quantitatively analyzed, and then pore regional division methods were established accordingly. The relationship between the damage mechanical properties of concrete material and its microstructure was analyzed under the coupling of sulfate attack and drying-wetting cycles.

2. Experimental Procedure

2.1. Experiment Materials and Specimen Preparation. The materials used to be of the Chinese medium-heat Ordinary Portland cement 42.5R produced by Shanxi Jidong Cement Limited Company were adopted. The coarse aggregate was Shanxi Province Yang gao crushed stone with a diameter within 5–30 mm; the fine aggregate was Shanxi Province Yang gao natural river sand with a diameter of 0–5 mm. A naphthalene-based superplasticizer was applied to produce fresh concrete with improving workability. The mix proportions by weight of concrete specimens were listed in Table 1. We made cubic specimens for compressive strength and for mass loss tests. Figure 1 shows the main concrete specimens. Cubic plain concrete specimens ($100 \times 100 \times 100$ mm) were cast in steel molds. After 24 h, all the specimens were demolded and cured at 25°C and 95% relative humidity for 28 days in a standard curing room.

2.2. Experiment Methods. In order to obtain the experimental results more quickly, the specimens were kept in Na_2SO_4 mixed solution with 15% concentration (by mass) and pH 3. Since the pH values of the solutions were changing with the conditioning continued, acids were measured with a pH meter every 8 hours to ensure a pH value constant, and acidity was recorded by DZS-708 Acidometer (Figure 2).

The concrete specimens were immersed in the solution for 60 hours and then placed in a drying oven for 10 hours at 60°C followed by another 4-hour cooling process (Figure 3). The specimens were then soaked in sulfate solution for 16 hours. This process was repeated over 3 days to complete one drying-wetting cycle. Drying-wetting cycles were continued for 63 days. The dimensions of specimens for the test of compressive strength, mass loss, and CT were $100 \times 100 \times 100$ mm. It was important to note that all experiments are performed on three specimen replicates.

TABLE 1: Mix proportion of concrete specimens.

Materials	Cement	Aggregate	Sand	Water	Superplasticizer
Content (kg/m^3)	311.11	1091.38	857.51	140	2.1

FIGURE 1: Concrete specimens exposed to drying-wetting cycles with sulfate attack.

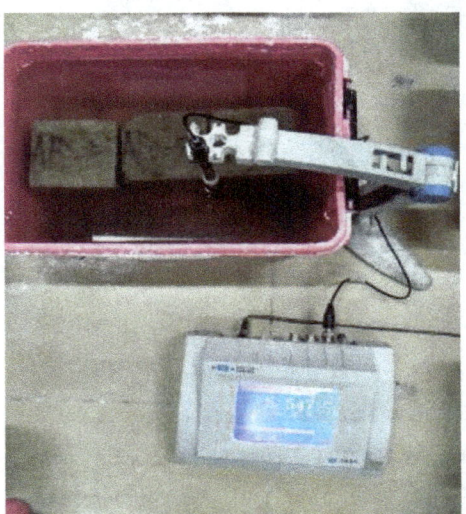

FIGURE 2: DZS-708 Acidometer.

2.2.1. Visual Appearance. The visual appearance of the concrete specimens immersed in sodium sulfate solution was periodically examined to check for spalling, cracking, expansion, and mass loss due to sulfate attack.

2.2.2. Mass Loss Test. According to the GB/T50082-2009 [20] method, mass of the specimens at each sulfate attack period was measured on an electronic scale with an accuracy of 0.01 g.

2.2.3. Uniaxial Compressive Strength Test. We also measured the average compressive strength of three specimens per batch subjected to sulfate solutions according to the GB/T50081-2002 [21] method. The compressive test was carried out on an electrohydraulic servo compressive testing system (WAW3100, Figure 4) with a capacity of 1000 kN. An

(a) (b)

FIGURE 3: Drying-wetting cycle process.

FIGURE 4: WAW3100 electrohydraulic servo testing system.

antifriction measure was taken by using a PTFE plastic board and carbon dust between loading plates. Tests were conducted in the stress-controlled mode at a loading rate of 0.5 MPa/s until failure.

2.2.4. X-Ray CT Test. The X-ray CT device used in this test was a Toshiba Aquilion One computed tomography scanner at Shaanxi Province Hospital (Figure 5). The internal structures of the concrete specimens (with the dimensions of $100 \times 100 \times 100$ mm) were tested at 0, 21, 42, and 63 days of sulfate attack. The specimens were scanned at 0.5 mm interval, and representative four scanning cross sections of the specimen were selected to analyze in this study (Figure 6). The dimensions of the gray images were 1024×1024 pixel^2, and the minimum resolution was 0.1 mm.

2.2.5. SEM and XRD Test. Scanning electron microscopy (SEM) and X-ray diffraction (XRD-Empyrean) were used to investigate and analyze the microstructure of the specimens

that underwent sulfates attack. The specimens used for SEM observations were 2 cm thick slices from the concrete prisms. The pressure in the specimen chamber was 50 Pa, and the accelerating voltage was 20 kV. After grinding into powder, the cement paste was investigated by XRD to analyze the change in reaction products after sulfate attack.

3. Results and Analysis

3.1. Visual Appearance. Figure 7 shows images of the surface damage characteristics of specimens exposed to sodium sulfate solutions for 63 days. After 21 days of sulfate attack, the concrete surface showed white sodium sulfate crystals so that the surface of the concrete precipitation was "frosted." This is because the saturated sodium sulfate solution directly crystallized, and the chemical reaction products were hydrated to produce crystal products accompanied by physical deterioration. The specimen surfaces were generally coarse at this point. After 42 days of sulfate attack, honeycomb pores were generated, and the edges and corners of specimens showed sanding phenomenon and pitting deterioration. After 54 days of sulfate attack, the pitting deterioration depth markedly increased, and several microcracks appeared on the specimen surface from the edges and corners. The cement was dissolved, and coarse aggregates were seen at the edges. After 63 days of sulfate attack, the specimens became severe loose and powdery.

The surfaces of the concrete specimens underwent sulfate attack, and drying-wetting cycle treatments were observed under a Zeiss Stemi 508 3D high-depth stereo microscope (Figure 8). This was enlarged 60-fold for analysis as shown in Figure 9. As expected, more crystals appeared on the specimen pores as the drying-wetting cycles progressed. At 42 days of sulfate attack, expansive products in the solution filled the pores of the material with a white precipitate. The cracks initiated and propagated from the pores. At 54 days of sulfate attack, the surface became rough, and the edges of the pores became blurred. After 63 days of sulfate attack, the aggregate was ejected from the surface, and cracks rapidly propagated from the pores.

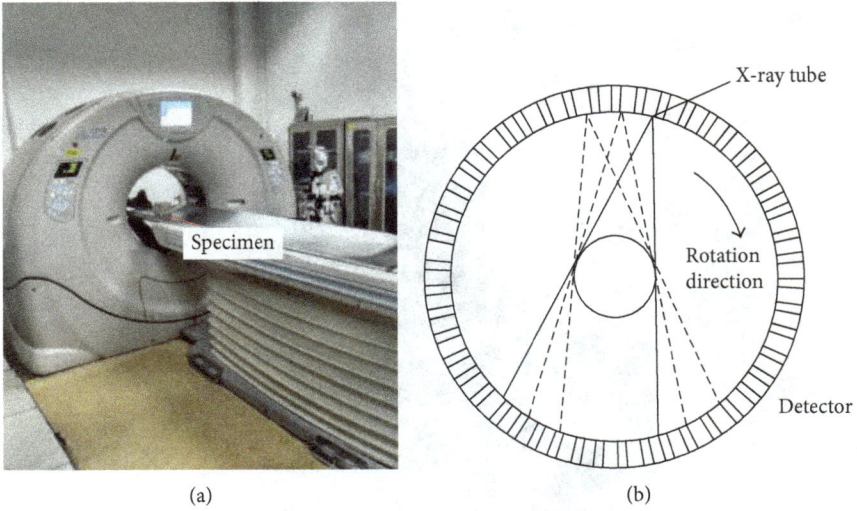

FIGURE 5: A schematic diagram of CT scanner process and the sketch of its working principle.

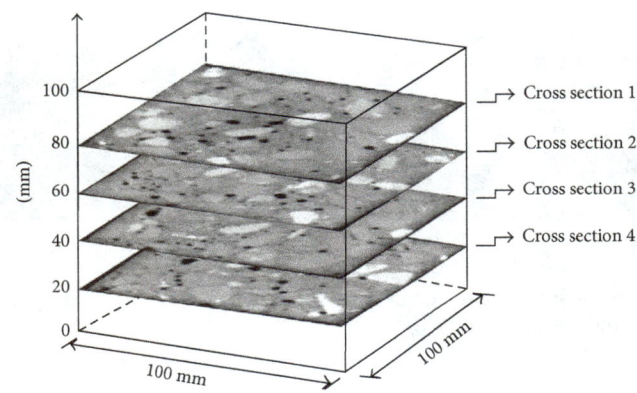

FIGURE 6: Location of scanning cross section.

FIGURE 7: Surface damage characteristics of specimens exposed to drying-wetting cycles with sulfate attack. (a) 0 days, (b) 21 days, (c) 42 days, (d) 54 days, and (e) 63 days.

3.2. Mass and Uniaxial Compressive. Figure 10 shows the variations in mass and compressive strength in concrete exposed to drying-wetting cycles with sulfate attack. It was shown that the variation in mass experienced two periods. In the early period of sulfate attack, the mass of the concrete specimens increased. This was because the solution reacted with the cement hydration products, filled the material's pores, and increased its mass. In the late period of sulfate attack, the mass continued to decrease as hydration products

such as C-S-H and calcium hydroxide gradually dissolved from the surface and specimens became severe loose and sanding.

The uniaxial compressive strength of the specimens' presented "up-down" trend as the sulfate attack period progressed. Initially, the hydration expansive products ettringite, gypsum, and sulfate crystals filled the pores to improve the specimen's density and strength. Then, the expansion force of ettringite and gypsum and the

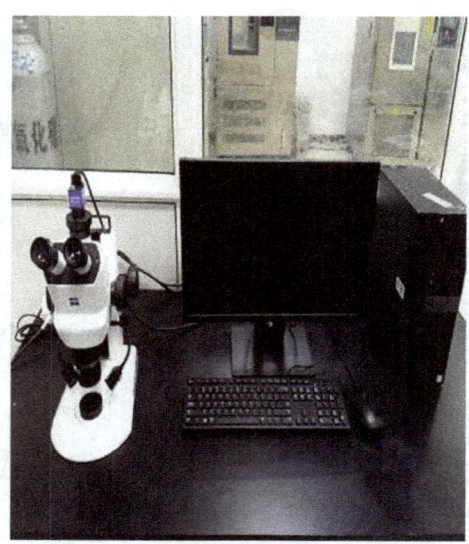

FIGURE 8: Stemi 508 stereo microscope.

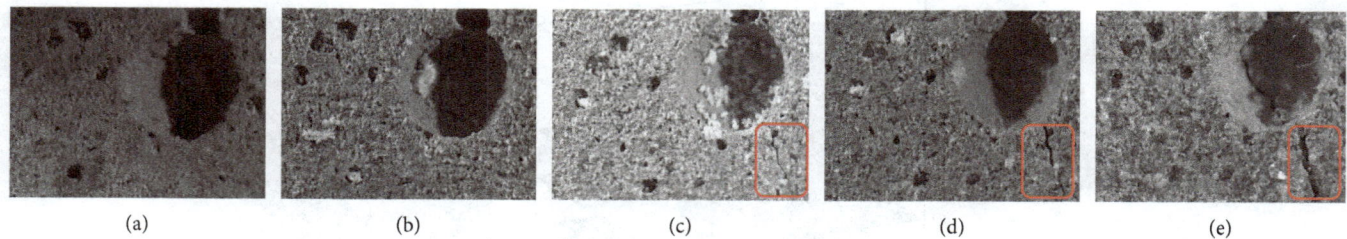

(a)   (b)   (c)   (d)   (e)

FIGURE 9: The microscope photographs of surface damage in concrete exposed to drying-wetting cycles with sulfate attack. (a) 0 days, (b) 21 days, (c) 42 days, (d) 54 days, and (e) 63 days.

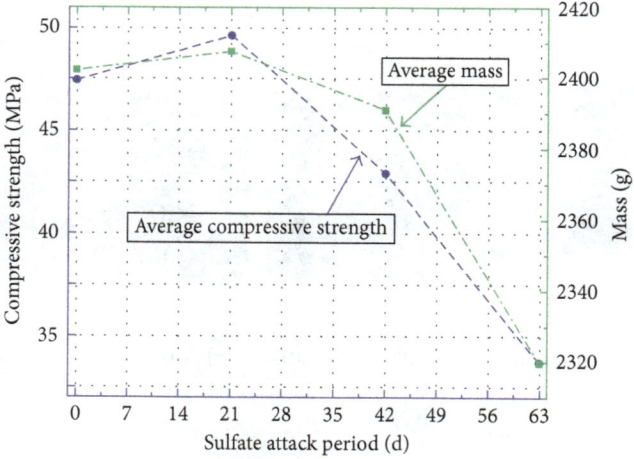

FIGURE 10: The variation of mass and compressive strength in concrete exposed to drying-wetting cycles with sulfate attack.

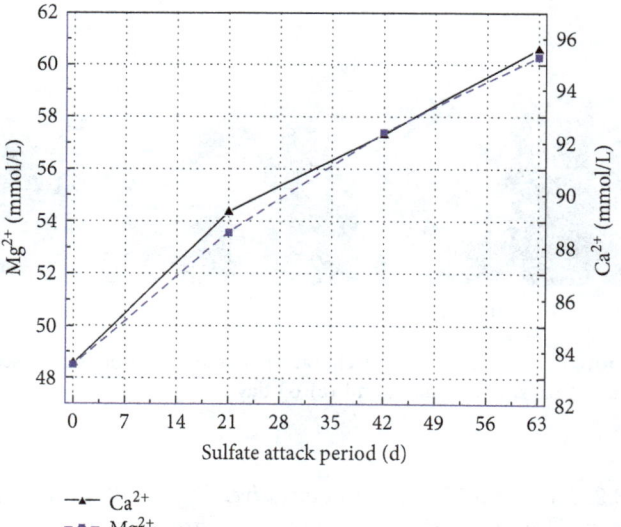

- ▲- Ca^{2+}
- ■- Mg^{2+}

FIGURE 11: Variation of Ca^{2+} and Mg^{2+} ion value in concrete exposed to drying-wetting cycles with sulfate attack.

crystallization pressure produced by sulfate crystallization then began to act on the pore walls. When the pressure of the pore wall exceeded the concrete's tensile strength, many internal microcracks appeared and decreased the uniaxial compressive strength. The deterioration of macroscopic mechanical properties was attributed to the combined action of specimens in the acidic sulfate solution and drying-wetting cycles.

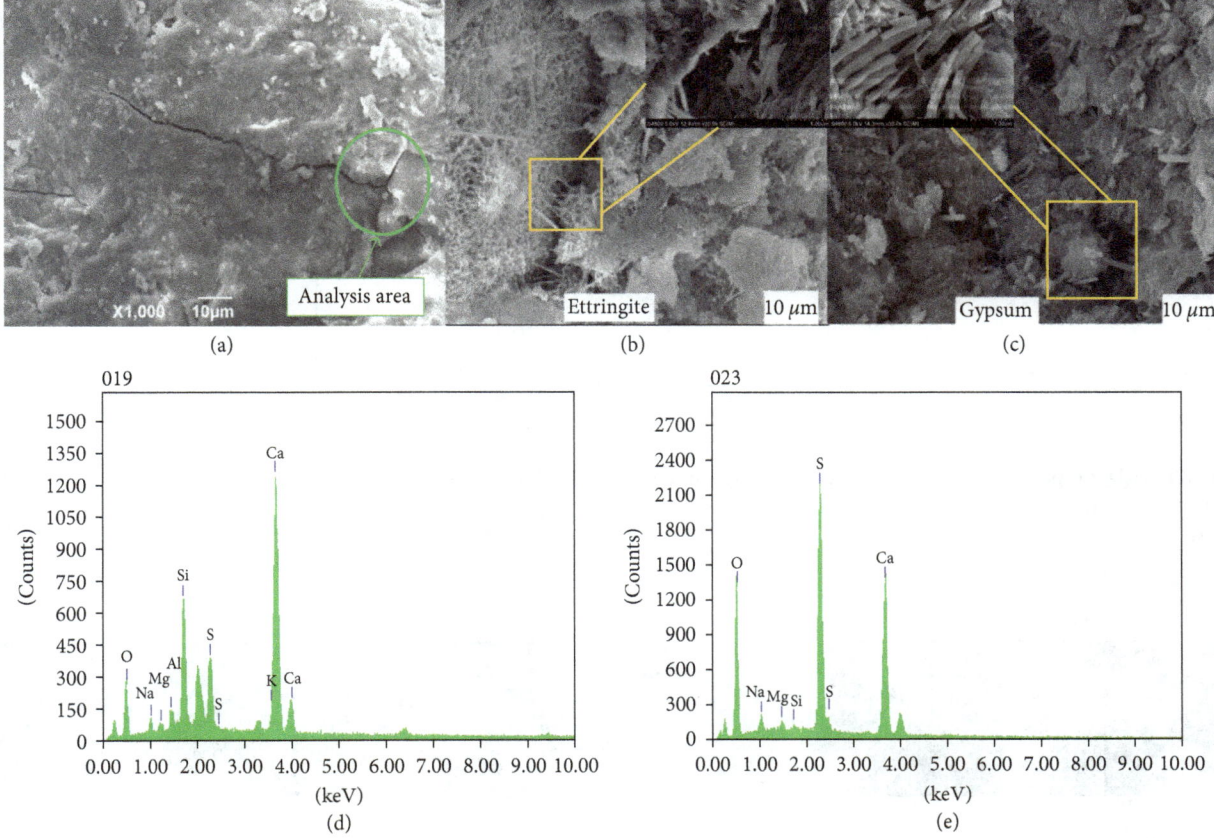

FIGURE 12: (a) SEM of thin sections from DW-1 specimen before chemical exposure, (b) SEM image from DW-1 specimen with ettringite, (c) SEM image from DW-1 specimen with gypsum in cracks and pores, (d) EDX analysis of squared area in (b) showing ettringite components, and (e) EDX analysis of squared area in (c) showing gypsum components.

3.3. Ion Concentration.

The concentrations of Ca^{2+} and Mg^{2+} in the solution dissolved due to a series of water chemistry reactions after 63 days of sulfate attack (Figure 11). These ions' concentrations were high at the initial period of sulfate attack, and then the calcium hydroxide and magnesium hydroxide on the surface of the concrete specimen rapidly dissolved in the acid solution. The dissolution rate of calcium and magnesium ions decreased slightly as the pores grew in size and number with better-connected infiltration pathways.

3.4. SEM Result Analyses.

Figures 12 and 13 shows SEM images of the concrete subjected to a sodium sulfate solution for 63 days. The needle-shaped ettringite and short columnar gypsum crystals of the sulfate attack products were found in the cracks and inner pores of the specimen. The formation of these products not only reduced the bond strength between the aggregates and mortar but also continually caused swelling, cracking, and spalling. The EDS spectra indicated that the expansive products were aluminum, sulfur, calcium, silicon, and other elements. The main elements were trace elements of gypsum and ettringite. The XRD patterns of the specimens show typical crystals and phases of hydrated cement. After sulfate attack, large amounts of expansive products such as gypsum and ettringite were produced, while the main hydration products such as C-S-H and $Ca(OH)_2$ gradually dissolved or decomposed—this destroyed the bonds between the aggregates and mortar and damaged the specimens. This reveals the presence of gypsum ($CaSO_4 \cdot 2H_2O$) and ettringite ($3CaO \cdot Al_2O_3 \cdot 3CaSO_4 \cdot 32H_2O$) that formed via the following equations:

$$Ca(OH)_2 + SO_4^{2-} + 2H_2O \rightarrow CaSO_4 \cdot 2H_2O + 2H_2O \quad (1)$$

$$3(CaSO_4 \cdot 2H_2O) + 4CaO \cdot Al_2O_3 \cdot 12H_2O + 14H_2O \\ = 3CaO \cdot Al_2O_3 \cdot 3CaSO_4 \cdot 32H_2O + Ca(OH)_2 \quad (2)$$

3.5. CT Test Result Analyses.

CT was used to characterize the changes in the internal structure of the concrete specimens at different sulfate attack periods. CT data were analyzed with the visualization software ENVI®. Through this strategy, the pore and crack distributions of different specimen cross sections as a function of sulfate attack period may be determined. Numerous 2D images were obtained from each specimen. Thus, we chose representative specimens labeled DW-1 and DW-2 images to analyze (Figure 14). It was noted that these image processing methods were not described in this study, and more research content can be found in our previous study [22].

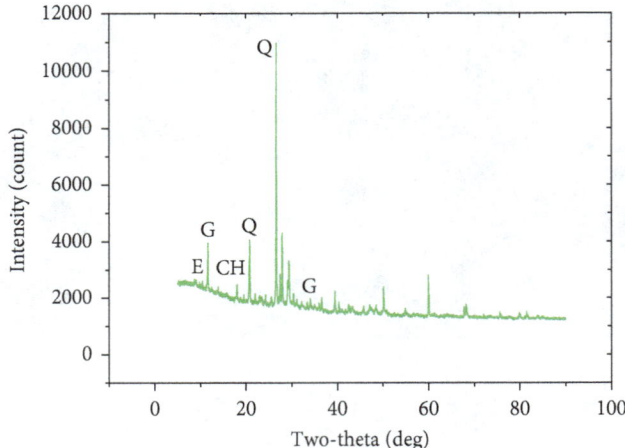

FIGURE 13: XRD of powder specimen from DW-1 specimen after 63 days sulfate attack. (E = ettringite, CH = calcium hydroxide, G = gypsum, Q = quartz).

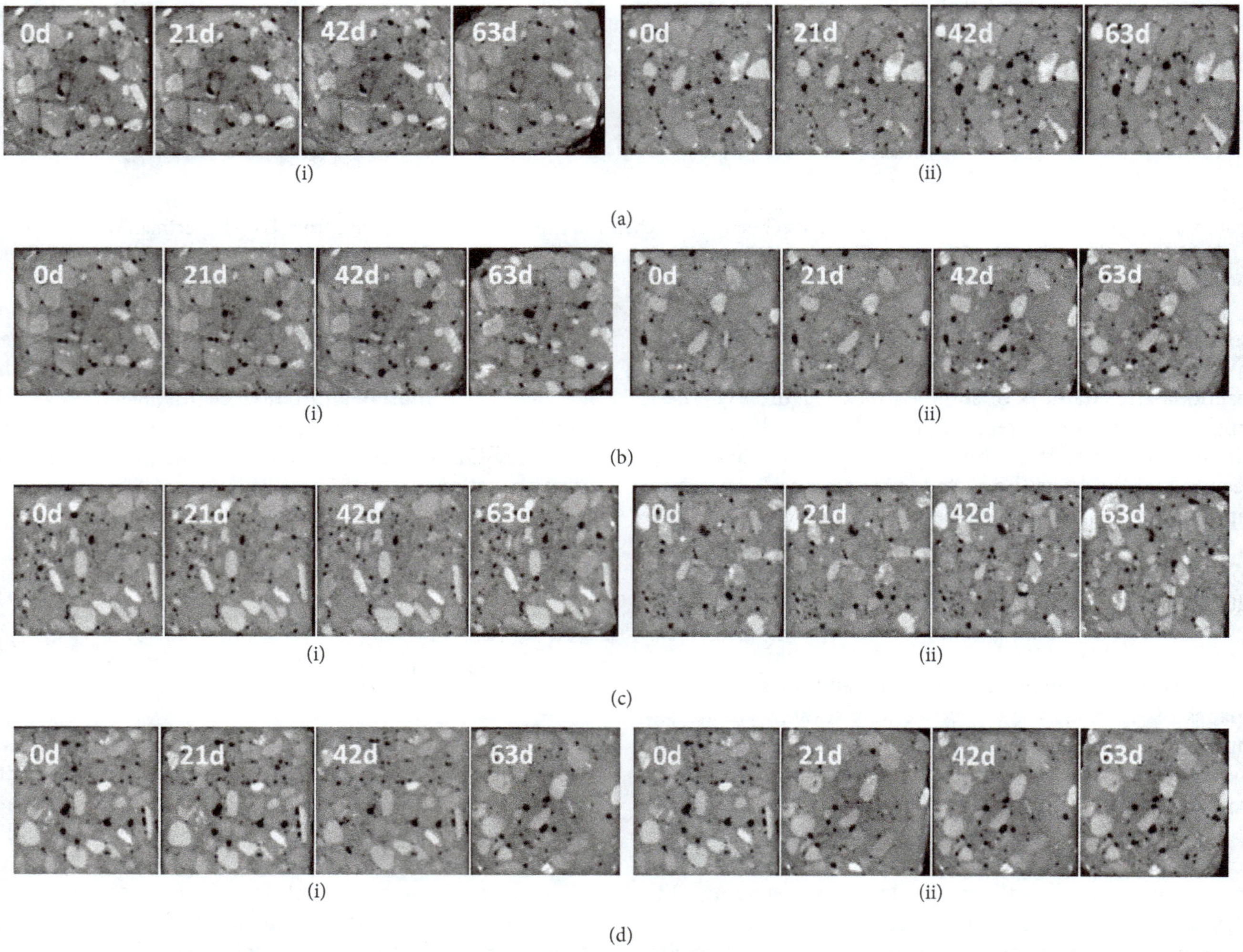

(a)

(i)   (ii)

(b)

(i)   (ii)

(c)

(i)   (ii)

(d)

(i)   (ii)

FIGURE 14: Cross-section CT images of DW-1 and DW-2 exposed to drying-wetting cycles with sulfate attack. (a) (i) Cross-section CT images of DW1-1 under different sulfate attack periods; (ii) cross-section CT images of DW2-1 under different sulfate attack periods. (b) (i) Cross-section CT images of DW1-2 under different sulfate attack periods; (ii) cross-section CT images of DW2-2 under different sulfate attack periods. (c) (i) Cross-section CT images of DW1-3 under different sulfate attack periods; (ii) cross-section CT images of DW2-3 under different sulfate attack periods. (d) (i) cross-section CT images of DW1-4 under different sulfate attack periods; (ii) cross-section CT images of DW2-4 under different sulfate attack periods.

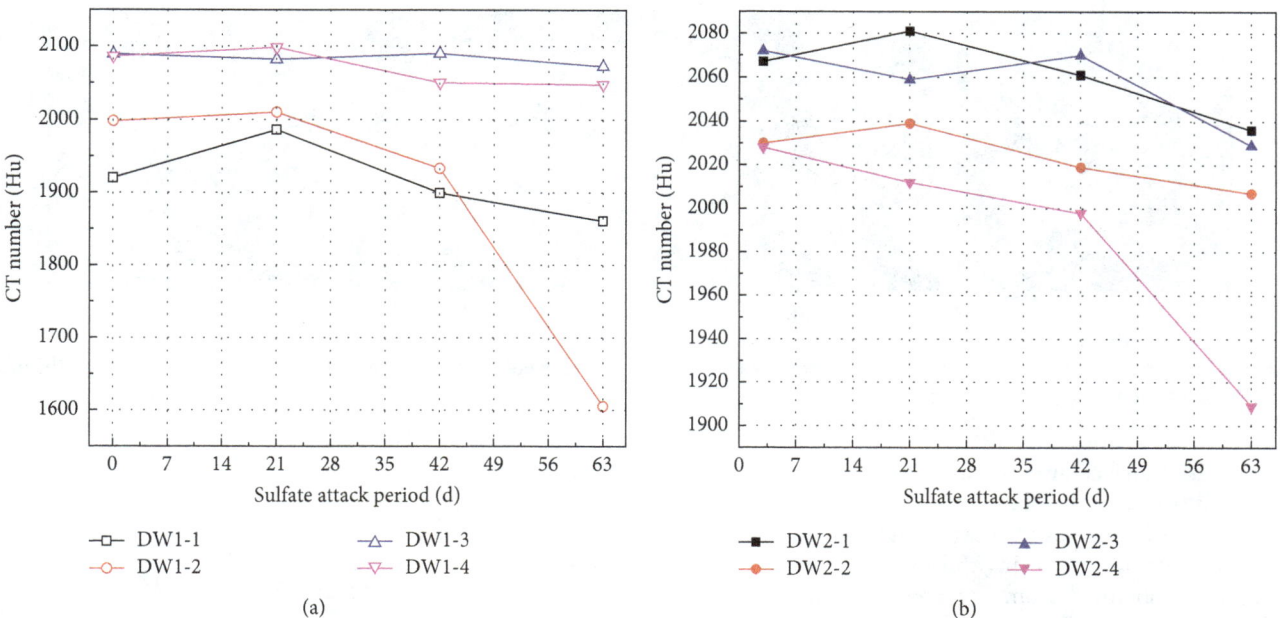

FIGURE 15: Variation of the average CT numbers of DW-1 and DW-2 exposed to drying-wetting cycles with sulfate attack.

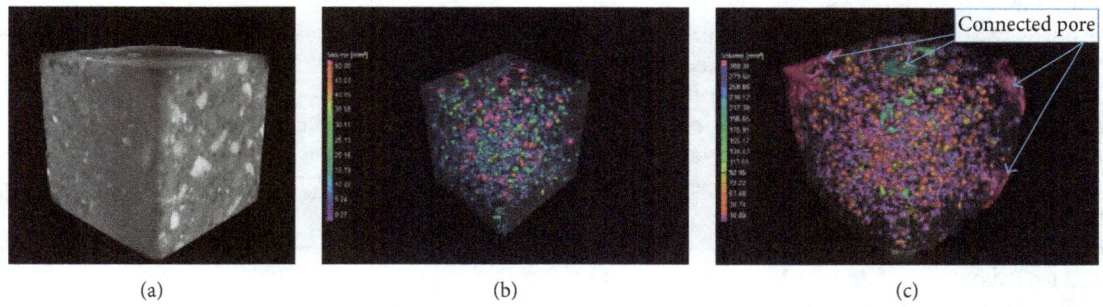

FIGURE 16: 3D reconstructed concrete and pores. (a) 3D concrete (aggregates in white and pores in black). (b) 3D pore without sulfate attack. (c) 3D pore at the 63-day sulfate attack.

TABLE 2: Statistical results of pore characteristics of DW-1 exposed to drying-wetting cycles with sulfate attack.

Specimen	D-W cycle	Volume porosity (%)	Pore number
DW-1	0	1.89	846
	21	1.67	896
	42	2.06	797
	63	2.16	917

TABLE 3: Statistical results of pore characteristics of DW-2 exposed to drying-wetting cycles with sulfate attack.

Specimen	D-W cycle	Volume porosity (%)	Pore number
DW-2	0	2.21	956
	21	2.07	855
	42	3.41	753
	63	3.67	987

3.5.1. CT Number Analysis Results.

CT number (CTN) in Hounsfield units represents the mean X-ray absorption which is related to the density of materials. The average CTN of the concrete samples' cross sections ranged from −1600 to 2100. The relationship curve between average CTN and sulfate attack period of DW-1 and DW-2 is shown in Figure 15. The average CTN increased throughout the early period (to day 21) of sulfate attack. Ettringite, gypsum, and sulfate crystals filled the specimen's micropores and increased the overall density. There were no significant changes in the CT images at 42 days of sulfate attack, but the CTNs did decrease as the swelling force of the sulfate attack products acted on the pore walls, and the pores size were slightly expanded. There was accumulating damage that accelerated the deterioration rate in the specimens. At 63 days of sulfate attack, the CTN decreased significantly. The values of the sections decreased by 19.6% and 5.86% for DW-1 and DW-2, respectively. The reason may be denser edges, and corners of the specimens were sanded by sulfate attack, which caused the CTN decreased.

3.5.2. Pore Characteristic Analysis Results.

The 3D pore structures were reconstructed by *VGStudio MAX 2.0*

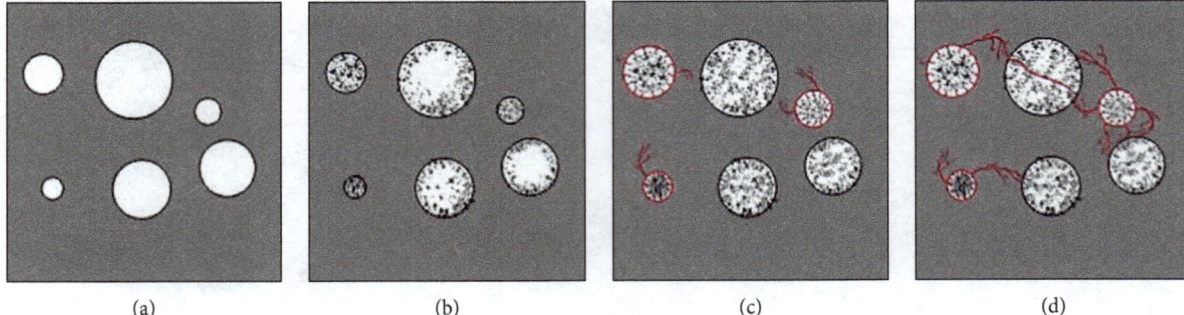

    (a)   (b)   (c)   (d)

FIGURE 17: The pore evolution rules of concrete specimens exposed to drying-wetting cycles with sulfate attack. (a) Initial stage, (b) filling stage, (c) expanded stage, and (d) connected stage.

software. The surface fit function was used to set the proper thresholds (the pore threshold was set as −831 to −1292 HU) and segment the images into two phases at which point we could calculate porosity using the volume analysis tool. The information obtained from the reconstruction model includes 3D pore distributions with color based on size or volume as shown in Figure 16.

3.5.3. Porosity.

Tables 2 and 3 show the effects of sulfate attack on the porosity of concrete. In the procedure of sulfate attack, the porosity of the concrete first decreased and then increased. This is consistent with the change in the average CT number. After 21 days, specimen porosity decreased by 11.6% and 6.3% for DW-1 and DW-2, respectively. For the reasons analyzed above, the porosity of the specimen decreased, which was attributed to the expansive products of the ettringite and gypsum as well as sulfate crystallization filling effect. At 42 days of sulfate attack, there was an increase in porosity because the swelling forces of the sulfate attack products and sulfate crystallization pressure acted on the pore walls. The resulting force of the two bodies exceeded the filling effect and drove the specimen porosity up to 2.06% and 3.41% for DW-1 and DW-2, respectively. After 63 days of sulfate attack, the pore connectivity was accelerated, and the volume of the pores increased eventually leading to microcracks being initiated and expanded. The specimen's internal porosity increased by 14.2% and 66%, and the specimen's overall structure became very loose. Tables 2 and 3 also indicate that the sulfate attack had a substantial impact on the number of internal micropores in the concrete. After 63 days, the total number of micropores increased by 15% and 31% for DW-1 and DW-2, respectively. The law of pore evolution is illustrated in Figure 17.

3.5.4. Pore Distribution Characteristics.

The pore-volume distributions of DW-1 and DW-2 were counted and are shown in Tables 4 and 5. Pores with a volume of 0.5–1 mm^3 dominated the specimens. Pores with volumes in the range of 0.5–1 mm^3 increased and then decreased throughout the experiment. The reason is due to ettringite, gypsum, and salt crystals filled the large pores and caused increase in small pores. With the continuous effect of swelling force and sulfate crystallization pressure, the small pores (0.5–1 mm^3)

TABLE 4: Pore distribution of DW-1 exposed to drying-wetting cycles with sulfate attack.

Pore volume (mm^3)	Sulfate attack period (day)			
	0 days	21 days	42 days	63 days
0–0.5	123	366	112	100
0.5–1.0	272	319	247	371
1.0–2.5	177	157	110	117
2.5–5.5	184	158	175	176
5.5–10	104	94	39	27

TABLE 5: Pore distribution of DW-2 exposed to drying-wetting cycles with sulfate attack.

Pore volume (mm^3)	Sulfate attack period (day)			
	0 days	21 days	42 days	63 days
0–0.5	266	103	64	81
0.5–1.0	239	233	137	237
1.0–2.5	137	173	176	183
2.5–5.5	162	166	173	186
5.5–10	140	155	171	140

began to connect and quickly aggregate into larger pores (5.5–10 mm^3) at 42 days of sulfate attack. After 63 days of sulfate attack, the pore volume with 0–5.5 mm^3 increased and pore volume with 5.5–10 mm^3 decreased. The reason for this may be as follows: (1) the microcracks expanded and cut through the larger pores significantly increasing the quantity of the small pores and (2) the small pores increased as expansion pressure caused by sulfate crystals created new pores.

3.5.5. Change in Porosity with Uniaxial Compressive Strength.

The relationship between porosity and uniaxial compressive strength of DW-1 and DW-2 specimens is plotted in Figure 18. As the sulfate attack progressed, the volume porosity decreased, and there was a negative correlation between the uniaxial compressive strength and porosity. The porosity of concrete has an important effect on concrete's mechanical properties and macrodamage characteristics. When the porosity was around 2%, the uniaxial

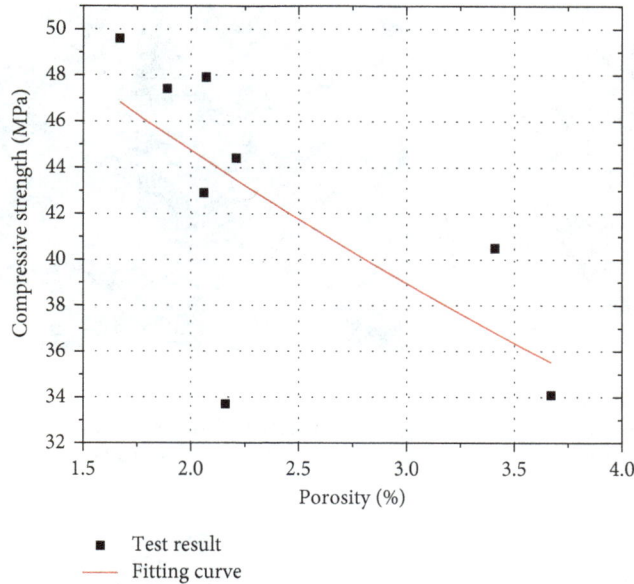

FIGURE 18: Relationship between porosity and uniaxial compressive strength.

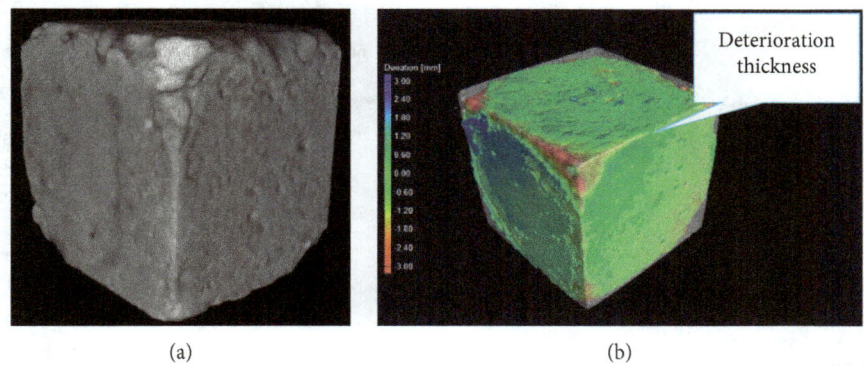

(a)   (b)

FIGURE 19: The deterioration thickness of concrete specimen (63 days). (a) Deterioration surface and (b) deterioration thickness.

compressive strength changed rapidly indicating that the expansion of the pore degraded the mechanical properties. The uniaxial compressive strength reached the minimum when the porosity was 3.76%. Equation (3) was obtained by fitting the test data:

$$f_c = \left[e^{(4.07-0.138p)} \right], \qquad (3)$$

where p is the porosity and f_c is the uniaxial compressive strength.

3.6. Pore Regional Division Characteristics.

The sulfate mainly acted on the surface of the concrete, which due to specimens' inner pores were mostly closed and the infiltration pathways were not connected. As Figure 19(a) displays, the significant sanding and pitting occurred on the surface of the specimen. It indicated the specimen's thickness (green color) after 63 days of sulfate attack, the thickness and depth of deterioration region near the surface of the specimen increased, and small cracks and connected pores formed during the sulfate attack period (Figure 19(b)). The solution more readily infiltrated the specimen over time to the point where it affected the internal pore regions of the specimen. So, we propose the pore regional division modeled in Figure 20(a): (1) Region 1 is one-fourth of the global region—a 25 mm × 25 mm cube central region; (2) Region 2 is one-half of the global region—a 50 mm × 50 mm cube area; and (3) Region 3 is the global region. Based on the pore regional division method, the 3D reconstruction of pores' structure in different divided regions is displayed in Figure 20(b).

3.6.1. Porosity of Regional Division.

Figure 21 and Table 6 display that the porosity in different regions of the specimens exposed to drying-wetting cycles during sulfate attack. Porosity first increased and then decreased in Region 3, but these changes were little complicated in Regions 1 and 2. In the early period of sulfate attack, sulfate ions entered the pores in the material and gradually infiltrated it. Region 3 experienced a significant decrease in porosity during the first 21 days of sulfate attack. This was mainly due to the formation of gypsum, ettringite, and other precipitation

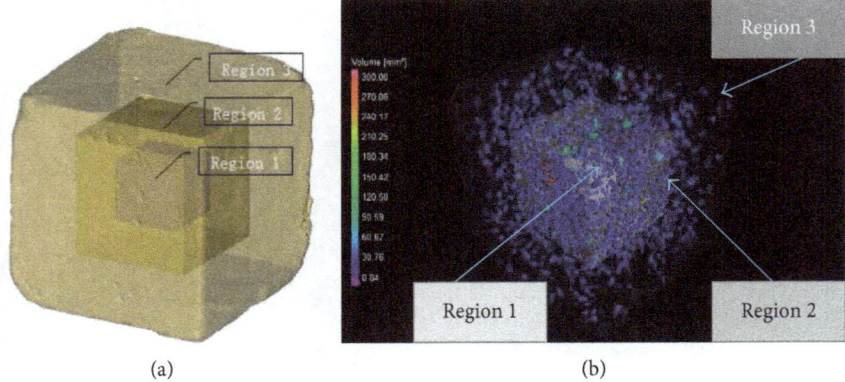

(a)   (b)

FIGURE 20: 3D view of pore's structure in different divided regions with color-coded pores according to size. (a) Pore divided region schemes and (b) 3D reconstruction of pore's structure in different divided regions (42 days of sulfate attack).

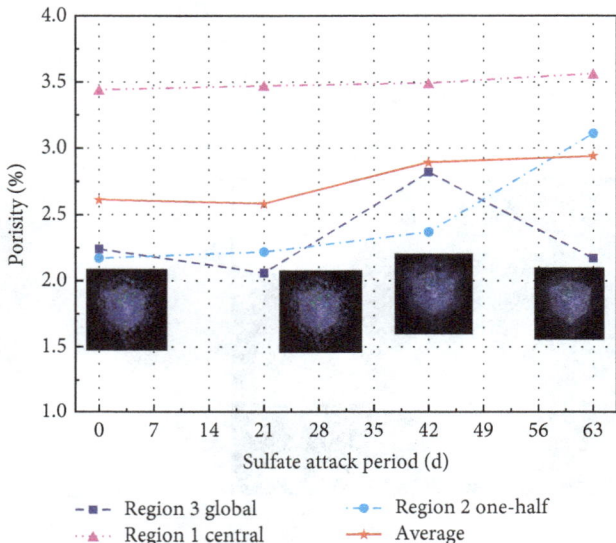

FIGURE 21: Porosity and pore volume in different divided regions in specimens exposed to drying-wetting cycles with sulfate attack.

TABLE 6: Porosity and pore volume in different divided regions in specimens exposed to drying-wetting cycles with sulfate attack.

	Total porosity (%)			
ROI	0 days	21 days	42 days	63 days
Region 3 global	2.24	2.06	2.82	2.17
Region 2 one-half	2.17	2.22	2.37	3.10
Region 1 central	3.43	3.46	3.48	3.55
Average	2.61	2.58	2.89	2.94

products as well as the salt crystal filling action. Regions 1 and 2 indicated slight increases in porosity, and the cause for this may be (1) pore water was converted to gas overflow through the drying-wetting cycles, which increased the size of the pores and thus their connectivity. (2) Regions 1 and 2 were not significantly affected by the sulfate solution as pore infiltration paths were not formed. After 42 days of sulfate attack, there was a significant increase in porosity in Region 3 due to the combined action of sulfate attack products'

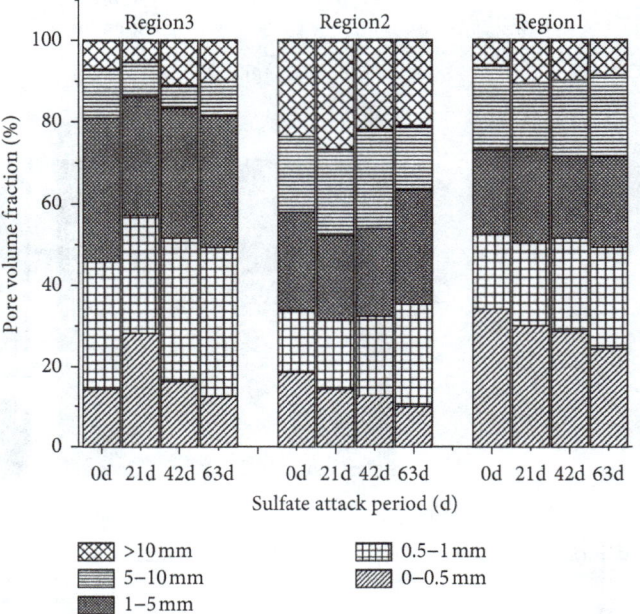

FIGURE 22: Pore volume of different regions with sulfate attack period in different divided regions.

swelling force and sulfate crystallization pressure. At this stage, cracks and pores in Region 3 allowed for solution infiltration leaving Region 2 susceptible to the sulfate as porosity continued to increase. After 63 days of sulfate attack, global porosity of Region 3 abruptly decreased due to local pitting and sanding. The porosity of Region 2 also increased at this point due to the constant expansion of the pores that were affected by solution infiltration. Region 1 central portions of the specimen were only affected by the drying-wetting cycle—their porosity continued to slightly increase throughout the experiment.

3.6.2. Pore Volume Distribution Characteristics of Regional Division. The pore-volume distributions of concrete are illustrated in Figure 22. The five characteristic ranges of pore volumes with corresponding distribution were observed: $0-0.5 \, mm^3$, $0.5-1 \, mm^3$, $1-5 \, mm^3$, $5-10 \, mm^3$, and $>10 \, mm^3$.

As shown in Figure 22, Region 3 of specimens was different in the pore distribution from Region 1 and Region 2.

In the Region 3 of specimens, the pores with volume 0.5–1 mm^3 were 36.7% of total pore volume, which was higher than the other pore-volume proportions. These ranges of pore volume first increased and then decreased during sulfate attack period. The reason is mainly due to the large pore space was filled, and thereby, the great number of small pores also emerged at the sulfate attack period. As the sulfate attack progressed, the small pores continuously emerged and developed, and the larger pores were formed. In Region 2 of specimens, the pores with volume more than 10 mm were mostly inside in this region. These pore volumes showed an increased and then decreased trend. As per the above analysis, when pore infiltration path was formed, the sulfate solution acted on the pore wall and made the pore volume increased. After 63 days of sulfate attack, the surface of specimens suffered macrodamage causing the large pore volume decrease. In Region 1 of specimens, the pores with volume 0-0.5 mm take up 34.2% higher than the other pore-volume proportions. These ranges of pore volume continued to decrease throughout the experiment. It indicates that this region is little significantly affected by the sulfate attack, and drying-wetting cycle was the main factor of the change in pore volume.

4. Conclusion

The behavior of concrete under sulfate attack and repeated drying-wetting cycles was investigated in this study. Our conclusions can be summarized as follows:

(1) Macroperformance was investigated via its apparent properties, mass loss, and compressive strength. Mass loss decreased initially due to the formation of expansive products by cement hydration. Mass loss increased in the middle stages of sulfate attack as hydration products such as C-S-H and calcium hydroxide gradually dissolved from the surface destroying the bonds between the aggregate and the mortar. The uniaxial compressive strength of the concrete specimen increased at the beginning of the drying-wetting cycles but then decreased over time.

(2) 3D pore structure of concrete specimens was reconstructed under the action of sulfate attack. The porosity and pore distributions changed over time in all specimens. After 21 days, the porosity decreased by 11.6% and 6.3% for DW-1 and DW-2 specimens, respectively. After 42 days, the porosity increased to 23.3% and 64.8% due to swelling force of the sulfate attack products and the sulfate crystallization pressure acting on the pore walls, respectively. After 63 days, the specimen's internal porosity increased by 4.8% and 7.6% for DW-1 and DW-2, respectively, and the specimen's structure became very loose. Pores with 0.5–1 mm^3 in volume dominated the specimens at this point. After 63 days, the pore volume decreased, and the quantity of micropores increased, indicating that the pressure caused by

crystal expansion created new pores. Microcracks also expanded and cut through larger pores.

(3) The pore partition analysis showed significant differences in porosity and pore-volume distribution across different regions. Porosity decreased and then increased in the global region, but these changes were more complicated inside the one-half region and the central region. Region 3 of specimens is different in the pore distribution from Region 1 and Region 2. In Region 3 of specimens, the pores with volume 0.5–1 mm^3 were higher than the other pores with volume proportions. In Region 2 of specimens, the pores with volume more than 10 mm were mostly inside in this region. These pore volumes showed an increased and then decreased trend. In Region 1 of specimens, the pores with volume more than 0–0.5 mm take up 34.2% higher than the other pore-volume proportions. These pore volumes continued to decrease throughout the experiment.

Conflicts of Interest

The authors declare that they have no conflicts of interest.

Acknowledgments

The authors would like to thank the National Natural Science Foundation of China (Grant nos. 51379015 and 51579013) and the Special Fund for Basic Scientific Research of Central Colleges, Chang'an University (Grant no. 310828172001) for the financial support for this project.

References

[1] P. K. Mehta, P. Schiessl, and M. R. aupach, "Performance and durability of concrete systems," in *Proceedings of 9th International Congress of the Chemistry of Cement*, pp. 571–659, New Delhi, India, 1992.

[2] C. D. Laurence, "Sulphate attack on concrete," *Magazine of Concrete Research*, vol. 42, no. 153, pp. 249–264, 1990.

[3] B. Lothenbach, B. Bary, P. Le Bescop, T. Schmidt, and N. Leterrier, "Sulfate ingress in Portland cement," *Cement and Concrete Research*, vol. 40, no. 8, pp. 1211–1225, 2010.

[4] J. Prasad, D. K. Jain, and A. K. Ahuja, "Factors influencing the sulphate resistance of cement concrete and mortar," *Asian Journal of Civil Engineering*, vol. 7, no. 3, pp. 259–268, 2006.

[5] N. Thaulow and S. Sahu, "Mechanism of concrete deterioration due to salt crystallization," *Materials Characterization*, vol. 53, no. 2–4, pp. 123–127, 2004.

[6] A. R. Suleiman, A. M. Soliman, and M. L. Nehdi, "Effect of surface treatment on durability of concrete subjected to physical sulfate attack," *Construction and Building Materials*, vol. 73, pp. 674–681, 2014.

[7] H. Okochi, H. Kameda, and S. I. Hasegawa, "Deterioration of concrete structures by acid deposition—an assessment of the role of rainwater on deterioration by laboratory and field exposure experiments using mortar specimens," *Atmospheric Environment*, vol. 34, no. 18, pp. 2937–2945, 2000.

[8] R. Vedalakshmi, V. Saraswathy, and A. K. Yong, "Performance evaluation of blended cement concretes under MgSO$_4$

attack," *Magazine of Concrete Research*, vol. 63, no. 9, pp. 669–681, 2011.

[9] T. H. Wee, A. K. Suyavanshi, S. F. Wong, and A. K. M. A. Rahman, "Sulfate resistance of concrete containing mineral admixtures," *Materials Journal*, vol. 97, no. 5, pp. 536–549, 2000.

[10] S. U. Al-Dulaijan, M. Maslehuddin, M. M. Al-Zahrani, A. M. Sharif, M. Shameem, and M. Ibrahim, "Sulfate resistance of plain and blended cements exposed to varying concentrations of sodium sulfate," *Cement and Concrete Composites*, vol. 25, no. 4-5, pp. 429–437, 2003.

[11] Z. Q. Jin, W. Sun, Y. S. Zhang, J. Y. Jiang, and J. Z. Lai, "Interaction between sulfate and chloride solution attack of concretes with and without fly ash," *Cement and Concrete Research*, vol. 37, no. 8, pp. 1223–1232, 2007.

[12] D. M. Yu, B. M. Guan, R. He, R. Xiong, and Z. Liu, "Sulfate attack of Portland cement concrete under dynamic flexural loading: a coupling function," *Construction and Building Materials*, vol. 115, pp. 478–485, 2016.

[13] L. Jiang and D. T Niu, "Study of deterioration of concrete subjected to different types of sulfate solutions under drying-wetting cycles," *Construction and Building Materials*, vol. 117, pp. 88–98, 2016.

[14] D. T. Niu, Y. D. Wang, R. Ma, J. B. Wang, and S. H. Xu, "Experiment study on the failure mechanism of dry-mix shotcrete under the combined actions of sulfate attack and drying-wetting cycles," *Construction and Building Materials*, vol. 81, pp. 74–80, 2015.

[15] J. Yuan, Y. Liu, Z. C. Tan, and B. K. Zhang, "Investigating the failure process of concrete under the coupled actions between sulfate attack and drying–wetting cycles by using X-ray CT," *Construction and Building Materials*, vol. 108, pp. 129–138, 2016.

[16] M. Santhanam, D. M. Cohen, and J. Olek, "Mechanism of sulfate attack: a fresh look part 2: proposed mechanisms," *Cement and Concrete Research*, vol. 33, no. 3, pp. 341–346, 2003.

[17] C. X. Qian, Y. F. Nie, and T. J. Cao, "Sulphate attack-induced damage and micro-mechanical properties of concrete characterized by nano-indentation coupled with X-ray computed tomography," *Structural Concrete*, vol. 17, no. 1, pp. 96–104, 2016.

[18] N. N. Naik, A. C. Jupe, S. R. Stock, A. P. Wilkinson, P. L. Lee, and K. E. Kurtis, "Sulfate attack monitored by micro CT and EDXRD: influence of cement type, water-to-cement ratio, and aggregate," *Cement and Concrete Research*, vol. 36, no. 1, pp. 144–159, 2006.

[19] R. El-Hachem, E. Rozière, F. Grondin, and A. Loukili, "Multi-criteria analysis of the mechanism of degradation of Portland cement based mortars exposed to external sulphate attack," *Cement and Concrete Research*, vol. 42, no. 10, pp. 1327–1335, 2012.

[20] National Standards of People's Republic of China (NSPRC), *Standard for Test Method of Long-Term Performance and Durability of Ordinary Concrete, GB/T 50082–2009*, China Architecture and Building Press (CABP), Beijing, China, 2009, in Chinese.

[21] National Standards of People's Republic of China (NSPRC), *Standard for Test Method of Mechanical Properties on Ordinary Concrete, GB/T 50081–2002*, China Architecture and Building Press (CABP), Beijing, China, 2002, in Chinese.

[22] W. Tian and N. Han, "Evaluation of damage in concrete suffered freeze-thaw cycles by CT technique," *Journal of Advanced Concrete Technology*, vol. 42, no. 11, pp. 679–690, 2016.

11

Mechanical Properties of Fiber-Reinforced Concrete Using Composite Binders

Roman Fediuk, Aleksey Smoliakov, and Aleksandr Muraviov

Far Eastern Federal University, Vladivostok 690950, Russia

Correspondence should be addressed to Roman Fediuk; roman44@yandex.ru

Academic Editor: Rishi Gupta

This paper investigates the creation of high-density impermeable concrete. The effect of the "cement, fly ash, and limestone" composite binders obtained by joint grinding with superplasticizer in the varioplanetary mill on the process of structure formation was studied. Compaction of structure on micro- and nanoscale levels was characterized by different techniques: X-ray diffraction, DTA-TGA, and electron microscopy. Results showed that the grinding of active mineral supplements allows crystallization centers to be created by ash particles as a result of the binding of $Ca(OH)_2$ during hardening alite, which intensifies the clinker minerals hydration process; the presence of fine grains limestone also leads to the hydrocarboaluminates calcium formation. The relation between cement stone neoplasms composition as well as fibrous concrete porosity and permeability of composite at nanoscale level for use of composite binders with polydispersed mineral supplements was revealed. The results are of potential importance in developing the wide range of fine-grained fiber-reinforced concrete with a compressive strength more than 100 MPa, with low permeability under actual operating conditions.

1. Introduction

Concrete on cementitious binder and natural aggregates is widely used as structural material in construction industry. Worldwide, the large-capacity ash waste and crushing of the rocks are generated as a result of activity of the fuel and energy sector and mining industry enterprises. It seems necessary to optimize the processes of concrete mixtures structure formation by using industrial waste. Traditional types of concrete have insufficient properties of gas permeability and vapor permeability. At the same time, it is necessary to improve the strength and deformability quality of composite to achieve fine-grained fiber-reinforced concrete.

Leading scientific schools in the field of building materials science have developed a number of concrete types with enhanced operational properties.

Textile-reinforced concrete is a type of reinforced concrete in which the usual steel reinforcing bars are replaced by textile materials. Materials with high tensile strengths with negligible elongation properties are reinforced with woven or nonwoven fabrics. The fibers used for making the fabric are of high tenacity like Jute, glass fiber, Kevlar, polypropylene, polyamides (nylon), and so forth.

Mechanical properties of high-strength concrete incorporate copper slag as a fine aggregate and concluded that less than 40% copper slag as sand substitution can achieve high-strength concrete comparable to or better than the control mix, beyond which, however, its behaviors decreased significantly [1–3].

Glass fiber-reinforced concrete consists of high-strength, alkali-resistant glass fiber embedded in a concrete matrix. In this form, both fibers and matrix retain their physical and chemical identities, while offering a synergistic combination of properties that cannot be achieved with either of the components acting alone.

High-performance fiber-reinforced cementitious composites (HPFRCCs) [4, 5] are a group of fiber-reinforced cement-based composites which possess the unique ability to flex and self-strengthen before fracturing. Strain hardening, the most coveted capability of HPFRCCs, occurs when a material is loaded past its elastic limit and begins to deform plastically. This stretching or "straining" action actually strengthens the material. The basis for the engineered design of different HPFRCCs varies considerably despite their similar compositions. For instance, the design of one type of HPFRCC called Engineered Cementitious Composite (ECC) stems from the principles of micromechanics. ECC, also called bendable concrete, is an easily molded mortar-based composite reinforced with specially selected short random fibers, usually polymer fibers [6, 7]. Unlike regular concrete, ECC has a strain capacity in the range of 3–7%, compared to 0.01% for ordinary Portland cement (OPC). ECC, therefore, acts more like a ductile metal than a brittle glass (as does OPC concrete), leading to a wide variety of applications [8, 9].

Ultrahigh-performance concrete (UHPC) is a new type of concrete that is being developed by agencies concerned with infrastructure protection [10–14]. UHPC is characterized by being a steel fiber-reinforced cement composite material with compressive strengths in excess of 150 MPa, up to and possibly exceeding 250 MPa. UHPC is also characterized by its constituent material make-up: typically fine-grained sand, silica fume, small steel fibers, and special blends of high-strength Portland cement. Note that there is no large aggregate. The current types in production differ from normal concrete in compression by their strain hardening, followed by sudden brittle failure. Ongoing research into UHPC failure via tensile and shear failure is being conducted by multiple government agencies and universities around the world.

The high-strength self-consolidating (self-compacting) concrete technology is made possible by the use of poly-carboxylates plasticizer instead of older naphthalene-based polymers and viscosity modifiers to address aggregate segregation [15, 16].

Combination of microsilica and nanosilica (colloidal silica) is considered to design high-strength self-consolidating concrete [1, 20, 21]. The results also revealed that 7% substitution of microsilica showed the same effect as 2% nanosilica replacement [1, 22, 23].

The fiber-matrix interfacial transition zone (ITZ) at nanoscale plays an important role in determining the mechanical performance of hybrid steel-polypropylene fiber-reinforced concrete at upper scales. This topic [24] presents the elastic behavior of the ITZ between steel/polypropylene fiber and pure cement paste through nanoindentation for different water/cement ratios.

Thus, the fine-grained structure with high homogeneity is characterized by increase of the integrated strength between aggregate and cement stone and decrease of specific stress in the contact area. Adhesion of sand component increases significantly with increase of contact area; these conditions were realized when creating the fine-grained concrete based on composite binders by using crushed granite from Wrangel deposit (Russian Far East). The aim of the study was to develop the concrete matrix dense structure with high gas, water, and vapor impermeability. Composite binders, obtained by cogrinding of Portland cement, fly ash, crushed limestone, and superplasticizer, have been proposed to achieve this aim. One of the ways to improve the properties of concrete and to reduce permeability parameters is the use of highly active additives of various compositions and genesis

at micro- and nanosized levels, which contribute to the optimization of structure formation processes by initiating the formation of hydrated compounds. The efficiency of use of the active mineral additives of nanostructured silica-modifier composition has been proven in topics [1, 20, 21, 23]. The possibility of the permeability reduction of the concrete by mechanically grinding the composite binders components was also studied previously [17]. However, the protective properties (impermeability parameters) and efficiency of high-density impermeable concrete (HDIC) produced on the basis of composite binder were not considered previously.

An assumption about the possibility of HDIC is obtained by varying the amount and type of additives, fineness of the composite binder components, and hardening conditions [18, 25]. The purpose of this work is to improve impermeability and strength characteristics of the fiber-reinforced concrete through use of composite binders obtained by cogrinding of Portland cement, superplasticizer, fly ash of a thermal power station, and crushing screenings limestone.

2. Materials and Methods

To achieve this goal, the following tasks were completed in this work:

(i) Study of mineral composition, particle size distribution, and physical and mechanical characteristics of the binders components and fillers for concrete

(ii) Research of the effect of mineral and organic additives on the properties of composite binders

(iii) Study of the properties of fiber-reinforced concrete, depending on the characteristics of the composite binder taking into account peculiarities of structure formation to improve the impermeability and strength characteristics, research on characteristics of water absorption and gas, water, and vapor permeability of developed concrete, and experimental industrial testing of the proposed compositions

The science work included grinding of composite binder components in a varioplanetary mill. Ordinary Portland cement (OPC), fly ash (FA), limestone crushing waste (LCW), and superplasticizer were used as ground components.

The chemical and mineralogical compositions of the raw materials obtained by means of X-ray florescence (on D8 ADVANCE powder diffractometer from Bruker AXS) are presented in Tables 1 and 2.

In the varioplanetary mill, the rotational speeds of the grinding jars and the support disc can be set completely independently. The movement and the trajectory of the grinding balls can be influenced by varying the gear ratio, so that the balls strike horizontally on the inner wall of the grinding jar (by high impact energy), approach tangentially (by high friction), or simply roll over the inner wall of the grinding jar (by centrifugal mills). All intermediate stages and combinations between pressure friction and impact can be freely set (Figure 1). Accordingly, grinding by the varioplanetary mills is more energy-efficient than that by the ball mills and the vibratory mills. In addition, due to the

TABLE 1: The chemical compositions of the fly ashes, Portland cement, and limestone crushing wastes.

The predominant type of coal	Fly ash				Portland cement	LCW
	Primorye TPP	Vladivostok TPP	Artem TPP	Partizansk TPP		
	Luchegorsk and Bikinsky	Primorsky brown (Pavlovsky section)	Coal	Neryungrinsky coal		
Content of elements in terms of oxides, %						
SiO_2	55.3	63	48.1	47.4	20.2–20.9	7.49
TiO_2	0.5	0.5	0	0.9		0.24
Al_2O_3	12.6	21.4	29.3	22.3	6.0–6.7	3.33
Fe_2O_3	10.7	7.5	6.5	19.6	3.5–4.0	0.24
CaO	12.5	3.4	9.7	4.8	66.2–67	44.21
MgO	3.5	2.1	1.8	2.8	1.4–2.0	2.57
K_2O	1	1.3	1.2	0.1		
Na_2O	0.4	0.3	0.2	0.4		
SO_3	3.4	0.6	2.3	1.62		
LOI	2.3	1.4	0.6	<5	0.18	38.71

TABLE 2: The mineral compositions of Spassky Portland cement.

Mineral composition	C_3S	C_2S	C_3A	C_4AF
Content [%]	58–67	8–15	10–12	10.5–12.5

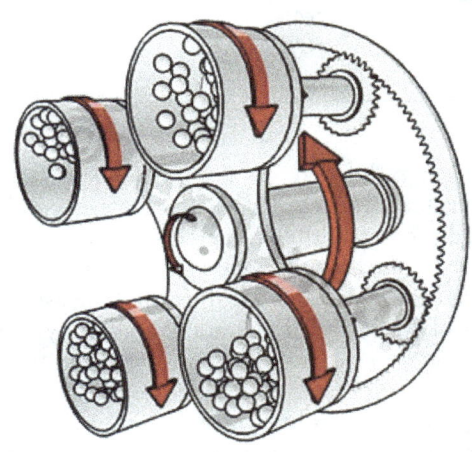

FIGURE 1: Operating principle of a varioplanetary mill.

joint action of shock, centrifugal shock, and abrasive forces, it becomes possible to achieve more highly disperse powders [26–29].

The reliability of results is provided by a systematic study with standard tools and methods for measuring, the mix of modern physical and chemical methods of analysis, X-ray diffraction and DTA, electron microscopy, and a sufficient amount of raw data and research results. The tasks of the scientific work were accomplished by implementing a systematic approach in the triad "composition (raw materials), structure, and properties" [30–32]. The studies were conducted with use of conventional physical-mechanical and physical-chemical

methods of quality assessment of raw materials and synthesized materials as well as finished products.

The fly ashes of largest thermal power plants (TPPs) of Russia (Vladivostok TPP (Figure 2), Artem TPP, Primorye TPP, and Partizansk TPP) were used as components of composite binders. The important factor is the possibility of dry ash selection, which is currently realized for these TPPs.

Taking into consideration the fact that the focus of the paper is the development and use of environmentally friendly materials, the ashes' radioactivity has been evaluated (Table 3).

Thermal studies of the ash showed that, in the range of low temperatures, physically bound water is removed from it. Exothermic effect with the maximum at about 400°C indicates burnout of organic substances and endotherm effect at 712°C indicates dissociation of calcite to CaO and CO_2, which was confirmed by X-ray diffraction data (Figure 3).

Optimization of structure formation processes at hydration of the composite binder components creates the matrix dense structure which is necessary for creating increased impermeability composites. This can be realized by cogrinding of the Portland cement and the polyfunctional mineral admixtures and by reducing the concrete mix water-cement ratio through use of superplasticizers.

To reduce the water demand of concrete mix, the superplasticizers were used. They were selected of six most common construction material markets in the Far East. Cement paste shrinkage was measured by Hagerman cone. The Spassky cement CEM I 42.5N was used as grout. Water-cement ratio was 0.3. Plasticizer dosage was 0.3%. Time of measurement of shrinkage cone is recorded after the end of the cement paste mixing.

Achievement of high values of shrinkage cone is marked in the raw mix of binder and superplasticizer PANTARHIT PC160 Plv (Table 4).

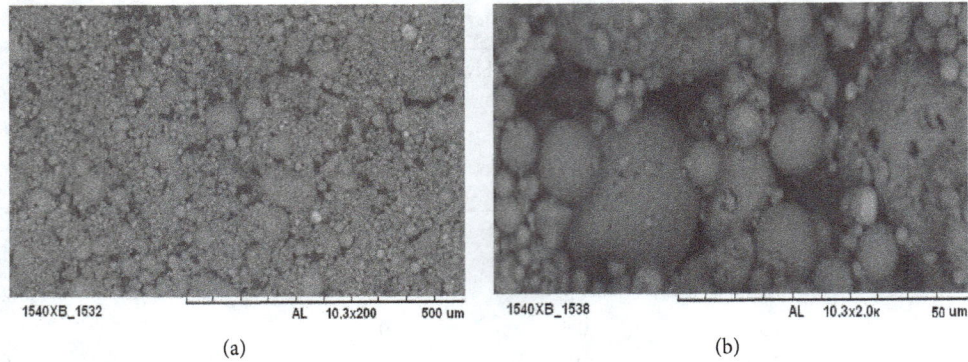

(a)   (b)

FIGURE 2: The Vladivostok TPP's fly ash micrographs: (a) ×200; (b) ×2000.

TABLE 3: Specific effective activity of fly ash depending on the composition [19].

Name of indicator	The measurement result (A) [Bq/kg]			
	Primorye TPP	Vladivostok TPP	Artem TPP	Partizansk TPP
Activity ^{40}K	496.9 ± 101	362 ± 89	342 ± 68	516.9 ± 101
Activity ^{232}Th	153.6 ± 20.3	31.5 ± 19.7	29.5 ± 15.7	193.2 ± 22.3
Activity ^{226}Ra	163.1 ± 9.36	37.63 ± 6.32	27.23 ± 5.93	113.1 ± 6.37
$A_{\text{eff}} = A_{\text{Ra}} + 1.31A_{\text{Th}} + 0.085A_{\text{K}}$	>398	80 ± 30	93 ± 20	>410

TABLE 4: Shrinkage of the cement paste with different superplasticizers.

Time [min]	Melflux 1641 F	Melflux 5581 F	PANTARHIT PC160 Plv	FOX™-8H	PC-1030	JK-04 PPM
			Shrinkage [mm]			
0	290	350	370	250	240	130
5	380	390	400	260	280	120
30	390	350	390	240	190	98

Time: the time since beginning of measurement.

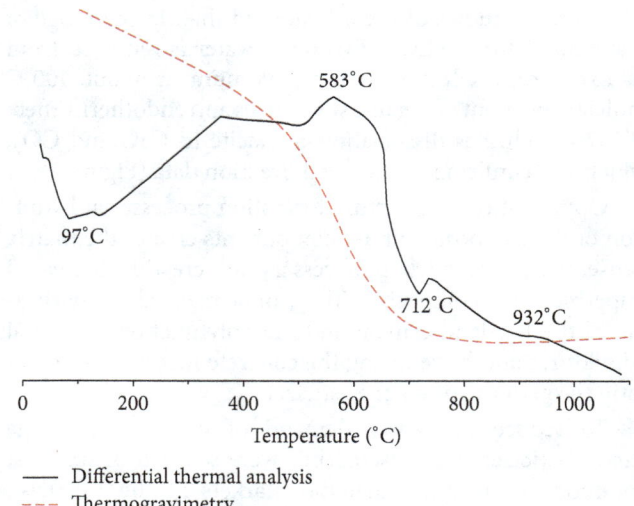

FIGURE 3: The DTA and TG results of the Vladivostok TPP fly ash [17–19].

The raw components were milled and mixed in various proportions (OPC 30–100%, FA 0–50%, LCW 0–20%, and superplasticizer 0.3%) in a varioplanetary mill for one hour.

The materials mixing was carried out using cyclic and continuous flow dispensers with automatic control with a maximum weighing cycle duration of 45–90 s, with a weighing accuracy of 1–3%. All raw materials were dosed by mass, with the exception of water and liquid additives (if any) dosed by volume. In our research, there were no liquid additives. Water was added at the last stage, during the mortar preparation.

The control samples were made from a composite paste without addition of sand and fiber. Number of specimens was fabricated to determine the optimum composition of the binder.

The flowability of the concrete mix was evaluated using Hagerman cone molded from a concrete mix. Water-binder ratio was 0.3. The inverted cone was filled with a freshly prepared concrete mix without sealing. 90 seconds after filling, the cone rose upward. Immediately the stopwatch was turned on. As the mixture reached a diameter of 500 mm and also after the spreading process was completed, the time was fixed. After the flow was completed, the maximum diameter of the spread of the concrete mix was determined.

The compressive strength and the modulus of elasticity of the specimens were researched on 70 mm cubes at the 28th day; however, 100 × 100 × 500 mm prisms were prepared for four-point bending to obtain flexural strength of the

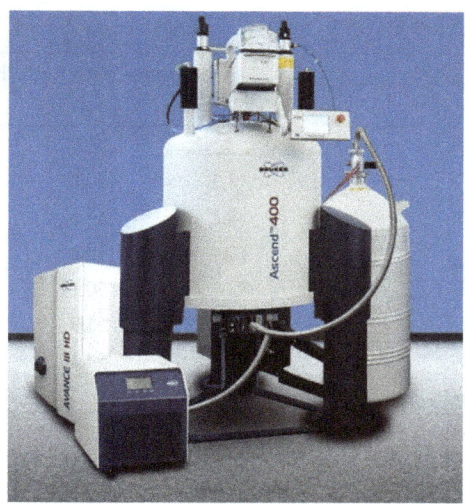

FIGURE 4: Bruker BioSpin NMR Spectroscopy.

FIGURE 5: Bruker D8 ADVANCE powder X-ray diffractometer.

specimens with effective span of 400 mm. Mechanical tests were performed with Servo-hydraulic Fatigue and Endurance Tester Shimadzu Servopulser U-type with capacity of 200 kN and as per BS EN 12390-3:2002. The compressive strength was calculated as the arithmetic average of the six samples.

The porosity of the hardened specimens was determined on $1 \times 1 \times 3$ cm and $3 \times 3 \times 3$ cm specimens. The structure of the cement stone was investigated at the age of 28 days. The porosity was determined by a number of mutually complementary methods, namely,

(i) proton magnetic resonance with a pore measurement range of $1 \times 10^{-3} \cdots 1 \times 10^{-1} \mu m$ in diameter (using Bruker BioSpin NMR Spectroscopy (Figure 4));

(ii) small-angle X-ray diffraction with a measurement range of $2 \times 10^{-3} \cdots 3 \times 10^{-1} \mu m$ (on D8 ADVANCE powder diffractometer from Bruker AXS (Figure 5));

(iii) mercury porosimetry with a measurable range of $1 \times 10^{-1} \cdots 4 \times 10 \mu m$ (using PoreMaster GT (Figure 6));

(iv) optical microscopy of thin sections with a measuring range of $4 \times 10 \cdots 1 \times 10^{3} \mu m$.

X-ray phase analysis determined the degree of cement hydration and the content of low-base calcium hydrosilicates, CSH (I). The phases were identified by the international JCPDS table. The degree of hydration was determined from the intensity of the main C_3S reflex. The amount of CSH (I) was established by comparing the intensity of the main β-CS reflex obtained on samples of hardened cement sintered at 1000°C with a standard sample (quartz).

3. Experimental Part

Seven binder composites were developed for further research (Table 5). Superplasticizer PANTARHIT PC160 Plv at quantity 0.3% was added to each of them. The binder : sand ratio is 1 to 3. To determine the optimal dosage of components in

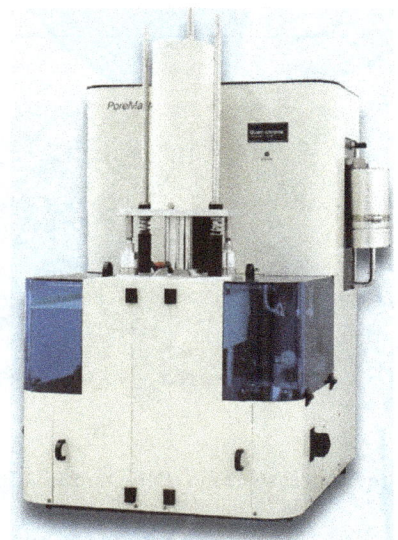

FIGURE 6: PoreMaster GT.

the "cement-limestone-ash" system, it was necessary to grind them to specific surface of 600 m^2/kg at various ratios.

According to Table 5, positive dynamics of strength growth of the composite binder under the joint influence of the ash fine constituents, limestone crushing wastes, and superplasticizer with maximum increase in the activity of the binder at 62% were found.

This is due to the fact that active mineral components of the composite binder contribute to the binding of Ca(OH)$_2$ produced during cement hydration, which results in formation of additional amount of hydrosilicate neoplasms. At the same time, optimizing the process of structure formation is achieved by composite polydispersed components. Highly dispersed spherical ash particles act as crystallization centers

TABLE 5: Compositions and strength of composite binders.

Number	Cement content, by weight [%]	Fly ash content, by weight [%]		Limestone content, by weight [%]	Compressive strength [MPa]		
		Vladivostok TPP	Artem TPP		3 d.	7 d.	28 d.
1	100	—	—	—	17	32.5	47.5
2	30	—	50	20	30.2	40.1	50.4
3	35	45	—	20	34.2	43.1	53.2
4	40	—	45	15	36.6	48.2	56.6
5	45	45	—	10	39.2	50.1	59.2
6	50	—	40	10	45.1	54.9	65.8
7	55	40	—	5	47.2	54.1	70.2
8	100	—	—	—	60.3	81	103.2

Note. The prototype is number 1 (without final grinding); compositions numbers 2–8 are ground to $S_{sp} = 600 \text{ m}^2/\text{kg}$.

TABLE 6: Bond between compressive strength [MPa] of the cement stone samples and the specific surface area of the composite binder [4].

Hardening time [d.]	Specific surface area of the composite binder [m²/kg]					
	500	550	600	700	800	900
3	46.1	**47.4**	**47.2**	46.0	45.6	45.5
7	50.3	**54.2**	**54.1**	49.1	48.6	48.4
28	68.1	**77.3**	**70.2**	65.8	55.0	65.0

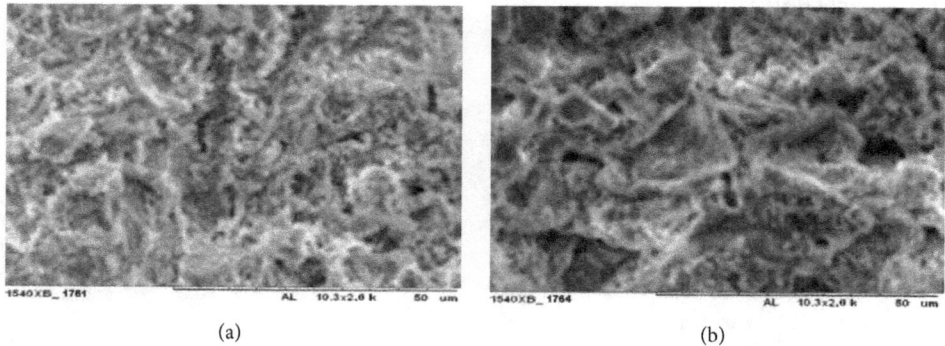

(a)   (b)

FIGURE 7: The cement stone microstructure: (a) CEM I 42.5N based; (b) composite binder based (composition number 7 in Table 5).

and are used as filler at the nano- and microlevels. In conjunction with the larger particles of the mineral component, denser filling of intergranular spaces is noted in concrete cement stone structure with reduction in number of pores and microcracks.

This is confirmed by micrographs of composite cement paste derived by joint grinding of clinker and industrial wastes of the Russian Far Eastern region. Cement stone structure is very dense packing of small grains in the crystalline neoplasms total mass (Figure 7). The additional amount of hydrated crystalline phases contributes to filling of the voids at the microlevel in the crystalline matrix of calcium hydrosilicates at the boundary of the contact area, increasing adhesion degree of binder with filler.

In order to determine the optimum particle size, the Portland cement, the superplasticizer, the ashes, and the limestone were ground (dosage according to composite number 7 of Table 5) to different specific surface area (S_{sp}): 500, 550, 600, 700, 800, and 900 m²/kg (Table 6).

According to Table 6, the 550–600 m²/kg specific surface area (S_{sp}) of binder is optimal. Increasing S_{sp} above these values does not lead to further significant increase in strength. Reduction of start setting time of binder to 35–40 minutes by intensifying the hydration process under the influence of highly active components of the composite [17–19] should be noted.

Thus, the optimum parameters chosen for binder composition are specific surface area of 550 m²/kg, the particle size of 0.15–500 microns, and average particle diameter of the grains of 0.65–11.2 mm [33, 34].

The most important task in creating HDIC is the rational formation and optimization of the pore space structure [35, 36]. In general, the overall reduction in porosity of compositions modified by the technogenic waste more than

TABLE 7: Influence of the composition of the composite binder to the cement stone porosity.

Composition according to Table 5	Porosity [%]			
	Technological (macroscopic level)	Capillary (microscopic level/submicroscopic level)	Gel (supramolecular level)	Total
1	1.2	4.6/2.3	8.2	16.3
2	2.6	1.7/4.5	1.6	10.4
3	1.3	1.1/5.0	3.5	10.8
4	1.4	1.9/2.3	4.4	9.6
5	3.6	1.7/2.5	1.6	9.4
6	3.2	1.1/1.0	3.5	8.8
7	1.0	0.9/1.8	4.4	8.1

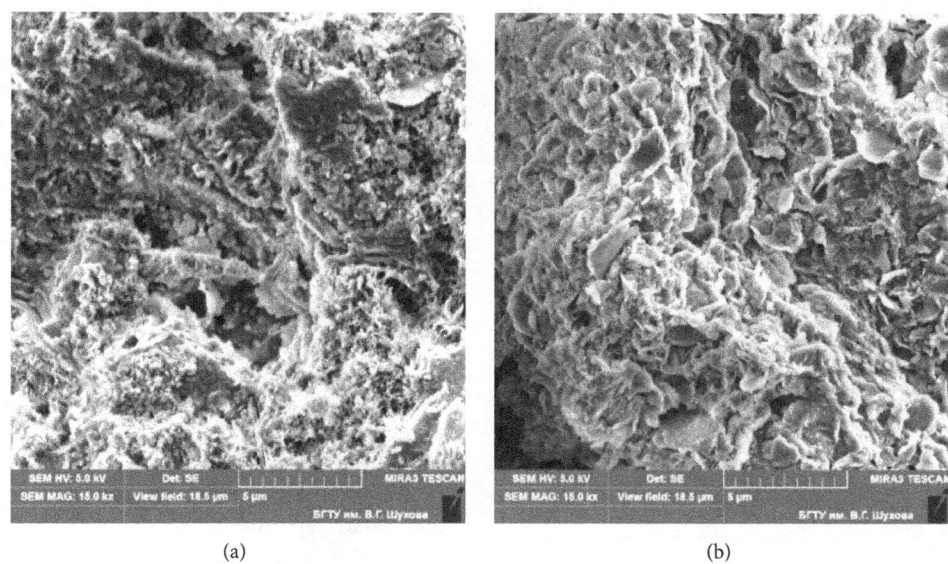

(a)   (b)

FIGURE 8: The neoplasms micrographs: (a) cement stone without any additives; (b) cement stone based on composite binders.

2 times (from 16.3% to 8%) should be noted. Fluctuations of different diameter pores which depend on nature of their formation should also be noted (Table 7).

Existence of a large quantity of hydrosilicate connections is confirmed by decreasing of the gel pores in crystalline bunch in conjunction with modified composites on molecular level with porosity maximum reduction more than 5 times [37, 38]. Although the maximum strength is 77.3 MPa in the optimal composition of binder (by grinding to specific area of 550 m^2/kg), the gel porosity of the composite fell almost in 2 times. In this case, high strength is influenced by the complex actions: reduction of capillary porosity due to the intensification of the processes of growth of primary crystals of hydrosilicate phases [39, 40], due to possible formation of secondary recrystallization and crystals creation, due to filling the space at the micro- and submicrolevels of structural organization composite with them, and in conjunction with reduction technological porosity on 17% due to the formation of dense packing of the grain structure at the macrolevel, with the participation of spherical fine components of fly ash and limestone crushing wastes.

Denser structure of the binder composition with lower porosity is confirmed by microstructural studies. In phase of the modified binder formation, the amount of gel-like hydrate new formation increases on the surface of the filler particles (Figure 8(b)), there are no visible portlandite crystals, and it shows a decline of its share in the total mass of ligament hydrosilicate [41, 42].

By varying the percentage of added ash, it is possible to control number and size of ettringite crystals, which further define the properties of composite binder and concrete [43, 44]. The carbonates also have close contacts with cement stone, which is explained by the emergence of bonds between the cement hydration products and limestone [45, 46]. Growth of crystals of "needle-like" and "stem-like," low-basic hydrosilicates is observed [47, 48]; there are also plate-like calcium crystals, allegedly hydrocarboaluminates (Figure 8(b)), in the structure of the modified binder. Synthesis of these compounds is the result of interaction during hydration of clinker minerals Ca(OH)$_2$ with active mineral ingredients of ash and limestone [49, 50]. Growth of needle-shaped crystals contributes to reinforcement of the composite structure on nano- and microlevels, reduction of porosity, and improvement of the composite complex strength [51, 52].

The greatest effect [53–55] is achieved through the synergistic impact of man-made pozzolanic additives (fly ash) and sedimentary-origin natural materials (limestone) at the content of OPC, 55 wt.%; LCW, 5 wt.%; and FA, 40 wt.%.

In this proportion, the composite material reaches the compressive strength of 77.3 MPa (by grinding to S_{sp} = 550 m^2/kg), at 45% replacement of cement with industrial waste. Thus, the effect of "cement, fly ash, and limestone" composite binder obtained by cogrinding with superplasticizer in the varioplanetary mill to structure formation process is determined. Ground active mineral supplements are the crystallization centers of neoplasms. Ash nanoparticles contribute to the binding of $Ca(OH)_2$, produced during the hydration of clinker minerals, intensifying binder hydration with forming needle-shaped crystals of low-basic hydrosilicate. Existence of the fine grains of limestone leads to forming hydrocarboaluminates. Implementation of the fiber-reinforced concrete potential is only possible by creating the material optimal structure, formation of which is determined by the following basic parameters: type and quality of raw materials, technology and preparation of concrete mixes, and quantitative relation between the components of fiber-reinforced concrete mixture.

4. Results and Discussions

Study of physical-mechanical properties of fine-grained concrete showed that use of the composite binder obtained by cogrinding of Portland cement, fly ash, limestone powder, and superplasticizer allowed increasing the compressive strength of fine concrete by 21%, while reducing almost in 2 times the proportion of cement. Prism strength and elastic modulus in the researched concrete types are significantly higher than in control samples (Table 8).

To optimize the fine-grained concrete structure forming on the macrolevel, steel fiber was used.

Taking into consideration previous studies [17–19], composition number 3 was adopted for the prototype (Table 8), to which fiber in an amount of 24 to 45 kg/m^3, that is, 2% of the total weight of the mixture in increments of 0.2%, was added. It was found that the structure optimization at the macrolevel improves the compressive strength by 24% (Table 9).

Addition of the domestically produced basalt fiber instead of the steel anchor fiber has not led to substantial improvement of physical and mechanical properties of concrete, so, for further investigations, fiber-reinforced concrete number 5 (1.6% reinforcement) is taken.

On basis of the research, it was revealed that the addition of the fly ash and the limestone waste to the composite binder promotes structural and phase changes in the formation of high-density composite impermeable structures.

The fine-grained fiber-reinforced types of concrete on the developed composite binder compositions number 2 and number 3 (Table 10), which achieve compressive strength of 100.2–100.9 MPa with the diffusion coefficient of 1.34 ⋅

10^{-4}–$1.39 \cdot 10^{-4}$ cm^2/s, have the best physical and mechanical properties (Table 11).

It is revealed that, for fine-grained structure of concrete, in addition to high homogeneity, it is also characteristic to reduce specific stresses in the contact zone and increase the integral adhesion force between the cement stone and aggregate. The structure-forming role of the fine aggregate is most evident with an increase in the interaction surface; these conditions are realized in fine-grained types of concrete with the use of screening of granite crushed stone on the basis of composite astringents which, due to the highly developed surface, allow intensifying the processes of structure formation and accelerating the strength of concrete and also consolidating the structure.

As can be seen from the test results (Tables 10 and 11), composite binder of cement, fly ash from thermal power plants, and limestone at all dosages reduces the water permeability and air permeability of concrete. Thus, there is a clearly defined relationship between the properties of concrete and the features of the structure of cement stone: increasing the number of low-basic calcium hydrosilicates as well as increasing gel content and, correspondingly, reducing capillary porosity, especially at the submicroscopic level, predetermine the increase in strength and decrease in the permeability of concrete.

The maximum decrease in impermeability parameters was found in composition number 2 with the replacement of the proportion of cement in the binder mixture by 45% with the technogenic waste (FA and LCW). The air permeability of concrete decreased by 2 times (to 0.0253 cm^3/s), which according to GOST 12730.5 (Russian regulatory requirements) corresponds to mark W14 on permeability. The fiber-reinforced concrete's dense structure provides humidity resistance and reduces water absorption by volume in almost 2.5 times. These patterns are reflected in the water vapor permeability characteristics, which reaches the limit of 0.021 mg/(m·h·Pa) in the humid climate. The concrete's diffusion permeability was determined on the basis of data on the concrete neutralization rate (carbonation) by carbon dioxide in the absence of gradient of common air-gas pressure at the difference between the concentration of carbon dioxide in the concrete and that in environment at the time when the neutralization process is limited by the speed of carbon dioxide diffusion into the concrete porous structure. The experimental procedure is intended for use in the technology development and the designing of concrete compositions that provide long-term maintenance-free operation in the construction in nonaggressive and aggressive gas-air environment.

When evaluating the diffusive permeability, the average value of neutralized concrete layer thickness was found for all developed compounds. It was found that the developed concrete has an effective diffusion coefficient of $D' = 1.34 \cdot 10^{-4}$ cm^2/s.

Thus, the clear link between the concrete properties and characteristics of the cement stone structure (the increase in the number of hydrosilicate neoplasms) at the complex reducing of gel and capillary porosity is revealed, which

TABLE 8: Physical and mechanical properties of fine-grained concrete depending on the binder composition [25].

| Composition numbers | Material consumption per 1 m^3 | | | | | | | Slump [cm] | Compressive strength [MPa] | Prism strength [MPa] | Elastic modulus [MPa] |
| | Cementitious binder [kg] | | | | Screenings of crushed granite [kg] | Sand [kg] | Water [l] | | | | |
	Cement	Fly ash	Limestone	Superplasticizer							
1*	550	—	—				220		107.5	86.3	61.2
2	288	235	27				240		83.7	59.5	43.8
3	275	246	29		1000	623	241	10–12	84.2	60.3	44.5
4	257	257	36	1.2			242		76.3	55.2	40.9
5	244	268	38				243		75.2	55.0	40.8
6	230	278	42				244		75.0	54.9	40.8
7**	550	—	—				215		63.1	42.3	36.2

*The binder of flow water ratio with the specific surface of 550 m^2/kg. **The binder based on Portland cement CEM I 42.5N.

TABLE 9: Dependence of the strength of fiber-reinforced concrete on the percentage of reinforcement.

| Composition numbers | Material consumption per 1 m³ | | | | Reinforcement [%] | R_{compr}, [MPa] |
	Binder [kg]	Water [l]	Aggregate [kg]	Fiber [kg]		
3-1*	550	240	1623	—	0	94.2
3-2	550	240	1623	23.97	1	96.1
3-3	550	240	1623	28.76	1.2	97.3
3-4	550	240	1623	33.56	1.4	99.8
3-5	550	240	1623	38.35	1.6	100.9
3-6	550	240	1623	43.15	1.8	99.5
3-7	550	240	1623	47.94	2	99.6

*Prototype composition corresponds to composition number 3 (according to Table 8).

TABLE 10: Compositions and strength characteristics of the fiber-reinforced concrete.

| Number | Material consumption per 1 m³ | | | | | | Slump [cm] | Prism strength [MPa] | Compressive strength [MPa] |
| | Cementitious binder [kg] | | | | Aggregates [kg] | Water [l] | | | |
	Cement	Fly ash	Limestone	Superplasticizer					
1	550	—	—		1623	220		66.3	115.5
2	288	235	27		1623	240		69.5	100.9
3	275	246	29	1.2	1623	241	10–12	70.3	100.2
4	257	257	36		1623	242		65.2	96.3
5	244	268	38		1623	243		65.0	95.2
6	230	278	42		1623	244		64.9	95.0

TABLE 11: The concrete performance characteristics depending on the binder composition.

| Number (according to Table 10) | Air permeability of concrete a_c [cm³/s] | Mark on water permeability, W | Effective diffusion coefficient [cm²/s] | Water absorption by volume [%] | Vapor permeability [mg/(m·h·Pa)] | |
					For dry climate	For wet climate
1	0.0565	W10	$1.56 \cdot 10^{-4}$	14.8	0.032	0.030
2	0.0253	W14	$1.34 \cdot 10^{-4}$	6.1	0.022	0.021
3	0.0289	W14	$1.39 \cdot 10^{-4}$	6.3	0.026	0.025
4	0.0402	W12	$1.64 \cdot 10^{-4}$	7.8	0.027	0.026
5	0.0465	W12	$1.79 \cdot 10^{-4}$	10.9	0.030	0.029
6	0.0423	W12	$1.82 \cdot 10^{-4}$	14.4	0.032	0.030

is especially observed at the molecular and submicroscopic levels determining the growth of strength and increase of concrete impermeability.

The approbation of theoretical and experimental studies is carried out on the example of monolithic fiber-reinforced concrete walls with permanent formwork developed by Fediuk et al. [18, 25, 56]. The thermal resistance of wall is $R_o = 4,223$ (m² ·°C)/W; the vapor permeability coefficient is $\mu = 0,021$ mg/(m·h·Pa). The fiber-reinforced types of concrete developed on basis of the composite binder can be used in the construction of high-rise buildings [57].

The technological circuit production of the high-density fiber-reinforced concrete is developed. It comprises the following steps: cogrinding of Portland cement, fly ash, and limestone crushing waste; two-stage mixing with the filler and the fiber; filling of formwork; and mechanical compaction of the concrete mix. This production line can be implemented in cement plants in different regions.

Thus, the possibility of reducing permeability of fiber-reinforced concrete by varying the amount and type of additives and fineness and taking into account the conditions of curing is studied. It allows creating materials for multilayer load-bearing structures with a compressive strength of 100 MPa, with low permeability under actual operating conditions. Implementation of the research results will help to improve the environmental situation of regions, as fiber-reinforced concrete comprises 50–60% of industrial waste.

5. Conclusion

Positive dynamics of strength growth of the composite binder under the joint influence of the ash fine constituents, limestone crushing wastes, and superplasticizer with maximum increase in the activity of the binder at 62% were found. This is due to the fact that active mineral components of the composite binder contribute to the binding of $Ca(OH)_2$

produced during cement hydration, which results in formation of additional amount of hydrosilicate neoplasms. The overall reduction in porosity of compositions modified by the technogenic waste more than 2 times (from 16.3% to 8%) should be noted.

High strength is influenced by the complex actions:

(i) Reduction of capillary porosity due to the intensification of the processes of growth of primary crystals of hydrosilicate phases

(ii) Possible formation of secondary recrystallization and crystals creation

(iii) Filling the space at the micro- and submicrolevels of structural organization composite with them and in conjunction with reduction technological porosity on 17%

(iv) The formation of dense packing of the grain structure at the macrolevel, with the participation of spherical fine components of fly ash and limestone crushing wastes

The greatest effect is achieved through the synergistic impact of man-made pozzolanic additives (fly ash) and sedimentary-origin natural materials (limestone) at the content of OPC, 55 wt.%; LCW, 5 wt.%; and FA, 40 wt.%.

The fine-grained fiber-reinforced types of concrete on the developed composite binder, which achieve compressive strength of 100.2–100.9 MPa with the diffusion coefficient of $1.34 \cdot 10^{-4}$–$1.39 \cdot 10^{-4}$ cm^2/s, have the best physical and mechanical properties. As can be seen from the test results, composite binder of cement, fly ash from thermal power plants, and limestone at all dosages reduces the water permeability and air permeability of concrete.

Conflicts of Interest

The authors declare that they have no conflicts of interest.

References

[1] S. Chithra, S. R. R. Senthil Kumar, and K. Chinnaraju, "The effect of Colloidal Nano-silica on workability, mechanical and durability properties of High Performance Concrete with Copper slag as partial fine aggregate," *Construction and Building Materials*, vol. 113, pp. 794–804, 2016.

[2] K. S. Al-Jabri, M. Hisada, S. K. Al-Oraimi, and A. H. Al-Saidy, "Copper slag as sand replacement for high performance concrete," *Cement and Concrete Composites*, vol. 31, no. 7, pp. 483–488, 2009.

[3] W. Wu, W. Zhang, and G. Ma, "Optimum content of copper slag as a fine aggregate in high strength concrete," *Materials and Corrosion*, vol. 31, no. 6, pp. 2878–2883, 2010.

[4] H. Stang and C. Pedersen, "HPFRCC - extruded pipes," in *Proceedings of the 1996 4th Materials Engineering Conference. Part 1 (of 2)*, pp. 261–270, November 1996.

[5] P. Tjiptobroto and W. Hansen, "Tensile strain hardening and multiple cracking in high-performance cement-based composites containing discontinuous fibers," *ACI Materials Journal*, vol. 90, no. 1, pp. 16–25, 1993.

[6] V. C. Li, "From mechanics to structural engineering: the design of cementitious composites for civil engineering applications," *Structural Engineering/Earthquake Engineering*, vol. 10, pp. 37–48, 1993.

[7] M. Li and V. C. Li, "Rheology, fiber dispersion, and robust properties of engineered cementitious composites," *Materials and Structures/Materiaux et Constructions*, vol. 46, no. 3, pp. 405–420, 2013.

[8] J. Qiu and E.-H. Yang, "Micromechanics-based investigation of fatigue deterioration of engineered cementitious composite (ECC)," *Cement and Concrete Research*, vol. 95, pp. 65–74, 2017.

[9] H. Luo, Y. Wu, A. Zhao et al., "Hydrothermally synthesized porous materials from municipal solid waste incineration bottom ash and their interfacial interactions with chloroaromatic compounds," *Journal of Cleaner Production*, vol. 162, pp. 411–419, 2017.

[10] D.-Y. Yoo and N. Banthia, "Mechanical properties of ultra-high-performance fiber-reinforced concrete: A review," *Cement and Concrete Composites*, vol. 73, pp. 267–280, 2016.

[11] D.-Y. Yoo, N. Banthia, and Y.-S. Yoon, "Predicting service deflection of ultra-high-performance fiber-reinforced concrete beams reinforced with GFRP bars," *Composites Part B: Engineering*, vol. 99, pp. 381–397, 2016.

[12] J. Yang, H. Shin, and D. Yoo, "Benefits of using amorphous metallic fibers in concrete pavement for long-term performance," *Archives of Civil and Mechanical Engineering*, vol. 17, no. 4, pp. 750–760, 2017.

[13] D. Yoo, I. You, and S. Lee, "Electrical properties of cement-based composites with carbon nanotubes, graphene, and graphite nanofibers," *Sensors*, vol. 17, no. 5, p. 1064, 2017.

[14] M. Castellote, I. Llorente, C. Andrade, and C. Alonso, "Accelerated leaching of ultra high performance concretes by application of electrical fields to simulate their natural degradation," *Materiaux et Constructions*, vol. 36, no. 256, pp. 81–90, 2003.

[15] M. Tabatabaeian, A. Khaloo, A. Joshaghani, and E. Hajibandeh, "Experimental investigation on effects of hybrid fibers on rheological, mechanical, and durability properties of high-strength SCC," *Construction and Building Materials*, vol. 147, pp. 497–509, 2017.

[16] M. Sahmaran and I. O. Yaman, "Hybrid fiber reinforced self-compacting concrete with a high-volume coarse fly ash," *Construction and Building Materials*, vol. 21, no. 1, pp. 150–156, 2007.

[17] R. S. Fediuk, "Mechanical Activation of Construction Binder Materials by Various Mills," in *Proceedings of the All-Russia Scientific and Practical Conference on Materials Treatment: Current Problems and Solutions*, vol. 125, November 2015.

[18] R. Fediuk and A. Yushin, "Composite binders for concrete with reduced permeability," in *Proceedings of the International Conference on Advanced Materials and New Technologies in Modern Materials Science 2015, AMNT 2015*, November 2015.

[19] R. S. Fediuk and A. M. Yushin, "The use of fly ash the thermal power plants in the construction," in *Proceedings of the 21st International Conference for Students and Young Scientists: Modern Technique and Technologies, MTT 2015*, vol. 93, October 2015.

[20] M. Sánchez, M. C. Alonso, and R. González, "Preliminary attempt of hardened mortar sealing by colloidal nanosilica migration," *Construction and Building Materials*, vol. 66, pp. 306–312, 2014.

[21] M. H. Mobini, A. Khaloo, P. Hosseini, and A. Esrafili, "Mechanical properties of fiber-reinforced high-performance concrete

incorporating pyrogenic nanosilica with different surface areas," *Construction and Building Materials*, vol. 101, pp. 130–140, 2015.

[22] N. Ranjbar, A. Behnia, B. Alsubari, P. Moradi Birgani, and M. Z. Jumaat, "Durability and mechanical properties of self-compacting concrete incorporating palm oil fuel ash," *Journal of Cleaner Production*, vol. 112, pp. 723–730, 2016.

[23] A. Hendi, H. Rahmani, D. Mostofinejad, A. Tavakolinia, and M. Khosravi, "Simultaneous effects of microsilica and nanosilica on self-consolidating concrete in a sulfuric acid medium," *Construction and Building Materials*, vol. 152, pp. 192–205, 2017.

[24] L. Xu, F. Deng, and Y. Chi, "Nano-mechanical behavior of the interfacial transition zone between steel-polypropylene fiber and cement paste," *Construction and Building Materials*, vol. 145, pp. 619–638, 2017.

[25] R. Fediuk, "High-strength fibrous concrete of Russian Far East natural materials," in *Proceedings of the International Conference on Advanced Materials and New Technologies in Modern Materials Science 2015, AMNT 2015*, vol. 116, November 2015.

[26] R. Ibragimov, "The influence of binder modification by means of the superplasticizer and mechanical activation on the mechanical properties of the high-density concrete," *ZKG International*, vol. 69, no. 6, pp. 34–39, 2016.

[27] L. H. Zagorodnjuk, V. S. Lesovik, A. A. Volodchenko, and V. T. Yerofeyev, "Optimization of mixing process for heat-insulating mixtures in a spiral blade mixer," *International Journal of Pharmacy and Technology*, vol. 8, no. 3, pp. 15146–15155, 2016.

[28] R. A. Ibragimov and S. I. Pimenov, "Influence of mechanochemical activation on the cement hydration features," *Magazine of Civil Engineering*, vol. 62, no. 2, pp. 3–12, 2016.

[29] E. Glagolev, L. Suleimanova, and V. Lesovik, "High reaction activity of nano-size phase of silica composite binder," *Journal of Environmental and Science Education*, vol. 11, no. 18, pp. 12383–12389, 2016.

[30] S. L. Buyantuyev, L. A. Urkhanova, S. A. Lkhasaranov, Y. Y. Stebenkova, A. B. Khmelev, and A. S. Kondratenko, "The methods of receiving coal water suspension and its use as the modifying additive in concrete," in *Proceedings of the 12th International Conference Radiation-Thermal Effects and Processes in Inorganic Materials*, vol. 168, September 2016.

[31] S. L. Buyantuev, L. A. Urkhanova, A. S. Kondratenko, S. Y. Shishulkin, S. A. Lkhasaranov, and A. B. Khmelev, "Processing of ash and slag waste of heating plants by arc plasma to produce construction materials and nanomodifiers," in *Proceedings of the 12th International Conference Radiation-Thermal Effects and Processes in Inorganic Materials*, vol. 168, September 2016.

[32] A. P. Semenov, N. N. Smirnyagina, L. A. Urkhanova et al., "Reception carbon nanomodifiers in arc discharge plasma and their application for modifying of building materials," in *Proceedings of the 12th International Conference Radiation-Thermal Effects and Processes in Inorganic Materials*, vol. 168, September 2016.

[33] R. Chihaoui, H. Khelafi, Y. Senhadji, and M. Mouli, "Potential use of natural perlite powder as a pozzolanic mineral admixture in Portland cement," *Journal of Adhesion Science and Technology*, vol. 30, no. 17, pp. 1930–1944, 2016.

[34] M. Koniorczyk, D. Gawin, and B. . Schrefler, "Modeling evolution of frost damage in fully saturated porous materials exposed to variable hygro-thermal conditions," *Computer Methods Applied Mechanics and Engineering*, vol. 297, pp. 38–61, 2015.

[35] K. Marcin, "Coupled heat and water transport in deformable porous materials considering phase change kinetics," *International Journal of Heat and Mass Transfer*, vol. 81, pp. 260–271, 2015.

[36] A. Peschard, A. Govin, P. Grosseau, B. Guilhot, and R. Guyonnet, "Effect of polysaccharides on the hydration of cement paste at early ages," *Cement and Concrete Research*, vol. 34, no. 11, pp. 2153–2158, 2004.

[37] R. S. Fediuk and D. A. Khramov, "Research on porosity of the cement stone of composite binders," *Int. Res. J*, vol. 1, pp. 77–79, 2016.

[38] Z. Liu, Y. Zhang, and Q. Jiang, "Continuous tracking of the relationship between resistivity and pore structure of cement pastes," *Construction and Building Materials*, vol. 53, pp. 26–31, 2014.

[39] Z. Liu, Y. Zhang, G. Sun, Q. Jiang, and W. Zhang, "Resistivity method for monitoring the early age pore structure evolution of cement paste," *Journal of Civil, Architectural and Environmental Engineering*, vol. 34, no. 5, pp. 148–153, 2012.

[40] M. Schmidt, H. Pöllmann, A. Egersdörfer, J. Göske, and S. Winter, "Investigations on the puzzolanic reactivity of a special glass meal in a cementitious system," in *Proceedings of the 32nd International Conference on Cement Microscopy 2010*, pp. 86–118, 2010.

[41] M. Schmidt, H. Pöllmann, A. Egersdörfer, J. Göske, and S. Winter, "Investigations on the use of a foam glass containing metakaolin in a lime binder system," in *Proceedings of the 33rd International Conference on Cement Microscopy 2011*, pp. 319–354, 2011.

[42] A. Sachdeva, M. J. Mccarthy, L. J. Csetenyi, and M. R. Jones, "Mechanisms of sulfate heave prevention in lime stabilized clays through pozzolanic additions," in *Proceedings of the International Symposium on Ground Improvement Technologies and Case Histories, ISGI'09*, pp. 555–560, December 2009.

[43] A. J. Puppala, E. Wattanasanticharoen, V. S. Dronamraju, and L. R. Hoyos, *Ettringite induced heaving and shrinking in kaolinite clay*, vol. 162 of *Geotechnical Special Publication*, 2007.

[44] S. Liu and P. Yan, "Hydration properties of limestone powder in complex binding material," *Journal of the Chinese Ceramic Society*, vol. 36, no. 10, pp. 1401–1405, 2008.

[45] S. Liu and L. Zeng, "Influence of new admixtures on the properties of hydraulic concrete," *Journal of Hydroelectric Engineering*, vol. 30, no. 2, pp. 118–122, 2011.

[46] K. Pushkarova, K. Kaverin, and D. Kalantaevskiy, "Research of high-strength cement compositions modified by complex organic-silica additives," *EasternEuropean Journal of Enterprise Technologies*, vol. 5, no. 5, pp. 42–51, 2015.

[47] E. V. Fomina, V. V. Strokova, and N. I. Kozhukhova, "Application of natural aluminosilicates in autoclave cellular concrete," *World Applied Sciences Journal*, vol. 25, no. 1, pp. 48–54, 2013.

[48] K. Ma, J. Feng, G. Long, and Y. Xie, "Effects of mineral admixtures on shear thickening of cement paste," *Construction and Building Materials*, vol. 126, pp. 609–616, 2016.

[49] P. Shafigh, M. A. Nomeli, U. J. Alengaram, H. B. Mahmud, and M. Z. Jumaat, "Engineering properties of lightweight aggregate concrete containing limestone powder and high volume fly ash," *Journal of Cleaner Production*, vol. 135, pp. 148–157, 2016.

[50] A. Balza, O. Corona, A. Alarcón, J. Echevarrieta, M. Goite, and G. González, "Microstructural study of Portland cement additivated with Nanomaterials," *Acta Microscopica*, vol. 25, no. 1, pp. 39–47, 2016.

[51] F. Faleschini, M. A. Zanini, K. Brunelli, and C. Pellegrino, "Valorization of co-combustion fly ash in concrete production," *Materials and Corrosion*, vol. 85, pp. 687–694, 2015.

[52] B. Boulekbache, M. Hamrat, M. Chemrouk, and S. Amziane, "Flexural behaviour of steel fibre-reinforced concrete under cyclic loading," *Construction and Building Materials*, vol. 126, pp. 253–262, 2016.

[53] M. Rudzki, M. Bugdol, and T. Ponikiewski, "An image processing approach to determination of steel fibers orientation in reinforced concrete," *Lecture Notes in Computer Science*, vol. 7339, pp. 143–150, 2012.

[54] T. Ponikiewski, J. Gołaszewski, M. Rudzki, and M. Bugdol, "Determination of steel fibres distribution in self-compacting concrete beams using X-ray computed tomography," *Archives of Civil and Mechanical Engineering*, vol. 15, no. 2, pp. 558–568, 2015.

[55] R. S. Fediuk and D. A. Khramov, "Physical equipment spectroscopic study of coal ash," *Modern Construction and Architecture*, vol. 1, pp. 57–60, 2016.

[56] R. S. Fediuk, A. K. Smoliakov, R. A. Timokhin, N. Y. Stoyushko, and N. A. Gladkova, *Fibrous Concrete with Reduced Permeability to Protect the Home Against the Fumes of Expanded Polystyrene*, vol. 66 of *Materials Science and Engineering*, 2017.

[57] M. Quintard, "Transfers in porous media," *Special Topics and Reviews in Porous Media*, vol. 6, no. 2, pp. 91–108, 2015.

12

Effects of Voids on Concrete Tensile Fracturing

Lei Xu and Yefei Huang

College of Water Conservancy and Hydropower Engineering, Hohai University, Nanjing 210098, China

Correspondence should be addressed to Lei Xu; leixu@hhu.edu.cn

Academic Editor: Francesco Caputo

A two-dimensional mesoscale modeling framework, which considers concrete as a four-phase material including voids, is developed for studying the effects of voids on concrete tensile fracturing under the plane stress condition. Aggregate is assumed to behave elastically, while a continuum damaged plasticity model is employed to describe the mechanical behaviors of mortar and ITZ. The effects of voids on the fracture mechanism of concrete under uniaxial tension are first detailed, followed by an extensive investigation of the effects of void volume fraction on concrete tensile fracturing. It is found that both the prepeak and postpeak mesoscale cracking in concrete are highly affected by voids, and there is not a straightforward relation between void volume fraction and the postpeak behavior due to the randomness of void distribution. The fracture pattern of concrete specimen with voids is controlled by both the aggregate arrangement and the distribution of voids, and two types of failure modes are identified for concrete specimens under uniaxial tension. It is suggested that voids should be explicitly modeled for the accurate fracturing simulation of concrete on the mesoscale.

1. Introduction

Concrete is widely used as a construction material and is traditionally treated as a homogeneous continuum on the structural scale (macroscale). This homogenization assumption can hold well as long as the mechanical response of concrete remains in the elastic regime [1, 2]. However, when fracturing occurs, the macroscale mechanical behavior of concrete is greatly controlled by its components and their interactions taking place on a finer scale (mesoscale) [3, 4], which means accurate modeling of concrete fracturing calls for the consideration of its mesostructure.

Up to date, several mesoscale models have been developed to provide tools for a better understanding of concrete fracturing. From the simulation strategy point of view, most of the existing concrete mesoscale models can be broadly grouped into two types: the continuum model and the lattice model. In the continuum model, concrete is usually characterized by a continuum composite material with each component discretized by finite elements, while, for the lattice model, a discrete system composed of lattice elements is used

to represent concrete. Moreover, the discrete element method (DEM) has been recently used to perform the mesoscale simulation of concrete [5], and it is shown that the discrete model requires a huge numerical effort that is necessary for this approach to obtain a reasonable representation of concrete mesostructure.

Several researchers studied the concrete fracturing by employing the continuum modeling strategy, and representative contributions can be found in [6–12]. The most recent investigations following this strategy were carried out by Du et al. [13] who studied the dynamic tensile fracturing of concrete by assuming concrete to be composed of aggregate and mortar matrix, by Huang et al. [14] who performed a 3D mesoscale fracturing simulation based on the actual concrete mesostructure, and by Wang et al. [15, 16] who developed a computational technology using the interface element with a cohesive law to perform Monte Carlo simulations of concrete fracturing and to study the 3D mesostructure effects on concrete damage and failure. Overall, the principal merit of the continuum model lies in the detailed representation of concrete mesostructure, which ensures the ability to realize reasonable simulations of cracking initiation on the mesoscale and coalescence of multiple distributed cracks into localized macroscale cracks and fracture propagation. However, it tends to be computationally intensive even for laboratory-scale specimens, especially for three-dimensional cases.

With respect to the lattice modeling strategy, representative studies were carried out in [17–20], and the most recent improvements are performed by Cusatis et al. [21, 22] who proposed a novel model named the lattice discrete particle model (LDPM) by exploiting the merits of both the lattice model and the discrete particle model. In contrast to the continuum model, the lattice model is considered computationally less demanding as concrete mesostructure is roughly represented by a discrete system with relatively less degrees of freedom and meanwhile can still possess the ability to capture the most important aspects of concrete fracturing. However, it is hard to investigate the interactions of concrete components in a real sense since the actual concrete mesostructure is not fully taken into account in the lattice model.

Voids (or pores) with different sizes always exist in concrete and typically take up 2–6% of the total volume, and the use of entrained air void system is a common approach in concrete technology to resist cyclic freezing and thawing degradation [23]. However, the effects of voids on concrete fracturing on the mesoscale are still not well understood. Wang et al. [15] built numerical concrete samples with pores using interface elements and studied the effects of porosity on concrete loading-carrying capability under uniaxial tension, but the fracturing mechanism on the mesoscale was not detailed. Huang et al. [14] reported the distribution of voids greatly influences the tensile strength and crack patterns based on the simulation results of a single 3D specimen. On the whole, it has been recognized that the existence of voids affects the concrete mechanical behavior to a large extent, but further research is needed to reveal the effects of voids on concrete fracturing.

With this in mind, a 2D finite element (FE) mesoscale modeling framework for concrete is proposed in this study in which concrete is considered as a four-phase material composed of aggregate, mortar, interfacial transitional zone (ITZ), and void, and the effects of voids on concrete tensile fracturing under the plane stress condition are detailed by performing several simulations. The rest of this paper is organized as follows: Section 2 presents the generation procedures of concrete mesostructure; the FE modeling methodology including mesh discretization, insertion of ITZ elements, and constitutive modeling of mortar and ITZ is described in Section 3; in Section 4, the effects of voids on concrete tensile fracturing are discussed in detail based on the simulation results of several concrete specimens with different mesostructures; finally, the study is summarized with conclusions in Section 5.

2. Generation of Concrete Mesostructure

In this study, concrete is treated as a four-phase composite material, that is, coarse aggregate, mortar composed of cement matrix and fine aggregate, interfacial transitional zone (ITZ), and void randomly distributed in the mortar. Regarding aggregate generation, gravel is idealized as circle, while crushed aggregate is considered as polygon. Mortar is assumed as a homogenous continuum, and the interface with a specified thickness between coarse aggregate (hereinafter referred to as aggregate) and mortar is used to represent ITZ. Moreover, void is viewed as circle for simplicity.

2.1. Size Distribution of Aggregates and Voids. The aggregate size distribution of concrete is described by Talbot's equation as

$$F_p(d_a) = \left(\frac{d_a}{d_{\max}}\right)^n, \tag{1}$$

where d_a is the size of aggregate, $d_{\max}$ is the maximum size of aggregate, $F_p(d_a)$ represents the ratio of aggregates by weight passing through a sieve of characteristic size equal to d_a, and n is the exponent of Talbot's equation. For $n = 0.5$, the corresponding curve is known as Fuller's curve extensively employed in concrete grading design for optimal packing properties.

For a concrete specimen with total volume V, the volume of aggregates within a grading segment $[d_i, d_{i+1}]$ can be calculated by

$$V_{\mathrm{agg}}[d_i, d_{i+1}] = \frac{F_p(d_{i+1}) - F_p(d_i)}{F_p(d_{\max}) - F_p(d_{\min})} \times A_F \times V, \tag{2}$$

where $d_{\min}$ is the minimum size of aggregate and A_F represents the aggregate volume fraction.

Currently, the size distribution of voids in concrete has not been detailed. In general, these voids can be broadly grouped into two types according to different formation ways and the resulting different sizes: the (smaller) entrained voids with typical sizes on the order of 0.1 mm and the (larger) entrapped voids with typical sizes commonly more than 1 mm. In this study void size is considered to be uniformly distributed, and the same assumption is also employed by other researchers [15, 16]. Thus, denoting the size range of void by $[d_{\min}^v, d_{\max}^v]$, the void size can be calculated by $d_v = d_{\min}^v + P \times (d_{\max}^v - d_{\min}^v)$ (P is a uniformly distributed random number between 0 and 1).

2.2. Generation and Placement of Aggregates and Voids. In order to build numerical concrete specimens automatically, a mesostructure generator for concrete (MGC) is developed using MATLAB based on the take-and-place method [24, 25].

In the take-process, aggregates and voids, which will be placed into the specimen volume in the place-process, are generated separately. For the aggregate generation, the aggregate volume for each grading segment is first calculated according to (2). Then, starting with the grading segment with the maximum average size, the aggregates are generated one by one for each grading segment. For a certain grading segment $[d_i, d_{i+1}]$, the generation of aggregates takes the following procedures.

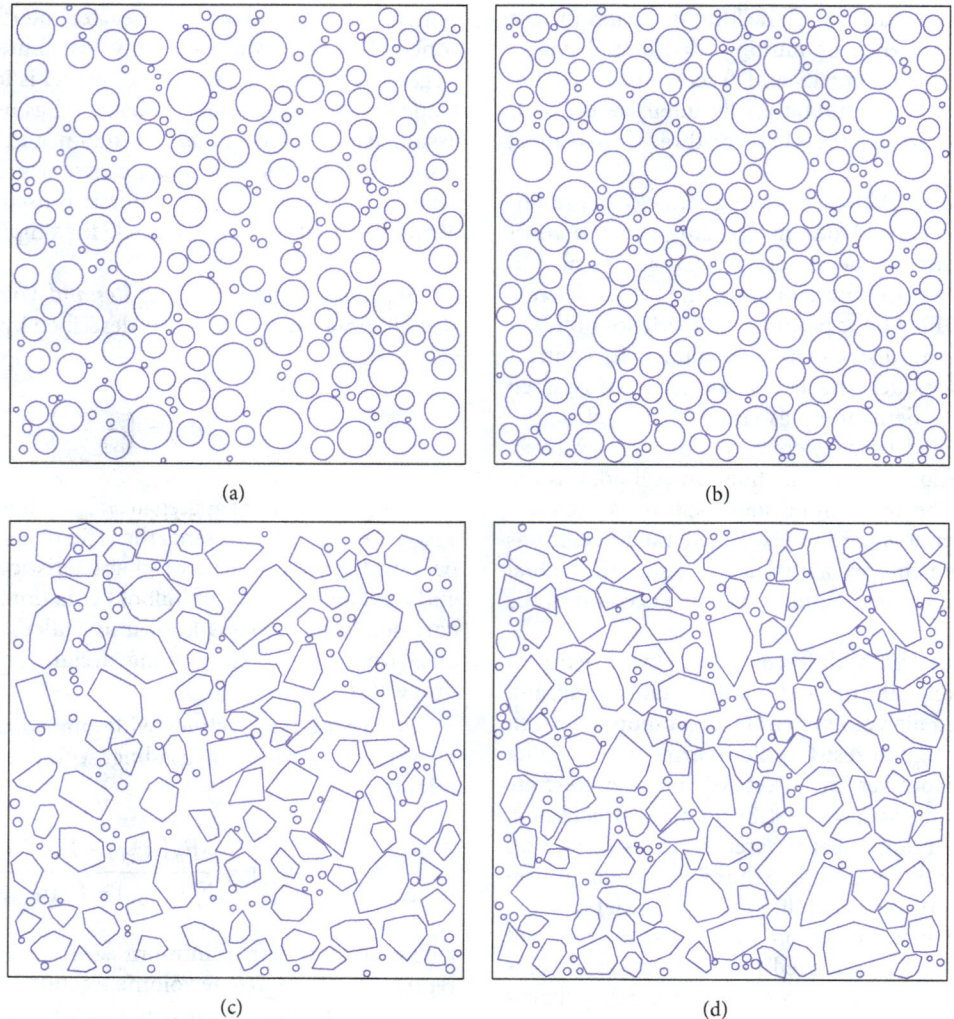

FIGURE 1: Numerical concrete specimens. (a) Circular aggregate ($A_F = 40\%$, $A_v = 2\%$). (b) Circular aggregate ($A_F = 50\%$, $A_v = 2\%$). (c) Polygonal aggregate ($A_F = 40\%$, $A_v = 2\%$). (d) Polygonal aggregate ($A_F = 50\%$, $A_v = 2\%$).

Step 1. Generate a random number representing the aggregate size d_a, which is assumed to follow a uniform distribution and therefore can be taken as $d_a = d_i + P \times (d_{i+1} - d_i)$.

Step 2. For gravel, a circle with radius of $d_a/2$ is defined to represent the aggregate, while, for crushed aggregate, a polygon with the random number of sides ranging from 4 to 10 and with the smallest width equal to d_a is generated to represent the aggregate (see [24] for more details). Then, the volume of the current generated aggregate is calculated.

Step 3. Repeat the previous two steps until the remaining volume left is less than $\pi d_a^2/4$, namely, not enough to generate a new aggregate.

Step 4. Transfer the remaining volume to the next grading segment.

Following the similar procedures for generating gravel aggregates, the generation of voids can be performed with ease provided by the given void volume fraction and size

range, which is followed by the placement of aggregates and voids (the place-process).

In the place-process, the generated aggregates and voids are first sorted according to their volume, respectively. Then, for the convenience of mesh discretization discussed in Section 3, the size of each aggregate is increased by a specified value (the thickness of ITZ, T_{ITZ}) to consider the surrounding ITZ, which means the aggregate finally placed consists of two parts (i.e., aggregate piece and the surrounding ITZ with a specified thickness). After the modification of aggregate size, all aggregates are placed into the specified specimen one by one starting with the aggregate with the largest volume, followed by the placement of voids starting from the biggest one. The procedures of the placement of aggregates and voids are detailed as follows.

Step 1. Define the shape of concrete specimen using X and Y coordinates of the boundary vertices numbered clockwise or anticlockwise, which will be used in Step 3 to check if an aggregate is inside the concrete specimen.

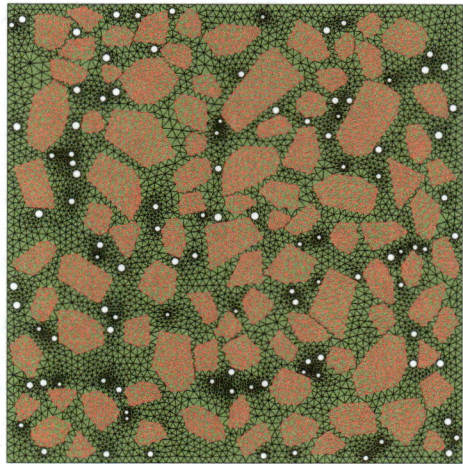

FIGURE 2: FE mesh discretization (polygonal aggregate).

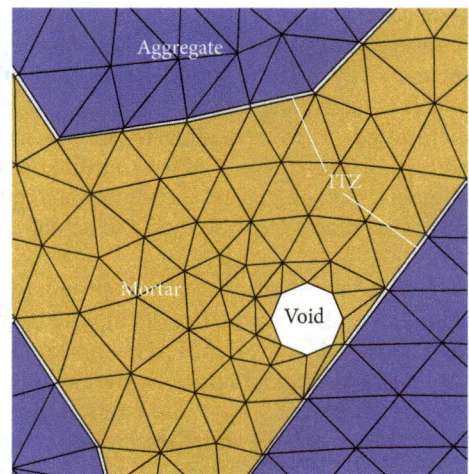

FIGURE 3: FE mesh discretization after inserting and adjusting.

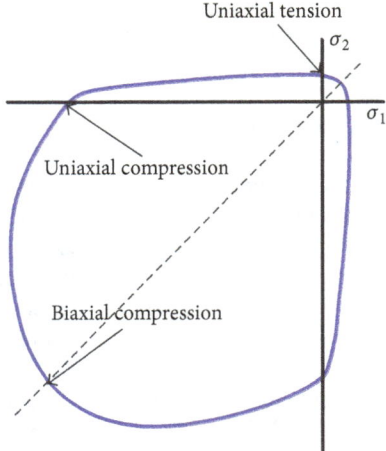

FIGURE 4: Yield surface in plane stress.

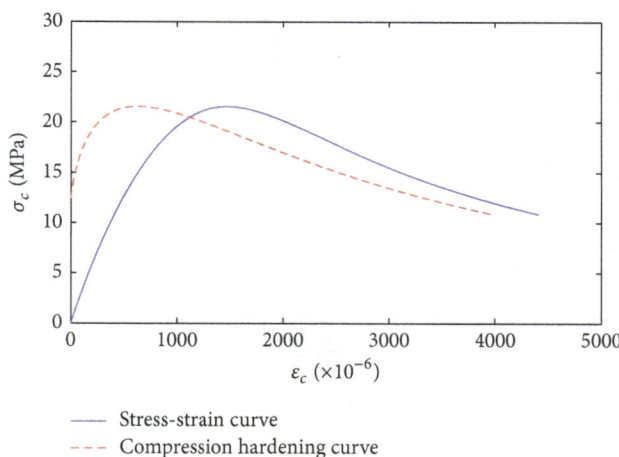

FIGURE 5: Total stress-strain relation under uniaxial compression and compression hardening curve of mortar.

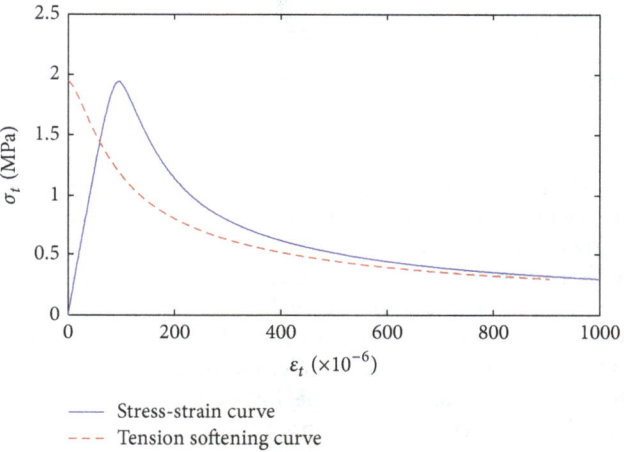

FIGURE 6: Total stress-strain relation under uniaxial tension and tension softening curve of mortar.

Step 2. Generate random numbers to define the position (and orientation if polygon is used to represent the crushed aggregate) of the aggregate using $X_{\min}$, $X_{\max}$, $Y_{\min}$, and $Y_{\max}$, which represent the minimum and maximum value of X and Y coordinates of all boundary vertices, respectively.

Step 3. Perform the aggregate placing. The placement is considered to be successful if the following four conditions are satisfied: (1) the whole aggregate should be within the concrete specimen; (2) no overlapping/intersection occurs between the aggregate to be placed and any existing aggregate; (3) a minimum distance (defined by r_1) should exist between the aggregate and the specimen boundary; and (4) a minimum gap (defined by r_2) should exist between any two aggregates. If any of the four conditions is violated, Step 2 is repeated to make a new attempt until the placement of the aggregate is completed.

Step 4. Repeat Steps 2-3 until all aggregates are placed inside the specimen.

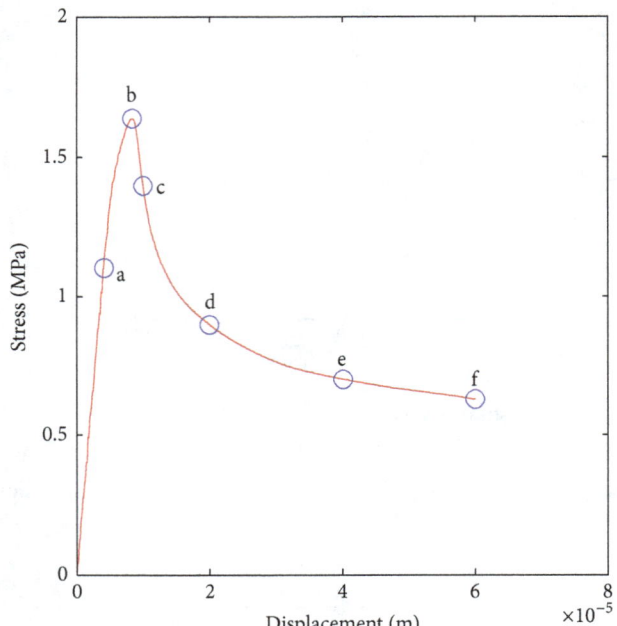

FIGURE 7: Stress-displacement curve (A_F = 40%, without void).

After the placement of aggregates, voids should also be placed into the specimen, which can be carried out by following the similar steps given above. It is worth noting that voids are considered to be embedded in mortar in this study.

Using MGC, numerical concrete specimens can be built with ease. The specimens shown in this paper are 100 mm squares, and the 4-segment Fuller curve is used to describe the aggregate grading for all specimens. For aggregates, d_{max} and d_{min} are set to 10 mm and 4 mm, respectively, while, for voids, d_{max}^v = 2 mm and d_{min}^v = 1 mm are used. In addition, T_{ITZ} is approximately set to 100 μm according to the experimental observation [1], and both r_1 and r_2 are taken as 0.1 times of the size of aggregate or void to be placed.

Figures 1(a) and 1(b) sketch two numerical samples using circular aggregates with the same void volume fraction (A_v = 2%) and the aggregate volume fraction A_F = 40% and 50%, respectively, whereas two numerical samples using polygonal aggregates with A_v = 2% and A_F = 40% and 50%, respectively, are plotted in Figures 1(c) and 1(d).

3. Finite Element Modeling Methodology

Once the concrete mesostructure is obtained, a corresponding FE model is required for performing numerical simulations. The details of the FE modeling methodology developed in this study are presented as follows.

3.1. Mesh Discretization. In order to automatically carry out the mesh discretization of the concrete specimen with complex mesostructure, a mesh generator is developed using MATLAB by exploiting the powerful preprocessing modules provided by the commercial finite element software ABAQUS. For a numerical concrete specimen with pregenerated mesostructure, a two-part python file, which defines

the boundary of the specimen together with the locations and shapes of aggregates and voids using the first part of the file and specifies the mesh discretization parameters using the second part of the file, is first generated using the mesh generator by taking concrete mesostructure as input data. Then, a FE mesh composed of linear triangular elements can be obtained by the mesh generator through calling ABAQUS/CAE kernel to execute the generated python file. An example of the FE mesh discretization with aggregate elements highlighted is shown in Figure 2. It must be noted that ITZ elements are not included in the original generated FE mesh as the tiny thickness of ITZ makes the mesh discretization harder and leads to a poor mesh quality, and therefore a modification of the original FE mesh is needed for obtaining the final FE mesh, which is discussed in detail in Section 3.2.

3.2. Insertion of ITZ Elements. As stated in Section 2, the size of an aggregate is increased by a specified value determined by the thickness of ITZ before placing this aggregate into the specimen volume. Hence, the aggregate elements in the FE mesh generated in Section 3.1 not only occupy the volume of aggregates but also take up the volume of ITZs. In order to explicitly represent the surrounding ITZs and the actual sizes of aggregates in the FE model, ITZ elements should be inserted between aggregate elements and the corresponding mortar elements and the coordinates of nodes on the boundaries of aggregate elements are required to be adjusted for accurately representing the actual aggregate sizes.

To this end, a four-step procedure is proposed. Firstly, the original nodes on the boundaries of amplified aggregates are identified, followed by the definition of new nodes based on the coordinates of original nodes on the boundaries and the given thickness of ITZ. Then, the connectivities of the aggregate elements associated with these nodes are redefined by replacing the number of the original nodes by the number of the corresponding new nodes. Subsequently, 4-noded ITZ elements are formulated one by one using the original nodes and the corresponding new nodes. Finally, an updated input file for ABAQUS, which contains final mesh data, is generated. An in-house MATLAB program, which follows the above procedure, is developed, and part of the final FE mesh discretization corresponding to Figure 2 is illustrated as an example in Figure 3.

3.3. Continuum Damaged Plasticity Model. Without considering voids, it is well recognized that ITZ is weaker than aggregate and mortar, and consequently mesoscale cracking in concrete under loading is commonly considered to first appear in ITZs. After that, mesoscale cracks propagate into mortar and additional cracks initiate within mortar with the further increase of loading [1]. In general, aggregates behave elastically during the process of concrete fracturing. Hence, the isotropic linear elastic model is employed to model the mechanical behavior of aggregates, while a continuum damaged plasticity (CDP) model [26, 27] implemented in ABAQUS is utilized to describe the mechanical behavior of both mortar and ITZ, which is briefly summarized as follows.

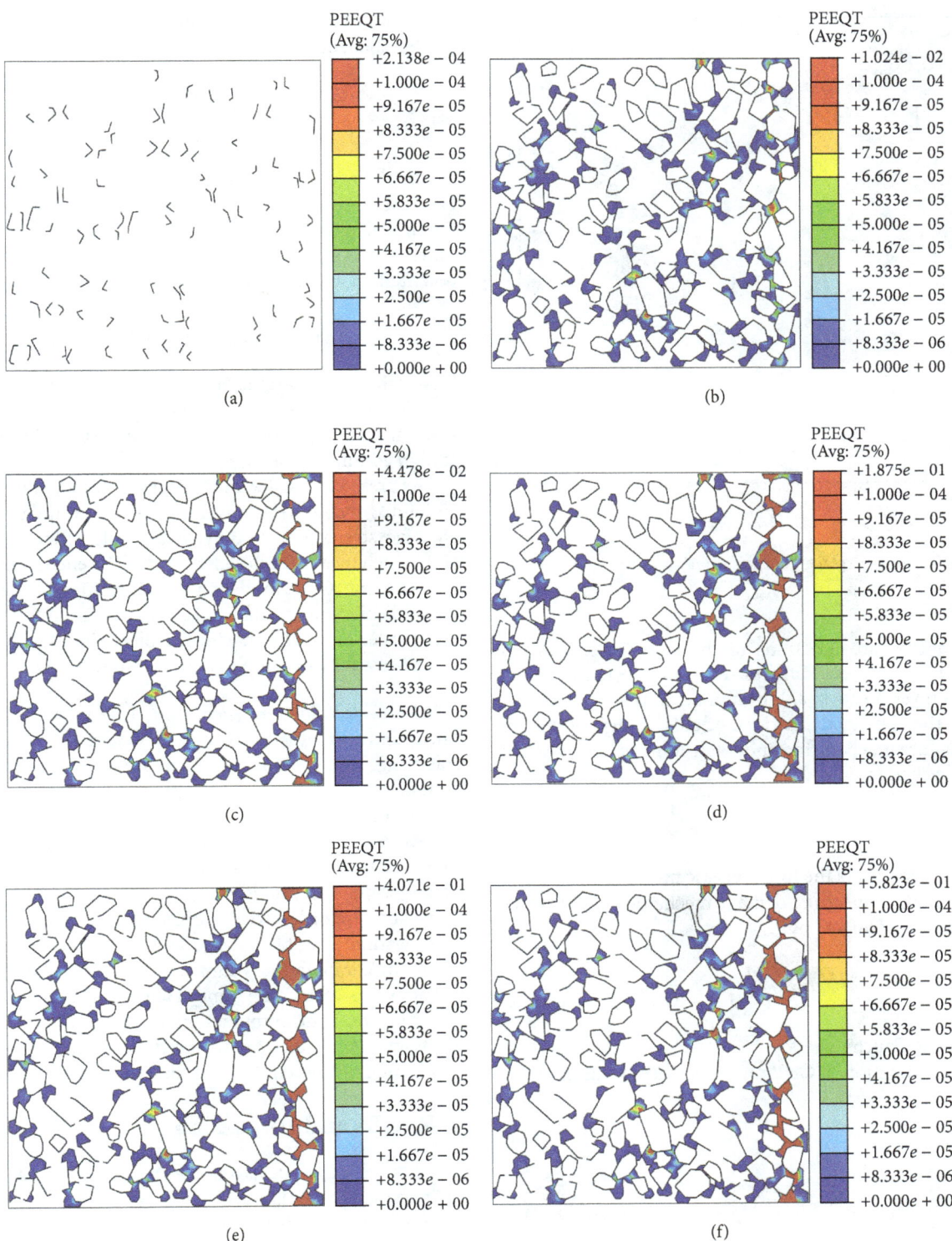

FIGURE 8: Mesoscale cracking development at different loading levels of the concrete specimen without void ($A_F = 40\%$). (a), (b), (c), (d), (e), and (f) show the equivalent tensile plastic strain distributions at points a–f, respectively.

In CDP model, two independent hardening variables, that is, equivalent compressive and tensile plastic strains $\widetilde{\varepsilon}_c^p$ and $\widetilde{\varepsilon}_t^p$, are introduced for considering compressive crushing and tensile cracking, respectively. Then, two independent damage variables, $d_c(\widetilde{\varepsilon}_c^p)$ and $d_t(\widetilde{\varepsilon}_t^p)$, are defined to represent the compressive and tensile damage states. Furthermore, in order to describe the overall damage in an isotropic manner, a scale variable d is defined as

$$d = 1 - \left(1 - s_t d_c\right)\left(1 - s_c d_t\right), \tag{3}$$

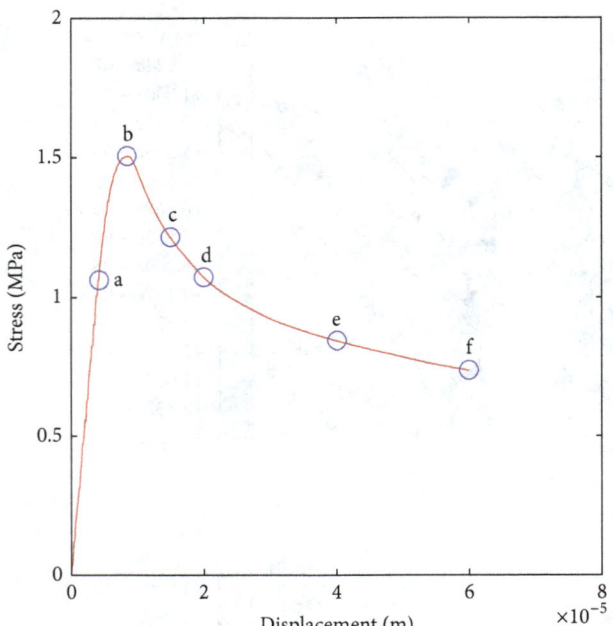

FIGURE 9: Stress-displacement curve ($A_F = 40\%$, $A_v = 2\%$).

where s_t and s_c are functions of the stress state that are introduced to represent stiffness recovery effects associated with stress reversals (see [27] for more details).

Then, the damaged elastic modulus E depending on different failure mechanisms under tension and compression can be described by

$$E = (1 - d) E_0 \qquad (4)$$

in which E_0 represents the initial elastic modulus.

Based on the concept of damage mechanics, the effective stress $\overline{\sigma}$ can be calculated by

$$\overline{\sigma} = \frac{\sigma}{1 - d}, \qquad (5)$$

where σ is the Cauchy stress.

The yield function of CDP model is given in the effective stress space as

$$F\left(\overline{\sigma}, \widetilde{\boldsymbol{\varepsilon}}^p\right)$$

$$= \frac{1}{1 - \alpha} \left(\overline{q} - 3\alpha\overline{p} + \beta\left(\widetilde{\boldsymbol{\varepsilon}}^p\right) \left\langle \widehat{\overline{\sigma}}_{\max} \right\rangle - \gamma \left\langle -\widehat{\overline{\sigma}}_{\max} \right\rangle\right) \qquad (6)$$

$$- \overline{\sigma}_c\left(\widetilde{\varepsilon}_c^p\right),$$

where $\widetilde{\boldsymbol{\varepsilon}}^p = [\widetilde{\varepsilon}_t^p \ \widetilde{\varepsilon}_c^p]^T$; $\overline{p}$ and $\overline{q}$ are the effective hydrostatic pressure and the effective Mises equivalent deviatoric stress, respectively; $\widehat{\overline{\sigma}}_{\max}$ is the algebraically maximum eigenvalue of $\overline{\sigma}$; the brackets $\langle \cdot \rangle$ are used in Macaulay sense; $\overline{\sigma}_c(\widetilde{\varepsilon}_c^p)$ is the uniaxial compressive effective strength; α and γ are dimensionless material constants, which can be determined by comparing the initial equibiaxial and uniaxial compressive yield stress and by comparing the yield conditions along the

tensile and compressive meridians, respectively; and $\beta(\widetilde{\boldsymbol{\varepsilon}}^p)$ can be calculated by

$$\beta\left(\widetilde{\boldsymbol{\varepsilon}}^p\right) = \frac{\overline{\sigma}_c\left(\widetilde{\varepsilon}_c^p\right)}{\overline{\sigma}_t\left(\widetilde{\varepsilon}_t^p\right)} (1 - \alpha) - (1 + \alpha) \qquad (7)$$

in which $\overline{\sigma}_t(\widetilde{\varepsilon}_t^p)$ is the uniaxial tensile effective strength. Figure 4 illustrates the yield surface in the case of plane stress.

In order to describe the dilatancy reasonably, nonassociated flow rule is employed in CDP model, and the flow potential takes the form as

$$G\left(\overline{\boldsymbol{\sigma}}\right) = \sqrt{\left(\epsilon\sigma_{t0} \tan\psi\right)^2 + \overline{q}^2} - \overline{p} \tan\left(\psi\right) \qquad (8)$$

in which ϵ is a parameter defining the rate at which the function approaches the asymptote; σ_{t0} is the uniaxial tensile stress at failure; ψ is the dilation angle measured in $\overline{p}$-$\overline{q}$ plane at high confining pressure.

As presented earlier, the material softening under tension is defined by the relationship between the uniaxial tensile effective strength and equivalent tensile plastic strain (see (7)), which means mesh sensitivity will be encountered when applying the CDP model in FE simulations. Therefore, a stress-displacement relation is used in this study to define the tensile softening behavior for alleviating the influence of mesh sensitivity on the simulation results.

3.4. Numerical Solution Algorithm. Due to the highly non-linear and softening behavior of concrete in the process of fracturing, the ABAQUS/Explicit solver is employed in all simulations with the aim of capturing the entire fracturing process of concrete.

As is well known, the dynamic effect inevitably exists in an explicit FE analysis, and its influence on the solution of a quasi-static problem should be small enough to be neglected. In order to minimise the dynamic effect, the loading time should be large enough, while, on the other hand, the computational effort increases proportionally with the increase of loading time. Hence, a balance has to be made between the computational efficiency and simulation accuracy, which can be achieved through comparing the results under different loading time (or loading rates).

4. Results and Discussion

4.1. Numerical Specimens and Mechanical Properties. Aiming to investigate the effects of voids on the tensile fracturing of concrete with different aggregate volume fractions, three sets of numerical concrete specimens with dimensions of 100 mm × 100 mm using polygonal aggregates are generated, and each set contains four specimens with the same aggregate arrangement and different A_v (0%, 2%, 4%, and 6%, resp.). For Set I, A_F is set to 30%, while $A_F = 40\%$ and 50% are chosen for Sets II and III, respectively. The typical element size is chosen as 0.4 times that of the minimum aggregate size; however the number of both elements and nodes of numerical specimens increases with the increase of A_v. For example, the number of elements and nodes for the specimen with $A_F = 40\%$ and $A_v = 2\%$ is 15796 and 9136, respectively,

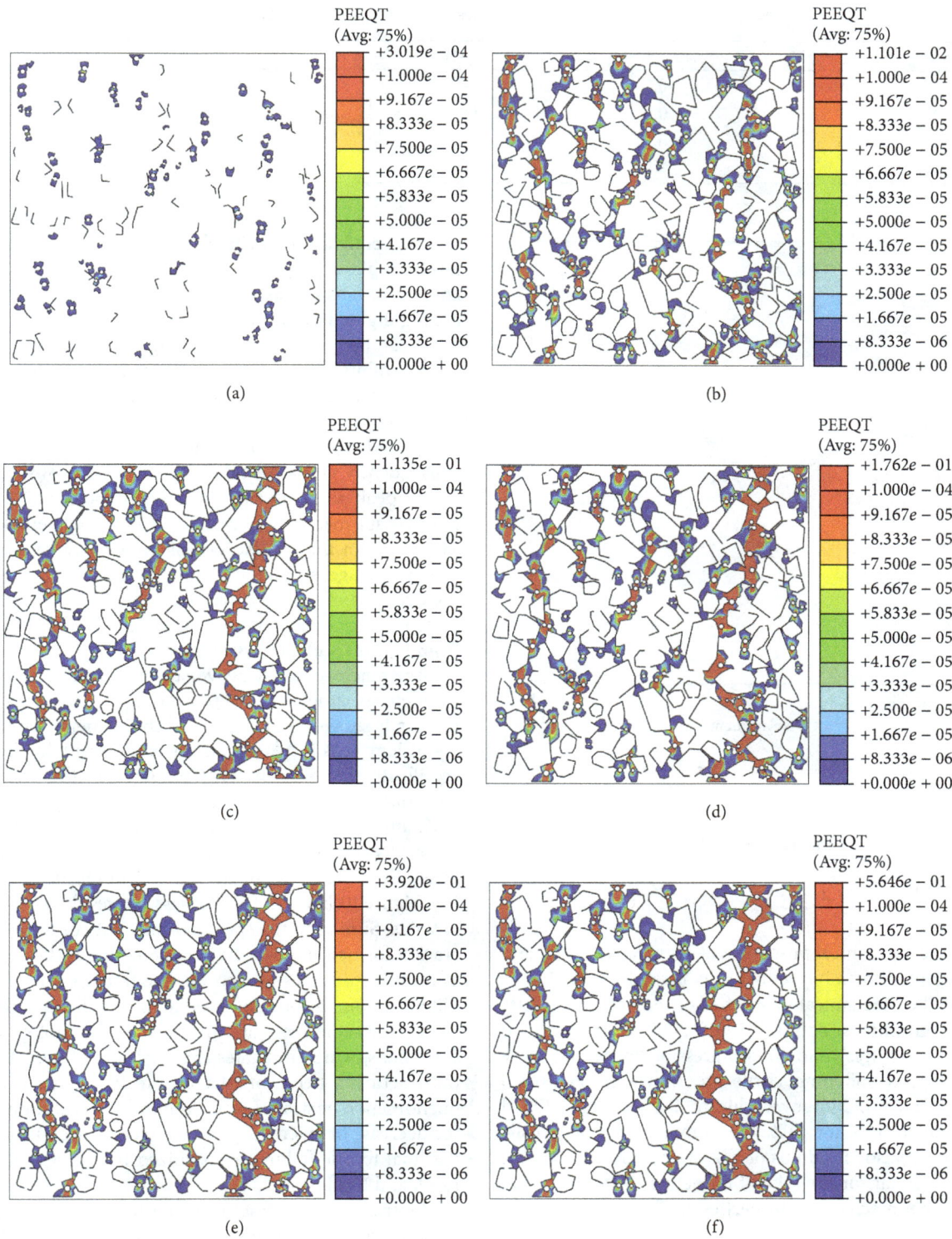

FIGURE 10: Mesoscale cracking development at different loading levels of concrete specimen with $A_F = 40\%$ and $A_v = 2\%$. (a), (b), (c), (d), (e), and (f) show the equivalent tensile plastic strain distributions at points a–f, respectively.

and the number of elements and nodes for the specimen with $A_F = 40\%$ and $A_v = 6\%$ is 20528 and 12126, respectively.

For each specimen, uniaxial tensile fracturing is simulated. In FE simulations, the left end of concrete specimen is fixed in the horizontal direction, while the opposite end is subjected to a uniformly distributed horizontal displacement up to 0.06 mm, namely, a displacement-controlled loading scheme is used. Following the strategy discussed in Section 3.4, the loading time is set to 0.012 s, which corresponds to a loading rate 5 mm/s.

TABLE 1: Mechanical properties of concrete components.

Material	E_0 (GPa)	Poisson's ratio (—)	σ_{t0} (MPa)	ϵ (—)	ψ (°)	α (—)	γ (—)
Aggregate	60	0.2	—	—	—	—	—
Mortar	20	0.2	1.94	0.1	35	0.12	2.0
ITZ	15	0.2	1.46	0.1	35	0.12	2.0
Void	—	—	—	—	—	—	—

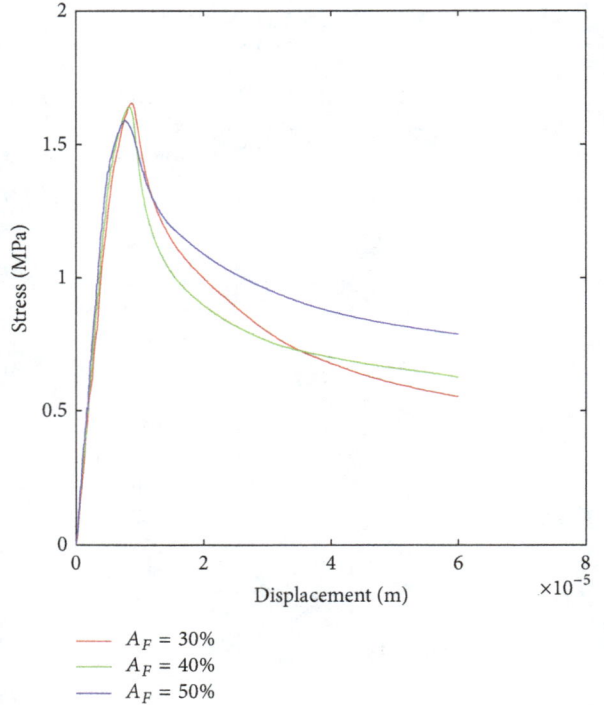

FIGURE 11: Stress-displacement curves of different aggregate volume fractions.

The same mechanical properties of aggregate, mortar, and ITZ are adopted for all specimens, as listed in Table 1. It is noted that the mechanical properties of mortar are directly obtained from Chinese code GB 50010-2002 (code for design of concrete structures), and the compression hardening curve and the tension softening curve are shown in Figures 5 and 6, respectively. Moreover, due to the lack of detailed experimental results for ITZ, the compressive and tensile strengths and elastic modulus of ITZ are assumed to be 75% of those of mortar for the sake of modeling the weaker ITZ, while the other mechanical properties of ITZ are taken as the same of mortar.

4.2. Effects of Voids on Concrete Tensile Fracturing Mechanism. In order to study the effects of voids on concrete tensile fracturing mechanism, the simulation results of two specimens in Set II, which include one without void and the other one with voids taking up 2% of the specimen volume, are discussed in detail in the following.

The macroscopic stress versus displacement curve of the concrete specimen without void ($A_F = 40\%$) under uniaxial tension is shown in Figure 7, and the elements with

nonzero equivalent tensile plastic strain at different loading levels are illustrated in Figure 8, in which the elements with equivalent tensile plastic strain bigger than 100 microstrains are highlighted in red color.

It can be observed that although concrete specimen initially exhibits elastic response on the macroscale, mesoscale cracking still occurs in ITZs (see Figure 8(a)) due to their lower fracture properties than mortar. With the increase of applied displacement, cracking develops in ITZs and subsequently propagates into mortar. The mesoscopic crack at the peak stress is shown in Figure 8(b), and it can be found that a macroscale crack close to the right end is formed as a result of the coalescence of mesoscopic cracks. In the softening stage, strain localization, which is accompanied by the decrease of macroscopic stress (see Figure 7) and finally leads to macroscopic tensile failure, can be clearly identified from Figures 8(c), 8(d), 8(e), and 8(f).

The macroscopic stress versus displacement curve of the concrete specimen ($A_F = 40\%$ and $A_v = 2\%$) under uniaxial tension is shown in Figure 9, and the elements with nonzero equivalent tensile plastic strain at different loading levels are illustrated in Figure 10. Compared to the case without void discussed above, mesoscale cracking appears in both ITZs and the mortar around voids in the macroscale elastic stage (see Figure 10(a)), and, additionally, cracked mortar generally suffers from bigger equivalent tensile plastic strain (see Figure 10(b)) at the peak stress as a result of the existence of voids. Similar to the specimen without void, a macroscale crack is formed at the peak stress point due to the coalescence of mesoscopic cracks. However, the fracture pattern is quite different from the one observed in the specimen without void even if the same aggregate arrangement is employed in both specimens, which indicates the existence of void dominates the fracturing behavior of concrete to a large extent. The phenomenon of strain localization, which is accompanied by the decrease of macroscopic stress (see Figure 9) and leads to the final tensile failure, is visualized in Figures 10(c), 10(d), 10(e), and 10(f).

Overall, different fracturing mechanisms can be observed for the two concrete specimens by comparing the development processes of mesoscale cracking shown in Figures 8 and 10. For the concrete specimen without void, mesoscale crack first appears in ITZ and then propagates into the neighboring mortar, while, for the concrete specimen with voids, mesoscale cracks are first found in both ITZ and mortar around voids and then coalesce in the fracturing process.

4.3. Effects of Void Volume Fraction. In this section, the effects of aggregate volume fraction (A_F) on concrete tensile

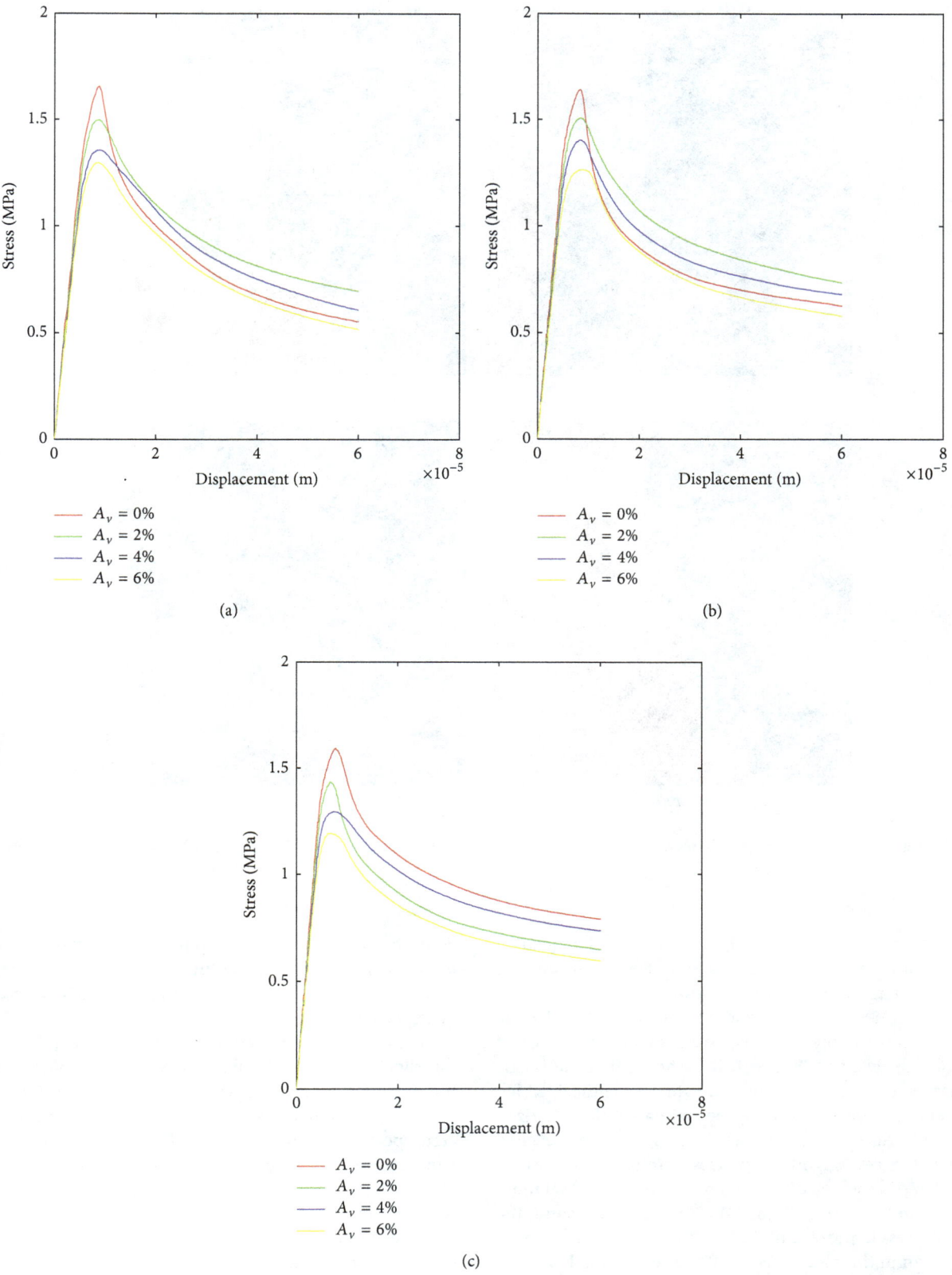

FIGURE 12: Stress-displacement curves for (a) $A_F = 30\%$; (b) $A_F = 40\%$; (c) $A_F = 50\%$.

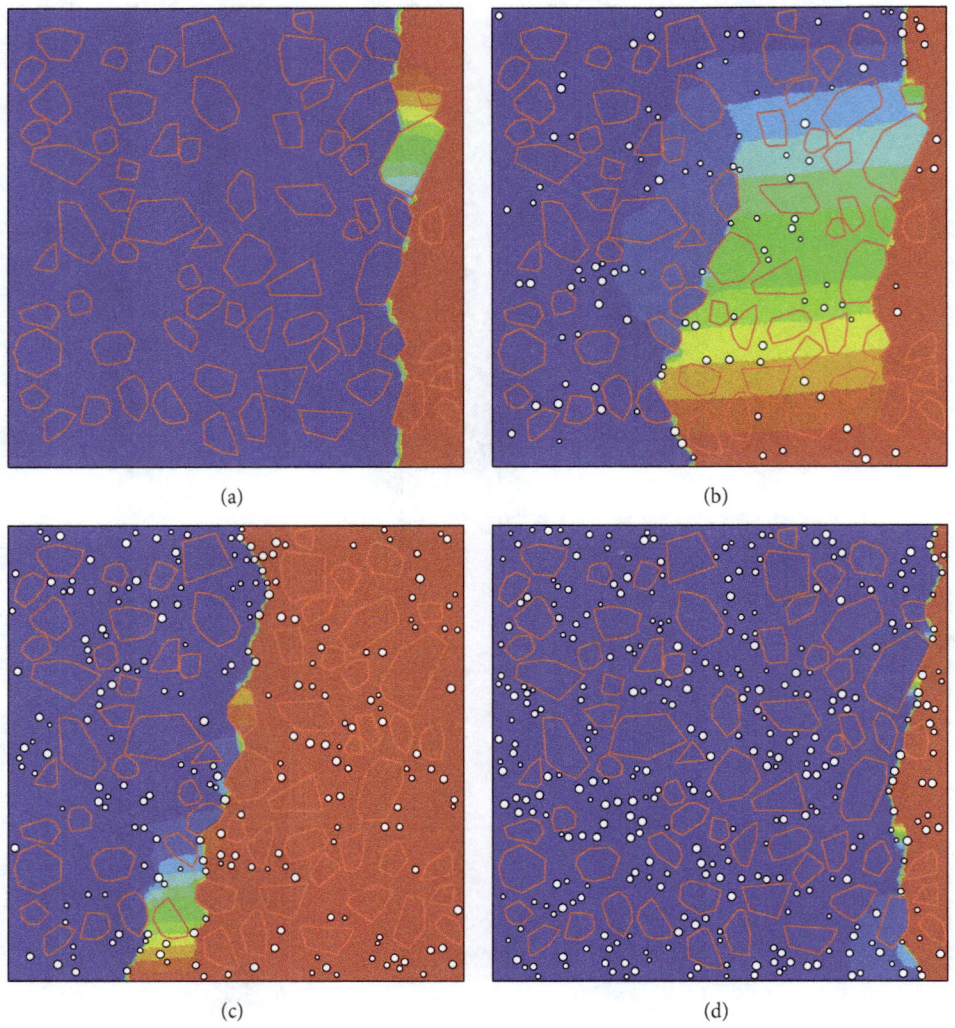

(a)   (b)

(c)   (d)

FIGURE 13: Fracture patterns of concrete specimens with A_F = 30% for (a) A_v = 0%; (b) A_v = 2%; (c) A_v = 4%; (d) A_v = 6%.

fracturing are first investigated, followed by the detailed discussion on the effects of void volume fraction on concrete with different aggregate volume fractions.

The specimens without void in Sets I, II, and III are simulated, and the macroscopic stress versus displacement curves are depicted in Figure 11. It is shown that the elastic responses of the three specimens are almost identical, which are in general linear and can be approximately characterized by the same elastic modulus. As A_F increases, the reached peak tensile stress slightly decreases (similar to the results reported in [15]), which can be attributed to the fact that more ITZs exist in the case of higher A_F. On the other hand, the postpeak stress in the case of A_F = 40% generally drops more quickly than in the case of A_F = 30%, as expected. However, in the case of A_F = 50%, the postpeak drops less quickly than in the other two cases, which may be due to the fact that this specimen finally fails with a different failure mode (two dominant macroscale cracks; see Figure 15(a)) from that of the other two specimens (one dominant macroscale crack; see Figures 13(a) and 14(a)), leading to higher fracture energy dissipated in tensile fracturing. From the above discussion,

it can be concluded that the postpeak behavior of concrete highly depends on the failure mode, namely, the more macroscale cracks appear, the higher fracture energy should be expected.

Provided the simulation results of Sets I, II, and III, the effects of void volume fraction (A_v) on concrete tensile fracturing are analyzed in the following. Figures 12(a), 12(b), and 12(c) present the stress-displacement relations of concrete specimens with A_F = 30%, 40%, and 50%, respectively, and Figures 13, 14, and 15 illustrate the fracture patterns of concrete specimens with A_F = 30%, 40%, and 50%, respectively. In Figures 12(a), 12(b), and 12(c), it is shown that the macroscale elastic modulus in prepeak stage can be considered to be independent of A_v, while the peak stress decreases as A_v increases. Moreover, the relation between A_v and the postpeak behavior is not straightforward as depicted. As shown in Figures 13, 14, and 15, the fracture patterns of concrete specimens with the same aggregate arrangement (represented by highlighted red polygons) and different A_v differ from each other. For concrete specimens without void, fracture pattern is mainly controlled by the distribution of

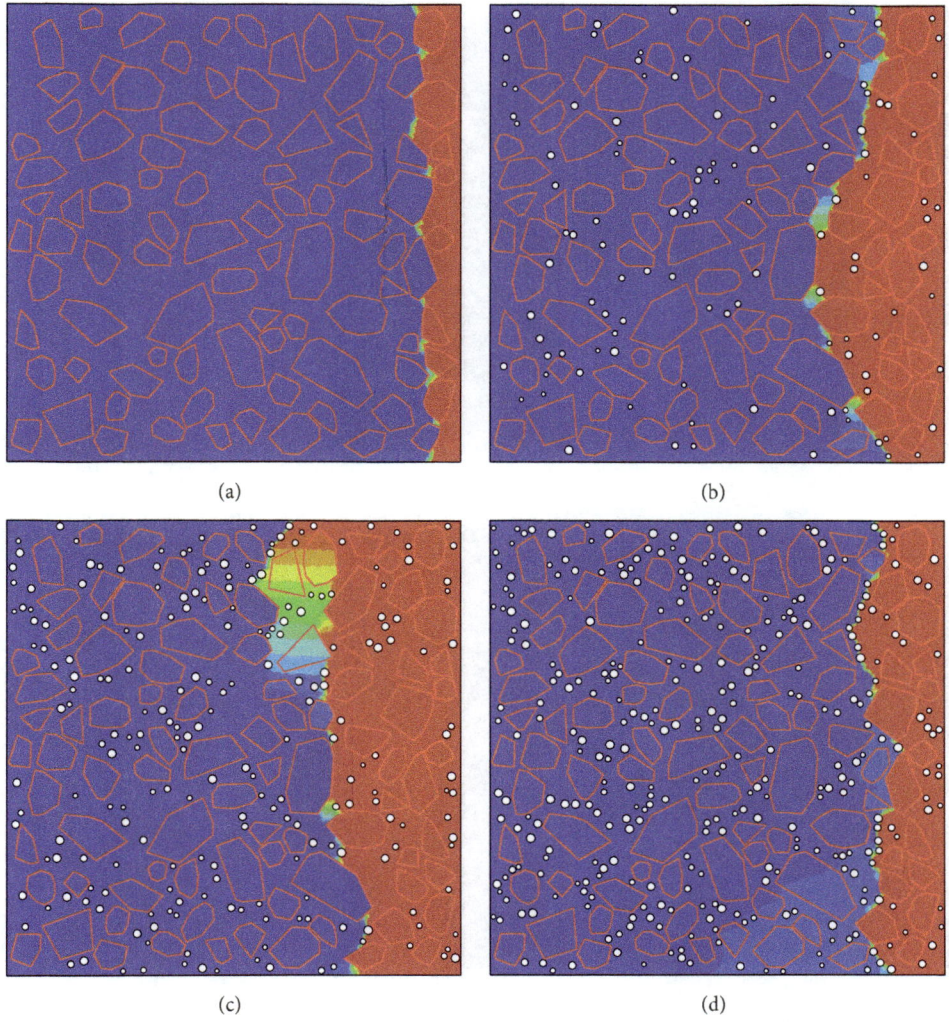

FIGURE 14: Fracture patterns of concrete specimens with $A_F = 40\%$ for (a) $A_v = 0\%$; (b) $A_v = 2\%$; (c) $A_v = 4\%$; (d) $A_v = 6\%$.

aggregates (or ITZs), whereas, for concrete specimens with voids, the fracture pattern is dominated not only by the aggregate arrangement but also by the distribution of voids in mortar, leading to the differences of fracture patterns illustrated in Figures 13, 14, and 15. Consequently, accurate fracturing analysis of concrete calls for the explicit modeling of voids. Furthermore, both the concrete specimen without void and the concrete specimens with voids present two types of failure modes: one is with a single macroscale crack (Type I, typically shown in Figure 15(b)), and the other is with two macroscale cracks (Type II; see Figure 13(b), e.g.). Owing to the longer crack length, the higher fracture energy is dissipated in Type II, and therefore the postpeak stress-displacement curve of Type II drops less quickly than that of Type I, which can be found in Figure 12.

5. Conclusions

A finite element modeling strategy of concrete with random mesostructure explicitly taking void into consideration has

been proposed in the present work. The tensile fracturing mechanism of concrete with voids is detailed on the mesoscale by comparing the simulation results of two specimens consisting of one without void and the other one with voids with the same aggregate arrangement. Then, several simulations are carried out with prime attention placed on the effects of void volume fraction on concrete tensile fracturing. The main conclusions are as follows:

(i) Different fracturing mechanisms are observed for the two concrete specimens with the same aggregate arrangement including one without void and the other one with voids, and the fracture pattern of concrete specimen with voids is controlled by both the aggregate arrangement and the distribution of voids.

(ii) Compared to aggregate volume fraction, void volume fraction has a larger influence on concrete tension strength. The elastic modulus of concrete in the prepeak stage can be considered to be independent

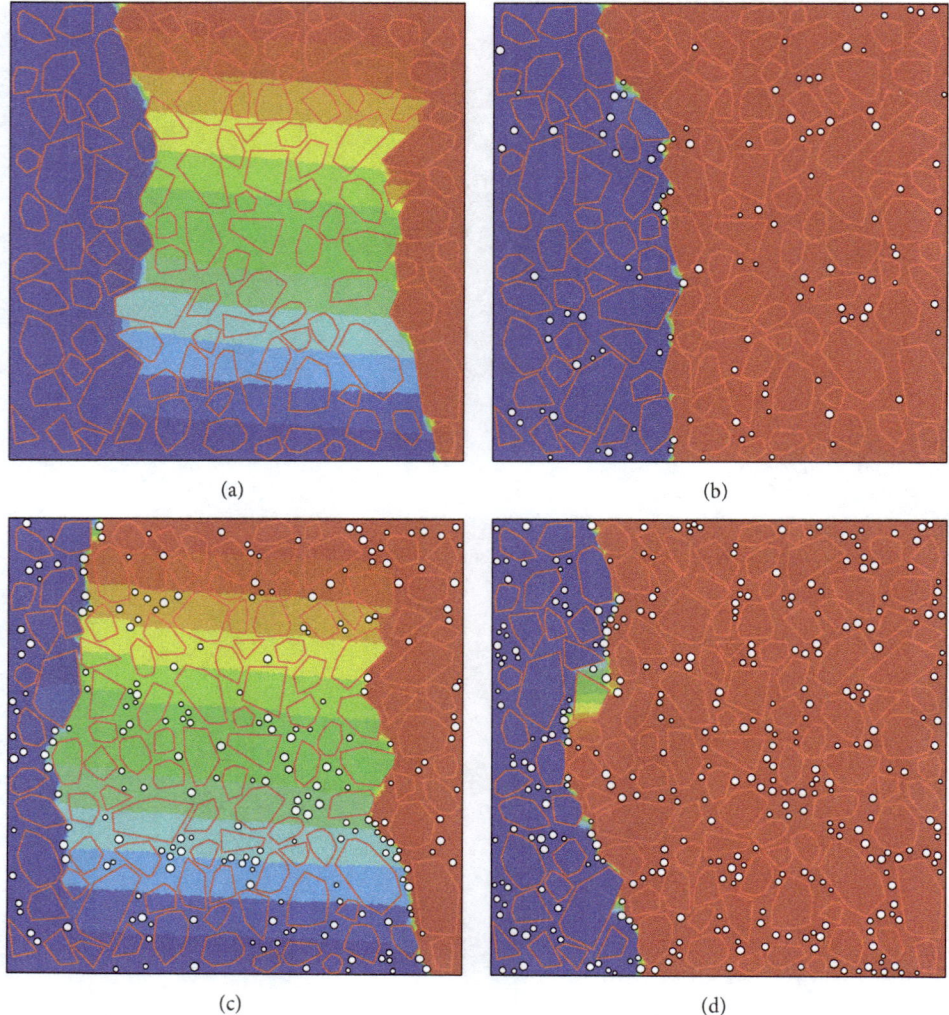

FIGURE 15: Fracture patterns of concrete specimens with $A_F = 50\%$ for (a) $A_v = 0\%$; (b) $A_v = 2\%$; (c) $A_v = 4\%$; (d) $A_v = 6\%$.

of both aggregate volume fraction and void volume fraction.

(iii) The relation between concrete postpeak behavior and void volume fraction is not straightforward due to the randomness of void distribution and the resulting fracture pattern.

(iv) Two types of failure modes are identified for concrete specimens under uniaxial tension, in which Type I is characterized by a single macroscale crack and Type II by two. Due to the longer crack length, the postpeak stress of Type II drops less quickly than that of Type I.

(v) It is necessary to model void explicitly for the accurate fracturing simulation of concrete on the mesoscale.

Conflicts of Interest

The authors declare that there are no conflicts of interest regarding the publication of this paper.

Acknowledgments

This work was supported by projects of the National Natural Science Foundation of China (Grants nos. 11132003 and 51109067).

References

[1] J. F. Unger and S. Eckardt, "Multiscale Modeling of Concrete," *Archives of Computational Methods in Engineering*, vol. 18, no. 3, pp. 341–393, 2011.

[2] I. M. Gitman, H. Askes, and L. J. Sluys, "Coupled-volume multi-scale modelling of quasi-brittle material," *European Journal of Mechanics-A/Solids*, vol. 27, no. 3, pp. 302–327, 2008.

[3] G. Cusatis, Z. P. Bažant, and L. Cedolin, "Confinement-shear lattice CSL model for fracture propagation in concrete," *Computer Methods in Applied Mechanics and Engineering*, vol. 195, no. 52, pp. 7154–7171, 2006.

[4] R. Rezakhani and G. Cusatis, "Asymptotic expansion homogenization of discrete fine-scale models with rotational degrees of freedom for the simulation of quasi-brittle materials," *Journal*

of the Mechanics and Physics of Solids, vol. 88, pp. 320–345, 2016.

[5] M. Nitka and J. Tejchman, "Modelling of concrete behaviour in uniaxial compression and tension with DEM," *Granular Matter*, vol. 17, no. 1, pp. 145–164, 2015.

[6] A. Caballero, I. Carol, and C. López, "A meso-level approach to the 3D numerical analysis of cracking and fracture of concrete materials," *Fatigue and Fracture of Engineering Materials and Structures*, vol. 29, no. 12, pp. 979–991, 2006.

[7] P. Wriggers and S. O. Moftah, "Mesoscale models for concrete: homogenisation and damage behaviour," *Finite Elements in Analysis and Design*, vol. 42, no. 7, pp. 623–636, 2006.

[8] C. M. López, I. Carol, and A. Aguado, "Meso-structural study of concrete fracture using interface elements. I: numerical model and tensile behavior," *Materials and Structures/Materiaux et Constructions*, vol. 41, no. 3, pp. 583–599, 2008.

[9] I. Comby-Peyrot, F. Bernard, P.-O. Bouchard, F. Bay, and E. Garcia-Diaz, "Development and validation of a 3D computational tool to describe concrete behaviour at mesoscale. Application to the alkali-silica reaction," *Computational Materials Science*, vol. 46, no. 4, pp. 1163–1177, 2009.

[10] P. Grassl and M. Jirásek, "Meso-scale approach to modelling the fracture process zone of concrete subjected to uniaxial tension," *International Journal of Solids and Structures*, vol. 47, no. 7-8, pp. 957–968, 2010.

[11] S. Shahbeyk, M. Hosseini, and M. Yaghoobi, "Mesoscale finite element prediction of concrete failure," *Computational Materials Science*, vol. 50, no. 7, pp. 1973–1990, 2011.

[12] L. Snozzi, A. Caballero, and J. F. Molinari, "Influence of the meso-structure in dynamic fracture simulation of concrete under tensile loading," *Cement and Concrete Research*, vol. 41, no. 11, pp. 1130–1142, 2011.

[13] X. Du, L. Jin, and G. Ma, "Numerical simulation of dynamic tensile-failure of concrete at meso-scale," *International Journal of Impact Engineering*, vol. 66, pp. 5–17, 2014.

[14] Y. Huang, Z. Yang, W. Ren, G. Liu, and C. Zhang, "3D meso-scale fracture modelling and validation of concrete based on in-situ X-ray Computed Tomography images using damage plasticity model," *International Journal of Solids and Structures*, vol. 67-68, pp. 340–352, 2015.

[15] X. F. Wang, Z. J. Yang, J. R. Yates, A. P. Jivkov, and C. Zhang, "Monte Carlo simulations of mesoscale fracture modelling of concrete with random aggregates and pores," *Construction and Building Materials*, vol. 75, pp. 35–45, 2015.

[16] X. Wang, M. Zhang, and A. P. Jivkov, "Computational technology for analysis of 3D meso-structure effects on damage and failure of concrete," *International Journal of Solids and Structures*, vol. 80, pp. 310–333, 2016.

[17] A. Zubelewicz and Z. P. Bažant, "Interface element modeling of fracture in aggregate composites," *Journal of Engineering Mechanics*, vol. 113, no. 11, pp. 1619–1630, 1987.

[18] Z. P. Bažant, M. R. Tabbara, M. T. Kazemi, and G. Pyaudier-Cabot, "Random particle model for fracture of aggregate or fiber composites," *Journal of Engineering Mechanics*, vol. 116, no. 8, pp. 1686–1705, 1990.

[19] E. Schlangen and J. G. M. van Mier, "Simple lattice model for numerical simulation of fracture of concrete materials and structures," *Materials and Structures*, vol. 25, no. 9, pp. 534–542, 1992.

[20] E. Schlangen and E. J. Garboczi, "Fracture simulations of concrete using lattice models: computational aspects," *Engineering Fracture Mechanics*, vol. 57, no. 2-3, pp. 319–332, 1997.

[21] G. Cusatis, D. Pelessone, and A. Mencarelli, "Lattice Discrete Particle Model (LDPM) for failure behavior of concrete. I: Theory," *Cement and Concrete Composites*, vol. 33, no. 9, pp. 881–890, 2011.

[22] G. Cusatis, A. Mencarelli, D. Pelessone, and J. Baylot, "Lattice Discrete Particle Model (LDPM) for failure behavior of concrete. II: calibration and validation," *Cement and Concrete Composites*, vol. 33, no. 9, pp. 891–905, 2011.

[23] F. Chen and P. Qiao, "Probabilistic damage modeling and service-life prediction of concrete under freeze–thaw action," *Materials and Structures/Materiaux et Constructions*, vol. 48, no. 8, pp. 2697–2711, 2014.

[24] Z. Wang, A. Kwan, and H. Chan, "Mesoscopic study of concrete I: generation of random aggregate structure and finite element mesh," *Computers and Structures*, vol. 70, no. 5, pp. 533–544, 1999.

[25] V. Palmieri and L. De Lorenzis, "Multiscale modeling of concrete and of the FRP-concrete interface," *Engineering Fracture Mechanics*, vol. 131, pp. 150–175, 2014.

[26] J. Lubliner, J. Oliver, S. Oller, and E. Oñate, "A plastic-damage model for concrete," *International Journal of Solids and Structures*, vol. 25, no. 3, pp. 299–326, 1989.

[27] J. Lee and G. L. Fenves, "Plastic-damage model for cyclic loading of concrete structures," *Journal of Engineering Mechanics*, vol. 124, no. 8, pp. 892–900, 1998.

13

Fatigue Performance of SFPSC under Hot-Wet Environments and Cyclic Bending Loads

Shanshan Luo [ID],[1] Peiyan Huang [ID],[1,2] Xinyan Guo [ID],[1] and Xiaohong Zheng [ID][1]

[1]*School of Civil Engineering and Transportation, South China University of Technology, Guangzhou 510640, China*
[2]*State Key Laboratory of Subtropical Building Science, South China University of Technology, Guangzhou 510640, China*

Correspondence should be addressed to Peiyan Huang; pyhuang@scut.edu.cn

Academic Editor: Antonio Gilson Barbosa de Lima

A new structural material named "steel fiber polymer structural concrete (SFPSC)" with features of both high strength and high toughness was developed by this research group and applied to the bridge superstructures in the hot-wet environments. In order to investigate the fatigue performance and durability of SFPSC under hot-wet environments, the environment and fatigue load uncoupling method and the coupling action of environment and fatigue load were used or developed. Three-point bending fatigue experiments with uncoupling action of environments and cyclic loads were carried out for SFPSC specimens which were pretreated under hot-wet environments, and the experiments with the coupling action of environments and cyclic loads for SFPSC specimens were carried out under hot-wet environments. Then, the effects of hot-wet environments and the experimental methods on the fatigue mechanism of SFPSC material were discussed, and the environmental fatigue equations of SFPSC material under coupling and uncoupling action of hot-wet environments and cyclic bending loads were established. The research results show that the fatigue limits of SFPSC under the coupling action of the environments and cyclic loads were lower about 15%. The proposed fatigue equations could be used to estimate the fatigue lives and fatigue limits of SFPSC material.

1. Introduction

The durability of bridge structures in service has been a frontier research topic in civil engineering field. As a significant part of the durability issues, the study on environmental fatigue performance of structural materials often has decisive significance for assuring the safety and durability of the bridge structures. With the increase of the span of concrete bridge, the common high-strength concrete materials could no longer satisfy the comprehensive requirements of load-bearing capacity, anticracking, antideformation, and durability [1–3], and there is an urgent need to develop a new material with features of high strength and high ductility.

Steel fiber reinforced concrete (SFRC) was a compound with high strength and high ductility developed in the 20th century. After modifications and developments in many decades, SFRC had been gradually applied in the various kinds of building structures [4–6]. In the transportation field, SFRC was used mostly for constructions such as the road pavements, bridge deck pavements, and airport runway pavements. However, in many cases, it had been discovered during the observations of damaged sections of the SFRC specimens that the steel fibers in the damaged sections were broken by being "pulled out" and very few by being "broken." This was due to insufficient cohesive force between the matrixes of the steel fibers and concrete and thus resulted in the low stress levels of the steel fibers during the damage of the specimens, and their tensile properties had not been fully realized. In order to further increase the bonding strength of the steel fiber and concrete and give full play to the good mechanical properties of the steel fibers, Luo et al. [7] of this research group added polymer latex into SFRC and developed the "steel fiber reinforced polymer concrete (SFRPC)." The experimental results showed that the new material had more superior antitensile, antibending, anti-fatigue, and antiimpact properties [8–10]. SFRPC belongs to medium to low strength modified concrete. Although it had been successfully applied in the construction of highway road pavements and bridge deck pavements [11], due to its insufficient strength, it was still unable to be used for main load-bearing members such as the bridge superstructures. Therefore, this research group conducted secondary development on SFRPC and developed a new high-strength concrete composite material named "steel fiber polymer structural concrete (SFPSC)." After the experiments of systematic mechanical properties and the optimum designs of the structures [12–15], SFPSC was successfully applied in the main girders (box girders) of three large-span concrete continuous rigid frame bridges on two highways in Guangdong Province, China [16, 17].

Regardless of SFRC, SFRPC, or SFPSC, there had been scarce reports about the research results of the environmental fatigue performance/durability of the concrete composite materials containing steel fibers. With regards to the bridge structures servicing in subtropical areas such as Guangdong Province, China, it is necessary to investigate its antifatigue performance/durability under hot-wet environments. Therefore, considering the actual weather conditions in the subtropical areas, the experimental studies on fatigue performance of SFPSC applied in bridges under different temperatures were carried out by this research group [14, 15], and a thermal fatigue equation for SFPSC was proposed. However, the effect of the humidity was not taken into the consideration.

In addition, for the bridge superstructures servicing in a hot-wet environment, the environment and vehicle loads are interacted, and thus their working conditions are different from the traditional environment and load uncoupling experiments. The differences in the environmental fatigue performance/durability of materials and components under uncoupling and coupling action are a scientific problem that needs to be proved but has not been proved. Therefore, in this paper, the superstructures of SFPSC bridges servicing in subtropical areas were taken as the research objects, and the environmental fatigue experimental researches and theoretical analyses were carried out based on the hot-wet environments and cyclic loads coupling and uncoupling (traditional) experimental methods, and the environmental fatigue performance/durability of SFPSC under different hot-wet conditions was discussed.

2. Fatigue Experiments under Hot-Wet Environments

In order to study the environmental fatigue performance/durability of SFPSC material under coupling and uncoupling action of hot-wet environments and cyclic loads and to prove the mechanism of the effect brought by hot-wet environments on the fatigue performance of SFPSC, in this research, three types of fatigue experiments were designed:

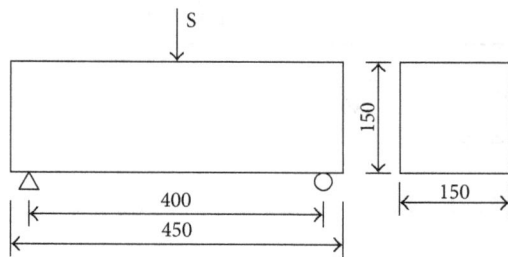

FIGURE 1: Three-point bending SFPSC beam for fatigue test (unit: mm).

(1) Fatigue experiments under room temperature and atmospheric environment. (2) Fatigue experiments under uncoupling action of hot-wet environments and cyclic loads. This is the traditional method for environmental fatigue tests. SFPSC specimens were pretreated in the hot-wet environments, and then the specimens were transferred to the test machine to carry out the fatigue experiments under room temperature and atmospheric environment. (3) Fatigue experiments under coupling action of hot-wet environments and cyclic loads. The experiments were carried out by the special testing device, and the specimens were treated by the hot-wet environments while the cyclic loads were applied.

2.1. Experimental Materials and Specimens. The specimens used for fatigue experiments were all the same and were three-point bending beams as shown in Figure 1. The specimen materials were based on the design requirements of a long-span bridge girder (box girder) of a highway in Guangdong Province and were designed for 2 types of materials: C55 concrete and SFPSC. The size of a specimen was length 450 mm × width 150 mm × height 150 mm, and the span length was $L = 400$ mm. There were total of 66 specimens produced, including five C55 concrete specimens and 61 SFPSC specimens. The usages of all specimens were as follows: five C55 concrete and five SFPSC specimens were used to test the static mechanical properties of the two types of materials. The remaining 56 SFPSC specimens were used in fatigue experiments, including 10 specimens used for the room temperature and atmospheric environmental tests (controlling experiments), 20 specimens used for the experiments under uncoupling action of hot-wet environments and cyclic loads, and 26 specimens used for the experiments under coupling action of hot-wet environments and cyclic loads.

Based on the design requirements of the superstructures (box girders) of the above bridges, the compressive strength of SFPSC should be slightly higher than that of C55 concrete. Therefore, the C55 concrete used in the experiments was designed with the standard mix ratio. The designs of the mix ratios of SFPSC should be based on the strength requirements of the structures (close to the static mechanical properties of C55 concrete); then, consider the quantity of the steel fiber to be added, the water/cement ratio, sand rate, water consumption of unit volumes, the amount of gel materials, and so on. Considerations should also be given to

TABLE 1: Static mechanical properties of SFPSC and C55 concrete.

Materials	Steel fiber (v%)	Polymer latex (wt.%)	Static mechanical properties		
			Compressive strength (MPa)	Flexural strength (MPa)	Ultimate bearing capacity (P_u/kN)
C55	0	0	69.0	5.36	36.0
SFPSC	0.580	1.30	70.1	8.68	40.4

the construction of the composite materials, such as liquidity (slump) and economy. Finally, the mix ratio of SFPSC was confirmed to be based on the standard mixture ratio, which was determined as "w(cement) : w(sand) : w(gravel) : w (water) : w(water reducer) = 1 : 1.493 : 2.432 : 0.321 : 0.012," and steel fiber of concrete volume rate of 0.58 v% and polymer latex of cement weight of 1.3 wt.% were mixed.

The details of experimental materials used were as follows: the cement was Huizhou Ta brand P.II42.5R cement; the sand was Dongjiang River natural washed-out sand, with a modulus of fineness of 2.68; the stone was 5~25 mm granite gravels; the steel fiber was the end-hook RS06073/35-800 indentation steel fiber manufactured by Shanghai Zhenjiang Fiber Co., Ltd., with the specification of 0.3 × 1 × 30 mm, the length/diameter ratio of 48, and the tensile strength of 800 MPa; for the polymer latex, the VINNAPAS RE5010N cement modified special styrene butadiene latex produced by German Wacker Co., Ltd., which was kind of vinyl acetate/ethylene copolymerized emulsion powder that can be redispersed when it comes into contact with water; and the water reducer was JB-ZSC polycarboxylate concrete admixtures.

The experimental results (average) of 28 d static mechanical properties of SFPSC and C55 concrete which were designed using the above mix ratios are shown in Table 1.

2.2. *Environmental Fatigue Experimental Methods.* There were three types of fatigue experiments with different environment. Fatigue experiments for group A specimens (controlling experiments) were carried out under room temperature and the atmospheric environment. In this type of the experiments, the experimental environment was recorded by measuring the temperature and humidity in the lab room during the experimental period and taking the average value (23°C and 78% RH). The methods of environmental treatment and the required devices were different for the fatigue experiments with uncoupling and coupling of hot-wet environments and cyclic loads. The two environmental treatment methods and the loading methods of all fatigue experiments were described as follows.

2.2.1. *Environmental Pretreatment for Uncoupling Fatigue Experiments.* For the environmental fatigue experiments, the determination of the hot-wet environments were based on the actually measured working temperature and humidity of servicing bridges in Guangdong region, China, and its higher value (severe environment) was taken. Therefore, the experimental environments were divided into 3 groups

B~D for the fatigue experiments under uncoupling action of the hot-wet environments and the cyclic loads. Among them, the experimental temperature and humidity for 7 specimens in group B was (50°C and 80% RH), for 6 specimens in group C was (50°C and 90% RH), and for 7 specimens in group D was (50°C and 95% RH).

Regarding the specimens of groups B~D, the environment pretreatment methods were conducted with reference to the relevant stipulations of Chinese national technical standard GB/T 2573-2008 "Test method for aging properties of glass fiber reinforced plastics" [18]. Firstly, the specimens were placed in temperature and humidity test chamber (Type Number Q8-901) and pretreated with constant temperature and humidity for 6 days (144 hours), and then the specimens were removed from the environmental box to be air-cooled for 2 days to ensure the dryness of the specimens. Finally, the fatigue experiments were carried out under room temperature and atmospheric environment.

2.2.2. *Hot-Wet Environments for Coupling Fatigue Experiments.* Regarding the fatigue experiments of coupling action of hot-wet environments and cyclic loads, the hot-wet environments used were the same with the above uncoupling fatigue experiments and were also divided into three groups. Among them, the temperature and humidity for 9 specimens in group E were 50°C and 80% RH, for 12 specimens in group F were 50°C and 95% RH, and for 5 specimens in group G were 35°C and 95% RH.

Regarding specimens in group E~G, the environmental treatments were completed by using the intelligent environment simulation and the controlling system, which was designed with the MTS810 material testing system by this research group (Figure 2) [19]. In order to ensure that the internal and external hot-wet conditions of the specimens were consistent at the beginning of loading action, the specimens of group E, F, and G were placed in the working box of the environment simulation and the controlling system in advance. Before the loading began, the environment simulation and controlling system was turned on for 4 hours to conduct pretreatment in hot-wet environment to reduce the influence of fatigue life by the difference of internal and external temperature and humidity of SFPSC specimen.

2.2.3. *Fatigue Loading Method*
  (1) Loading way: the loading ways were the same for all fatigue experiments. Three-point bending loading

FIGURE 2: Intelligent environment simulation and controlling system.

way, stress controlling method, and sinusoid wave were applied; the stress ratio was $R = 0.1$. However, the loading frequencies were different in various experimental methods. Among them, the loading frequency for room temperature and atmospheric condition and for the uncoupling action of hot-wet environments and cyclic loads was fixed at 10 Hz. For the fatigue experiments with coupling action of hot-wet environments and cyclic loads, the loading frequency was reduced to 2 Hz for better coupling action of the environments and loads.

Based on the design specifications of the common concrete structures [19], if the number of loading cycles N reached 2×10^6 times, and the specimen remained unfailed, the fatigue experiment would be stopped. In other words, if the specimen still remained unfailed after 2×10^6 times of loading cycles, then it would be assumed that the specimen could endure infinite times of cyclic loads, and thus the fatigue life would be assumed to be infinite.

(2) Loading (stress) level: in order to obtain the complete fatigue curve ($S \sim N$ curve), the loading (stress) levels S_R of group A specimens were divided into three grades ($S_R = P_{max}/P_u = 0.70, 0.75,$ and 0.80). Among them, P_{max} was the maximum cyclic load, and P_u was the ultimate bearing capacity of SFPSC specimens (Table 2).

For the specimens of uncoupling action of environments and loads, the loading (stress) level S_R was set to 3 levels. Among them, group B was $S_R = 0.65, 0.70,$ and 0.75, group C was $S_R = 0.60, 0.65,$ and 0.75, and group D was $S_R = 0.60, 0.64,$ and 0.68.

For the specimens of coupling action of environments and loads, the loading (stress) level S_R was set to 4 levels. Among them, group E was $S_R = 0.6, 0.65, 0.70,$ and 0.75, group F was $S_R = 0.60, 0.67, 0.72,$ and 0.77, and group G was $S_R = 0.60, 0.70, 0.72,$ and 0.75.

The experimental conditions of group A~D specimens are shown in Table 2, and the experimental conditions of group E~G group specimens are shown in Table 3.

2.2.4. Data Acquisition and Recording. In the fatigue experiments, environmental data such as temperature and humidity were collected and recorded by the intelligent environment simulation and controlling system and are shown in Figure 2. The maximum and minimum loads, the deflections of the specimens (displacements), loading cycle number N, and other experimental data were automatically collected and recorded by the MTS810 testing system, and 8 sets of data were recorded for each loading cycle. The strain data of concrete were automatically measured and recorded by the Wavebook 516E dynamic strain device with 4 strain gages attached in the middle and bottom of the center and the lower edges of the sides of the specimen, and the sampling frequency was 100 Hz.

3. Fatigue Performance under Hot-Wet Environments

3.1. Experimental S~N Curves

3.1.1. Experimental S~N Curves under Uncoupling Action. According to the experimental conditions and the experimental methods described in Section 2, the fatigue experiments were carried out under the uncoupling action of hot-wet environments and cyclic loads for SFPSC specimens of group B~D. The controlling experiments were also carried out for group A specimens under room temperature and atmospheric environment. The experimental results of group A~D specimens are shown in Table 2. Based on the fatigue test data, the fatigue experimental curves ($S_R \sim N$ curves) represented by the loading level (stress level) S_R were obtained, as shown in Figure 3.

By using the least square method, the experimental data of the specimens of each group were fitted, respectively, and the equations of $S \sim N$ curves could be determined from the logarithmic coordinates:

$$S_R = A_i + B_i \lg(N) \quad (i = 1, 2, 3, 4), \quad (1)$$

where $S_R = P_{max}/P_u$; i indicates the group number of the specimens, and $i = 1~4$, respectively, represents group A~D;

TABLE 2: Experimental conditions and results of group A~D specimens.

Specimen number	Temperature, T (°C)	Relative humidity, H (%RH)	Stress level, $S_R = P_{\max}/P_u$	Loading frequency, f (Hz)	Fatigue lives, N_f/cycles
A1					>2000000
A2			0.70		>2000000
A3					825899
A4					293301
A5			0.75		328359
A6	23	78		10	349997
A7					>2000000
A8					24099
A9			0.80		35677
A10					36071
B1			0.65		>2000000
B2					1800031
B3					63470
B4	50	80	0.70	10	207900
B5					157413
B6			0.75		92486
B7					50936
C1			0.60		>2000000
C2					1784833
C3	50	90	0.65	10	166591
C4					186596
C5			0.75		7940
C6					1960
D1			0.60		>2000000
D2					>2000000
D3			0.64		22236
D4	50	95		10	62746
D5					21355
D6			0.68		3647
D7					15173

and A_i and B_i were constant coefficients, and their values are shown in Table 4.

As shown in Figure 3, the fatigue performance of SFPSC was affected significantly by the pretreatment under hot-wet environments. In contrast to the specimens under room temperature and atmospheric environment (group A), the fatigue lives of all SFPSC specimens (group B~D) which were pretreated with hot-wet environments have been remarkably decreased. Furthermore, the more severe the hot-wet environment (the higher the temperature and humidity), the less the fatigue life was shown, that is, under the same stress level, the shorter the fatigue lives of the specimens. However, for the specimens of group C (50°C and 90% RH) and group D (50°C and 95% RH), the differences of the experimental S_R~N curves were small. The reason could be that in hot-wet environments, the fatigue performance of SFPSC specimens was mainly influenced by high temperature (50°C), and the relative humidity above 90% RH could cause the saturation of the moisture absorption ability of SFPSC specimens.

3.1.2. Experimental S~N Curves under Coupling Action. According to the experimental conditions and experimental methods described in Section 2, the fatigue experiments were carried out under coupling action of hot-wet environment and cyclic loads for group E~G SFPSC specimens, and the results are shown in Table 3. Based on the fatigue test data of each group, the fatigue experimental curves (S_R~N curves) represented by the loading level (stress level) S_R were obtained, as shown in Figure 4. For the convenience of comparison, the experimental results of SFPSC specimens under room temperature and atmospheric environment (group A) were also shown in the same figure.

Using (1) and paying attention to $i = 1, 2, 3$, and using the least square method, the experimental data of the specimens of each group E~G were fitted respectively. The coefficients A_i and B_i of each S_R~N equation could be determined, and their values are shown in Table 5.

With regard to the fatigue performance of SFPSC material under coupling action of hot-wet environments

TABLE 3: Experimental conditions and results of group E~G specimens.

Specimen number	Temperature, T (°C)	Relative humidity, H (%RH)	Stress level, $S_R = P_{max}/P_u$	Loading frequency, f (Hz)	Fatigue lives, N_f/cycles
E1					1678342
E2			0.60	2	1048011
E3					410854
E4					470951
E5	50	80	0.65	2	338499
E6					336204
E7			0.70	2	249136
E8			0.75	2	3082
E9					1733
F1			0.60	2	>2000000
F2					>2000000
F3					3106
F4			0.67	2	13237
F5					18525
F6	50	95			50766
F7					4266
F8			0.72	2	15037
F9					16241
F10					223
F11			0.77	2	906
F12					151
G1			0.60		>2000000
G2			0.70		>2000000
G3	35	95	0.72	2	39535
G4					27171
G5			0.75		1549

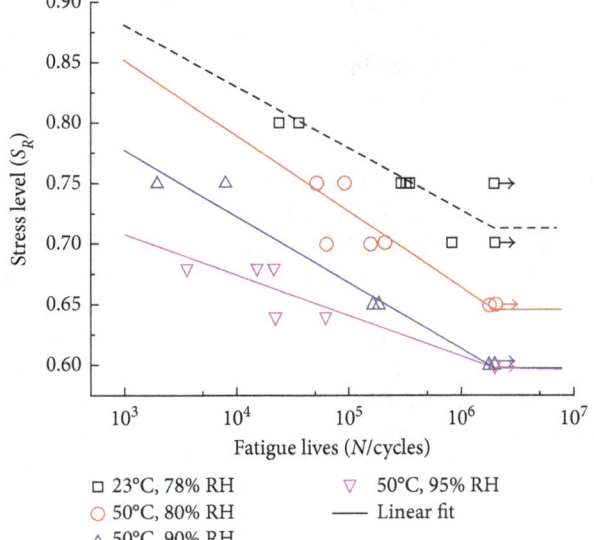

FIGURE 3: Experimental S_R~N curves under uncoupling of hot-wet environments and cyclic loads.

- □ 23°C, 78% RH
- ○ 50°C, 80% RH
- △ 50°C, 90% RH
- ▽ 50°C, 95% RH
- —— Linear fit

TABLE 4: Coefficients A_i and B_i for group A~D specimens.

Coefficients	i (group)			
	1 (group A)	2 (group B)	3 (group C)	4 (group D)
A_i	1.03	1.04	0.940	0.808
B_i	−0.0510	−0.0621	−0.0544	−0.0336

action also had shorter fatigue lives than those under room temperature and atmospheric environment (group A). Moreover, the higher the temperature and humidity, the shorter the fatigue life of SFPSC under the same stress level. In the same high humidity environment (95% RH), the fatigue life of SFPSC decreased significantly as the temperature increased. In the same high temperature (50°C) environment, the higher the humidity was, the shorter the fatigue life of SFPSC observed. However, with the increase of environmental treatment time, the decrease of fatigue lives of specimens became gradual. When the fatigue life of specimen approached to its fatigue limit, humidity had less influence on fatigue performance of SFPSC.

It is shown in Figure 4 that the influence of temperature on fatigue lives of SFPSC specimens was dominant, and the influence of humidity was secondary.

and cyclic loads, as shown in Figure 4, the experimental results were similar to that under the uncoupling action, and the specimens of each groups under the coupling

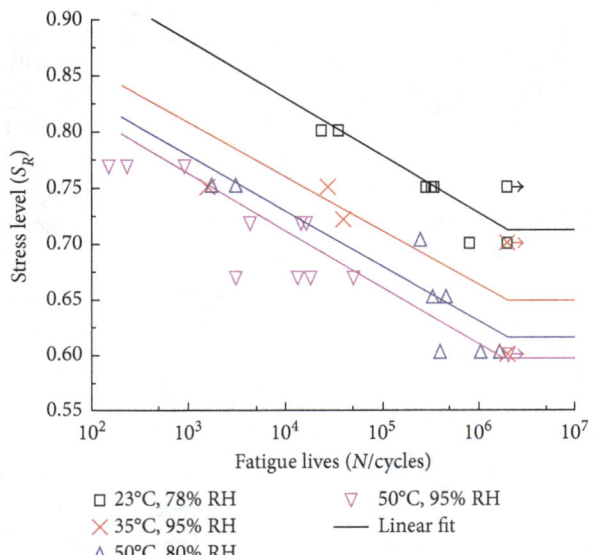

FIGURE 4: Experimental $S_R \sim N$ curves under coupling of hot-wet environments and cyclic loads.

TABLE 5: Coefficients A_i and B_i for group E~G specimens.

Coefficients	i (group)		
	1 (group E)	2 (group F)	3 (group G)
A_i	0.953	0.926	0.914
B_i	−0.0482	−0.0492	−0.0508

3.2. Analysis of Failure Mechanism of SFPSC in Hot-Wet Environments.

As shown in the previous sections regarding $S_R \sim N$ curves, fatigue experiment results of SFPSC specimens under both the coupling/uncoupling action of hot-wet environments and cyclic loads showed that the environmental fatigue lives of the specimens were shorter than those in room temperature and atmospheric environment. Furthermore, the more severe the environments were and the higher the temperature and humidity were, the shorter the fatigue life of SFPSC was, under the same stress level. The fatigue failure mechanism of SFPSC in hot-wet environments could be analyzed as follows:

(1) The influences of the temperature and humidity on the interfaces between steel fibers and concrete matrixes. When the hot-wet environment was severe, especially when there was a big difference between the environmental humidity and the internal humidity of the concrete, the transmission of water from the surfaces to internal through the concrete pores would be intensified. Consequently, a higher surface tensile force would be generated on the liquid surfaces of the internal pores of the concrete, and such surface tensile force would intensify the crack growth within the concrete and lead to decrease of the cohesive force of the interfaces between the steel fibers and the concrete matrixes.

(2) The influences of humidity on structural compactness of SFPSC. Figure 5(a) shows that there are a lot

of needle-like crystals of calcium vanadium stone and flocculation of C-S-H gel in the hydration products of the concrete without polymer latex, which have many pores with very little connections. The combination of the hydration products was relatively loose with some large hole in the combination. Some crossed microcrack can be observed on the surface of the slurry. After mixing of the polymer latex, as shown in Figure 5(b), the gaps or the cavities of the cement hydration products would be filled up or joined by the membranous substances formed by the polymer latex, which made the hydration products in the concrete more compact, caused reduction of the microcracks, and improved the cracking resistance of the material [20]. In addition, the concrete would show the characteristics of dry shrinkage and wet expansion; the higher the strength of the concrete, the more obvious such phenomenon is. Therefore, during the pretreatment process in the hot-wet environment, when the humidity was high, the gaps and cavities in the concrete, which was originally filled by the latex membrane substance, would expand due to wet expansion, resulting in the decrease of the structural compactness, and the fatigue performance became poor. Moreover, the concrete strength used in this research was higher (C55), and such phenomenon was more obvious.

(3) The influences of temperature on the mechanical properties of polymer latex. The reactions on the mechanical properties of the polymer to temperature were rather sensitive. It was showed by existing research [14] that polymer latex would be softened as the temperature increased, and when the temperature reached to a certain value, the supporting capacities of the polymer would be lost, and thus caused greater influences on the mechanical properties of the polymer concrete.

3.3. Influence of Experimental Methods on Fatigue Performance.

To study the influence of environmental fatigue test methods on the fatigue performance of SFPSC material, the fatigue experimental data of group B and E (50°C and 80% RH), as well as group C and F (50°C and 95% RH) in Tables 2 and 3 were taken and are shown in Figures 6(a) and 6(b), respectively. Figure 6(a) shows that under lower relatively humidity (80% RH), compared to the experimental method of uncoupling action of environment and loads, the fatigue lives of SFPSC would be lower for the experimental data under coupling action of the hot-wet environment and the cyclic loads, as long as they were under the same loading level regardless of high or low loading level. The whole $S_R \sim N$ curve under coupling action would be below the former.

Under the same temperature (50°C), when the relative humidity was higher (95% RH), the influence rules of environmental fatigue experimental method of SFPSC to fatigue life would be changed. In such hot-wet environment,

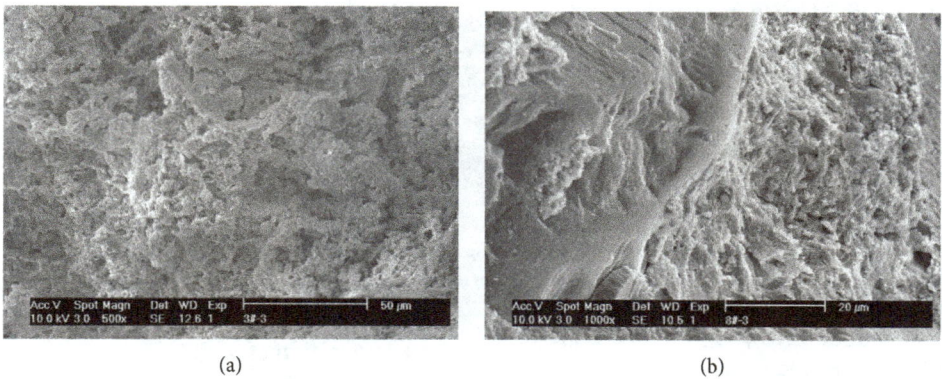

(a)   (b)

FIGURE 5: Effect of the polymer latex on the hydraulic products (SEM photos). (a) Specimens without polymer latex. (b) Specimens with polymer latex.

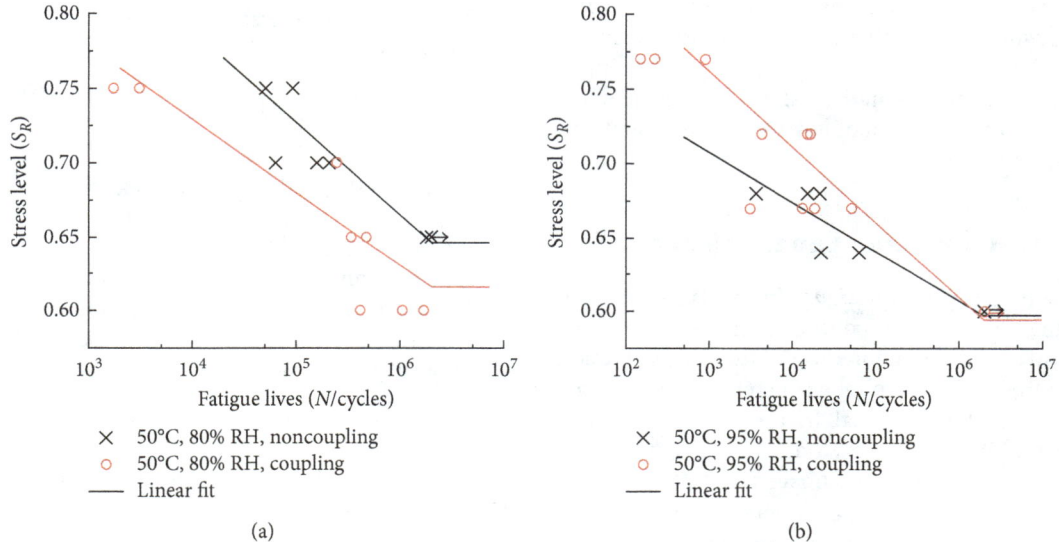

(a)   (b)

FIGURE 6: Influence of the experimental method on S_R~N curves. (a) 50°C and 80% RH. (b) 50°C and 95% RH.

when the loading level was high, the fatigue lives under the uncoupling action were lower. When the loading level was low and especially close to the fatigue limit of the material, the fatigue lives of SFPSC under coupling action were lower.

For the analysis of the experimental results shown in Figure 6, the effect mechanism of environmental fatigue test methods on the fatigue performance of SFPSC could be considered as follows:

(i) The temperature and humidity set in these environmental fatigue experiments were higher than that in room temperature and atmospheric environment as shown in Tables 2 and 3. For the case of uncoupling action of hot-wet environments and loads, the specimens were pretreated in the environmental box and placed in room temperature and atmospheric environment to dry. Then, the fatigue experiments were carried out. During air cooling and drying, the moisture originally absorbed by concrete material of the specimens in the hot-wet

environments (environmental box) gradually vaporized, causing the gaps produced under hot-wet environments to reduce or vanish as the temperature and humidity decreased. The fatigue performance would be partially recovered.

(ii) Previous studies [20] showed that, under the conditions of coupling action of hot-wet environments and loads, the deformation of the specimens caused acceleration of moisture and heat transmission rate into the interior of concrete material and led to a decrease in fatigue performance.

Based on the analysis of (i) and (ii), under common hot-wet environments, the experimental condition was more severe under coupling action compared to uncoupling action, and it would lead to a decrease in fatigue lives, as shown in Figure 6(a).

(iii) For high temperature and high humidity environment (50°C and 95% RH), when the traditional experimental method with uncoupling action of environment and loads was used, as described in (i), the fatigue performance of the specimens would be partially recovered if the specimens were placed in room temperature and atmospheric environment to be dried after the hot-wet pretreatment. The comparison to fatigue performance of the specimens under coupling action of environment and loads could be divided into the following two parts:

Firstly, when the experimental period was short (high loading level and low fatigue lives), high temperature and high humidity would not be able to have sufficient time to cause erosion of the specimens. Therefore, the fatigue lives of specimens were longer under coupling action during this interval.

Secondly, when the experimental time was long enough (low loading level especially close to the fatigue limit and long fatigue lives of the specimens), high temperature and high humidity would have sufficient time to cause erosion of the specimens. Due to the reason described in (ii), the fatigue lives of SFPSC specimens would be shorter than the specimens under uncoupling action of hot-wet environment and loads during this interval.

4. Environmental Fatigue Equations of SFPSC

4.1. Environmental Fatigue Equations. In the components of SFPSC, polymer latex has a consolidation action on the concrete gel and a bonding action on steel fiber. The adhesion and fixation of the steel fiber and the concrete gel were caused by the adhesion and mechanical friction of their contact interface. When SFPSC was cracked due to external loading, the steel fibers dispersed in the damaged section would be pulled out by the two ends of the concrete. However, this adhesion force would form resistance to the cracking and significantly disperse the tensile stress that was originally concentrated. Meanwhile, it would also continuously absorb the fracture energy and cause a dramatic improvement of the anticracking property of SFPSC. When the polymer latex and steel fiber in SFPSC came into contact, the surface of steel fiber would adhere to the hydrophilic base of the surfactant. As the cement hydration effect increased, water was consumed and a large amount of heat was released, and a chemical adhesion of the surface of hydrophilic base and steel fibers could be caused. Furthermore, as the temperature increased and water decreased, the adhesion effect would also increase.

When studying the effect mechanism of serving environmental temperature to the fatigue performance of SFPSC material, this research group [14, 15] considered the influence of temperature would be mainly presented in (1) thermal stress caused by different thermal expansion coefficient of each phase in the composites, (2) influence of temperature on polymer latex performance, (3) the influence of the change in polymer latex properties on the mechanical properties of the interface between steel fiber and the base matrix, and so on.

For the influence of serving environmental humidity to the mechanical properties of SFPSC, this research suggested

that (1) the high humidity environment would weaken the adhesion action of polymer latex and steel fibers and reduce the strength; (2) in high humidity environments, moisture absorption stress would be generated in the concrete, and the destruction would be accelerated, and (3) high humidity could accelerate the speed of corrosion reaction for steel fibers, leading to a decrease of its mechanical properties.

For SFPSC materials serving in the hot-wet environment, the effects of temperature and humidity on fatigue performance mentioned above should be taken into consideration. Therefore, when establishing the environmental fatigue equation of SFPSC in this paper, the previous studies by this research group on the thermal fatigue equation [21] and hot-wet fatigue equation [20] for reinforced concrete (RC) beams strengthened with carbon fiber reinforced polymer (CFRP) were used as references. It suggested that for experimental methods of both coupling and uncoupling action of hot-wet environments and cyclic loads, fatigue life $f(N)$ of SFPSC in hot-wet environments under bending loads could be presented by the following formula:

$$f(N) = \frac{[S_R + C_1 + C_2 f(T, H)]}{[C_3 + C_4 f(T, H)]}, \qquad (2)$$

where S_R was the loading level (also known as the stress level, $S_R = P_{\max}/P_u$) and $f(T, H)$ was the hot-wet function.

Based on the analysis of the influence mechanism of temperature and humidity on the fatigue performance of SFPSC materials and the mathematical description of the hot-wet function generated by this research group, the expression of $f(T, H)$ in this paper was taken as follows:

$$f(T, H) = C_5 e^{C_7 T} + C_6 e^{C_8 H}. \qquad (3)$$

In (2) and (3), $C_1 \sim C_8$ were constant coefficients, determined by experiments.

For (2), the fatigue life function $f(N)$ could be described in single logarithmic coordinates, considering the fatigue curves ($S \sim N$ curves) of the material. The expression of $f(N)$ was as follows:

$$f(N) = \lg N. \qquad (4)$$

The fatigue equation of SFPSC under hot-wet environments, which was also called the hot-wet fatigue equation, was generated by ((2)~(4)).

By using these equations, the bending fatigue lives and fatigue limits of the SFPSC specimens under coupling and uncoupling action of the hot-wet environments and loads could be conveniently and accurately estimated.

4.2. Environmental Fatigue Equation under Uncoupling Action. Using the experimental data of group B~D specimens shown in Table 2 and the least squares method, the data were fitted in ((2)~(4)), and the coefficients $C_1 \sim C_8$ could be determined. Then, the expression for fatigue lives of SFPSC materials under uncoupling action of the hot-wet environments and three-point bending cyclic loads presented by loading level S_R was determined. The environmental fatigue equation of SFPSC was as follows:

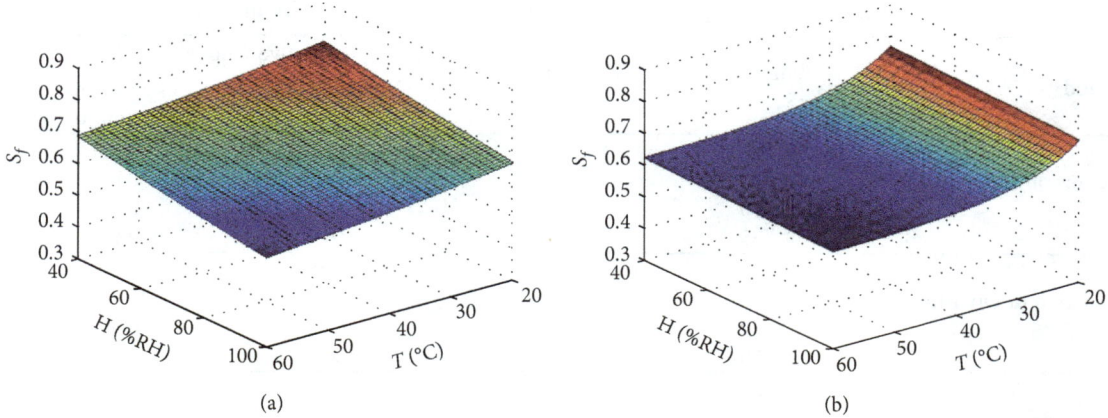

FIGURE 7: The influence of hot-wet environments on the relative fatigue limit S_f. (a) S_f under the uncoupling action. (b) S_f under the coupling action.

$$S_R = 0.343 + 0.238e^{-0.0174T} + 0.851e^{-0.619H}$$
$$-\left(0.0137 - 0.00219e^{-0.0174T} + 0.0657e^{-0.619H}\right)\lg(N),$$

$$(5)$$

where the unit of temperature T is presented in (°C) and humidity H is presented in (% RH).

By using (5), the fatigue lives of the SFPSC materials under uncoupling action of hot-wet environments and cyclic bending loads and the traditional environmental fatigue experimental method could be conveniently and accurately estimated.

4.3. Environmental Fatigue Equation under Coupling Action.
Similar to the analysis method shown in Section 4.2, the experimental data of group A, E, F, and G specimens and the least square method were used to fit in ((2)~(4)), and the coefficients of $C_1 \sim C_8$ could be determined. Then, the expression for fatigue lives of SFPSC materials under coupling action of hot-wet environments and cyclic bending loads presented by loading level S_R was determined. The environmental fatigue equation of SFPSC was as follows:

$$S_R = 0.543 + 1.12e^{-0.0998T} + 0.412e^{-0.114H}$$
$$-\left(0.0294 + 0.0101e^{-0.0998T} + 0.0221e^{-0.114H}\right)\lg(N).$$

$$(6)$$

By using (6), the fatigue lives of the SFPSC materials under coupling action of hot-wet environments and cyclic bending loads could be conveniently and accurately estimated.

4.4. Fatigue Limits of SFPSC under Hot-Wet Environments.
For general concrete structures, the corresponding loading value of $N = 2 \times 10^6$ cycles could be defined as the fatigue limit (depending on the specifications, it can also be defined as $N = 4 \times 10^6$ cycles and 1×10^7 cycles). $N = 2 \times 10^6$ cycles was substituted into (5) and (6), and the relations of fatigue limits S_f of the bending SFPSC material under coupling and uncoupling action of the environments and cyclic loads

presented by S_R to the temperature T and humidity H could be obtained and the expressions were

$$S_f = 0.257 + 0.252e^{-0.0174T}$$
$$+ 0.437e^{-0.619H} \text{ (uncoupling action)},$$

$$(7)$$

$$S_f = 0.358 + 1.06e^{-0.0998T} + 0.273e^{-0.114H} \text{ (coupling action)}.$$

$$(8)$$

Using (7) and (8), the change rules of the fatigue limits S_f of SFPSC material under different hot-wet environments and cyclic bending loads could be obtained and are shown in Figure 7. As shown in Figure 7, when the environmental temperature was increased, the relative fatigue limit S_f decreased; when the environment humidity was increased, the relative fatigue limit S_f would also be decreased. In addition, comparing Figures 7(a) and 7(b), it showed that regardless of coupling or uncoupling action of environments and loads, the rules of changes in relative fatigue limit S_f of SFPSC materials estimated by the environmental fatigue equations due to the shift in temperature T and humidity H remained congruity. However, in general, the relative fatigue limit of SFPSC material under coupling action of environments and loads was lower than that under uncoupling action. In addition, in contrast to room temperature and atmospheric environment, under the high temperature and high humidity (50°C and 95% RH) environment, the fatigue limit S_f of SFPSC would be reduced by about 15%. Therefore, the effects of the hot-wet environments on fatigue lives of SFPSC were very significant. Meanwhile, this also showed that, for the antifatigue/durability designs of the bridge structures using SFPSC material, the adverse effects of the hot-wet environments must be considered.

In order to verify the effectiveness and feasibility of (5) and (6), the experimental conditions of the hot-wet environments shown in Tables 2 and 3 were substituted into (7) and (8) separately, and the environmental fatigue limits of SFPSC under bending loads were obtained respectively, as shown in Table 6. According to Table 6, the actual experimental estimation of relative fatigue limits S_f of SFPSC

TABLE 6: Relative fatigue limits of SFPSC S_f in the hot-wet environments.

Testing method	Temperature, T (°C)	Relative humidity, H (%RH)	Relative fatigue limit S_f		Relative error (%)
			Calculated values by (7) or (8)	Experimental estimation by (1)	
Coupling action of environments and loads	23	78	0.715	0.709	0.846
	50	80	0.614	0.616	0.325
	50	95	0.611	0.594	2.86
	35	95	0.635	0.649	2.16
Uncoupling action of environments and loads	23	78	0.696	0.709	1.83
	50	80	0.629	0.649	3.08
	50	90	0.613	0.597	2.68
	50	95	0.605	0.596	1.51

obtained from two types of experimental methods well matched with the calculated values, and the average relative errors were presented as 2.28% (uncoupling action) and 1.55% (coupling action), respectively. This explained that using (5) and (6) to estimate the hot-wet environmental fatigue lives and fatigue limits of SFPSC material under coupling and uncoupling action of hot-wet environments and bending loads was effective and feasible in this research.

5. Conclusions

The experimental studies for environmental fatigue/durability of "steel fiber polymer structural concrete (SFPSC)" were conducted using the conventional uncoupling and coupling methods of hot-wet environments and cyclic bending loads. The experiments were carried out under three different hot-wet conditions (50°C and 80% RH, 50°C and 90% RH, and 50°C and 95% RH) for uncoupling action and three different hot-wet conditions (50°C and 80% RH, 50°C and 95% RH, and 35°C and 95% RH) for coupling action. The experimental results were discussed in comparison with the results obtained under room temperature and atmospheric environment (23°C and 78% RH), and conclusions were obtained as follows:

(1) The effects of hot-wet environments on the bending fatigue performance/durability of SFPSC were obvious. Within the range of testing conditions in this paper, when the environmental temperatures were the same, fatigue lives and fatigue limits of SFPSC material would decrease along with the increase of the humidity; when the humidity remained unchanged, the fatigue lives and fatigue limits would decrease along with the increase of the temperature. In comparison to the room temperature and atmospheric environment, the fatigue limit S_f would be reduced by about 15% under high temperature and high humidity (50°C and 95% RH) environment.

(2) Compared with the experimental results of traditional uncoupling environmental fatigue experiments, the fatigue lives and especially the fatigue limits of SFPSC were reduced under coupling action of hot-wet environments and cyclic loads. Therefore, in order to study the environmental

fatigue performance of the structures in service and under coupling action such as main load-bearing structures of bridges, it would be necessary to conduct the experiments under a condition close to the actual service environment, and testing method of coupling action should be used.

(3) Environmental fatigue life equations of SFPSC under the uncoupling/coupling action of hot-wet environments and cyclic bending loads (i.e., the hot-wet fatigue equations) were proposed based on the data of environmental fatigue experiments and in combination with the classical theories of fatigue strength. By using the fatigue equations, the fatigue lives and fatigue limits of the SFPSC materials in service under hot-wet environments could be conveniently and accurately estimated.

(4) This research results showed that the adverse effects of the hot-wet environments must be considered for designing of antifatigue/durability of the bridge structures of SFPSC in the subtropical areas in order to avoid safety hazards.

It is necessary to point out that the environmental fatigue equations proposed in this paper should be verified on a greater environmental scale, and more experimental data should be collected to verify its accuracy of calculation and applicability.

Conflicts of Interest

The authors declare that they have no conflicts of interest.

Acknowledgments

The project is supported by the National Key R&D Program of China (no. 2017YFC0806000) and the National Natural Science Foundation of China (nos. 11627802, 51678249, 11132004, and 51508202).

References

[1] F.-W. Wang and X.-F. Shi, "The research about how to control long-term deflection of large span prestressed concrete girder bridges," *Road*, vol. 8, pp. 72–76, 2006.

[2] H. S. Chiu, J. C. Chern, and K. C. Chang, "Long-term deflection control in cantilever prestressed concrete bridges I:

control method," *Journal of Engineering Mechanics*, vol. 122, no. 6, pp. 489–494, 1996.

[3] I. N. Robertson, "Prediction of vertical deflections for a long-span prestressed concrete bridge structure," *Engineering Structures*, vol. 27, no. 12, pp. 1820–1827, 2005.

[4] O. C. Choi and C. Lee, "Flexural performance of ring-type steel fiber-reinforced concrete," *Cement and Concrete Research*, vol. 33, no. 6, pp. 841–849, 2003.

[5] L. Daniel and A. Loukili, "Behavior of high-strength fiber-reinforced concrete beams under cyclic loading," *ACI Structural Journal*, vol. 99, no. 3, pp. 248–256, 2002.

[6] N. Banthia and M. Sappakittipakorn, "Toughness enhancement in steel fiber reinforced concrete through fiber hybridization," *Cement and Concrete Research*, vol. 37, no. 9, pp. 1366–1372, 2007.

[7] L.-F. Luo, M. Zhong, P.-Y. Huang et al., "Steel fibers reinforced polymers concrete and the method of it's mix ratio," China Patent for Invention ZL200410026780, 2006.

[8] L.-F. Luo, J.-C. Zhou, and P.-Y. Huang, "Reinforced mechanism with the polymer latex added in the steel fiber reinforced concrete," *Acta Materiae Compositae Sinica*, vol. 19, no. 3, pp. 46–50, 2002.

[9] L.-F. Luo, "Impact behavior of steel fiber reinforced polymer modified concrete," *China Journal of Highway and Transportation*, vol. 19, no. 5, pp. 71–76, 2006.

[10] Y.-P. Liu, L.-Q. Tang, and X.-Q. Huang, "Fatigue damage behavior of steel fiber reinforced polymer modified concrete," *Journal of South China University of Technology*, vol. 35, no. 2, pp. 18–22, 2007.

[11] L.-F. Luo, M. Zhong, and C.-Z. Huang, *Construction Technology of SFRPC Bridge Deck Pavement*, Press of South China University of Technology, Guangzhou, China, 2004.

[12] S.-C. Zheng, *Study on the Mechanical Property of Steel Fiber Polymer High Strength Structural Concrete*, South China University of Technology, Guangzhou, China, 2010.

[13] S.-C. Zheng, P.-Y. Huang, and X.-Y. Guo, "Experimental study on fatigue performance of steel fiber polymer high strength concrete," *Journal of Experimental Mechanics*, vol. 26, no. 1, pp. 1–7, 2011.

[14] S.-C. Zheng, P.-Y. Huang, X.-Y. Guo et al., "Temperature fatigue performance of steel fiber polymer high strength concrete," *Journal of South China University of Technology*, vol. 39, no. 9, pp. 88–92, 2011.

[15] H. Hu, *Temperature Fatigue Performance of Steel Fiber Polymer High Strength Concrete*, South China University of Technology, Guangzhou, China, 2011.

[16] Z.-S. Chen, *Experimental Study on Environmental Fatigue Behavior of Steel Fiber Polymer Structural Concrete*, South China University of Technology, Guangzhou, China, 2014.

[17] Z.-G. Chen, *Study on Key Mechanical Properties of Steel Fiber Polymer Structural Concrete*, South China University of Technology, Guangzhou, China, 2016.

[18] China National Technology Standard GB/T 2573-2008, *Test Method for Aging Properties of Glass Fiber Reinforced Plastics*, China National Technology Standard, 2008, in Chinese.

[19] P.-Y. Huang, H. Zhou, H.-Y. Wang, and X.-Y. Guo, "Fatigue lives of RC beams strengthened with CFRP at different temperatures under cyclic bending loads," *Fatigue and Fracture of Engineering Materials and Structures*, vol. 34, no. 9, pp. 708–716, 2011.

[20] G. Qin, P. Huang, H. Zhou, X. Guo, and X. Zheng, "Fatigue and durability behavior of RC beams strengthened with CFRP under hot-wet environment," *Construction and Building Materials*, vol. 111, pp. 735–742, 2016.

[21] P.-Y. Huang, S.-C. Zheng, W. Guo, and J. Deng, "Effect of admixtures on mechanical properties of steel fibre reinforced polymer high-strength concrete," in *Proceedings of SPIE, (ICEM 2008)*, vol. 7375, pp. 7375 6L-1–7375 6L-6, Nanjing, China, August 2008.

14

Evaluation of the Flexural Performance and CO_2 Emissions of the Voided Slab

Seungho Cho[1] and Seunguk Na [ID][2]

[1]*Architectural Engineering Department, Seoul National University of Science and Technology, Seoul, Republic of Korea*
[2]*Architectural Engineering Department, College of Architecture, Dankook University, Yongin-si, Gyeonggi-do, Republic of Korea*

Correspondence should be addressed to Seunguk Na; naseunguk@dankook.ac.kr

Academic Editor: Fernando Lusquiños

Reinforced concrete is regarded as one of the ideal structural materials which comprises concrete with high compressive strength and reinforcing bars with high tensile strength. However, concrete has been pointed out that it consumes a large volume of energy and emits a lot of carbon dioxide during its manufacturing. In order to lower such environmental burdens of concrete structures, a number of studies and approaches have been carried out. The voided slab is also suggested as a new method to reduce the environmental burden since voided section of the slab would use less concrete compared with the normal reinforced concrete slab. However, no studies have evaluated the CO_2 emissions and environmental performance of voided slabs. The purpose of this study was to evaluate the structural performance of voided slabs and empirically corroborate their environmental influence. The flexural performance test was carried out based on the variables of the depth of slab, types of the void former materials, and the hollowness ratio. In addition, comparison of the emission of CO_2 was also performed by considering the hollowness ratio and types of void former materials over the normal reinforced concrete slab. The structural performance of the voided slab was similar or slightly higher than the normal reinforced concrete slab. The yield strength of specimens was increased approximately 10~30% over the anticipated yield strength. Based on this result, it is considered that the voided slab would be sufficient to structural performance and beneficial to plane planning in buildings. In general, it is considered that the voided slab would be beneficial to both structural and environmental aspects. However, the test results in this research showed that the voided slab would emit more carbon dioxide emissions compared to the normal reinforced concrete slab. The main source of more CO_2 emissions in the voided slab was the anchoring materials. In this research, wires were used to fix the void former materials to the reinforcing bars. In order for the voided slab to become a more eco-friendly and sustainable material, new anchoring methods such as use of recycled materials, new void former materials without anchoring, or other eco-friendly materials should be applied to reduce the emission of CO_2.

1. Introduction

Reinforced concrete is regarded as one of the ideal structural materials and is commonly used in the architectural, engineering, and construction (AEC) industry [1, 2]. Reinforced concrete is a composite structure which comprises concrete with high compressive strength and reinforcing bars with high tensile strength. Reinforced concrete is highly useful not only because it can be structured in any form that the architect intends but also as it ensures a highly durable structure. Moreover, concrete is relatively inexpensive compared to steel and other construction materials, although there are various advantages such as a low corrosion rate, high fire resistance, and high water resistance. For these reasons, reinforced concrete has been chosen as a suitable material and used for a long time in the AEC industry.

However, studies have shown that concrete consumes a large volume of energy and emits large quantities of carbon dioxide [3–6]. In order to reduce the environmental burdens from concrete structures, a number of studies have been carried out and approaches tested. Kim et al. [7] suggested the I-slab system, which incorporates polystyrene forms in precast concrete panels to reduce the amount of concrete.

Various other researchers have also proposed new methods to reduce the amount of concrete, such as replacing normal-strength materials with high-strength materials, using by-products and recycled materials, and designing optimal structural systems to minimise construction materials [8–14]. For example, using high-strength concrete with added by-products such as silica fume or other by-products is one of frequently suggested approaches to enhance environmental performance of concrete. In addition, design and selection of optimal structural system would be one of the frequently suggested approaches to reduce the emission of CO_2 in the AEC industry [8, 15–18]. Baek et al. [15] propose that changing the block-type bearing wall system which is common to building apartment housing in South Korea to a column and beam system would reduce a significant amount of CO_2 emission during the construction stage. In a similar vein, Han and Kim [8] researched the emission of carbon dioxide from reinforced concrete structure and steel structure building. In this study, they showed that a steel structure building would emit relatively less CO_2 over the reinforced concrete building. Penadés-Plà et al. [16] indicate that optimal design of structural system would make it possible to lower the environmental impact even though the manufacturing stage of concrete box-girder would emit a large volume of carbon dioxide. Moreover, Molina-Moreno et al. [17] highlight the significance of low-carbon design for three aspects, which are structural performance, economic costs, and environmental impacts. Like other methods to minimise the amount of material used in the AEC industry, the voided slab is also suggested as a new method to reduce the environmental burden since voided section of the slab would use less concrete compared to the normal reinforced concrete slab [19–23]. Despite a number of studies regarding the voided slabs systems describing that the environmental friendliness of the voided slabs, it is sparse to corroborate the carbon dioxide emission of the voided slab by an empirical manner. The voided slab has also been suggested as a new method to reduce the environmental burden, since the voided section of the slab would use less concrete than a normal reinforced concrete slab [8, 15–18]. However, it is sparse that studies have evaluated the CO_2 emissions and environmental performance of voided slabs. The purpose of this study was to evaluate the structural performance of the voided slab by testing the flexural experiment and empirically corroborate their environmental influence.

2. Literature Review

The AEC industry is regarded as one of the main actors emitting large volumes of carbon dioxide and consuming a significant amount of energy [24]. The life cycle of a building can be divided into four distinctive phases: design, construction, operation and maintenance, and demolition or decomposition. Throughout the life cycle, the CO_2 emissions of a building are divided into embodied carbon and operating carbon. As a lot of new technologies have been developed and applied to newly constructed and existing buildings, the emissions of carbon dioxide from operating

carbon have gradually shown a decreasing tendency [25, 26]. As a result of this, the proportion of embodied carbon in buildings is experiencing a relatively incremental trend. In order to reduce embodied carbon during building construction, various approaches have been suggested and researched, such as the using high-performance materials, material replacement, implementation of optimal design, and the application of structural system alternatives [5, 25–28].

A number of academics have proposed that an effective approach to reduce CO_2 emissions in the AEC industry is to replace conventional materials with low-carbon ones [11, 12, 28]. Enhancing the strength of materials, applying recycled substances or by-products, and planning the optimal design are popular methods for replacing materials to reduce CO_2. Cho and Chae [11] suggested application of recycled materials or industrial by-products and shortening the manufacturing process of materials to lower CO_2 emissions during construction. According to Kim et al. [28], using 60 MPa high-strength concrete would incur 1.5 times less CO_2 emissions compared with 24 MPa normal-strength concrete. In this research, when buildings with same area were constructed, the total amount of CO_2 emissions was reduced when using high-strength concrete since the amount of materials required decreased. The authors suggested three ways to reduce CO_2 emissions from construction: firstly, using high-strength concrete, which would lower the required quantity of concrete and rebars; secondly, an optimal mixture design, using admixtures such as blast-furnace slag; and lastly, taking the location of ready-mixed concrete into consideration, as this would lower emissions from transportation. Cho and Na [12] indicated that the application of high-strength reinforcing bars would reduce carbon dioxide emissions in reinforced concrete structure buildings. Their research showed that high-strength rebars lead to a reduction in the amount of materials needed, even though there was a slight increase in splice and development in connections.

Many researchers have emphasised the effectiveness of reducing CO_2 emissions through application of high-strength concrete. Tae et al. [5] evaluated the environmental impact of high-strength concrete by comparing energy consumption and CO_2 emissions in existing apartment housing in South Korea. They concluded that the application of high-strength concrete has a significant impact on the reduction of carbon dioxide emissions compared to normal-strength concrete. According to this research, the application of high-strength concrete would lead to the lowering of rebars in concrete, as well as cross section reduction of vertical members. They assert that pursuing material strength enhancement is an effective approach to minimise emission of carbon dioxide and energy consumption.

Additionally, Park et al. [29] analysed the CO_2 emissions of concrete based on different compressive strengths and seasonal attributes, and they proposed a formula to estimate the emissions of carbon dioxide based on the compressive strength of concrete. These researchers also considered seasonal characteristics, which would have a crucial impact on the emissions of CO_2 during the construction stage.

They suggested that construction during the winter season would emit much more CO_2 compared with other seasons in South Korea. According to them, the reason for this is the reduced application of admixture, increased inputting of cement, and longer curing time.

Along with the implementation of high-strength concrete, some researchers have proposed ensuring the optimal design of concrete and material selection in the design stage. González and Navarro [9] highlighted the significance of the early design stage for crucial decision-making about construction materials. Their research shows that appropriate materials selection would reduce approximately 28% of CO_2 emissions, compared with a building that was built without considering carbon dioxide emissions. In addition, Chau et al. [10] examined ten buildings to assess the relationship between construction materials and their environmental impacts in Hong Kong. According to them, concrete and reinforcing bars are the first and second contributors to the environmental impact of buildings throughout their life cycle. Since these are two major building materials, they concluded that it would be necessary to account for the environmental effects of concrete and rebars in building construction. They asserted that designers would have a significant role in reducing carbon dioxide emissions during building construction by selecting low-emission building materials, thus improving sustainability.

There are also other approaches to reduce CO_2 such as recycling and developing optimal design programmes. Lee et al. [30] considered the life cycle of apartment housing and suggested CO_2 reduction strategies in South Korea. They proposed not only the extension of the service life of the building but also recycling the concrete as sub-base material or aggregates after demolition of the building. Similar to Lee et al. [30], Yan et al. [31] compared the emissions of carbon dioxide between residential and commercial buildings. In their research, they suggested that utilisation of recyclable, rather than nonrecyclable, materials would be ideal for reduction of CO_2 emissions.

While CO_2 reduction strategies such as selection of alternative materials and optimizing design might be considered a micro-perspective, there is also a "macro-approach," which considers the entire building system rather than its individual parts or materials. Nadoushani and Akbarnezhad [14] studied the relationships between different structural systems and CO_2 emissions. In this research, they indicated the significance of comprehensive assessment of embodied carbon and operating carbon for choosing the optimal structural materials during structural design. Moreover, Han and Kim [8] examined the total carbon dioxide emissions of different structural types for apartment and office buildings in South Korea. The test results showed that steel structure buildings emitted less carbon dioxide than reinforced concrete structures. Moreover, the authors proposed that using high-strength materials (e.g., high-strength concrete and reinforcing bars) would be an effective method to minimise CO_2 emissions in reinforced concrete structures.

Cole [32] examined [33] not only energy consumption and greenhouse gas emissions of different structural systems but also energy consumption during transportation and from construction personnel. In this study, the author showed that concrete structures consumed the highest energy as well as emitted the largest volume of CO_2 during construction. In contrast, steel structures consumed the lowest energy and emitted the lowest CO_2 among concrete, wood, and steel structures. The author inferred that "the more labour intensive a process, the greater the amount of worker transportation." Baek et al. [15] compared different structural types of apartment buildings in South Korea. According to them, structural system has a significant influence on lowering carbon dioxide emissions in the construction phase. In this research, apartment houses built with a column and beam system emitted less CO_2 than bearing wall systems. They also suggested that the application of high-strength concrete and adding blast furnace slag would be beneficial to reduce CO_2 emissions.

3. Materials and Methods

3.1. Assessment of Structural Performance of the Voided Slabs

3.1.1. Flexural Performance Test. In this study, the depth of slab, type and presence of void former materials, and hollowness ratio were used as the main variables to test the flexural strength of concrete slab specimens. The details of the dimensions and properties of the specimens are summarised in Table 1, and Figure 1 shows cross-sectional diagrams of the specimens. The tested specimens were 4230 mm in length and 700 mm in width. The depth of slabs varied between 169 and 210 mm.

To evaluate the flexural performance of the voided slab specimens, the test specimens were simply supported as shown in Figure 2, and four-point bending tests were conducted. The applied load to the specimens was monotonic loading which used an actuator with a capacity of 100 kN resisted by a steel frame with a capacity of 200 kN. The applied load was at a relatively low deflection rate of 2.0 mm/min for accurate examination of the initial flexural behavior before and immediately after the initial flexural cracklings. A load cell with a capacity of 100 kN was used to obtain the load data. In order to receive the displacement of the slabs, a linear variable differential transformer (LVDT) was installed under the specimens. The LVDT in this research was able to measure up ranges of 200 mm of the displacement. The data were gathered by a data logger (Tokyo Sokkie).

All the specimens were reinforced by 10 mm and 13 mm deformed bars, and $\phi 6$ wires were used as a means to anchor the void formers to the upper and lower reinforcing bars. Details of the void former materials and their anchoring are depicted in Figure 3.

3.1.2. Properties of Materials. The detailed characteristics of the materials used in the voided slab are shown in Table 2. The concrete had a designed compressive strength of 27 MPa and comprised 333 kg/m^3 ordinary Portland cement, 821 kg/m^3 fine aggregate, 1086 kg/m^3 coarse aggregate, and 108 kg/m^3

TABLE 1: Properties of specimens.

Specimens	Type of slab	Depth of slab (mm)	Specification of void former materials
S1	Normal reinforced concrete slab	210	—
S2	Voided slab	210	Sphere shape Diameter 100 mm
S3	Voided slab	210	Oblate shape Diameter 170 mm Height 110 mm
S4	Normal reinforced concrete slab	169	—

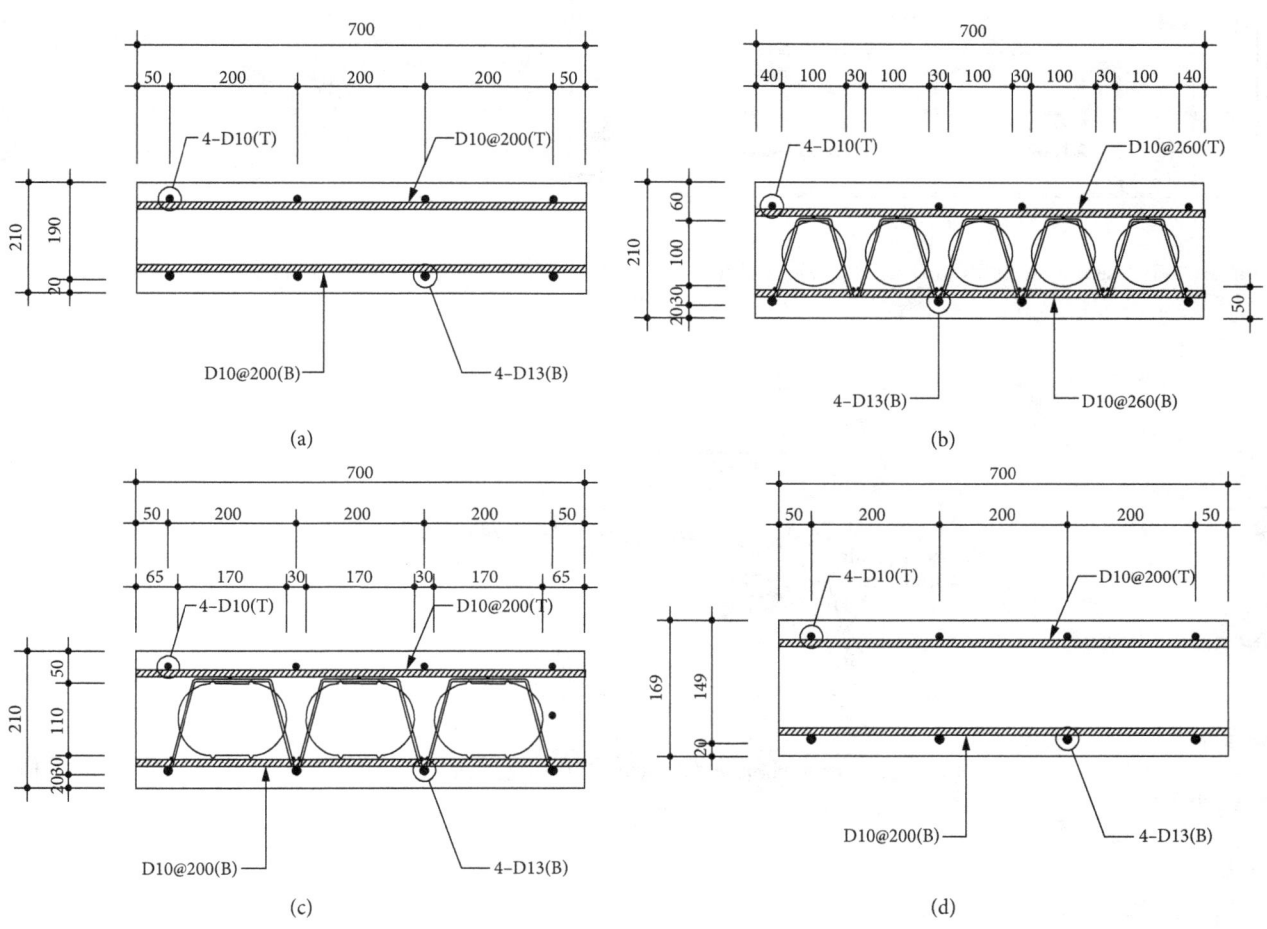

FIGURE 1: Details of the specimens. (a) S1. (b) S2. (c) S3. (d) S4.

water. The tested compressive strength was 35.16 MPa after 28 days curing in a water bath. The slump and the entrapped air void were 80 mm and 2%, respectively (Table 2).

The reinforcing bars used in the specimens were named D10 and D13 based on the diameter; their nominal yield strength was $f_y = 400$ MPa. According to the direct tension test specified in KS B 0802 [29], the ultimate tensile strength of the reinforcing bars used in this research was 630–651 MPa (Table 3).

3.2. Evaluation of CO_2 Emissions in the Voided Slabs

3.2.1. A System Boundary. It is necessary to calculate the life cycle of voided slabs to evaluate their CO_2 emissions.

Based on ISO 14044 [34] and ISO 21930 [35], the system boundary of voided slabs was determined as the product stage of voided slabs (i.e., cradle to gate). The production stage of voided slabs is divided into three broad stages, which are the raw materials, transportation, and manufacturing stages. CO_2 assessment of the first stage requires evaluating the CO_2 of each ingredient in the voided slab. The raw materials of voided slabs are the mixed ingredients of concrete (i.e., cement, aggregates, admixture, and water), reinforcing bars, and void former materials (i.e., high-density polyethylene and expanded polystyrene). Each material was weighed in a certain ratio in compliance with the mix design of concrete and casted into cement moulds to produce voided slabs.

The transportation stage refers to CO_2 emissions during transporting the ingredients of voided slabs to the

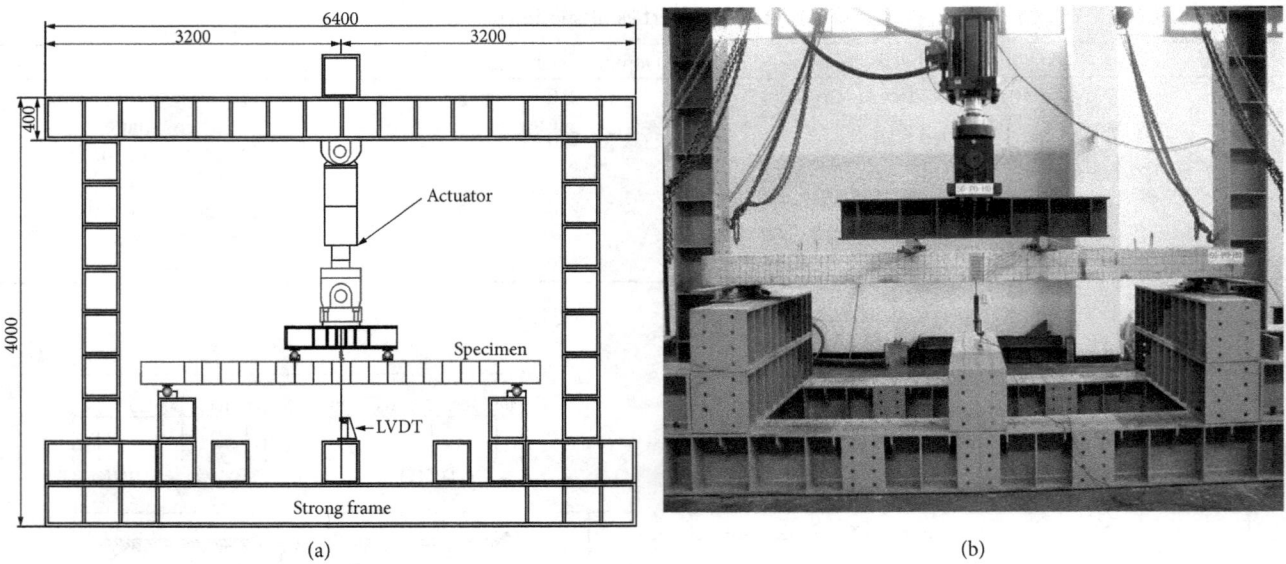

FIGURE 2: Test setting of the voided slab. (a) Flexural test setup diagram (unit: mm). (b) Test setting and loading configuration.

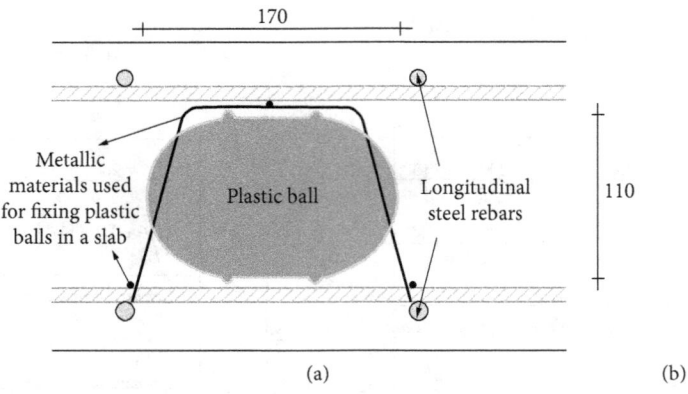

FIGURE 3: Details of the void former materials and anchoring. (a) Details of the void former materials and anchoring. (b) Photo of the void former materials

TABLE 2: Design of concrete mixture.

Designed compressive strength (MPa)	w/c (%)	s/a (%)	Unite content (kg/m^3)					Air content (%)
			Water	Cement	Fine aggregate	Coarse aggregate	Admixture	
27	45.7	44.3	108	333	821	1086	3.33	2.0

Note: w/c is the water/cement ratio, and s/a is the sand/aggregate ratio.

TABLE 3: Specifications and test results of reinforcing bars.

Types	Diameter (mm)	Yield strength (MPa)	Ultimate strength (mm)
D10	10	519	630
D13	13	531	651

manufacturing site. These occur during the manufacturing of the voided slabs from the electricity, gas, oil, etc., used in the manufacturing plant. The system boundary of life cycle CO_2 emissions of voided slabs is depicted in Figure 4.

3.2.2. Raw Materials Stage. Carbon dioxide emissions from the raw materials of voided slabs were calculated as the sum of the quantity of individual components, which were concrete, reinforcing bars, and void formers. The CO_2 emissions of concrete were evaluated through the sum of multiplication of the quantity of each ingredient utilised for producing 1 m^3 of concrete and the CO_2 emission factor for producing concrete. The following equation was used for computation of the CO_2 emissions of a unit of concrete:

$$CO_2M = \sum (M(i) \times CO_2 \text{emission factor } M), \quad (1)$$

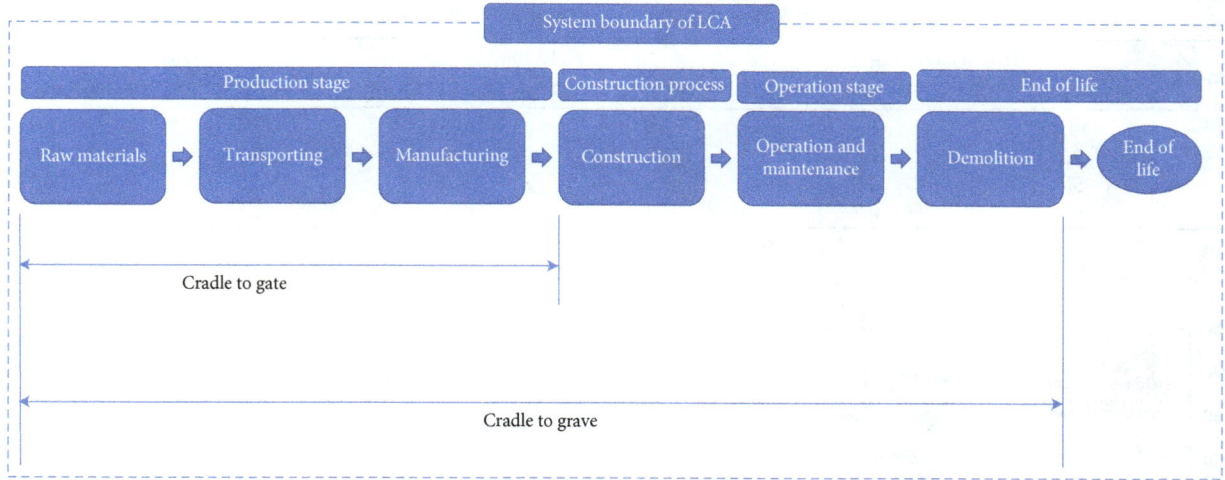

FIGURE 4: The system boundary of the voided slabs.

in which CO_2M (kg-CO_2/m^3) is the quantity of CO_2 emissions during the raw materials stage; M ($i = 1$: cement, 2: aggregate, 3: chemical admixture, 4: water, 5: reinforcing bar, 6: HDPE) is the amount of each material used to produce the concrete (kg/m^3); and CO_2 emission factor M (kg-CO_2/kg) is the CO_2 emission factor of each material used during the production of concrete.

The CO_2 emission factors of cement, aggregates, and water given in the Korea Life Cycle Database (LCI DB) [36] were applied. Since the CO_2 emission factor for chemical admixture was not listed in the Korea LCI DB, the overseas LCI DB [37] was applied to the chemical admixture. Table 4 shows the source of LCI DB for the individual ingredients of concrete.

3.2.3. Transportation Stage.
The CO_2 emissions of transportation occur as individual components of the voided slabs are delivered to the manufacturing site. The number of vehicles, distance from the origin to the manufacturing plant, and fuel efficiency of each vehicle were considered when assessing CO_2 emissions. The speed of the vehicles and the traffic situation were not considered. Equation (2) was used to calculate the CO_2 emissions from the transportation stage of the voided slabs:

$$CO_2T = \sum\left[\left(\frac{M(i)}{L_t}\right) \times \left(\frac{d}{e}\right) \times CO_2 \text{ emission factor } T\right]. \tag{2}$$

Here, the quantity of CO_2 emitted from transporting a unit of manufactured voided slab is CO_2T (kg-CO_2/m^3); the quantity of material applied in the voided slab is M ($i = 1$: cement, 2: aggregate, 3: chemical admixture, 4: water, 5: reinforcing bar, 6: HDPE) (kg/m^3); the transported load is L_t (tons); the transportation distance is d (km); e is the fuel efficiency of the transportation method (km/L); and CO_2 emission factor T (kg-CO_2/kg) is the CO_2 emission factor of the energy resource consumed by the transportation method.

3.2.4. Manufacturing Stage.
The CO_2 emission of voided slabs from the manufacturing process is the sum of

TABLE 4: Reference of life cycle inventory database.

Material	Unit	Source
Ordinary Portland cement	kg	The Korea LCI DB (South Korea)
Coarse aggregate	kg	The Korea LCI DB (South Korea)
Fine aggregate	kg	The Korea LCI DB (South Korea)
Chemical admixture	kg	Overseas LCI DB (ecoinvent)
Water	kg	The Korea LCI DB (South Korea)
HDPE	kg	The Korea LCI DB (South Korea)
Iron wire	Kg	The Korea LCI DB (South Korea)

consumed energy and unloading of raw materials and manufacturing of reinforcing bars and HDPE. The energy consumption of the manufacturing process was estimated based on the process of manufacturing which was divided into loading, storage, transportation, and mixing. The types of energy sources used in the voided slabs manufacturing process were electricity, diesel, liquefied natural gas (LNG), and water. The calculation of CO_2 emissions during the manufacturing is shown in the following equation:

$$CO_2F = \sum\left[\left(\frac{E(i)}{R}\right) \times CO_2 \text{ emission factor } F\right], \tag{3}$$

here, CO_2F is the amount of CO_2 occurring from a unit of voided slab manufacturing stage (kg-CO_2/m^3); $E(i)$ represents the annual energy usage (unit/year); R (m^3/year) denotes the annual production of concrete; and CO_2 emission factor F is the CO_2 emission factor of each energy resource (kg-CO_2/kg).

4. Results

4.1. Structural Analysis of the Voided Slabs.
The flexural test results of the voided slab specimens are summarised in Table 5 and Figure 5. The overall trend of cracking in the

TABLE 5: Test results of flexural performance.

Specimens	Initial cracking		Yield load		Ultimate load	
	Load (kN)	Displacement (mm)	Load (kN)	Displacement (mm)	Load (kN)	Displacement (mm)
S1	14.4	1.5	53.9	24.3	73.3	216.4
S2	9.2	1.0	60.7	34.0	70.3	141.8
S3	5.6	0.7	46.0	29.2	54.5	175.1
S4	5.4	0.7	38.1	40.4	45.6	213.2

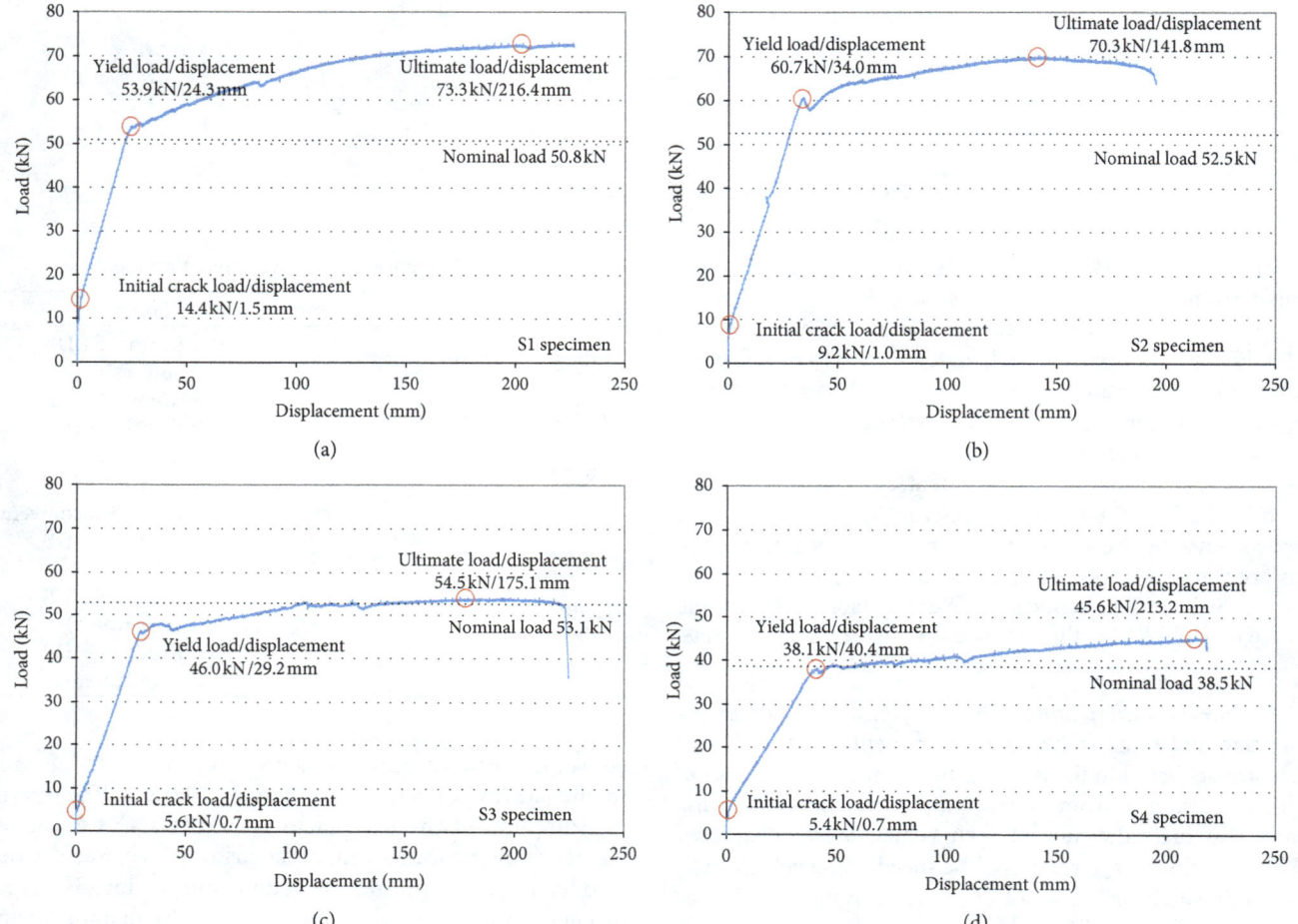

FIGURE 5: Load-displacement curves of specimens. (a) S1 specimen. (b) S2 specimen. (c) S3 specimen. (d) S4 specimen.

specimens indicated that, after initial flexural cracking occurred in all specimens, cracking gradually expanded to both ends as the loads increased. When flexural yielding occurred, the strength of specimens increased approximately 10–30% compared to the reference specimen S1, followed by flexural fracture in all specimens.

The comparison between the expected value of cracking strength and yield strength, calculated in accordance with the Structural Concrete Design Code and Commentary in South Korea [38], and the actual experimental values of the voided slabs are summarised in Table 6. According to the test results, the experimental value of each specimen was higher than the anticipated load. The experimental ultimate strength was about 115–139% of the yield strength of the specimens. A likely reason that the ultimate strength observed in the experiment exceeded the expected strength is that the wires that anchored the void formers influenced the strength of the voided slab.

4.2. Analysis of CO_2 Emission of the Voided Slabs. The CO_2 emissions of each component of the voided slab were 334.0749 kg-CO_2/m³ for concrete, 2.5266 kg-CO_2/kg for D10 reinforcing bars, 2.4858 kg-CO_2/kg for D13 reinforcing bars, 2.034 kg-CO_2/kg for wires, and 2.06 8kg-CO_2/kg for HDPE (Tables 7 and 8). The CO_2 emissions of reinforcing bars were calculated using input-output analysis.

TABLE 6: Comparison between anticipated and actual values.

Specimen	A Anticipated strength (kN)	B Anticipated strength with anchoring (kN)	C Ultimate strength (kN)	C/A	C/B
S1	50.8	—	73.3	1.443	—
S2	52.5	59.6	70.3	1.339	1.180
S3	53.1	57.5	54.5	1.026	0.948
S4	38.5	—	45.6	1.184	—

TABLE 7: CO_2 emission of concrete in product stage.

Items	Raw materials			Transportation			
	A	B	C = A * B	D	E	F	G
	kg/unit	kg-CO_2/kg	kg-CO_2/unit	Location	km	kg-CO_2/kg-km	kg-CO_2/kg
Concrete							
Ordinary Portland cement	333.000	0.948000	315.6840	Damyang	277.00	$6.06 * 10^{-5}$	5.590
Fine aggregate	821.000	0.000152	0.124792	Hadong	322.00	$1.16 * 10^{-5}$	3.067
Coarse aggregate	1086.00	0.007740	8.405640	Gyeonggi, Gwangju	37.60	$1.16 * 10^{-5}$	0.474
Chemical admixture	3.330	0.002050	0.006826	Gyeonggi, Anseong	66.000	$1.16 * 10^{-5}$	0.003
Sub sum			324.232274				9.133
Manufacturing	1 m^3	0.71 kg-CO_2/FU	0.710000				
Total			334.078975				

TABLE 8: CO_2 emission of voided slabs components in product stage.

Items	CO_2 emission	Transportation			D = A * B * C
		A Location	B Distance	C Factor	
HDPE	1.875	Shihwa	70.40	$2.735 * 10^{-3}$	0.193
		CO_2 emission + D			2.068
Wire	1.732	Dangjin	110.54	$2.735 * 10^{-3}$	0.302
		CO_2 emission + D			2.034
Rebar D10	2.5266	Dangjin	110.54	$2.735 * 10^{-3}$	0.302
		CO_2 emission + D			2.8286
Rebar D13	2.4858	Dangjin	110.54	$2.735 * 10^{-3}$	0.302
		CO_2 emission + D			2.788

4.2.1. CO_2 Emissions of the Voided Slabs with Consideration of Anchoring

(1) Overview of CO_2 Emissions. The CO_2 emissions of all specimens are summarised in Table 9. The total amount of concrete used in specimen S1 was 0.21 m^3, and the amounts of reinforcing bars used in S1 were 8.38 kg of D10 and 4.97 kg of D13. Based on the materials used in the specimen S1, the total emissions of carbon dioxide from S1 were 107.15 kg-CO_2/FU. In the case of S1, concrete accounted for the highest proportion of the volume. The amount of CO_2 emissions from concrete was 69.59 kg-CO_2, which accounted for 69.94% of the total CO_2 emissions (Figure 6).

The specimen S2 was made of voided slab of approximately 16% hollowness ratio. In this specimen, 64 spherical void formers of 100 mm diameter were used to fill the hollow sections of the slab. The volume of concrete used was 0.17 m^3, approximately 0.04 m^3 less than that used in the reference model S1. Reinforcing bars applied in the S2

TABLE 9: The CO_2 emissions of specimens (unit: kg-CO_2/FU).

Ingredients of the voided slab	S1	S2	S3	S4
Concrete				
Volume (m^3)	0.21	0.17	0.16	0.17
CO_2 emissions (kg-CO_2)	69.59	57.66	53.36	55.89
Rebars				
D10				
Weight (kg)	8.38	6.71	8.38	8.38
CO_2 emissions (kg-CO_2)	23.72	18.97	23.72	23.72
D13				
Weight (kg)	4.97	3.97	4.97	4.97
CO_2 emissions (kg-CO_2)	13.85	11.08	13.85	13.85
Wires				
Weight (kg)	—	19.86	7.76	—
CO_2 emissions (kg-CO_2)	—	34.42	13.44	—
HDPE				
Weight (kg)	—	3.90	3.98	—
CO_2 emissions (kg-CO_2)	—	7.31	7.45	—
Total emissions (kg-CO_2/FU)	107.15	129.44	111.82	93.45

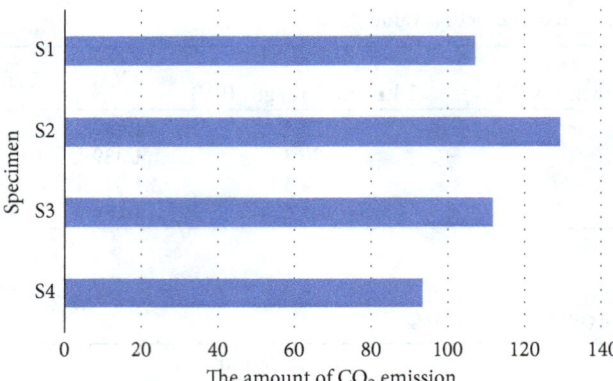

FIGURE 6: Total CO_2 emissions with anchoring.

specimen were 6.71 kg of D10 and 3.97 kg of D13. In addition, 19.86 kg of wires (ϕ6) were also used to address buoyancy of the void formers in the specimens. The CO_2 emissions of the specimen S2 were 129.44 kg-CO_2/FU. In this specimen, quite a large amount of wires was applied to anchor the void formers, which is the reason that the CO_2 emissions were higher than the ordinary reinforced concrete slab specimen S1.

The specimen S3 was voided slab model with about a 22% hollowness ratio, including 25 oblate shape void formers. The oblate shape void formers were 170 mm in length and 110 mm in height, and each component had a volume of 1,903,594 mm^3. The concrete applied in the S3 specimen was 0.16 m^3, and the quantity of CO_2 emitted from this concrete was 53.36 kg-CO_2. The rebars used in the S3 specimen were 8.38 kg for D10 and 4.97 kg for D13. The CO_2 emissions of each type of reinforcing bars in specimen S3 were 23.72 kg-CO_2 for D10 and 13.85 kg-CO_2 for D13. The total CO_2 emissions from deformed bars were 37.57 kg-CO_2. Furthermore, 7.76 kg of wires (ϕ6) was used to anchor the void formers in the section of the slab, from which the amount of carbon dioxide emissions was 13.44 kg-CO_2. Based on these data, the total amount of carbon dioxide emissions from the S3 specimen was 111.82 kg-CO_2/FU.

The S4 specimen was made of 0.17 m^3 of concrete which was comparable with the amount of concrete used in the S2 and S3 specimens. It was designed to compare the structural performance with the voided slabs. The mass of the reinforcing bars applied to the S4 specimen was 8.38 kg for D10 and 4.97 kg for D13. The carbon dioxide emitted by the S4 specimen was 93.45 kg-CO_2/FU.

(2) Comparisons of the CO_2 Emissions with Anchoring. The hollowness ratio is one of the most significant characteristics regarding the voided slab. In this study, the hollowness ratio of the specimens was established 0, 15.96, and 22.66% for the S1, S2, and S3 specimens, respectively. The CO_2 emissions from each sample showed distinctive features dependent upon the ratio of hollowness. Figure 7 shows the overall tendency of CO_2 emissions of the specimens based on the hollowness ratio of the slabs.

The first specimen, S1, was an ordinary slab which is normally applied to reinforced concrete buildings in South

Korea. The depth of this slab was 210 mm, which is the minimum depth of slab for prevention of noise complaints between floors of apartment housing in South Korea. The emitted carbon dioxide from the S1 specimen was 107.15 kg-CO_2/FU. Concrete was the highest carbon dioxide emitter among all of the components of this reinforced concrete slab. Reinforcing bars were the second largest source of CO_2 emissions in this specimen.

The S2 and S3 models were voided slab specimens that showed 129.44 kg-CO_2/FU and 111.82 kg kg-CO_2/FU, respectively. The S2 specimen had a 15.96% ratio of hollowness to the total volume of concrete. The hollow section of the slab was filled with void formers made from HDPE. The void formers were spherical, with 100 mm diameter, and 64 balls were inserted in the specimen. Twenty-five oblate shape void formers were buried in the S3 slab to create a ratio of 22.66% hollowness to concrete. The size of each component was 170 mm in length and 110 mm in height. All the void formers in both specimens were anchored by wires to prevent buoyancy or separation between reinforcing bars during the placing and curing of concrete. The amount of concrete used in the specimens was 0.17 m^3 for S2 0.16 m^3 for S3, respectively. Based on the reduction of concrete use in the slab, approximately 12 and 16 kg-CO_2/FU less CO_2 were emitted from the S2 and S3 specimens, respectively. As the CO_2 emissions from concrete decreased as the concrete use was reduced, the emissions from concrete also decreased in response to this tendency. However, the total CO_2 emissions from specimen S2 and S3 were higher than those in the S1 specimen. The S2 specimen showed about 21% increase of carbon dioxide emissions compared to the S1 specimen.

In addition, the S3 voided slab indicated approximately a 4% rise of CO_2 emissions over the S1 specimen. The reason for the increase of CO_2 emissions in the voided slabs might be the application of wires for anchoring the void formers. It is unavoidable for voided slabs to utilise anchoring materials such as wires, deck-plate, and wire mesh; for example, in this research, wires were utilised for the anchoring component of the voided slab. The amount of wires applied in both voided slab specimens was quite large: 19.86 kg for S2 and 7.76 kg for S2. Based on these data, the emitted carbon dioxide from the S2 and S3 specimens was 34.42 kg-CO_2 and 13.44 kg-CO_2, respectively. These results show that about 27 and 12% more carbon dioxide was emitted from anchoring wires, and anchoring wires accounted for a significant proportion of CO_2 emissions in the voided slab (Figure 8). In other words, wires for anchoring void formers to reinforcing bars were a source of large volumes of CO_2 emissions in the voided slab specimens.

Furthermore, the CO_2 emissions of the voided slab exhibited different features depending upon the shape of the void formers. Two different shapes of materials were applied to fill the hollow section of the concrete slab in this research. The S2 specimen, into which spherical void formers were inserted, showed higher CO_2 emissions than the S3 specimen, into which oblate shape ones were inserted. Oblate shape materials emitted approximately 18 kg-CO_2/FU less carbon dioxide than spherical ones. The reason for higher emissions of CO_2 from spherical materials might be that the

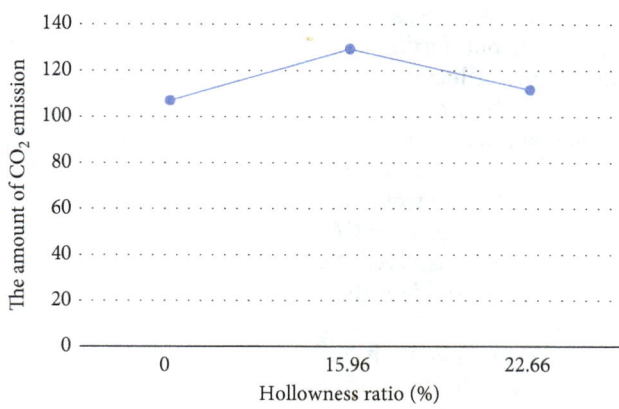

FIGURE 7: CO_2 emission variation based on hollowness ratio.

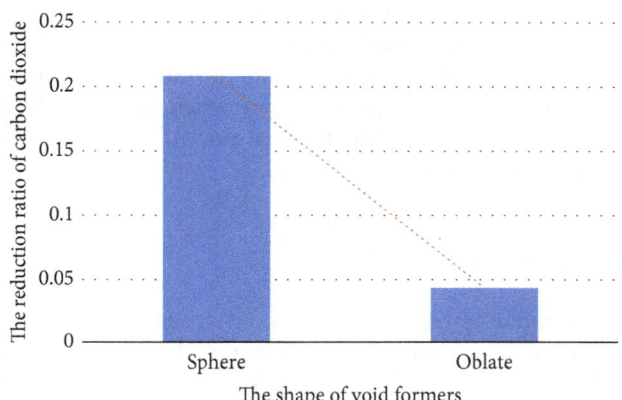

FIGURE 8: CO_2 emissions reduction ratio by the shape of void former materials.

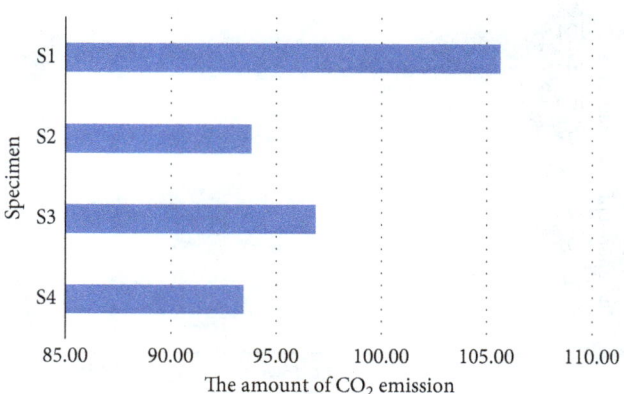

FIGURE 9: Total CO_2 emissions without anchoring.

individual spherical void formers were smaller than the oblate shape ones. As the size of each former was smaller, more anchoring wires may have been required to harness the formers in the voided slabs overall. As shown in Table 9, about 19.86 kg of wires were used in the S2 specimen, which was about 13 kg more than that in the S3 specimen. Moreover, the carbon dioxide emission of reinforcing bars and wires is different depending on their size. The CO_2 emissions of steel materials in construction works such as reinforcing bars, wires, wire mesh, and so forth might increase as the diameter of rebars or wires decreases [12, 39]. In this research, quite a large amount of smaller size wires was applied to anchor, and this might be a significant reason for the observed increase in carbon dioxide emissions.

(3) Comparisons of the CO_2 Emissions without Anchoring. Considering the CO_2 emissions in terms of the pure hollowness, which refers to the void slab without considering the utilisation of anchoring materials, the amount and characteristics of carbon dioxide emissions were considerably different than when anchoring materials were considered. Figure 9 shows total carbon dioxide emissions of the specimens without consideration of anchoring wires.

The CO_2 emissions of S2, S3, and S4 specimens were 95.0241, 98.3753, and 93.4536 kg-CO_2/FU, respectively. The

reduced amount of CO_2 compared to the reference model S1 was 12.1265 kg-CO_2/FU for S2, 8.7754 kg-CO_2/FU for S3, and 13.6971 kg-CO_2/FU for S4 (Figure 9). When the CO_2 reduction ratio was considered, S2 and S3 showed 11.32 and 8.19% reduced amounts of CO_2 versus the reference model, S1 (Figure 10). Since concrete was one of the main sources of CO_2 emission in reinforced concrete structures, removal of concrete would seem to have a significant impact on the reduction of carbon dioxide emissions. While the void formers accounted for quite a large volume in the voided slab, the CO_2 emissions from the void formers was of smaller proportion because the thickness of each material was only 20 mm.

In summary, the CO_2 emissions of higher void ratios were lower in this research. The reason for this tendency is the application of more anchoring materials, since the size of the individual components in the higher hollowness ratio was smaller than in the lower hollowness ratio. Thus, more utilisation of materials led to the occurrence of more carbon dioxide emissions in this research.

5. Discussion and Limitations

It is generally known that voided slabs are environmentally friendly and beneficial for a sustainable environment. In addition, utilisation of less concrete would lead to lower energy consumption and carbon dioxide emissions from voided slabs. However, the carbon dioxide emissions of voided slabs showed higher CO_2 emissions compared to normal reinforced concrete slabs in this research.

The structural performance of the voided slab proved to be comparable to a normal reinforced concrete slab. In this study, the flexural performance of the voided slab was similar or slightly higher than the normal reinforced concrete slab. From a structural perspective, the voided slab would be appropriate to apply to long span structures due to its light weight, as well as to prevent noise complaints between floors in apartment housing in South Korea. Moreover, the application of the voided slab would make it possible to remove beams in reinforced concrete structure. The voided slab is thus an alternative to lightweight materials.

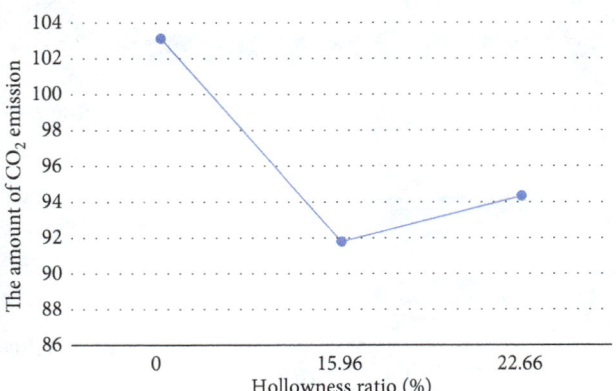

FIGURE 10: CO_2 variation by void ratio without anchoring.

When comparing the carbon dioxide emissions of normal reinforced concrete slab and voided slab, the CO_2 emissions of the voided slab were higher than those of the normal reinforced concrete slab, considering all the materials used in the voided slab such as concrete, void former materials, reinforcing bars, and wires for anchoring. In particular, the amount of anchoring wires to manage buoyancy occupied a considerable proportion of the total amount of carbon dioxide emissions, approximately 25% in each specimen. Based on these results, replacing or reducing current anchoring materials to more environmentally friendly ones should be considered to lower the amount of CO_2 emissions in the voided slab.

Additionally, the variation of carbon dioxide emissions by the application of different types of void formers (i.e., spherical and oblate shape) was studied. In this study, oblate shape void formers emitted 4.35% in S2 and 20.80% in S3 additional carbon dioxide. Since the overall size of the spherical type was smaller than the oblate one, more wires were used to anchor the void formers to the reinforcing bars. The anchoring wires were one of the main sources of CO_2 emissions in the voided slab.

It is normally considered that plastics or other petrochemical products might be less effective than other materials for environmental performance. The CO_2 emission per unit weight of petrochemical material (i.e., HDPE in this study, which was used for void materials) was higher than other components of voided slab. However, the amount of HDPE applied in the specimens was smaller than other materials such as concrete, reinforcing bars, and wires. Thus, the influence of carbon dioxide emissions of HDPE in the voided slab was not significant compared to other materials.

In order for the voided slab to meet both environmental and structural performance requirements, current anchoring methods using wires, deck-plate, cable ties, and so forth should be replaced with more eco-friendly or recycled materials. Additionally, other methods to connect void former materials directly to reinforcing bars should be developed, in order to reduce the application of anchoring materials in the voided slab. Based on such research approaches, further studies should consider the optimal design of the voided slab with using optimal materials which would meet both structural and environmental performance.

This research was limited to the slab component of a building, but further studies should be carried out to confirm the reduction of construction materials and variation of carbon dioxide emissions in the entire building through application of the voided slab. Although this study only focused on the case of South Korean construction industry, this study would also provide a useful reference for assessing the emission of CO_2 from the voided slab to other countries. It is considered that further studies should be also carried out to validate the cases of voided slab in other countries.

Moreover, this study only dealt with the emission of carbon dioxide for the voided slab during the manufacturing phase. However, it is recently considered that the significant impact of operational carbon of a building's life cycle. As for this reason, further research should be conducted to assess the influence of operational carbon for the voided slab system compared to the normal reinforced concrete structure. Along with the consideration of operational carbon for the voided slab system, other factors such as geographical aspect which would impact the transportation distance and energy consumptions.

6. Conclusions

The purpose of this research was to confirm the flexural performance as well as the variation of carbon dioxide emissions from voided slabs. The flexural performance was assessed based on the variables of the depth of slab, types of the void former materials, and hollowness ratio. In addition, CO_2 emission comparisons were also conducted considering the hollowness ratio and types of void formers compared to the normal reinforced concrete slab. The system boundary of the voided slab was limited to the production stage, known as cradle to gate, in accordance with ISO 14044 [30].

The structural performance of the voided slab was similar or slightly better than the normal reinforced concrete slab. The yield strength of specimens increased approximately 10–30% over the anticipated yield strength. Based on this result, it is considered that the voided slab has sufficiently good structural performance and would be beneficial to planning in buildings.

The results of assessment of the CO_2 emissions showed that the voided slabs emitted more carbon dioxide compared to the normal reinforced concrete slab, regardless of the hollowness ratio and the types of void formers. In this research, the slab with a higher hollowness ratio emitted less carbon dioxide emissions than that with a lower hollowness ratio. Additionally, the voided slab would require additional materials to anchor the void formers to the reinforcing bars in order to prevent buoyancy of the formers during the placing and curing of concrete.

In general, a number of studies have shown that voided slabs are beneficial both structurally and environmentally. However, the results of this study showed that voided slabs emitted more carbon dioxide compared to normal reinforced concrete slabs. The main source of this additional CO_2 from the voided slab might be the materials anchoring the void formers to the reinforcing bars. In this research, wires were used to fix the void formers, and they were a reasonably large source of CO_2 emissions. In order for the voided slab to become a more eco-friendly and sustainable material in buildings, new anchoring methods such as use of recycled materials, new void formers without anchoring, or other eco-friendly materials should be developed to reduce the emission of CO_2.

Conflicts of Interest

The authors declare that they have no conflicts of interest.

Acknowledgments

This research was supported by the Basic Science Research Program through the National Research Foundation of Korea (NRF) funded by the Ministry of Education (2017R1D1A1B03032279).

References

[1] Korea Concrete Institute, *Concrete and Environment*, Korea Concrete Institute, Seoul, Korea, 2nd edition, 2016.

[2] A. Akbarnezhad and J. Xiao, "Estimation and minimization of embodied carbon of buildings: a review," *Buildings*, vol. 7, no. 4, p. 5, 2017.

[3] P. Purnell, "The carbon footprint of reinforced concrete," *Advances in Cement Research*, vol. 25, no. 6, pp. 362–368, 2013.

[4] S. H. Cho and C.-U. Chae, "The comparative study on the environmental impact assessment of construction material through the application of carbon reducing element-focused on global warming potential of concrete products," *International Journal of The Korea Institute of Ecological Architecture and Environment*, vol. 33, pp. 149–156, 2015.

[5] S. Tae, C. Baek, and S. Shin, "Life cycle CO_2 evaluation on reinforced concrete structures with high-strength concrete," *Environmental Impact Assessment Review*, vol. 31, no. 3, pp. 253–260, 2011.

[6] S. Tae, S. Shin, J. Woo, and S. Roh, "The development of apartment house life cycle CO_2 simple assessment system using standard apartment houses of South Korea," *Renewable and Sustainable Energy Reviews*, vol. 15, no. 3, pp. 1454–1467, 2011.

[7] S.-H. Kim, I.-S. Kang, and H.-S. Lee, "Experimental study on the flexural behavior of I-slab," *Proceedings of Korea Concrete Institute*, vol. 19, no. 2, pp. 5–8, 2007.

[8] Y.-S. Han and S.-D. Kim, "A comparative study on CO_2 amount of construction material in structural design," *Journal of the Architectural Institute of Korea*, vol. 25, no. 1, pp. 203–206, 2005.

[9] M. J. González and J. G. Navarro, "Assessment of the decrease of CO_2 emissions in the construction field through the selection of materials: practical case study of three houses of low environmental impact," *Building and Environment*, vol. 41, no. 7, pp. 902–909, 2006.

[10] C. K. Chau, F. W. H. Yik, W. K. Hui, H. C. Liu, and H. K. Yu, "Environmental impacts of building materials and building services components for commercial buildings in Hong Kong," *Journal of Cleaner Production*, vol. 15, no. 18, pp. 1840–1851, 2007.

[11] S. H. Cho and C. U. Chae, "A study on life cycle CO_2 emissions of low-carbon buildings in South Korea," *Sustainability*, vol. 8, no. 6, p. 579, 2016.

[12] S. Cho and S. Na, "The reduction of CO_2 emissions by application of high-strength reinforcing bars to three different structural systems in South Korea," *Sustainability*, vol. 9, 2017.

[13] D. Y. Moon, "Flexural behavior of concrete beams reinforced with high-strength steel bars," *Journal of Korean Society of Disaster and Security*, vol. 13, no. 6, pp. 107–113, 2013.

[14] Z. S. M. Nadoushani and A. Akbarnezhad, "Effects of structural system on the life cycle carbon footprint of buildings," *Energy and Buildings*, vol. 102, pp. 337–346, 2015.

[15] C. Baek, S. Tae, R. Kim, and S. Shin, "Life cycle CO_2 assessment by block type changes of apartment housing," *Sustainability*, vol. 8, no. 8, p. 752, 2016.

[16] V. Penasadés-Plà, J. V. Martí, T. García-Segura, and V. Yepes, "Life-cycle assessment: a comparison between two optimal post-tensioned concrete box-girder road bridges," *Sustainability*, vol. 9, no. 10, pp. 1–21, 2017.

[17] F. Molina-Moreno, J. V. Martí, and V. Yepes, "Carbon embodied optimization for buttressed earth-retaining walls: implications for low-carbon conceptual designs," *Journal of Cleaner Production*, vol. 164, pp. 872–884, 2017.

[18] J. V. Martí, T. García-Segura, and V. Yepes, "Structural design of precast-prestressed concrete U-beam road bridges based on embodied energy," *Journal of Cleaner Production*, vol. 120, pp. 231–240, 2016.

[19] M. Schnellenbach-Held and K. Pfeffer, "Punching behavior of biaxial hollow slabs," *Cement & Concrete Composites*, vol. 24, no. 6, pp. 551–556, 2002.

[20] H. Hwang, "Structural performance of hollow reinforced concrete half slabs," *Proceedings of Korea Concrete Institute*, vol. 20, no. 1, pp. 45–48, 2008.

[21] J. Kang, S. Park, and G. Kim, "Application of hollow tube system to concrete slab for improvement of impact sound insulation," *Proceedings of Architectural Institute of Korea*, vol. 24, no. 12, pp. 75–82, 2008.

[22] M. Aldejohann and M. Schnellenbach-Held, "Investigations on the shear capacity of biaxial hollow slabs-Test results and evaluation," *Darmstadt Concrete*, vol. 18, pp. 1–11, 2003.

[23] I. S. Ibrahim, K. S. Elliott, R. Abdullah, A. B. H. Kueh, and N. N. Sarbini, "Experimental study on the shear behaviour of precast concrete hollow core slabs with concrete topping," *Engineering Structures*, vol. 125, pp. 80–90, 2016.

[24] J. Wang and V. W. Y. Tam, "Construction industry carbon dioxide emissions in Shenzhen, China," *Proceedings of the Institute of Civil Engineers–Waste and Resource anagement*, vol. 169, no. 3, pp. 114–122, 2016.

[25] RICS, *Methodology to Calculate Embodied Carbon*, Royal Institution of Chartered Surveyors, London, UK, 2014.

[26] G. P. Hammond and C. I. Jones, "Embodied energy and carbon in construction materials," *Proceedings of the Institution of Civil Engineers-Energy*, vol. 161, no. 2, pp. 87–98, 2008.

[27] G. Baiocchi and J. C. Minx, "Understanding changes in the UK's CO_2 emissions: a global perspective," *Environmental Science and Technology*, vol. 44, no. 4, pp. 1177–1184, 2010.

[28] T. H. Kim, C. Chae, G. Kim, and H. Jang, "Analysis of CO_2 emission characteristics of concrete used at construction sites," *Sustainability*, vol. 8, no. 4, p. 348, 2016.

[29] J. Park, S. Tae, and T. Kim, "Life cycle CO_2 assessment of concrete by compressive strength on construction site in Korea," *Renewable and Sustainable Energy Reviews*, vol. 16, no. 5, pp. 2940–2946, 2012.

[30] S. Lee, W. Park, and H. Lee, "Life cycle CO_2 assessment method for concrete using CO_2 balance and suggestion to decrease $LCCO_2$ of concrete in South-Korean apartment," *Energy and Buildings*, vol. 58, pp. 93–102, 2013.

[31] H. Yan, Q. Shen, L. C. H. Fan, Y. Wang, and L. Zhang, "Greenhouse gas emissions in building construction: a case study of One Peking in Hong Kong," *Building and Environment*, vol. 45, no. 4, pp. 949–955, 2010.

[32] R. J. Cole, "Energy and greenhouse gas emissions associated with the construction of alternative structural systems," *Building and Environment*, vol. 34, no. 3, pp. 335–348, 1998.

[33] Korean Standards Association, *Korea Standard Information Centre, KS B 0802*, Korea Standards Association, Seoul, Korea, 2003.

[34] International Organization for Standardization, *ISO14044: Life Cycle Assessment (Requirement and Guidelines)*, ISO, Geneva, Switzerland, 2006.

[35] International Organization for Standardization, *ISO:21930: Environmental Declaration of Building Product*, ISO, Geneva, Switzeland, 2007.

[36] The Korea Environmental Industry & Technology Institute (KEITI), "Korea LCI DB Information Network," 2017, http://www.epd.or.kr/en/lci/lci_intro.asp.

[37] Ecoinvent, "Ecoinvent version 3.01 database," 2013, https://ecoquery.ecoinvent.org/.

[38] Korea Institute of Concrete, *Structural Concrete Design Code and Commentary*, Korea Institute of Concrete, Seoul, Korea, 2012.

[39] T.-H. Hong, C.-Y. Ji, and M.-H. Jang, "An analysis on CO_2 emission of structural steel materials by strength using input-output analysis," *Korean Journal of Construction Engineering and Management*, vol. 13, no. 4, pp. 132–140, 2012.

15

Identification of Key Indicators for Sustainable Construction Materials

Humphrey Danso ⓘ

Department of Construction and Wood Technology, University of Education Winneba, P.O. Box 1277, Kumasi, Ghana

Correspondence should be addressed to Humphrey Danso; dansohumphrey@yahoo.co.uk

Academic Editor: Nadezda Stevulova

Studies on sustainable construction materials are on the rise with their environmental, social, and economic benefits. This study identifies the key indicators for measuring sustainable construction materials. The design used for the study was that of a survey which relied on a questionnaire with five-point Likert scale to generate data for the analysis. For this purpose, 25 indicators from the three dimensions (environmental, social, and economic) identified from the literature were presented to the respondents in a structured questionnaire, and responses were collected and analysed using SPSS. The study identified three key environmental indicators for measuring sustainable construction materials, and these indicators are human toxicity, climate change, and solid waste. Furthermore, adaptability, thermal comfort, local resources, and housing for all were identified as the four key social indicators for sustainable construction materials. In addition, maintenance cost, operational cost, initial cost, long-term savings, and life span were found to be the five key economic indicators for measuring sustainable construction materials. The study therefore suggests that these twelve indicators should be considered in future studies that seek to measure sustainable construction materials.

1. Introduction

The choice of construction materials has wide-reaching economic, environmental, and social consequence on any structure. The total cost of most structures depends greatly on the types of materials used. The energy consumption of any building has a significant relationship with the materials used for construction and the design of the building. Adaptability of some buildings by the society sometimes has bearing on the types of materials used. This therefore makes construction material an important factor in sustainable development. According to Ness et al. [1], sustainable development has been integrated into many aspects of the society in recent years. This has made the concept of sustainable development an important aspect of many industries worldwide [2]. According to Mailler [3], sustainability is an integrated concept, whereby environmental, social, and economic factors are interrelated. Sustainable development is a concept that is applicable in almost all fields of study and industry.

In the field of construction, sustainability is of great concern due to huge capital injunction and environmental factors coupled with societal adaptability. Construction materials and methodologies applied in building of structures have great influence in the sustainable development issues in the field of construction. Sustainable buildings take advantage of the natural resources available and depend on a "green" choice of materials [4]. According to Buildabroad [4], the availability of sustainable construction materials is on the rise, with new innovations and sourcing of materials that are not detrimental to the environment and are designed to enhance the energy efficiency of buildings. Sustainable construction materials are materials that are cost-effective and socially accepted and reduce negative environmental impacts, today and into the future during manufacturing and building structures. Peach [5] mentioned that sustainable construction materials include wool bricks, solar tiles, sustainable concrete, paper insulation, and triple-glazed windows. In low-income communities, earth-based technologies (such as rammed earth, compressed

TABLE 1: Some studies on the properties of sustainable construction materials produced with waste.

Material	Reference
Boron waste in soil bricks	[12]
Cassava peel in soil blocks	[13]
Coconut (coir) fibre in soil blocks	[14–17]
Coconut fibre in concrete	[18]
Waste paper in concrete	[19]
Waste paper in sandcrete blocks	[20]
Date palm in soil blocks	[21]
Flax (harakeke) in soil blocks	[22]
Hemp in concrete	[23]
Kenaf in soil blocks	[24]
Oil palm fibre in soil blocks	[25–27]
Pineapple leaves fibre in soil blocks	[28]
Plastic in soil blocks	[29–31]
Plastic in concrete	[32]
Sawdust in soil blocks	[33]
Scrap tire (crumb) rubber in soil blocks	[34, 35]
Seaweed fibre in soil blocks	[36]
Sheep wool in soil blocks	[37, 38]
Sisal fibre in soil blocks	[39]
Straw in soil blocks	[40–43]
Sugarcane bagasse ash in soil blocks	[44]
Sugarcane bagasse fibre in soil blocks	[14, 25]
Waste phosphogypsum and natural gypsum in soil blocks	[45]
Waste tea residue in soil bricks	[46]

earth blocks, and stabilised soil blocks) are the commonly used sustainable construction materials.

A number of studies [6–11] have reviewed the volume of literature in earth-based sustainable construction materials. Danso et al. [6] reviewed the existing published works on the effect of stabilisers (fibres and binders) on technical performance of soil blocks or bricks using parameters such as compressive strength, water absorption, and flexural strength. Delgado and Guerrero [7] also reviewed the state of use of the earth building in Spain which presented research organisations, modern projects carried out, and the existing manufacturers for compressed earth blocks, and also examined a pair of nonregulatory guides that could act as national reference documents. Hejazi et al. [8] reviewed the history, benefits, application, and possible executive problems of using different types of natural and/or synthetic fibres in soil reinforcement. Maniatidis and Walker [10] made a review of rammed earth construction by considering national codes, materials, structural design, construction, quality control, foundations, and maintenance. Pacheco-Torgal and Jalali [11] also reviewed some of the environmental benefits associated with earth construction including an overview about its past and present. They also included a review of economic issues, nonrenewable resource consumption, waste generation, energy consumption, carbon dioxide emissions, and indoor air quality.

The use of waste in producing sustainable construction materials is generating research interest in the last decade. Studies have investigated the properties of sustainable construction materials produced with waste in concrete or blocks. Table 1 outlines some of these studies and the types of sustainable construction materials they investigated.

It can be seen from Table 1 that some studies have investigated the properties of sustainable concrete with waste materials such as paper, plastic, and natural fibres. Earth-based construction materials have rather seen a large volume of research works. While rammed earth has also seen a number of studies into its properties, earth/soil blocks/bricks have recorded a massive investigation with different waste materials such as natural fibres, sawdust, scrap tire rubber, plastics, and animal wool fibres. All these studies contribute to the literature on sustainable construction materials.

Identifying and measuring the indicators for sustainable construction materials is relevant in placing construction material studies in the context of sustainable development. Danso [2] reviewed the general and relevant available indicators for measuring sustainable construction materials. However, there is lack of information on the most important sustainable indicators in relation with construction materials. This paper therefore identifies the key indicators for measuring sustainable construction materials.

2. Methodology

The design used for this study was that of a survey which relied on a questionnaire to generate data for the analysis. The questionnaire was developed, pretested to a sample of three experts in construction materials, and then modifications were made to obtain a more efficient instrument. A five-point Likert scale was used to measure the key indicators for sustainable construction materials. The Likert scale ranged from unimportant (1) to very important (5). Three main dimensions used are environmental, social, and economic. The environmental dimension consisted of eleven indicators, social dimension had eight indicators, and economic dimension had six indicators, making a total of 25 indicators. These indicators were adopted from the previous study [2] which reviewed the general and relevant available indicators for measuring sustainable construction materials. The indicators under their various dimensions are shown in Figure 1 which presents the hierarchical structure for construction material sustainability assessment. The questionnaire also asked the respondents to add any indicators they consider important and make any suggestions.

The population for the study was the professionals from the construction industry and academics who have researched and published in the area of construction and building materials. A purposive sampling technique was adopted to select the professional for data collection. The professionals were identified through their published materials (articles, books, and reports), and their email addresses noted. The questionnaire was designed with Adobe Acrobat Professional version 11, which made it easy to be emailed as an attachment to the respondents. In the email, the respondents were told to download, click preferred responses, save, and return by email. The questionnaire was administered to the construction professionals through their email addresses. After a month, a reminder was sent to those who have not responded. It must be noted that few emails

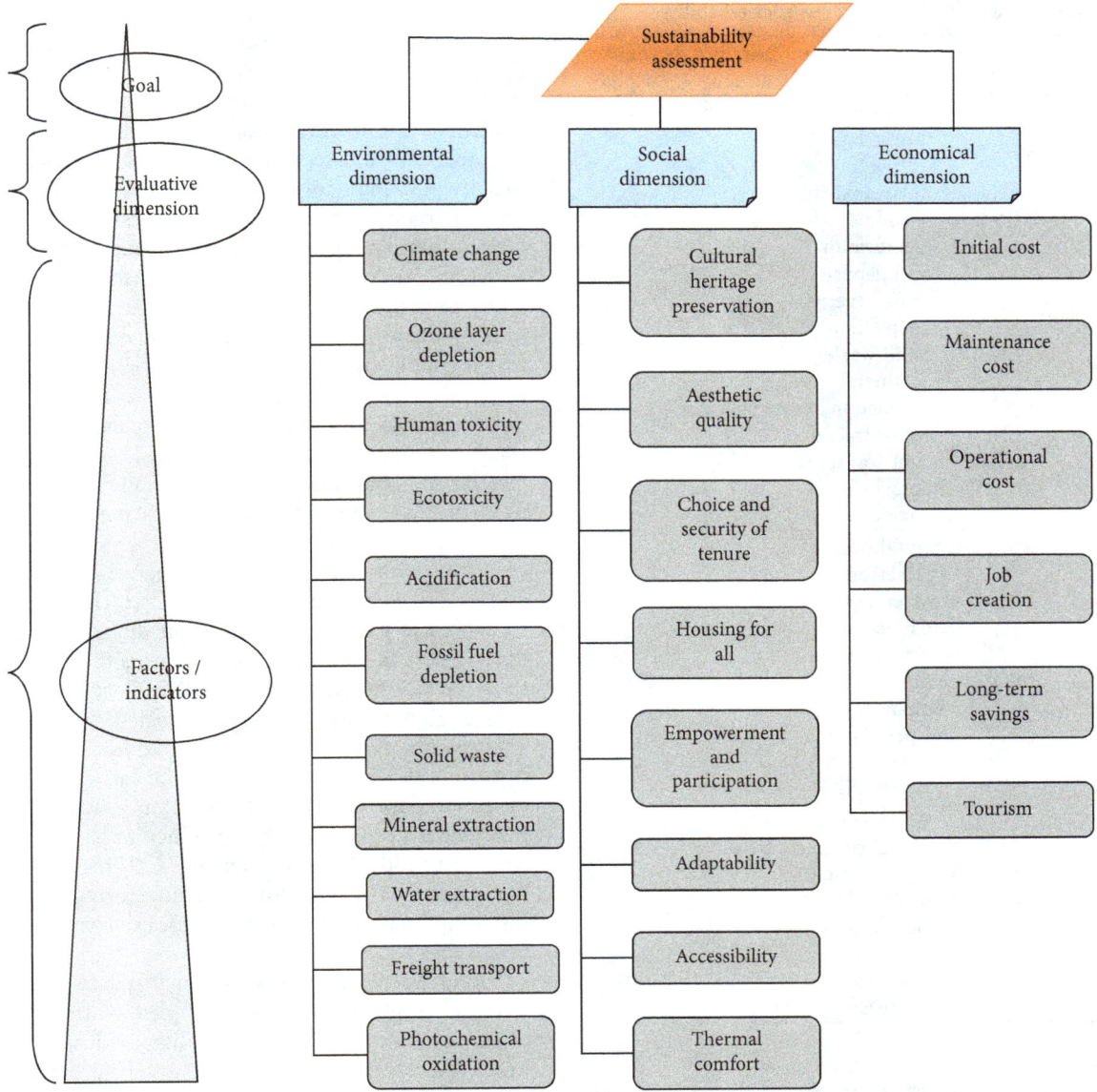

FIGURE 1: Hierarchical structure for construction material sustainability assessment [2].

were returned as error which might be due to invalid and wrong addresses.

The completed and returned questionnaires were downloaded, printed, and coded. The data obtained were analysed using the statistical package for social sciences (SPSS) version 22. The data from the questionnaire were inputted into SPSS and analysed by the use of descriptive statistics for obtaining the mean and standard deviation. In determining the key indicators for each dimension, a hypothesized mean of 4.000 and above was set as a critical cutoff point. It means that any indicator that recorded a mean of 4.000 or above was considered a key indicator, which the study identified as measuring sustainable construction materials. This represented the respondents' choice of indicators ranked between important and very important.

3. Results and Discussion

Out of the 115 questionnaires administered, 45 usable questionnaires were obtained, representing 39% response

rate. Construct validity was also ensured by critically developing the indicators within an established theoretical framework. The Cronbach alpha reliability test for the items was above the recommended 0.7 [47]. From Table 2, all of the constructs have item loadings higher than the recommended 0.70. Data were analysed using descriptive analysis. The responses of the respondents are presented under three dimensions: (1) environmental, (2) social, and (3) economic.

3.1. Key Environmental Indicators. Table 3 presents the analysis of the environmental indicators of sustainable construction materials. The descriptive analysis shows that three items recorded mean values of 4.000 or more, which implies that the respondents rated the items between important and very important. The first ranked indicator for the environmental dimension is "human toxicity" with a mean value of 4.267. The following two indicators are "climate change" and "solid waste," both with a mean value of 4.000. The

TABLE 2: Item loading and construct reliability.

Dimension	Code	Indicators	Number of items	Cronbach alpha
Environmental	EN1	Climate change	11	0.734
	EN2	Ozone layer depletion		
	EN3	Human toxicity		
	EN4	Ecotoxicity		
	EN5	Acidification (acid deposit)		
	EN6	Fossil fuel depletion		
	EN7	Solid waste		
	EN8	Mineral extraction		
	EN9	Water extraction		
	EN10	Freight transport		
	EN11	Photochemical oxidation		
Social	SO1	Cultural heritage preservation	8	0.875
	SO2	Aesthetic quality		
	SO3	Choice and security of tenure		
	SO4	Housing for all		
	SO5	Empowerment and participation		
	SO6	Adaptability		
	SO7	Accessibility		
	SO8	Thermal comfort		
Economic	EC1	Initial cost	6	0.820
	EC2	Maintenance cost		
	EC3	Operational cost		
	EC4	Job creation		
	EC5	Long-term savings		
	EC6	Tourism		

TABLE 3: Descriptive statistics on environmental indicators ($n = 45$).

Indicator	$\overline{x}$	σ	R
Human toxicity	4.267*	1.009	1
Climate change	4.000*	1.108	2
Solid waste	4.000*	0.739	2
Ecotoxicity	3.933	0.780	3
Water extraction	3.867	0.815	4
Photochemical oxidation	3.800	1.057	5
Acidification	3.667	1.023	6
Fossil fuel depletion	3.667	1.023	6
Ozone layer depletion	3.667	1.087	6
Mineral extraction	3.600	0.963	7
Freight transport	3.000	0.826	8

*Key indicator, mean ≥ 4.000.

other environmental indicators (ecotoxicity, water extraction, photochemical oxidation, acidification, fossil fuel depletion, ozone layer depletion, mineral extraction, and freight transport) recorded values less than the hypothesized mean.

From the results, it can be seen that three indicators were identified as the key environmental sustainable construction materials. These indicators are human toxicity, climate change, and solid waste. Human toxicity aims to quantify the degree to which a particular substance causes damage to living organisms. Assessments of toxicity are based on guidelines for tolerable concentrations in air and water, tolerable daily intake, and acceptable daily intake for human toxicity [48]. Many of the chemicals used in the society have not undergone a risk assessment, and assessment techniques are still developing, something that manufacturers of construction products should also be aware of [48]. Issues relating to toxicity generate much debate in the construction industry, especially manufacturing of construction materials and their disposal after demolition. Users of construction materials should carefully review the material supplier's guidance and note any relevant regulations, codes, and standards appropriately. They should also consider the context and application within which the materials are to be used to ensure that industry-specific regulations or standards are complied with.

Climate change refers to the change in global temperature caused by the greenhouse effect by the release of "greenhouse gases" such as carbon dioxide by human activity. There is now scientific consensus that the increase in these emissions is having a noticeable effect on climate. Raised global temperature is expected to cause climatic disturbance, desertification, rising sea levels, and spread of disease [49]. Embodied carbon is the carbon dioxide (CO_2) or greenhouse gas (GHG) emissions associated with the manufacture and use of a product or service. For construction products, this means the CO_2 or GHG emissions associated with extraction, manufacturing, transporting, installing, maintaining, and disposing of construction materials and products [48].

Solid waste represents the environmental issues associated with the loss of resource implied by the final disposal of waste, and any waste that is disposed of in landfill or incinerated without energy recovery. The characterisation methodologies, for example, the Dutch EcoIndicator30 and the Swiss Ecopoints31, and the characterisation factors are based on the mass of solid waste [49]. Solid waste generated by construction materials is huge, from manufacturing of the materials, application of the materials, and disposal of the demolished structures.

3.2. Key Social Indicators. The analysis of the social indicators of sustainable construction materials can be seen in Table 4. The result indicates that four indicators were rated by the respondents as the key indicators within the theoretical mean of 4.000. "Adaptability" was the first ranked indicator with a mean value of 4.133. The second ranked indicator was "thermal comfort" with a mean value of 4.067. The next ranked indicators were "local resources" and "housing for all," both with a mean value of 4.000. Indicators such as cultural heritage, social value, choice and security of tenure, accessibility, empowerment and participation, and aesthetic quality had values less than the hypothetical mean.

Adaptable buildings are defined as "dynamic systems that carry the capacity to accommodate a set of evolving demands

TABLE 4: Descriptive statistics on social indicators ($n = 45$).

Indicator	$\overline{x}$	σ	R
Adaptability	4.133*	0.815	1
Thermal comfort	4.067*	0.939	2
Local resources	4.000*	0.977	3
Housing for all	4.000*	1.108	3
Cultural heritage	3.867	1.217	4
Social value	3.733	0.688	5
Choice and security of tenure	3.733	0.939	5
Accessibility	3.600	1.095	6
Empowerment and participation	3.600	1.156	6
Aesthetic quality	3.533	0.815	7

*Key indicator, mean ≥ 4.000.

TABLE 5: Descriptive statistics on economic indicators ($n = 45$).

Indicator	$\overline{x}$	σ	R
Maintenance cost	4.400*	0.889	1
Operational cost	4.400*	0.889	1
Initial cost	4.400*	0.889	1
Long-term savings	4.333*	0.879	2
Life span	4.067*	1.136	3
Job creation	3.667	1.087	4
Tourism	2.800	0.919	5

*Key indicator, mean ≥ 4.000.

regarding space, function, and components [50]". To ensure sustainability, building materials must build in adaptability to both its existing and new buildings. Buildings are more likely to be occupied and reused if they can be easily adapted to meet changing needs [51]. Buildings which are unable to adapt with such changing needs will become obsolete or require substantial refurbishment or demolition, where neither option may create a sustainable built environment [52].

Thermal comfort is the condition of mind which expresses satisfaction with the thermal environment [53], that is, the condition when someone is not feeling either too hot or too cold [54]. Building materials contribute greatly to thermal comfort of any structure. Adaptive thermal comfort broadens understanding of the human comfort zone by taking into account the ways that people's perceptions of their environment change based on seasonal expectations of temperature and humidity [55]. It is an undeniable fact (especially in warm climate areas) that houses built with local materials have cool room temperature particularly with houses built with soil or earth and thatches [56]. The consideration of thermal comforts is therefore important when selecting construction materials.

Local resources (materials) are usually available and affordable in most localities. Nature has provided mankind with some wonderful materials to build houses, and these materials require little processing or transporting and the costs are low [57]. It is important to identify and use locally manufactured and available materials (sand, stones, grass, thatches, clay, timber, clay bricks, and clay blocks) in providing houses especially in the developing countries in order to meet their housing demand [56].

"Housing for all" is a concept that describes the situation of ensuring that housing becomes affordable and available to every individual. Housing for all is therefore associated with affordable or low-cost housing. UN Habitat [58] defined affordable housing as a house that is adequate in quality and location and does not cost such that it prohibits its occupants from meeting other living costs and threatens their enjoyment of basic human rights. Low-cost housing on the other hand is a housing concept whose total cost for purchase or rent is deemed affordable for those in the median income bracket [59, 60]. The factors that contribute to housing for all include cost of materials, cost of labour, and cost of land, among others. The cost of construction materials constitutes between 60% and 70% of the cost of a building in developing countries [61]. Therefore, reduction of the material cost will invariably help in promoting housing for all.

3.3. Key Economic Indicators. Descriptive analysis of the economic indicators of sustainable construction materials is shown in Table 5. Five indicators were identified by the respondents as key economic factors of sustainable construction materials. These are maintenance cost, operational cost, initial cost, long-term savings, and life span, which recorded a mean value above 4.000 by the respondents. Indicators such as job creation and tourism obtained mean values below the hypothesized mean.

The maintenance cost of building components is the cost involved in the process of sustaining the performance of a building in accordance with the documented design and the operational needs. This process involves a set of activities that help in sustaining a building's components. Sahely et al. [62] opined that economic indicators such as operation and maintenance costs continue to play an important role in decision making as part of a larger set of indicators. Maintenance cost involves the cost of caring for the components of building to ensure reliability and prevention of failure.

The operational cost is the cost incurred in the day-to-day running or operation of the building. The operational cost includes mortgage payments, building insurance, taxes, maintenance and repair cost, and general and administrative expenses [54].

Initial costs are the costs that are incurred during the design and construction process. The initial cost of construction project includes the cost at the following stages of construction: planning, preliminary, design, and construction. The initial cost includes the cost of acquiring land and cost of materials for constructing the structures.

Long-term savings is critical in reducing the running cost of any structure. Construction materials and components with a long-term savings and low-maintenance effort reduce investments for maintenance, replacement, and renovation. While some sustainable materials may require greater upfront costs, they bring with them long-term savings due to reduced energy and transportation costs, as well as being beneficial for the environment in the long term [4].

4. Summary and Conclusion

The aim of this paper was to identify the key indicators for measuring sustainable construction materials. This study was based on the three main dimensions of sustainable development, which are environmental, social, and economic. The study adopted 25 indicators from the literature out of which the respondents identified twelve (12) as key indicators for measuring sustainable construction materials. These include three key environmental indicators for measuring sustainable construction materials, which are human toxicity, climate change, and solid waste. Furthermore, adaptability, thermal comfort, local resources, and housing for all were identified as the four key social indicators for sustainable construction materials. In addition, maintenance cost, operational cost, initial cost, long-term savings, and life span were found to be the five key economic indicators for measuring sustainable construction materials. The study therefore suggests that these twelve indicators should be considered in future for any study that seeks to measure sustainable construction materials.

Conflicts of Interest

The authors declare that they have no conflicts of interest.

References

[1] B. Ness, E. Urbel-Piirsalu, S. Anderberg, and L. Olsson, "Categorising tools for sustainability assessment," *Ecological Economics*, vol. 60, no. 3, pp. 498–508, 2007.

[2] H. Danso, "Dimensions and indicators for sustainable construction materials: a review," *Research and Development in Material Science*, vol. 3, no. 4, pp. 1–9, 2018.

[3] M. Mailler, "Sustainability assessment of an IAIA educational & networking forum," in *Proceedings of 28th Annual Conference of the International Association for Impact Assessment*, Perth, Australia, May 2008.

[4] Buildabroad, *Sustainable Materials: The Backbone of Any Sustainable Building Project*, 2017, https://buildabroad.org/2017/08/15/sustainable-materials.

[5] J. Peach, *Five Sustainable Building Materials that Could Transform Construction*, 2010, http://thisbigcity.net/five-sustainable-building-materials-that-could-transform-construction.

[6] H. Danso, B. Martinson, M. Ali, and C. Mant, "Performance characteristics of enhanced soil blocks: a quantitative review," *Building Research and Information*, vol. 43, no. 2, pp. 253–262, 2015.

[7] M. C. J. Delgado and I. C. Guerrero, "Earth building in Spain," *Construction and Building Materials*, vol. 20, no. 9, pp. 679–690, 2006.

[8] S. M. Hejazi, M. Sheikhzadeh, S. M. Abtahi, and A. Zadhoush, "A simple review of soil reinforcement by using natural and synthetic fibres," *Construction and Building Materials*, vol. 30, pp. 100–116, 2012.

[9] J. M. Khatib, *Sustainability of Construction Materials*, Woodhead Publishing, CRC Press, Cambridge, UK, 2009.

[10] V. Maniatidis and P. Walker, *A Review of Rammed Earth Construction*, University of Bath, Bath, UK, 2003.

[11] F. Pacheco-Torgal and S. Jalali, "Earth construction: lessons from the past for future eco-efficient construction," *Construction and Building Materials*, vol. 29, pp. 512–519, 2012.

[12] T. Kavas, "Use of boron waste as a fluxing agent in production of red mud brick," *Building and Environment*, vol. 41, no. 12, pp. 1779–1783, 2006.

[13] M. C. N. Villamizar, V. S. Araque, C. A. R. Reyes, and R. S. Silva, "Effect of the addition of coal-ash and cassava peels on the engineering properties of compressed earth blocks," *Construction and Building Materials*, vol. 36, pp. 276–286, 2012.

[14] H. Danso, B. Martinson, M. Ali, and J. B. Williams, "Effect of sugarcane bagasse fibre on the strength properties of soil blocks," in *Proceedings of 1st International Conference on Bio-based Building Materials*, Clermont-Ferrand, France, June 2015.

[15] B. Gaw and S. Zamora, *Soil Reinforcement with Natural Fibers for Low-Income Housing Communities*, MSc thesis, Worcester Polytechnic Institute, Worcester, MA, USA, 2011.

[16] J. I. Aguwa, "Study of coir reinforced laterite blocks for buildings," *Journal of Civil Engineering and Construction Technology*, vol. 4, pp. 110–115, 2013.

[17] M. G. Sreekumar and D. G. Nair, "Stabilized lateritic blocks reinforced with fibrous coir wastes," *International Journal of Sustainable Construction Engineering and Technology*, vol. 4, pp. 23–32, 2013.

[18] P. P. Yalley, *Use of Waste and Low Energy Materials in Construction*, Lambert Academic Publishing, Saarbrücken, Germany, 2012.

[19] I. I. Akinwumi, O. M. Olatunbosun, O. M. Olofinnade, and P. O. Awoyera, "Structural evaluation of lightweight concrete produced using waste newspaper and office paper," *Civil and Environmental Research*, vol. 6, pp. 160–167, 2014.

[20] H. Yun, H. Jung, and C. Choi, "Mechanical properties of papercrete containing waste paper," in *Proceedings of 18th International Conference on Composite Materials*, Jeju Province, Republic of Korea, August 2011.

[21] B. Taallah, A. Guettala, S. Guettala, and A. Kriker, "Mechanical properties and hygroscopicity behavior of compressed earth block filled by date palm fibers," *Construction and Building Materials*, vol. 59, pp. 161–168, 2014.

[22] T. M. Le and K. L. Pickering, "The potential of harakeke fibre as reinforcement in polymer matrix composites including modelling of long harakeke fibre composite strength," *Composites Part A: Applied Science and Manufacturing*, vol. 76, pp. 44–53, 2015.

[23] S. Elfordy, F. Lucas, F. Tancret, Y. Scudeller, and L. Goudet, "Mechanical and thermal properties of lime and hemp concrete ("hempcrete") manufactured by a projection process," *Construction and Building Materials*, vol. 22, no. 10, pp. 2116–2123, 2008.

[24] Y. Millogo, J. E. Aubert, E. Hamard, and J. C. Morel, "How properties of kenaf fibers from Burkina Faso contribute to the reinforcement of earth blocks," *Materials*, vol. 8, no. 5, pp. 2332–2345, 2015.

[25] H. Danso, D. B. Martinson, M. Ali, and J. Williams, "Effect of fibre aspect ratio on mechanical properties of soil building blocks," *Construction and Building Materials*, vol. 83, pp. 314–319, 2015.

[26] S. Ismail and Z. Yaacob, "Properties of laterite brick reinforced with oil palm empty fruit bunch fibres," *Pertanika Journal of Science and Technology*, vol. 19, pp. 33–43, 2011.

[27] S. M. Marandi, M. H. Bagheripour, R. Rahgozar, and H. Zare, "Strength and ductility of randomly distributed palm fibers

reinforced silty-sand soils," *American Journal of Applied Sciences*, vol. 5, no. 3, pp. 209–220, 2008.

[28] C. M. Chan, "Effect of natural fibers inclusion in clay bricks: physico-mechanical properties," *International Journal of Civil and Environmental Engineering*, vol. 3, pp. 51–57, 2011.

[29] P. P. Yalley and A. S. K. Kwan, "Use of waste and low energy materials in building block construction," in *Proceedings of 25th Conference on Passive and Low Energy Architecture (PLEA)*, Dublin, Ireland, June 2008.

[30] C. K. Subramaniaprasad, B. M. Abraham, and E. K. K. Nambiar, "Sorption characteristics of stabilised soil blocks embedded with waste plastic fibres," *Construction and Building Materials*, vol. 63, pp. 25–32, 2014.

[31] J. Cid-Falceto, F. R. Mazarrón, and I. Cañas, "Assessment of compressed earth blocks made in Spain: international durability tests," *Construction and Building Materials*, vol. 37, pp. 738–745, 2012.

[32] M. Raghatate Atul, "Use of plastic in a concrete to improve its properties," *International Journal of Advanced Engineering Research and Studies*, vol. 1, pp. 109–111, 2012.

[33] B. R. T. Vilane, "Assessment of stabilisation of adobes by confined compression tests," *Biosystems Engineering*, vol. 106, no. 4, pp. 551–558, 2010.

[34] S. Akbulut, S. Arasan, and E. Kalkan, "Modification of clayey soils using scrap tire rubber and synthetic fibers," *Applied Clay Science*, vol. 38, no. 1-2, pp. 23–32, 2007.

[35] P. Turgut and B. Yesilata, "Physico-mechanical and thermal performances of newly developed rubber-added bricks," *Energy and Buildings*, vol. 40, no. 5, pp. 679–688, 2008.

[36] M. Achenza and L. Fenu, "On earth stabilization with natural polymers for earth masonry construction," *Materials and Structures*, vol. 39, no. 1, pp. 21–27, 2006.

[37] F. Aymerich, L. Fenu, and P. Meloni, "Effect of reinforcing wool fibres on fracture and energy absorption properties of an earthen material," *Construction and Building Materials*, vol. 27, no. 1, pp. 66–72, 2012.

[38] C. Galán-Marín, C. Rivera-Gómez, and J. Petric, "Clay-based composite stabilised with natural polymer and fibre," *Construction and Building Materials*, vol. 24, no. 8, pp. 1462–1468, 2010.

[39] S. S. Namango and D. S. Madara, "Compressed earth blocks reinforced with sisal fibres," *Journal of Agricultural Pure Applied Science and Technology*, vol. 19, pp. 10–22, 2014.

[40] W. Quagliarini and S. Lenci, "The influence of natural stabilisers and natural fibres on the mechanical properties of ancient Roman adobe bricks," *Journal of Cultural Heritage*, vol. 11, no. 3, pp. 309–314, 2010.

[41] S. Yetgin, O. Cavdar, and A. Cavdar, "The effects of the fiber contents on the mechanic properties of the adobes," *Construction and Building Materials*, vol. 22, no. 3, pp. 222–227, 2008.

[42] F. Parisi, D. Asprone, L. Fenu, and A. Prota, "Experimental characterization of Italian composite adobe bricks reinforced with straw fibers," *Composite Structures*, vol. 122, pp. 300–307, 2015.

[43] A. Laborel-Préneron, J.-E. Aubert, C. Magniont, P. Maillard, and C. Poirier, "Effect of plant aggregates on mechanical properties of earth bricks," *American Society of Civil Engineers*, vol. 29, no. 12, p. 04017244, 2017.

[44] S. A. Lima, H. Varum, A. Sales, and V. F. Neto, "Analysis of the mechanical properties of compressed earth block masonry using the sugarcane bagasse ash," *Construction and Building Materials*, vol. 35, pp. 829–837, 2012.

[45] N. Degirmenci, "The using of waste phosphogypsum and natural gypsum in adobe stabilization," *Construction and Building Materials*, vol. 22, no. 6, pp. 1220–1224, 2008.

[46] I. Demir, "An investigation on the production of construction brick with processed waste tea," *Building and Environment*, vol. 41, no. 9, pp. 1274–1278, 2006.

[47] D. Straub, M. Boudreau, and D. Gefen, "Validation guidelines for IS positivist research," *Communications of the Association for Information Systems*, vol. 13, pp. 380–427, 2004.

[48] J. Anderson and J. Thornback, *A Guide to Understanding the Embodied Impacts of Construction Products*, Construction Products Association, London, UK, 2012.

[49] BRE, *Methodology for Environmental Profiles of Construction Products: Product Category Rules for Type III Environmental Product Declaration of Construction Products (Draft)*, 2007.

[50] Adaptable Futures, *Homepage of adaptable futures*, 2012, http://www.adaptablefutures.com.

[51] E. Annex, *Buildings–Adaptability, Durability and Materials*, 2015, http://democracy.york.gov.uk/documents/s4858/Annex%20E%20Sustainable%20Development.pdf.

[52] A. Manewa, M. Siriwardena, A. Ross, and U. Madanayake, "Adaptable Buildings for sustainable built environment," *Built Environment Project and Asset Management*, vol. 6, no. 2, pp. 139–158, 2016.

[53] BS EN ISO 7730, *Ergonomics of the thermal environment, Analytical determination and interpretation of thermal comfort using calculation of the PMV and PPD indices and local thermal comfort criteria*, 2005.

[54] Designing Buildings Wiki, *Thermal comfort in buildings*, 2016, https://www.designingbuildings.co.uk/wiki/Thermal_comfort_in_buildings.

[55] A. Ward, J. Boehland, and N. Malin, *Thermal Comfort and Building Management*, BuildingGreen, 2018, http://fmlink.com/articles/thermal-comfort-and-building-management.

[56] H. Danso, "Building houses with locally available materials in Ghana: benefits and problems," *International Journal of Science and Technology*, vol. 2, no. 2, pp. 225–231, 2013.

[57] GreenHomeBuilding, *Use Local Materials*, 2016, http://www.greenhomebuilding.com/localmaterials.htm.

[58] UN Habitat, *Affordable land and housing in Africa United Nations Human Settlement Programme*, 2011.

[59] B. Bhatta, "Analysis of urban growth and sprawl from remote sensing data," in *Advances in Geographic Information Science*, pp. 23–42, Springer, Berlin, Germany, 2010.

[60] H. Danso, *Use of agricultural waste fibres as enhancement of soil blocks for low-cost housing in Ghana*, Ph.D. thesis, School of Civil Engineering and Surveying, University of Portsmouth, 2015.

[61] H. Danso and D. Menu, "High cost of materials and land acquisition problems in the construction industry in Ghana," *International Journal of Research in Engineering and Applied Sciences*, vol. 3, no. 3, pp. 18–33, 2013.

[62] H. R. Sahely, C. A. Kennedy, and B. J. Adams, "Developing sustainability criteria for urban infrastructure systems," *Canadian Journal of Civil Engineering*, vol. 32, no. 1, pp. 72–85, 2005.

16

Investigation on the Deterioration Mechanism of Recycled Plaster

Zhixin Li ⓘ,[1] **Kaidong Xu,**[1] **Jiahui Peng,**[2] **Jina Wang,**[1] **Xianwei Ma ⓘ,**[1] **and Jishou Niu**[1]

[1]*School of Material and Chemistry Engineering, Henan University of Urban Construction, Pingdingshan 467036, China*
[2]*College of Materials Science and Engineering, Chongqing University, Chongqing 400045, China*

Correspondence should be addressed to Zhixin Li; li.zhixin1989@163.com

Academic Editor: Ana S. Guimarães

The deterioration mechanism of recycled plaster (R-P) was studied. The large specific surface area (SSA), improper preparation temperature, increased water requirement of R-P, and microstructure of its hardened body were analyzed by particle size distribution (PSD), Blaine method, differential scanning calorimetry (DSC), scanning electron microscopy (SEM), and nitrogen adsorption porosimetry. The results indicated that the properties of R-P were deteriorated, but its strength decreases from 50% at the same manufacturing process to 30%–40% at similar specific surface area. The analysis shows that the large SSA, poor morphology, narrow PSD, and increased internal detects give rise to increase of water requirement. In addition, the deterioration properties are caused by unsuitable temperature of preparation, loose structure, and large average pore diameter in hardened R-P as well.

1. Introduction

The development of economy has brought about the wide use of plaster of Paris (POP) in construction [1–3] and in ceramic factories [4, 5] as the raw materials of slip casting models, thus producing brazen increase of waste gypsum (WG) after their utilization. According to Suárez, the output of gypsum wallboards produced annually is 80 million tons, and the amount of gypsum dumped in landfills is 15 million tons per year [6]. It not only wastes the gypsum resources but also arouses environment problems, endangering human health [7–9]. Hence, recycling WG is necessary for offering cost reduction and environment protection.

Great efforts have been exerted in the utilization of WG and several approaches are also tried, such as using WG as a retarder in making Portland cement [6], soft clay [10–13], ceramic products [14], new drywalls, and non-load-bearing bricks [15]. However, in most of the related application mentioned above, the utilization of WG is limited, and the problems brought by WG can not be effectively solved. Large plate gypsum is also prepared from WG via the wet process [16], which is unfavorable considering the complex process and may not be feasible in the industrial production on a large scale. β-Hemihydrate of calcined WG (R-P) as a low-energy material can be produced by the reversible reaction between gypsum dihydrate and gypsum hemihydrate in an electric oven at 180°C [16]. Nevertheless, the utilization is still very little owing to the deterioration performances of R-P compared with POP. So, mastering the deterioration mechanism of R-P is of great importance and urgency for its utilization. There have been several investigations on the deterioration mechanism of R-P, and the deterioration is commonly ascribed to its large specific surface area [17]. However, only limited research on other possible factors inducing R-P deterioration except for the large specific surface areas (SSAs) was published in the common sources by the scholars, which seriously hinders the utilization of R-P. Our study found that the unsuitable preparation temperature, increase of water requirement, and microstructure changes of its hardened R-P were also important factors. Thus, deterioration mechanism of R-P studied in our work is systematic.

In this paper, the above mentioned factors have been fully analyzed to investigate the deterioration mechanism of R-P comprehensively. The authors expect the research will lay a ground work for the utilization of R-P.

TABLE 1: Chemical composition of VG used (%).

Virgin gypsum	SO_3	CaO	SiO_2	Al_2O_3	Fe_2O_3	K_2O	SrO
Chemical composition	47.57	38.73	1.15	0.42	0.13	0.07	0.04

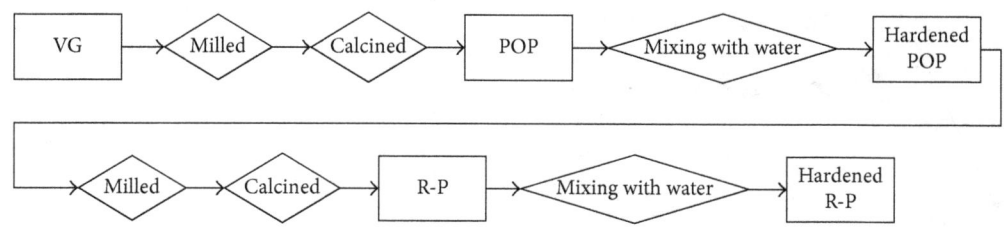

FIGURE 1: The flowchart of preparing R-P and its hardened R-P.

TABLE 2: Properties of POP and R-P under the similar SSA.

Gypsum	SSA ($m^2 \cdot kg^{-1}$)	W/P	Setting time (min) Initial	Final	2 h strength (MPa) Flexural strength	Compressive strength	Dry strength (MPa) Flexural strength	Compressive strength	Water absorption (%)
POP	1098	0.68	26	30	2.47	4.59	4.50	6.98	38.89
R-P	1114	0.85	33	46	1.66	2.94	2.70	4.80	40.65

2. Experimental Details

2.1. Materials and Preparation of R-P.
The virgin gypsum (VG) was purchased from Yingcheng in China. The chemical composition of VG determined by X-ray fluorescence (XRF) is shown in Table 1. Preparation of R-P is shown in Figure 1, and their measurement of water-plaster ratio (W/P), setting time, and strength were done according to GB/T 9776-2008 [18]. The experimental details have been introduced in our earlier research [19], and the SSA was tested according to GB/T 8074-2008 [20].

2.2. Experimental Techniques.
To eliminate the effect of SSA of R-P, it is necessary to mill POP into the approximate SSA with R-P (Table 2). Thus, POP with SSA of 1098 m^2/kg and R-P with SSA of 1114 m^2/kg were employed to analyze possible influencing factors that can induce the deterioration of R-P.

In order to study the effect of unsuitable preparation temperature on the deterioration of R-P, the relationship between strength and calcination temperatures was established. Moreover, the calcination temperatures are 130°C, 150°C, 160°C, 165°C, 170°C, 175°C, 180°C, and 200°C (Figure 2), respectively.

To facilitate understanding the going of the work, a diagram for experimental work is presented in Figure 3. Briefly, the properties (water requirement, setting time, strength, and water absorption) were measured first, and then the deterioration mechanism was determined from the large SSA, unsuitable preparation temperature, reasons for increased water requirement, and microstructure changes of its hardened R-P.

2.3. Materials Characterization.
The morphological investigations were observed by the scanning electron microscope (TESCAN VEGA III LMH). The internal detects were measured via simultaneous DSC/TG instrument (SMP/PF7548/MET/400W) [21]. The particle size distribution was analyzed via a laser particle size analyzer (Mastersizer 2000) after dispersing POP and R-P in anhydrous ethanol with an ultrasonic bath. Pore size distribution and porosity of pastes were investigated by an adsorption meter (ASAP 2020) with nitrogen full adsorption.

3. Results and Discussions

3.1. Large Specific Surface Area.
Under the same preparation process as POP, the properties of R-P were deteriorated seriously, which is displayed in Table 3. Clearly, the W/P {ml water/300 g plaster powder (water-plaster ratio)}, setting time, and water absorption were all increased, whereas the strength of R-P was decreased by approximately 50% (Figure 4). At this time, SSA of R-P was increased from 452 m^2/kg to 1114 m^2/kg; thus, the deterioration properties of R-P were caused by its large SSA, which is in harmony with the previous findings [17].

Table 2 gives the properties of R-P and POP under the similar SSA. This table was overwhelmingly proving that the properties of R-P were also deteriorated despite approximate SSA obtained by POP and R-P. To our surprise, the strength decreasing rate was reduced from 50% to 30%–40% (Figure 4). Therefore, it could be concluded that there were, in addition to large SSA brought about, other reasons for property deterioration of R-P.

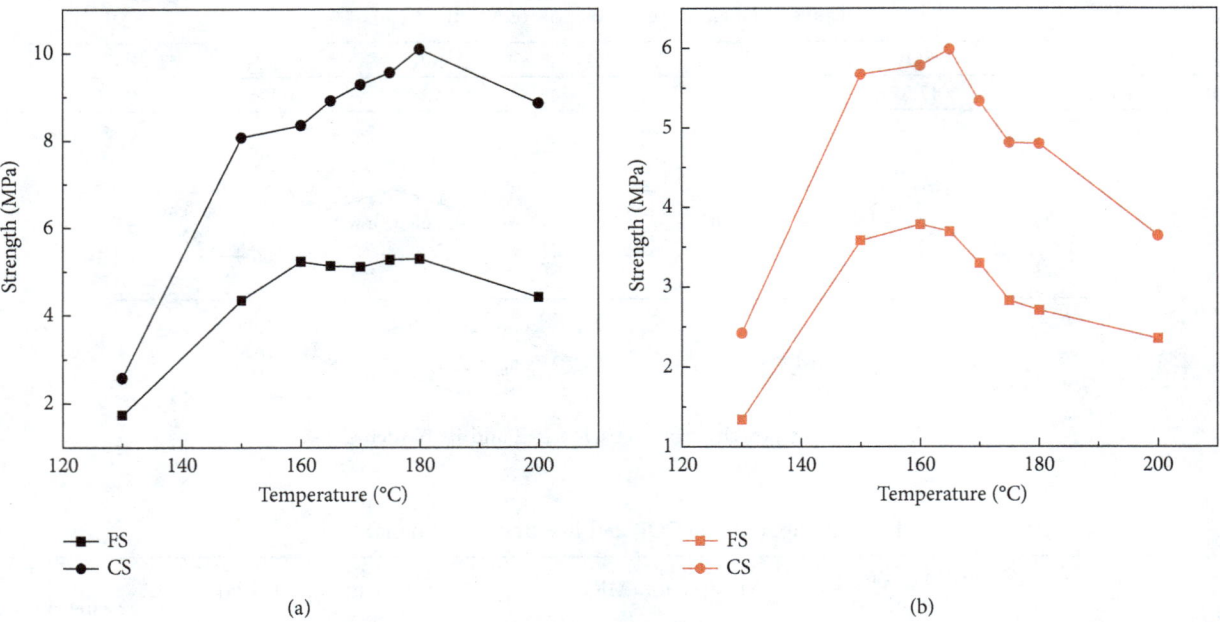

FIGURE 2: Effect of preparation temperatures on the strength of POP and R-P.

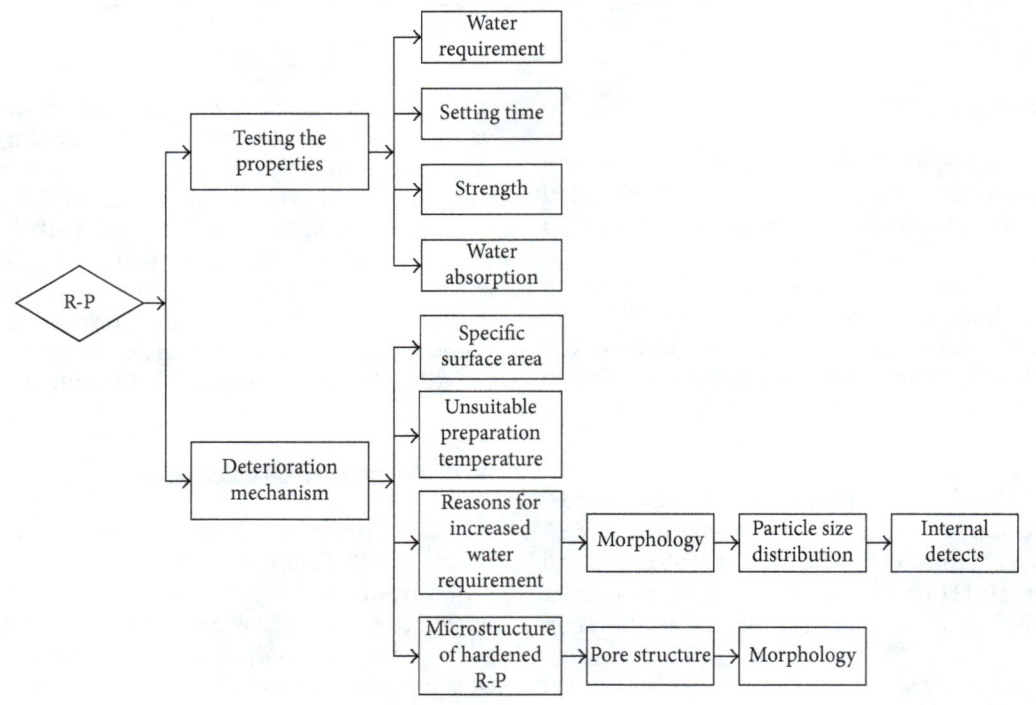

FIGURE 3: The test program.

TABLE 3: Properties of POP and R-P under the same preparation process.

Gypsum	SSA (m²·kg⁻¹)	W/P	Setting time (min)		2 h strength (MPa)		Dry strength (MPa)		Water absorption (%)
			Initial	Final	Flexural	Compressive	Flexural	Compressive	
POP	452	0.63	8.5	13.5	2.94	5.95	5.29	10.08	29.50
R-P	1114	0.85	33	46	1.66	2.94	2.70	4.80	40.65

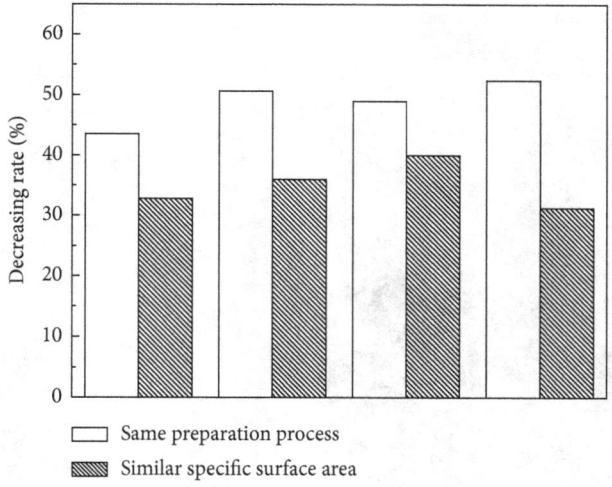

FIGURE 4: The decreasing rate in the strength of R-P.

3.2. Unsuitable Preparation Temperature.

3.2. Unsuitable Preparation Temperature. It has been reported that under constant experimental conditions, gypsums of different origin have different dehydration characteristics and physical properties [22], and the reactivity of hemihydrate depends largely on the temperature of preparation [23]. So, we conclude that various physical properties of R-P can be observed under different calcination temperatures since it is known that strength is the comprehensive aspect charactering the properties of plaster. Therefore, an attempt has been made to correlate the strength of R-P and POP with different temperatures of preparation so as to find out the effect of preparation temperature on properties of R-P, which is seen in Figure 2.

As the preparation temperature increases, the strength changes of POP and R-P are shown in Figures 2(a) and 2(b), respectively. From the figures, it is clear that the compressive strength (CS) of POP increased rapidly to 10.08 MPa around 180°C and then dropped to 8.85 MPa at 200°C. The flexural strength (FS) of POP rose dramatically to 5.11 MPa around 160°C, and this number remained slightly increased until 180°C but then suffered a sharp drop. Therefore, the suitable preparation temperature of POP is 180°C; at this time, the initial setting time and final setting time are 8.5 min and 13.5 min, respectively, meeting the standard of GB 9776-2008 [18]. While for R-P, the compressive strength and flexural strength were at a peak around 165°C and 160°C, respectively. The setting time can satisfy the standard requirement when the calcination temperature was 165°C. Hence, it could be drawn that the unsuitable preparation temperature led to the deterioration properties of R-P as well.

3.3. Reasons for Increased Water Requirement.

3.3. Reasons for Increased Water Requirement. The morphology [24], particle size distribution (PSD) [25], and internal detects [21] are of critical factors associated with the water demand property of plaster. Hence, the morphology, PSD, and internal detects of POP and R-P were measured via SEM, laser particle size analyzer, and DSC curves, which are presented in Figure 5.

Figures 5(a) and 5(b) show the crystal morphology of POP and R-P, respectively. It can be seen that the morphology of R-P is totally different from that of POP, depending on their origin of gypsum. The morphology of plaster, made of needle-shaped gypsum crystals, changes with the properties of gypsum. The origin gypsum of POP was relatively dense with almost no pores (Figure 6(a)). However, the origin gypsum of R-P was loose with much pores (Figure 6(b)), which contributed to the morphology changes. Clearly, POP possesses many spheroidal particles and includes some rode-like crystals with an aspect ratio of 1-2. While the aspect ratio of R-P crystals increases, the morphology are changed to acicular-like crystals with an aspect ratio of 5-6, which make R-P crystals occupy acicular-like and rode-like crystals. According to the principle enunciated by Li et al. [19] and Peng et al. [24], the water requirement of plasters are increased with the increase of aspect ratio. The needle-shaped crystal has poor fluidity and can increase the water requirement of plaster greatly, and the ideal crystal has an aspect ratio of 1 : 1 for reducing the water demand. Hence, it can be concluded that the water requirement of R-P is increased by its poor crystals.

Simultaneously, particle size distribution (PSD) results presented in Figure 5(c) further confirm the increase of water demand of R-P. It is noticed that the grading of R-P is, though two peaks are both occupied, obviously different from POP. A broad peak of POP is observed, where the full width at half maximum (FWHM) of the first peak and second peak are 18.75 and 22.50, respectively, if the width of POP is 100. While a narrow peak is seen in R-P, FWHM of the first peak and second peak are 16.25 and 18.75, respectively. Besides, its average particle size of R-P decreases from 54.614 μm to 12.882 μm. The coarse plasters require little water while plasters with narrow PSD reduce the packing density, thus requiring much water for standard consistency [26]. Therefore, much water of R-P is acquired to satisfy the fluidity.

To gain further insight into the reasons for the increase of water requirement, DSC analysis was carried out to evaluate the internal detects of R-P, which is presented in

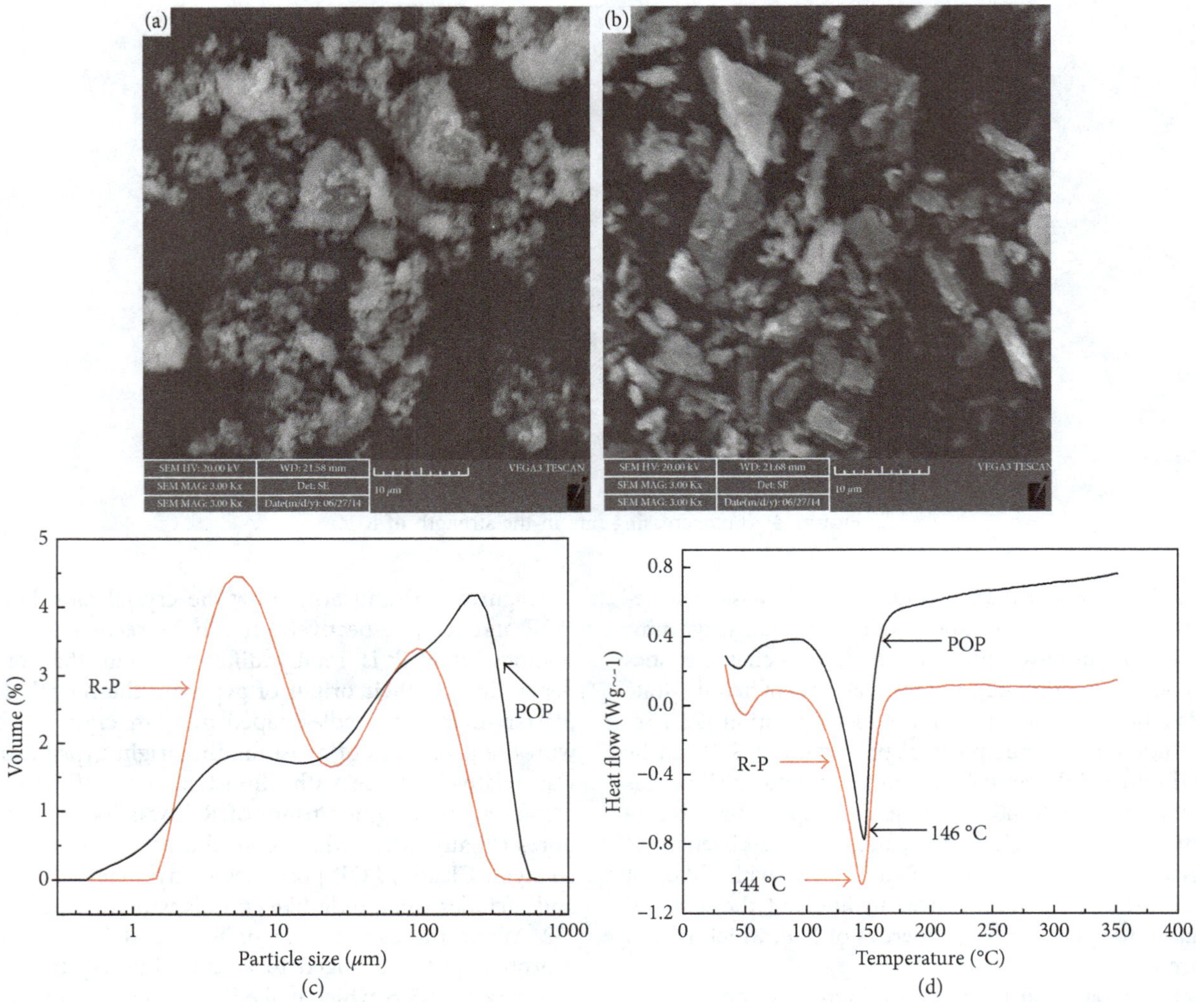

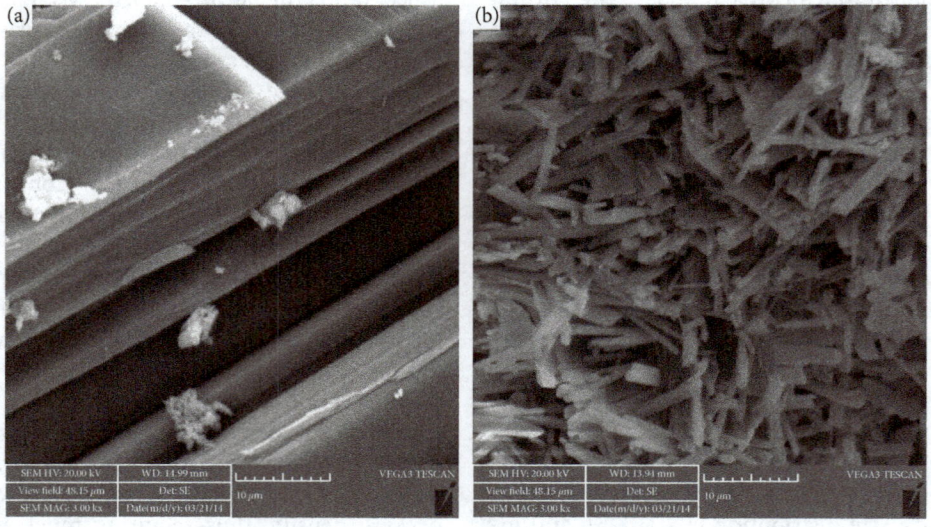

FIGURE 5: Microstructure, particle size distribution, and DSC curves of POP and R-P.

FIGURE 6: Microstructure of origin gypsum of POP and R-P.

Figure 5(d). Clearly, the decomposition temperature of R-P is 144°C, a little lower than that of POP. Some reported studies have indicated that plasters with little internal detects gain good crystallization while plasters with more internal detects have low decomposition temperature; thus, little internal detects acquire high decomposition temperature [21].

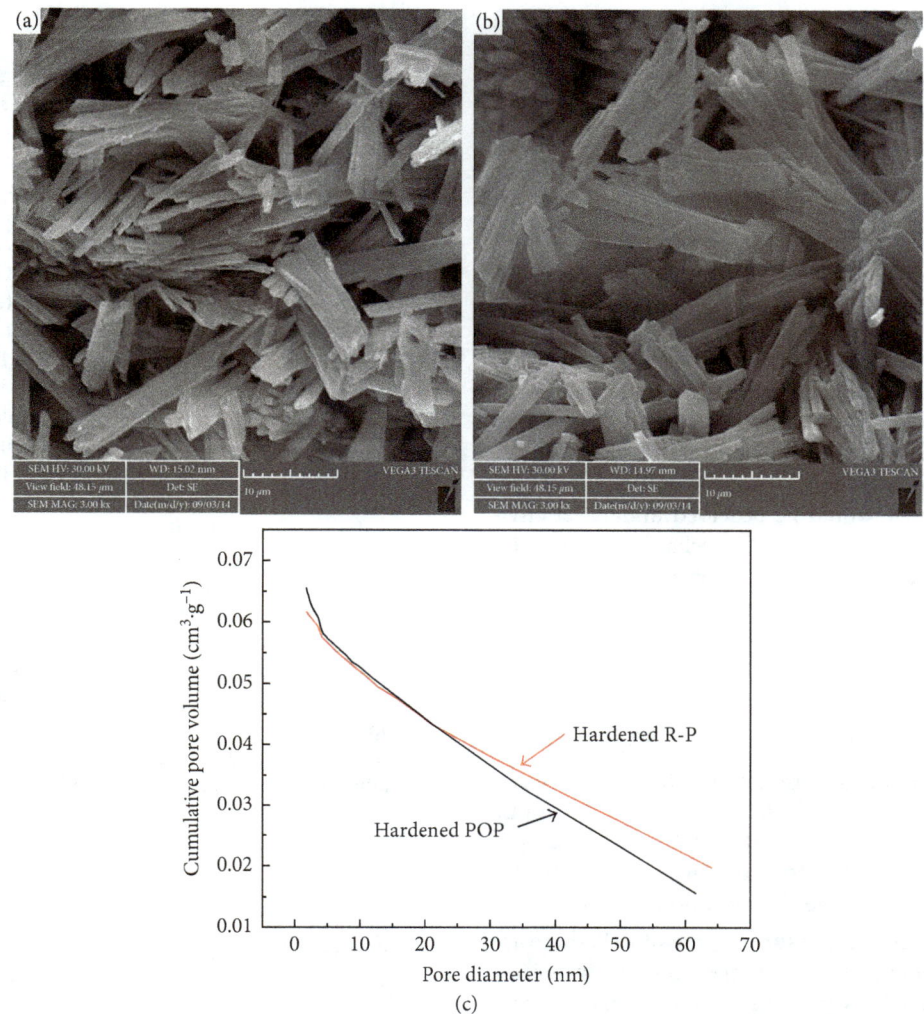

FIGURE 7: The morphology and pore size distribution of hardened POP and hardened R-P.

TABLE 4: Pore characteristics of hardened POP and hardened R-P.

Sample	Average pore diameter (nm)	Cumulative pore volume (cm$^3 \cdot$g^{-1})
Hardened POP	11.5892	0.066
Hardened R-P	14.6917	0.062

So, it could be concluded that more internal detects are obtained by R-P, increasing its water requirement, which is in good agreement with the SEM and PSD results.

Although the increased water requirement of R-P is not conducive for use in construction, it is of great importance for the ground improvement projects. According to Kamei et al., the improvement in strength when R-P was added to the soil was mainly depending on the potential of R-P to absorb water from the tested soil. The presence of R-P in soil mixture had a significant effect on the reduction of natural water content, subsequently the strength was improved.

3.4. Microstructure of Hardened R-P.

The strength of R-P paste depends largely on the characteristics of the microstructure, such as crystal morphology and size, and characteristics of matrix joints and pore structure [25].

Therefore, the morphology and pore structure were measured by SEM and nitrogen adsorption porosimetry, respectively, which are shown in Figure 7 and Table 4.

The SEM images of hardened POP and hardened R-P are presented in Figures 7(a) and 7(b), where significant difference can be observed. Owing to the high W/P and low reaction rate, a loose structure with many tufted crystals was shown in hardened R-P. Crystals of short aspect ratio can be clearly observed as well, reducing the overlapping of crystals; consequently, the strength of R-P is weakened. While for hardened POP, the crystals are slightly refined, little rode-like crystals of short aspect ratio can be readily detected. It displays a relatively compacted network with much needle-like crystals of long aspect ratio interweaving together horizontally and vertically, resulting in the preferable development and lapping tightly of crystals attributing high strength of POP.

It is widely known that the characteristics of pore structure are of essential importance for understanding the strength and wettability of hardened R-P [23]. So the porosity and pore size distribution of hardened POP and hardened R-P are measured, which are displayed in Table 4 and Figure 7(c), respectively. Clearly, the average pore diameter of hardened R-P, though the close cumulative pore volumes with hardened POP are obtained, increases drastically. The high quantity of coarse pores existed in hardened R-P is also confirmed by the pore size distribution results in Figure 7(c). Fine pores acquire high strength while pores with big diameter obtain low strength. The pore structure results are good consistent with SEM. The attainment of low strength and high wettability is associated with its coarse pores. It is well known that wettability is an important parameter for construction. R-P presents high water absorption. Therefore, low moisture resistance would be observed under wet environment, and radical decrease of mechanical properties would appear with the increasing moisture as well, which limits its utilization in building construction.

4. Conclusions

The following conclusions can be drawn from this study:

(1) The properties of R-P are deteriorated; nevertheless, the strength decreases from 50% at the same manufacturing process to 30%–40% at similar specific surface area. Therefore, the large specific surface area contributes to the deterioration of R-P.

(2) Except for large specific surface area of R-P, there are other influencing factors increasing the water requirement, such as poor morphology, narrow particle size distribution, and its incremental internal detects.

(3) The suitable preparation temperature of R-P is reduced to 165°C; thus, the properties of R-P are deteriorated severely when unsuitable temperature of preparation is adopted.

(4) Hardened R-P possesses a loose structure and coarse pores, thus decreasing its strength.

(5) In principle, the research on recycled plaster will provide a theoretical basis of efficient utilization of waste gypsum.

Conflicts of Interest

The authors declare that they have no conflicts of interest.

Acknowledgments

The authors thank Research Fund for the Key Program of Higher Education of Henan (18B430002), Program of Science and Technology Development of Henan (182102310928), and Research Foundation of Henan University of Urban Construction (2017YY014). The authors thank Electron Microscopy Center of Chongqing University for materials characterizations.

References

[1] J. L. Davy, T. J. Phillips, and J. R. Pearse, "The damping of gypsum plaster board wooden stud cavity walls," *Applied Acoustics*, vol. 88, pp. 52–56, 2015.

[2] J. Zhou, Z. Shu, L. Tiantian, Y. Dongxue, Z. Sheng, and Y. Wang, "Novel fabrication route for non-fired ceramic tiles only using gypsum," *Ceramics International*, vol. 41, no. 7, pp. 9193–9198, 2015.

[3] J. Zhou, Z. Sheng, T. Li, Z. Shu, Y. Chen, and Y. Wang, "Preparation of hardened tiles from waste phosphogypsum by a new intermittent pressing hydration," *Ceramics International*, vol. 42, no. 6, pp. 7237–7245, 2016.

[4] H. C. Wu, Y. M. Xia, X. Y. Hu, and X. Liu, "Improvement on mechanical strength and water absorption of gypsum modeling material with synthetic polymers," *Ceramics International*, vol. 40, no. 9, pp. 14899–14906, 2014.

[5] N. B. Singh and B. Middendorf, "Calcium sulphate hemihydrate hydration leading to gypsum crystallization," *Progress in Crystal Growth and Characterization of Materials*, vol. 53, no. 1, pp. 57–77, 2007.

[6] S. Suárez, X. Roca, and S. Gasso, "Product-specific life cycle assessment of recycled gypsum as a replacement for natural gypsum in ordinary Portland cement: application to the Spanish context," *Journal of Cleaner Production*, vol. 117, pp. 150–159, 2016.

[7] O. Bergersen and K. Haarstad, "Treating landfill gas hydrogen sulphide with mineral wool waste (MWW) and rod mill waste (RMW)," *Waste Management*, vol. 34, no. 1, pp. 141–147, 2014.

[8] A. López and A. Lobo, "Emissions of C&D refuse in landfills: a European case," *Waste Management*, vol. 34, no. 8, pp. 1446–1454, 2014.

[9] T. Raghavendra and B. C. Udayashankar, "Engineering properties of controlled low strength materials using fly ash and waste gypsum wall boards," *Construction and Building Materials*, vol. 101, pp. 548–557, 2015.

[10] A. Ahmed, K. Ugai, and T. Kamei, "Investigation of recycled gypsum in conjunction with waste plastic trays for ground improvement," *Construction and Building Materials*, vol. 25, no. 1, pp. 208–217, 2011.

[11] A. Ahmed and K. Ugai, "Environmental effects on durability of soil stabilized with recycled gypsum," *Cold Regions Science and Technology*, vol. 66, no. 2-3, pp. 84–92, 2011.

[12] T. Kamei, A. Ahmed, and T. Shibi, "Effect of freeze–thaw cycles on durability and strength of very soft clay soil stabilised with recycled Bassanite," *Cold Regions Science and Technology*, vol. 82, no. 8, pp. 124–129, 2012.

[13] T. Kamei, A. Ahmed, and T. Shibi, "The use of recycled bassanite and coal ash to enhance the strength of very soft clay in dry and wet environmental conditions," *Construction and Building Materials*, vol. 38, pp. 224–235, 2013.

[14] A. P. Godinho-Castro, R. C. Testolin, L. Janke, A. X. R. Corrêa, and C. M. Radetski, "Incorporation of gypsum waste in ceramic block production: proposal for a minimal battery of tests to evaluate technical and environmental viability of this recycling process," *Waste Management*, vol. 32, no. 1, pp. 153–157, 2012.

[15] K. F. Ragab, *Recycling of Waste Gypsum Board to Produce New Drywalls and Non-Load Bearing Bricks*, American University in Cairo, Cairo, Egypt, 2014.

[16] Y. Kojima and T. Yasue, "Synthesis of large plate-like gypsum dihydrate from waste gypsum board," *Journal of the European Ceramic Society*, vol. 26, no. 4-5, pp. 777–783, 2006.

[17] L. N. Ren, *Reclaimed Gypsum Properties Change Law and Mechanism*, Standardization Administration of China, Beijing, China, 2014.

[18] GB/T 9776-2008, *Calcined Gypsum*, Standardization Administration of China, Beijing, China, 2008.

[19] Z. Li, J. Peng, and X. Qiu, "Effect of different ways of STPP retarder addition on properties of recycled gypsum," *Journal of Wuhan University of Technology*, vol. 32, no. 5, pp. 1125–1131, 2017.

[20] GB/T 8074-2008, *Testing Method for Specific Surface of Cement-Blaine Method*, Standardization Administration of China, Beijing, China, 2008.

[21] M. Guodong, "A study on the phase-transformation of the bassanites during hydration," *Journal of the Chinese Ceramic Society*, vol. 30, no. 4, pp. 532–536, 2002.

[22] R. A. Kuntze, "Differential thermal analysis of calcium sulphate dihydrate," *Nature*, vol. 193, no. 4817, pp. 429–432, 1962.

[23] M. J. Ridge and H. Surkevicius, "Influence of some conditions of calcination on the reactivity of calcium sulphate hemihydrate," *Journal of Chemistry Technology and Biotechnology*, vol. 12, no. 10, pp. 425–432, 1962.

[24] J. Peng, J. Qu, J. Zhang, C. Zou, and M. Chen, "Adsorption characteristics of EDTA on of α-hemihydrate desulfurization gypsum surface and its influence on crystal morphology," *Chinese Journal of Material Research*, vol. 25, no. 6, pp. 566–572, 2011.

[25] B. H. Guan, Q. Ye, Z. Wu, W. Lou, and L. Yang, "Analysis of the relationship between particle size distribution of α-calcium sulfate hemihydrate and compressive strength of set plaster—using grey model," *Powder Technology*, vol. 200, no. 3, pp. 136–143, 2010.

[26] Q. Ye, B. Guan, W. Lou, L. Yang, and B. Kong, "Effect of particle size distribution on the hydration and compressive strength development of α-calcium sulfate hemihydrate paste," *Powder Technology*, vol. 207, no. 1–3, pp. 208–214, 2011.

17

Chloride Transport in OPC Concrete Subjected to the Freeze and Thaw Damage

Ki Yong Ann,[1] Min Jae Kim,[1] Jun Pil Hwang,[1] Chang-geun Cho,[2] and Ki Hwan Kim[1]

[1]Department of Civil and Environmental Engineering, Hanyang University, Ansan 15588, Republic of Korea
[2]School of Architecture, Chosun University, Gwangju 61452, Republic of Korea

Correspondence should be addressed to Ki Yong Ann; kann@hanyang.ac.kr

Academic Editor: Rishi Gupta

To predict the durability of a concrete structure under the coupling degradation consisting of the frosting and chloride attack, microstructural analysis of the concrete pore structure should be accompanied. In this study, the correlation between the pore structure and chloride migration for OPC concrete was evaluated at the different cement content in the concrete mix accounting for 300, 350, and 400 kg/m^3 at 0.45 of a free water cement ratio. The influence of frosting damage on the rate of chloride transport was assessed by testing with concrete specimens subjected to a rapid freezing and thawing cyclic environment. As a result, it was found that chloride transport was accelerated by frost damage, which was more influential at the lower cement content. The microscopic examination of the pore structure showed that the freezing environment increased the volume of the large capillary pore in the concrete matrix.

1. Introduction

Concrete structures such as concrete pavements and concrete bridge decks in the cold climate often experience the deicer to mitigate the slipperiness of the pavement surface arising from snowing. Sodium chloride (NaCl) salt and calcium chloride (CaCl$_2$) are the representative deicers, which however imposes a potential risk of chloride-induced corrosion of steel in the concrete structure. Steel embedment in concrete is usually protected from the risk of corrosion by a thin oxide layered film, the so-called "γ-Fe$_2$O$_3$ passive film" generated in the alkaline concrete pore solution [1]. Notwithstanding, chloride ions present in the concrete, pore solution may activate the electrochemical reaction then to corrode steel [2]. Once corrosion starts on the steel surface, the volume of the steel rebar expands about 3–8 times of the original volume [3], leading to a loss of the tensile capacity of steel and spalling of cover concrete, which substantially could bring a devastating failure of the structure. Despite the risk of steel corrosion, the deicing salt is an unavoidable manner to secure the traffic condition against snowing.

Moreover, the concrete pore structure may be damaged by freezing and thawing in the cold climate. Water confined in the pore system of cement matrix, when freezing, exposes the pore walls to stresses, which might cause intense damage to the structure. In particular, in the presence of the deicing salt, the destruction of the pore structure can be accelerated. Ice bodies, which have already been formed in coarser pores, are able to attract unfrozen water from finer capillary pores and gel pores, due to the difference in free energy between ice and water [4]. Then, there would be a water transfer to the freezing site, thereby a growth of the ice body in the pores and a breakdown of cement paste surrounding the pore system (i.e., a generation of microcracking). Substantially, the frost damage may provide a better path for chloride ions to reach at the depth of the steel in concrete, as being associated with further corrosion process and thus the severer deterioration of the concrete structure may occur. However, the combined deterioration, especially about a degradation of physical property and accelerated chloride penetration induced by frost damage, has not been fully clarified.

TABLE 1: Oxide composition of use material.

Binder type	Oxide compositions (%)							
	CaO	SiO$_2$	Al$_2$O$_3$	Fe$_2$O$_3$	MgO	Na$_2$O	K$_2$O	SO$_3$
OPC	64.7	20.7	4.6	3.0	1.0	0.13	0.65	3.0

In this reason, the present study is concerned about the combined effect of the two deteriorations (i.e., chloride attack and frost damage) by measuring relative dynamic modulus of elasticity and the degree of chloride transport. Moreover, microstructure analysis was also conducted to verify the change of pore system during repetitive freezing and thawing condition.

2. Experimental Work

Concrete specimens were fabricated in a cylindrical (Ø100 mm × 200 mm) and cuboidal (100 × 100 × 400 mm) mould for chloride migration test and relative dynamic modulus of elasticity test, respectively. The specific gravity of the fine aggregate was 2.60 and coarse aggregate was 2.65. The maximum size of the coarse aggregate used in a concrete mix was always 10 mm to minimize the disruption of ionic transport. The ratio of the fine aggregate to total was 0.4 for all mixes. Ordinary Portland cement (OPC) was used for a binder and its oxide composition is given in Table 1.

The binder content was ranged from 300, 350 and 400 kg/m^3. The freezing and thawing test specimens generally manufactured by less than 0.35 of water-cement ratio. However, since the purpose of this study is to observe the combined deterioration induced by frost damage and chloride attack, the water-cement ratio was kept at 0.45 for all cases. Moreover, no chemical admixture such as air entrainer was used to avoid unexpected chemical reaction which can influence the test result. All concrete specimens were demoulded 24 hours after casting and cured in a chamber with 95% RH at 20 ± 2°C for 500 days of which hydration effect become negligible during the test. After curing, the concrete specimens were directly subjected to the freeze-thaw condition, followed by tests for chloride migration and microscopic investigation. For all tests, the replications were always five to cover the variation in the values. The details of each test procedure are given in the following.

2.1. Testing for Frost Damage. To emulate the frost damage on concrete, the concrete specimens were exposed to a cyclic freeze and thaw environment. The concrete specimens after 500 days of curing were allowed to rapidly freeze in the atmospheric condition and then to thaw in water in an insulting chamber. The temperature variation in the chamber ranged from 5 to −18°C. For the freezing period, the temperature of the chamber was kept at −18°C for 1.5 hours and then still water was supplied to thaw the specimens by increasing the temperature back to 5°C within 5.0 h. The freezing and thawing rate were kept at 10°C/h. The cyclic freeze and thaw were repeated up to 60, 120, and 180 cycles. The details of testing procedure are given elsewhere [5].

The relative dynamic modulus of elasticity of concrete specimens was measured to detect internal degradation by applying ultrasonic pulse. The test was conducted at 60, 120, and 180 cycles of freeze and thaw, accompanied with chloride migration test and MIP.

2.2. Testing for Chloride Migration. Cylindrical concrete specimens sawed off to produce a disc sample with 50 mm in the thickness, after the completion of curing for 500 days. The specimen was connected to two chambers filled with 10.0% NaCl and with 0.3 M sodium hydroxide (NaOH) solutions, respectively, to form an electric cell. A direct current at 30 V was applied to the cell for 30 hours.

After applying the current, the specimens were axially split into two pieces and 0.1 M silver nitrate (AgNO$_3$) solution was sprayed on the surface of the split section to form the visible precipitation of silver chloride (AgCl). The depth of the precipitation of silver chloride was measured by a ruler four times and then the average value was chosen as the chloride penetration depth. Then, the migration coefficient was calculated by (1), derived by the Nernst-Plank equation. The detailed procedure for measuring the chloride migration coefficient is given elsewhere [6].

$$D_{\text{mig}} = \frac{0.0239\,(273 + T)\,L}{(U-2)\,t} \left\{ X_d - 0.0238 \sqrt{\frac{(273+T)\,LX_d}{U-2}} \right\},$$

(1)

where D_{mig} is migration coefficient, $\times 10^{-12}\,\text{m}^2/\text{s}$; U is applied voltage, V; T is temperature of the cell, °C; L is thickness of the specimen, mm; X_d is chloride penetration depth, mm; t is test duration, hour.

2.3. Testing for Microscopic Examination. Deformation of pore structure after severe freeze-thaw environment was examined using mortar sample, not to reflect the distortion effect of coarse aggregate. The mortar samples, after 500 days of curing, were crushed to obtain the middle of specimen, which may form a homogeneous pore matrix. Then, the sample was dried out in an oven at 50°C for 7 days to liberate water content inside of the material that might otherwise give undesired test result. The sample was initially evacuated to about 50 μm mercury (Hg) and the low pressure was generated up to 0.21 MPa by nitrogen gas, and then the pressure was gradually increased to 117.21 × 10^3 MPa at the rate of 9.1 × 10^3 kPa/s. The pressure was converted to the

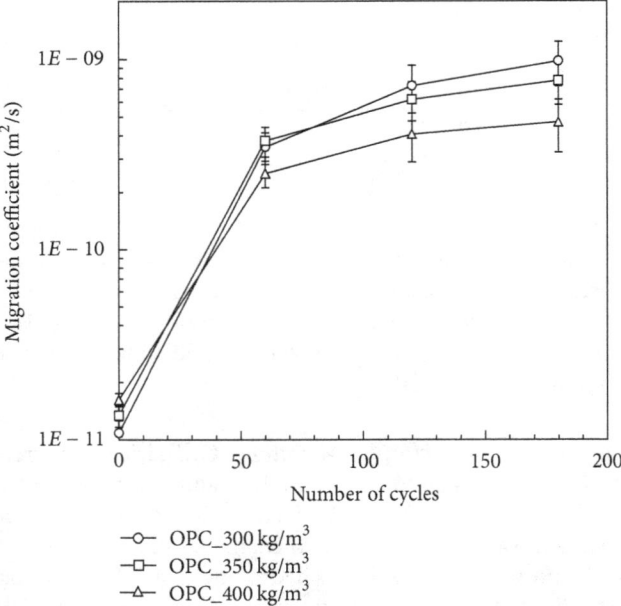

FIGURE 1: Migration coefficient for concrete after frosting depending on binder content.

equivalent pore diameter using the Washburn equation as given in (2). Then the pore volume distribution at a given pore diameter was obtained. The pore volume was adjusted to the percentage of the volume of each sample.

$$d = \frac{-4\gamma \cos\theta}{P}, \tag{2}$$

where d is pore diameter, m; γ is surface tension, N; θ is contact angle, °; P is pressure, MPa.

3. Result and Discussion

3.1. Frost Damage. After an exposure of concrete specimens to different cyclic durations for freeze and thaw (i.e., 0, 60, 120 and 180 cycles), the migration coefficient of chloride ions of the specimens was calculated by (1), as given in Figure 1. The initial migration coefficient for 300, 350, and 400 kg of cement content was 0.109×10^{-10}, 0.134×10^{-10}, and 0.162×10^{-10} m²/s, respectively, before being subjected to a freeze and thaw cyclic environment. The migration coefficients increased with the number of cycles, irrespective of cement content. This implies that the pore structure was presumed to turn more porous in the process of freeze and thaw, due to the internal pressure generated by ice formation in the pore solution, which could, in fact, enhance the connectivity between capillary pores including even isolated pores. Thus, continuous pore fractions can be used for chloride ions to increase the mobility through the pores. Furthermore, a reduction of the concrete surface in the process of freeze and thaw could produce further porosities, in terms of cracking of cover concrete, which provides additional paths for chloride transport. It is notable that, at a high number of freeze and thaw cycles, the rate of chloride penetration was less

increased. At the initial cycles, up to 60, the migration coefficient was more rapidly increased, compared to longer exposure period, from 60 to 180. For example, the migration coefficient for 300 kg of cement content was 3.47×10^{-10} m²/s at 60 cycles and reached 9.77×10^{-10} m²/s after 180 cycles. This may be attributed to limited deterioration of cover concrete arising from freeze and thaw cyclic conditions. An increase in the pore connectivity and cracking-originated pores may result from frost damage more or less, and however, after a certain level of deterioration, the pore structure may be stably sustained with no increase in the capillary pores. Then, ionic transport would be less influenced by the number of freeze and thaw cycles in a long term.

It is evident that increased cement content in the concrete mix was more resistant to chloride migration under frost damage. It was observed that an increase in the cement content resulted in a lower migration coefficient with cycles of freeze and thaw. In fact, there was a marginal variation in the migration coefficient with cement content at no exposure to freeze and thaw condition. Then, with increasing the cycles of freeze and thaw, the higher cement content in mix produced the lower migration coefficient. For example, at 180 cycles, the migration coefficient for 300, 350, and 400 kg/m³ of cement accounted for 9.77×10^{-10}, 7.73×10^{-10}, and 4.70×10^{-10} m²/s, respectively. It may reflect that the higher cement content may resist the frost damage. As a given free water cement ratio (0.45) was used in concrete mixes, no possibility of the variation in the pore structure could be achieved, as long as the pores are assumed to be produced in the cement matrix. However, the higher cement content, in turn, implies the lower aggregate content and thus lowers interfacial porosity between cement paste and aggregate, which is usually vulnerable to frost damage. Thus, the higher cement content in the concrete mix may be more resistant to frost damage in terms of increased porosities, in particular, connectivity between capillary pores.

This phenomenon can be confirmed by measuring the relative dynamic modulus of elasticity of concrete subjected to freeze and thaw cycles, as given in Figure 2. An increase in the cycle numbers of freeze and thaw resulted in a decrease in the relative dynamic modulus of elasticity, irrespective of cement content. However, concrete at the lower cement content produced a more rapid reduction of the relative dynamic modulus of elasticity at a given number of the freeze and thaw cycles. For example, for 300 kg/m³ of cement content, the relative dynamic modulus of elasticity accounted for 88.29, 72.32, and 49.39% of elasticity at 60, 120, and 180 cycles, whilst 400 kg/m³ produced 87.55, 75.28, and 58.59% at the equivalent cycles. Thus it can be said that the higher cement content may benefit in raising the resistance to frost damage at a given free water cement ratio, except for the use of air-entraining agents. Notwithstanding, all specimens did not maintain the minimum level for the relative dynamic modulus of elasticity to secure the safety of the structure, accounting for 60% at 300 cycles; in fact the relative dynamic modulus of elasticity was lower than 60% at 180 cycles due to the absence of air entraining agent and the higher water-cement ratio rather than guided values.

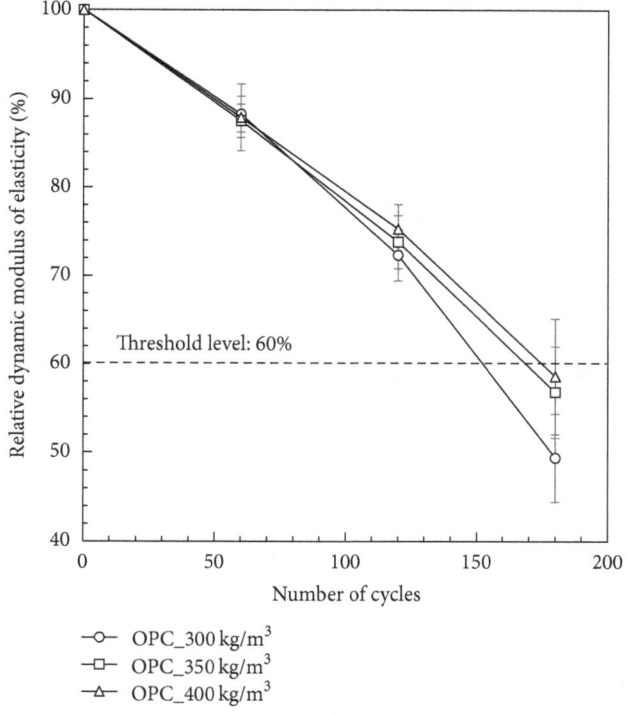

FIGURE 2: The relative dynamic modulus of elasticity under freeze-thaw condition.

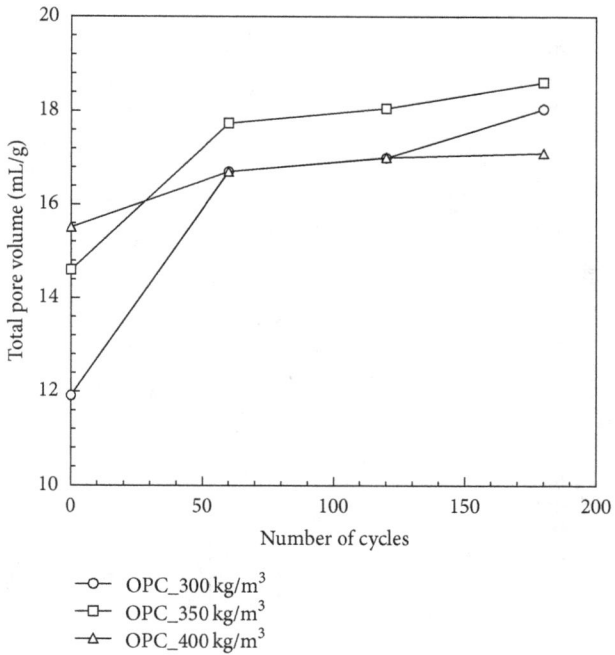

FIGURE 3: Total pore volume for 300, 350, and 400 kg/m^3 of cement content at 0, 60, 120, and 180 cycles.

3.2. Pore Structure Analysis.

To ensure a modification of the porosity under freeze and thaw damage, the pore structure was examined by the mercury intrusion porosimetry (MIP). The total pore volume was obtained at each cycle and cement content as given in Figure 3. It is evident that cement content was crucial in increasing the total pore volume; the initial porosity was 11.91, 14.60, and 15.51 mL/g for 300, 350, and 400 kg/m^3 of cement content, respectively. However, after 180 cycles of freezing and thawing, the total pore volume was changed to 18.04, 18.61, and 17.10 for 300, 350, and 400 kg/m^3 of cement content, respectively. This may be attributed to the resistance of porosity or/and air bubble to frost damage. In fact, increased porosity may buffer internal stress arising from freezing the pore solution in the pores and thus frost damage could be mitigated. The higher porosity may benefit in lowering breakage of pore network, resulting from freeze and thaw cycles. Additionally, it was observed that an increase in the porosity was observed at 60 cycles rather than subsequent cycles, at which there was no significant change in the total pore volume for 350 and 400 kg/m^3 of cement content and only marginal increase was observed for 300 kg/m^3 of cement content. It implies that the crack-induced porosity would be formed at an early stage of freeze and thaw cycles, which would buffer subsequent internal stress from freezing the pore solution and thus no further formation of cracking was formed.

The pore size distribution was used to determine modification of pore structure change after frost damage, as given Figure 4. The pore diameter of pore at the highest peak (i.e., critical diameter) was determined, and simultaneously, pores were classified by the size as small capillary ($d < 0.1\,\mu$m), large capillary ($0.1 < d < 10\,\mu$m), and macropores ($d > 10\,\mu$m). At 0 cycle, the peak of porosity was observed at about 0.5–0.7 μm in the diameter. It was seen that the pore volume was strongly dependent on large capillary pore, of which the volume accounted for 4.32, 5.12, and 6.95 mL/g, respectively, and 300, 350, and 400 kg/m^3 of cement with no exposure to freeze and thaw cycles. With increasing cycles of freeze and thaw, the volume of large capillary pore was significantly increased, compared to other sizes of pores. This may be due to the fact that crack formed in the process of freeze and thaw would be equated to the large capillary pores in size. The other pores were also modified by the freeze and thaw. For example, the breakage of small capillary pore was reduced by suspended pressure in the pore network, which in turn gave relatively lower expansive force [7]. It is notable that freeze and thaw cycles were influencing critical pore diameter, which was, for example, 0.045, 0.072, 0.088, and 0.095 μm at 0, 60, 120, and 180 cycles of freeze and thaw, respectively, for 300 kg/m^3 of cement. The shift of peak diameter was commonly observed in all cement contents. Thus, small capillary pore has only marginal impact from freeze and thaw in the total volume. The macropore may reflect the resistance to frost damage in terms of spacing factor [8], which can be defined as average length of connected small pores with voids. Thus, ice formation may largely affect the macro pore wall but the degree of degradation become reduced as capillary pore fraction increases.

3.3. Combined Degradation.

As a parametric value of frost damage, the volume of large capillary pore at each cycle was used to relate the migration coefficient as given in Figure 5.

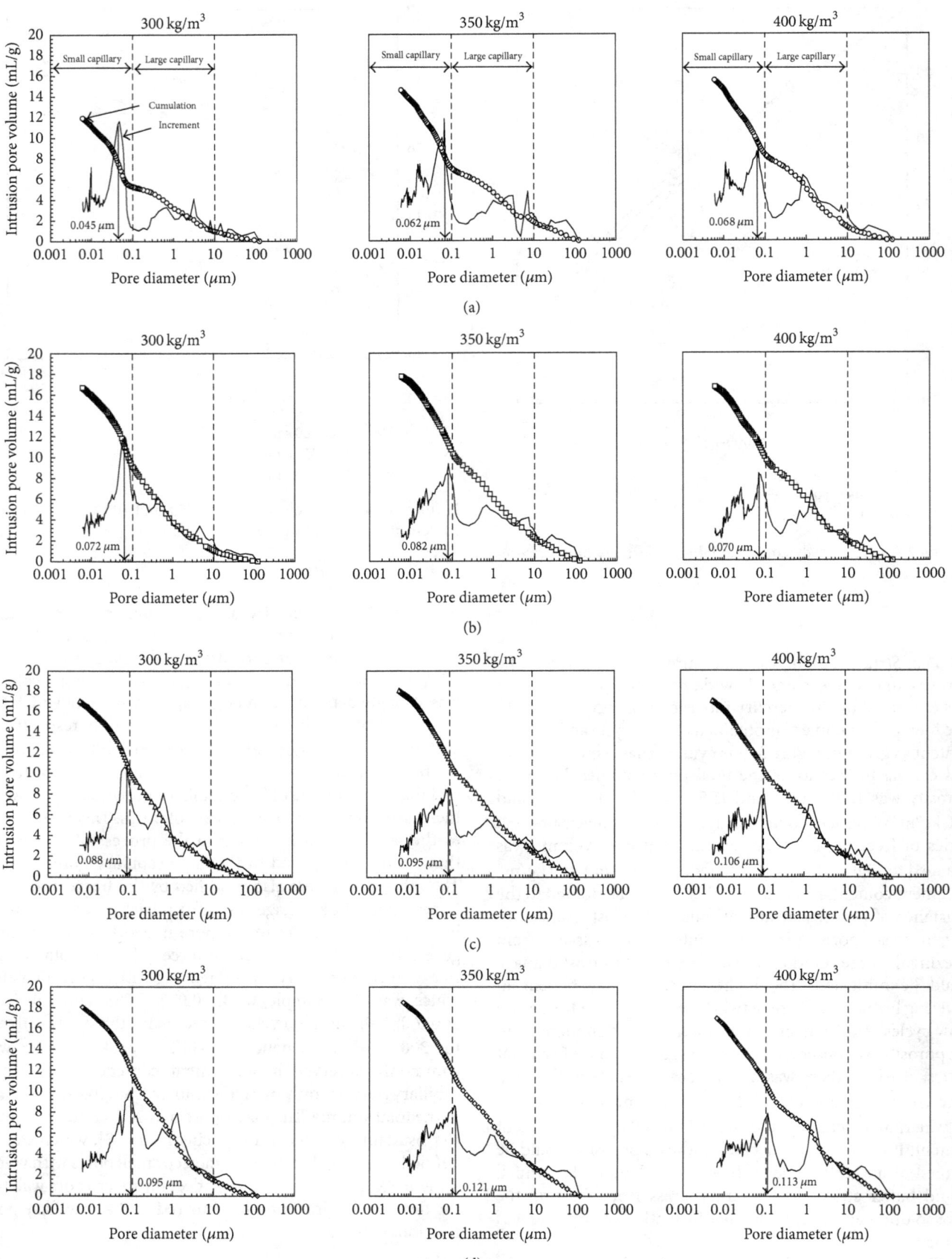

FIGURE 4: Pore-size distribution for 300, 350, and 400 kg/m³ of cement content at (a) 0, (b) 60, (c) 120, and (d) 180 cycles.

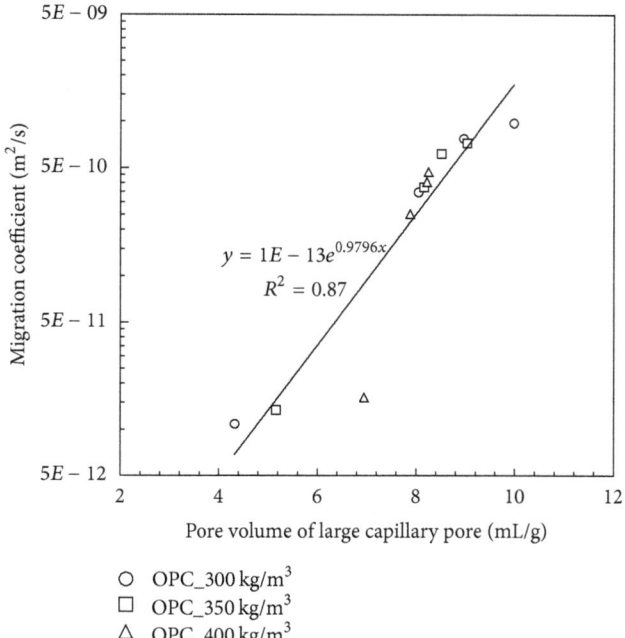

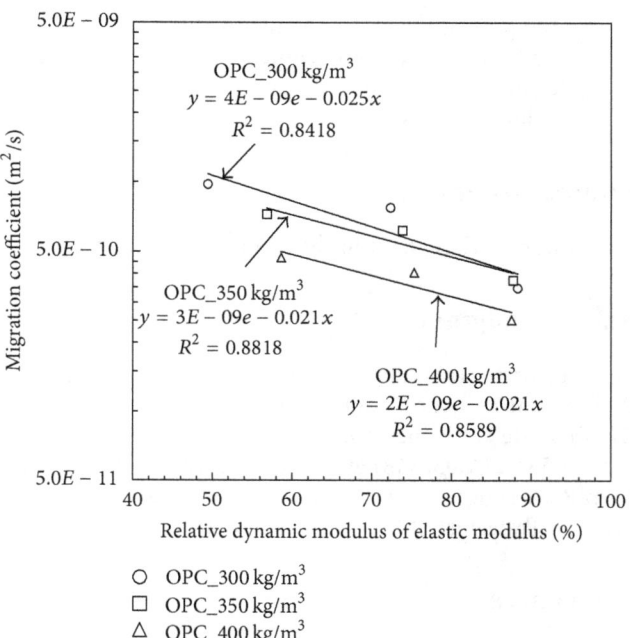

FIGURE 5: The relation between chloride migration coefficient and pore volume of large capillary pores for three different cement content in the concrete mix at freeze-thaw environment.

FIGURE 6: The relation between chloride migration coefficient and relative dynamic modulus of elasticity.

It is seen that there was a linear relation between the large capillary pore volume and migration coefficient. It has an important implication in predicting the corrosion risk of concrete under coupling degradation of frost damage and chloride attack. The pore distribution is usually influenced by a free water cement ratio and binder type (i.e., highly grained pozzolanic materials). However, in this study, the pore volume, in particular, large capillary pores, were further generated by the freeze and thaw cyclic condition. Substantially, an increase in the large capillary pores resulted in an increase in the chloride migration. Once a certain level of capillary pores was generated/produced by the freeze and thaw condition, the cement content had a marginal influence on chloride migration. The critical pore is regarded as the smallest pore that can complete the connection of pores to form pore network. Thus, a further formation and generation of pores in terms of increased critical pore diameter during freeze and thaw cycles may increase the penetration of chloride ion. Additionally, the critical pore diameter is crucial on concrete permeability [9, 10].

The relation between the chloride migration coefficient and relative dynamic modulus of elasticity was depicted in Figure 6. The relative dynamic modulus of elasticity was strongly influenced by the cement content; an increase in the cement content resulted in an increase in the relative dynamic modulus of elasticity at a given cycle of freeze and thaw. Substantially, a reduction of the relative dynamic modulus of elasticity was accompanied by increased migration of chloride ions. It is evident that the higher cement content concrete imposed the low increasing rate of chloride migration coefficient. Thus, to protect steel from corrosion in concrete, the relative dynamic modulus of elasticity also

must be kept high against freezing climate, for example, the concrete pavements or bridge deck exposed to the deicing salts in the cold climate. It suggests that the higher cement content could benefit in lowering frost damage and thus mitigating the rate of chloride transport after the frost damage.

4. Conclusion

In this study, the coupling degradation of concrete, when exposed to a chloride and frost environment, was assessed by experimental works using OPC. Concrete and mortar specimens were exposed to a cyclic freeze-thaw environment to induce the frost damage, and then the chloride migration and pore structure were examined. The freeze and thaw was repeated up to 180 cycles to meet the quality of concrete specimens for chloride transport. The conclusion obtained from experiments is given as follows:

(1) With no frost damage, the rate of chloride transport in terms of migration was increased by cement content at a given free water cement ratio, which could increase the porosity of the concrete matrix, arising from an increase in the amount of cement paste.

(2) When concrete is subjected to chloride permeation and frost damage, the rate of chloride transport was increased by the number of cycles of freeze and thaw; especially a rapid increase was sighted up to 60 cycles, presumably due to increased large capillary pores produced by microcracking in the cement matrix.

(3) Furthermore, the higher cement content lowered the rate of chloride transport at a given number of cycles of freeze and thaw, since the cement contents increase

the porosity to resist against frost damage. Substantially, higher cement contents in concrete benefits against the freeze and thaw environment and thus chloride permeation.

Competing Interests

The authors declare that they have no competing interests.

Acknowledgments

This research was supported by Basic Science Research Program through the National Research Foundation of Korea (NRF) funded by Ministry of Science, ICT & Future Planning (no. 2015R1A5A1037548) and a grant (16RDRP-B076268-03) from R&D program funded by Ministry of Land Infrastructure and Transport of Korea government.

References

[1] C. L. Page, "Mechanism of corrosion protection in reinforced concrete marine structures," *Nature*, vol. 258, no. 5535, pp. 514–515, 1975.

[2] K. Y. Ann and H.-W. Song, "Chloride threshold level for corrosion of steel in concrete," *Corrosion Science*, vol. 49, no. 11, pp. 4113–4133, 2007.

[3] A. M. Neville, *Properties of Concrete*, Longman Group Ltd, Harlow, UK, 4th edition, 1997.

[4] G. Fagerlund, "On the service life of concrete exposed to frost action," in *Freeze-Thaw Durability of Concrete*, J. Marchand, M. Pigeon, and M. Setzer, Eds., pp. 23–41, E & FN Spon, London, UK, 1997.

[5] ASTM C 666, "Standard test method for resistance of concrete to rapid freezing and thawing," ASTM International, 2003.

[6] NT Build 492, "Concrete mortar and cement-based repair materials," Nordtest method, Nordtest, Espoo, Finland, 1999.

[7] B. Zuber and J. Marchand, "Modeling the deterioration of hydrated cement systems exposed to frost action—Part 1: description of the mathematical model," *Cement and Concrete Research*, vol. 30, no. 12, pp. 1929–1939, 2000.

[8] K. A. Kalliopi, *Pore Structure of Cement Based Materials Testing, Interpretation and Requirement*, Taylor & Francis, Oxfordshire, UK, 1st edition, 2006.

[9] L. Cui and J. H. Cahyadi, "Permeability and pore structure of OPC paste," *Cement and Concrete Research*, vol. 31, no. 2, pp. 277–282, 2001.

[10] P. Halamickova, R. J. Detwiler, D. P. Bentz, and E. J. Garboczi, "Water permeability and chloride ion diffusion in portland cement mortars: relationship to sand content and critical pore diameter," *Cement and Concrete Research*, vol. 25, no. 4, pp. 790–802, 1995.

18

Influence of Steam Curing Method on the Performance of Concrete Containing a Large Portion of Mineral Admixtures

Mengyuan Li, Qiang Wang, and Jun Yang

Department of Civil Engineering, Tsinghua University, Beijing, China

Correspondence should be addressed to Qiang Wang; w-qiang@tsinghua.edu.cn

Academic Editor: Yao Luan

A comparison was made between the impact of raising the thermostatic temperature and the impact of prolonging the thermostatic time on the performance of steam-cured concrete containing a large portion of fly ash (FA) or ground granulated blast furnace slag (GGBS) by analysing the form removal strength, chemically combined water content, reaction degree, strength development, chloride permeability, and volume stability. For the materials and test conditions reported in this study, raising the thermostatic temperature is more favourable for concrete containing FA, as indicated by the significantly higher form removal strength and the higher growth of reaction degree of FA compared with prolonging the thermostatic time. With an increase in the thermostatic temperature, the hydration degree of a binder containing FA or GGBS initially increases and subsequently decreases. Although concrete containing FA can obtain satisfactory form removal strength with steam curing at 80°C, the late strength development of concrete containing FA is slow for the same curing conditions. The effect of the late performance of resistance to chloride ion permeability improved by FA is better than the effect improved by GGBS. The risk of destroying the structure of concrete containing a large portion of FA or GGBS due to delayed ettringite formation (DEF) is minimal when specimens were steam-cured at 80°C.

1. Introduction

Concrete is one of the most common construction materials. Cast-in-situ concrete and precast concrete are two techniques that housing developers and construction workers often adopt. However, precast concrete members have been increasingly utilized in civil engineering construction in recent years due to their advantages: reliable quality assurance, simple production process, faster construction speed, and environmentally friendly building operations [1–3].

Currently available information indicates that the technique of steam curing is the most frequently employed technique among various production processes of prefabricated members [4, 5]. The benefits of steam curing are as follows: simple process, convenient operation, production with high early strength, short production cycle, and superior economic benefits [5–7]. The steam curing process includes the following four stages: the precuring stage, the heating stage, the thermostatic stage, and the cooling stage [8, 9]. The thermostatic temperature is generally less than 60°C during steam curing because of two reasons. Firstly, the growth of the late performance of concrete will be small if the thermostatic temperature is excessive during steam curing [10–12]. Secondly, the formation of ettringite produced by cement hydration in the case of a minimum curing temperature of 70°C is decomposed during steam curing and reformed during the service life. This process is called delayed ettringite formation (DEF), and it can substantially weaken the late performance of concrete [13–15].

Mineral admixtures are extensively applied in blended cement and concrete; this process is a substantial contribution to the field of civil engineering. The technology of steam curing has been primarily employed for pure cement concrete rather than concrete with a large portion of mineral admixtures as many researchers and housing developers have expressed their concern that the early compressive strength of concrete with a large portion of mineral admixtures is low [16–18], which make it difficult to satisfy the requirements of form removal strength of steam-cured concrete at the end of steam curing. This is an obstacle to the use of steam curing for concrete that incorporates a large portion of mineral admixtures.

TABLE 1: Chemical compositions and specific surface areas of OPC, GGBS, and FA.

	OPC	GGBS	FA
SiO_2 (%)	21.10	31.76	53.33
Al_2O_3 (%)	6.33	14.84	27.65
Fe_2O_3 (%)	4.22	0.60	6.04
CaO (%)	54.86	36.44	2.86
MgO (%)	2.60	9.08	1.35
SO_3 (%)	2.66	1.94	0.45
Na_2O_{eq} (%)	0.53	0.56	0.64
Loss on ignition (%)	2.42	0.86	4.71
Specific surface area (m^2/kg)	376	430	358

$Na_2O_{eq} = Na_2O + 0.658K_2O$.

TABLE 2: Compositions of the pastes (%).

Sample	Binder			Water/binder ratio
	OPC	GGBS	FA	
CC	100	0	0	0.4
FF	60	0	40	0.4
BB	60	40	0	0.4

FIGURE 1: The measurement of volume stability of concrete.

Although the early strength of concrete with a large portion of mineral admixtures is low at room temperature, high temperatures can promote the early hydration of a binder. By adjusting the thermostatic time and the thermostatic temperature under steam curing conditions, concrete with a large portion of mineral admixtures may achieve the required form removal strength. To address the problem of form removal strength and promote the high performance of steam-cured concrete that incorporates mineral admixtures, this paper addresses the influence of prolonging the thermostatic time and improving the thermostatic temperature on the form removal strength of concrete that incorporates a large portion of mineral admixtures. Our study also focused on a comparison between the impact of prolonging the thermostatic time and the impact of raising the thermostatic temperature on the hydration degree of a binder, the strength development and the resistance to the chloride ion permeability of concrete, and the volume stability of steam-cured concrete with a large portion of mineral admixtures.

2. Raw Materials and Test Methods

2.1. Raw Materials. P.O 42.5 ordinary Portland cement (OPC), ground granulated blast furnace slag (GGBS), and fly ash (FA) were employed in this study. The chemical compositions and specific surface areas of these powder materials are shown in Table 1. The fine aggregates consisted of natural river sands with particle sizes that ranged from 0.08 to 5 mm. The coarse aggregates consisted of crushed limestone smaller than 25 mm.

2.2. Test Methods. Table 2 exhibits the mix proportions of pastes that were used to measure the hydration property of a composite binder. Table 3 exhibits the mix proportions of concrete. Pastes in the plastic tube were prepared. Concrete samples with the dimensions of $10 \times 10 \times 10 cm^3$ were prepared for an experiment on compressive strength and the chloride ion permeability of concrete. Concrete samples with the dimensions of $10 \times 10 \times 300 cm^3$ were prepared for an experiment on volume stability of concrete.

The precuring time for steam curing was three hours (20°C). The heating and cooling rate was 15 ± 1°C/h. Eight thermostatic times were adopted: 8 h, 9 h, 10 h, 11 h, 12 h, 13 h, 14 h, and 16 h. Four thermostatic temperatures were adopted: 60°C, 70°C, 80°C, and 90°C. Concrete that was used to measure the compressive strength and chloride ion permeability were cured at 20°C and a relative humidity greater than 95% after steam curing.

The chemically combined water (w_c) content of hydration products was tested by the difference of weight between a sample dried at 80°C and a sample heated at 1060°C, which were standardized by the weight after drying at 80°C, and by subtracting the loss of ignition of the raw materials. The permeability of chloride ion of the concrete was obtained in accordance with ASTM C1202 "Standard Test Method for Electrical Indication of Concrete's Ability to Resist Chloride Ion Penetration." In order to control the quality of steam-cured concrete, the deviation of strength was restricted to be less than 10%. The determination of the degree of reaction of fly ash was based on a selective dissolution procedure using concentrated hydrochloric acid and water [19, 20]. The determination of the degree of reaction of GGBS was based on a selective dissolution procedure using salicylic acid-methanol-acetone solution [21].

This study involves an experiment on the volume stability of concrete. Concrete that was used to measure volume stability was cured in a saturated $Ca(OH)_2$ solution at 20°C after steam curing. As water is a necessity to DEF, the specimens after steam curing were placed in a saturated $Ca(OH)_2$ solution to keep concrete completely wet during curing. This type of practice can prevent $Ca(OH)_2$ dissolution and drying shrinkage. The consequence of the volume stability analysis was confirmed by measuring the lengths of the concrete specimens using a comparator at scheduled ages. Test probes were installed in advance on both ends of the concrete specimens. Concrete that was used to measure volume stability and the process of measuring volume stability are shown in Figure 1.

3. Results and Discussion

3.1. Form Removal Strength of Concrete. The influence of thermostatic time and thermostatic temperature on the form removal strength of steam-cured concrete is presented in

TABLE 3: Mix proportions of the concrete (kg/m^3).

Sample	OPC	GGBS	FA	Fine aggregates	Coarse aggregates	Water
C	350	0	0	812	1077	161
F30	245	0	105	812	1077	161
F40	210	0	140	812	1077	161
F50	175	0	175	812	1077	161
B30	245	105	0	812	1077	161
B40	210	140	0	812	1077	161
B50	175	175	0	812	1077	161
B60	140	210	0	812	1077	161

TABLE 4: Removal strengths under different curing conditions.

Samples	Thermostatic temperature/°C	Thermostatic time/h	Removal strength/MPa
C	60	9	28.7
F30	60	11	23.7
	80	9	35.7
F40	60	11	17.7
	80	9	29.7
F50	60	13	10.4
	80	11	26.8
B30	60	11	27.8
	80	9	31.0
B40	60	11	32.7
	80	9	27.2
B50	60	11	20.0
	80	9	22.6
B60	60	11	19.4
	80	10	27.8

Table 4. The form removal strength of pure cement concrete that is cured at 60°C for 9 h in steam curing is established as the control group. When a thermostatic temperature of 60°C is maintained, the form removal strength of concrete F30 (thermostatic time: 11 h), F40 (thermostatic time: 11 h), and F50 (thermostatic time: 13 h) is lower than the form removal strength of the control group. When a thermostatic temperature of 80°C is maintained, the form removal strength of concrete F30 and F40 (thermostatic time: 9 h) is higher than the form removal strength of the control group. When the thermostatic temperature is increased to 80°C and the thermostatic time is simultaneously prolonged to 11 h, the form removal strength of concrete F50 is also similar to the form removal strength of the control group. It is an indication that the degree of influence of raising the thermostatic temperature on the form removal strength of concrete containing FA is superior to the degree of influence of prolonging the thermostatic time on the form removal strength of concrete containing FA. Raising the thermostatic temperature to 80°C can attain satisfactory form removal strength.

By utilizing the form removal strength of pure cement concrete under steam curing at 60°C for 9 h as the reference, the form removal strength of concrete B30 and B40 cured at 80°C for 9 h is similar to the form removal strength of the control group. The form removal strength of concrete B50

that was cured at 80°C for 9 h is obviously lower than the form removal strength of the control group. However, by prolonging the thermostatic time to 10 h, the form removal strength of concrete B60 that was cured at 80°C is similar to the form removal strength of the control group. By prolonging the thermostatic time to 11 h and controlling the thermostatic temperature (60°C) as a constant, the form removal strengths of concrete B30 and B40 are similar to the form removal strength of the control group and the form removal strengths of concrete B50 and B60 are lower than the form removal strength of the control group. Both raising the thermostatic temperature and prolonging the thermostatic time enhance the form removal strength of concrete that incorporates a large portion of GGBS. When the content of GGBS in cementing materials exceeds 50%, methods of prolonging the thermostatic time and raising the thermostatic temperature need to be simultaneously employed to obtain the satisfactory form removal strength as the influence of prolonging the thermostatic time on enhancing the form removal strength is limited.

3.2. Chemically Combined Water Content. w_c content of hydration products reveals the hydration degree of the same binder. The influences of the thermostatic time on w_c content of cement paste, the paste containing a large portion of GGBS,

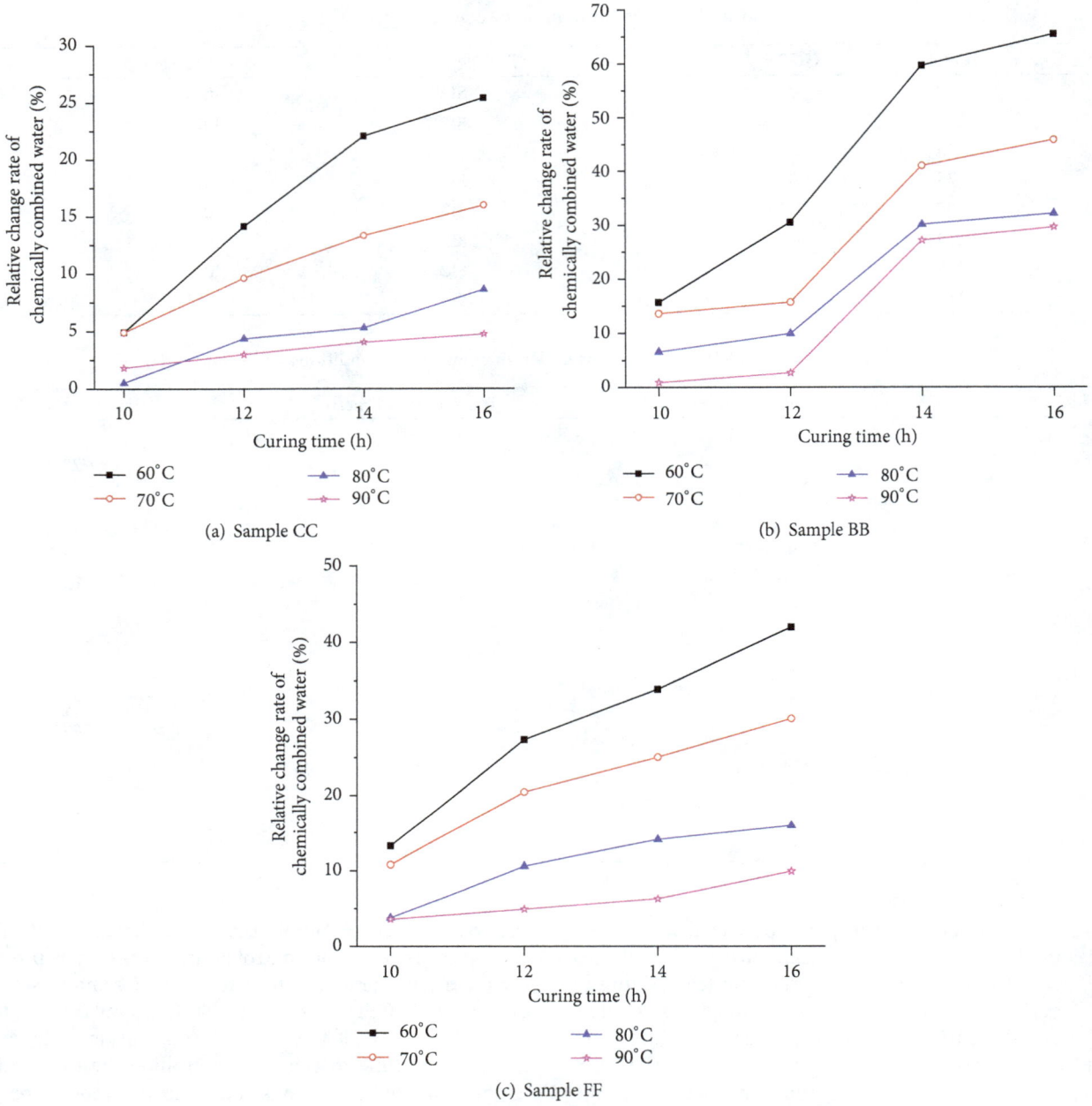

FIGURE 2: The influence of thermostatic time on w_c content.

and the paste containing a large portion of FA are illustrated in Figures 2(a), 2(b), and 2(c), respectively, at the end of the steam curing process. The ordinates in Figure 2 denote the relative change rate of w_c content of pastes that were cured for 10 h, 12 h, 14 h, and 16 h relative to the paste cured for 8 h.

Figure 2(a) indicates that w_c content of pure cement paste increases with an increase in thermostatic time; however, if the curing temperature in the thermostatic period is low, the growth rate of w_c content is high with the extension of thermostatic time. This finding may be attributable to this reason: 8 h of curing in the thermostatic period is sufficient for the hydration degree of cement achieving a high level when the temperature in the thermostatic period is sufficiently high. Few contributions to improving the

hydration degree of cement have been achieved by prolonging the thermostatic time under this circumstance.

Conclusions can be drawn from Figures 2(b) and 2(c). (1) When the curing temperature in the thermostatic period is low, the growth rates of w_c content of the binder with a large portion of GGBS and the binder with a large portion of FA are high due to the extension of thermostatic time, which is similar to the pure cement paste. (2) The order of the degree of influence of prolonging the thermostatic time on improving w_c content of binders is the binder containing large portion mineral admixtures > the pure cement paste at the same thermostatic temperature. This finding may be attributable to this reason: when considering the binder containing a large portion of GGBS or FA, the thermostatic time may

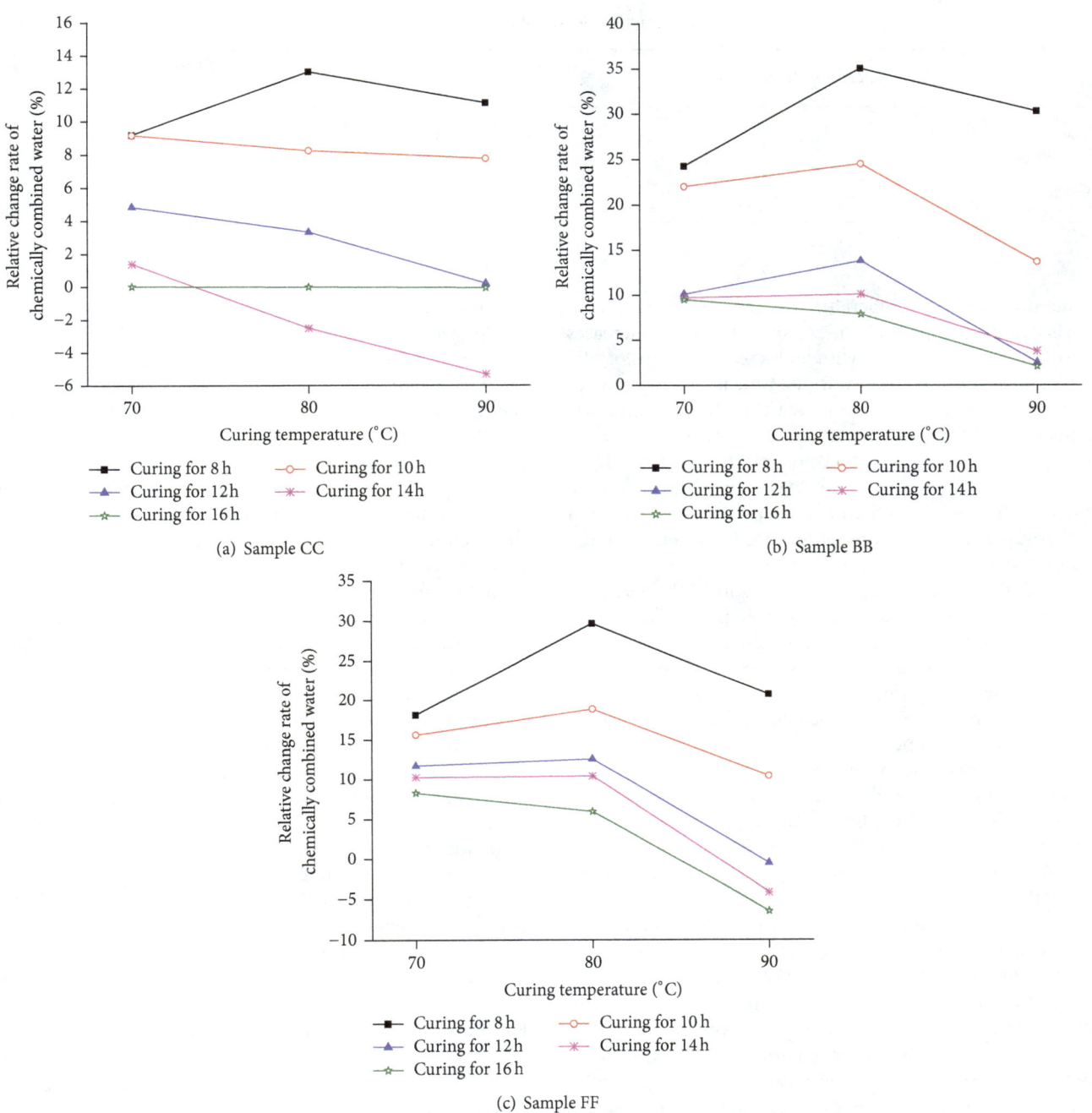

(a) Sample CC

(b) Sample BB

(c) Sample FF

FIGURE 3: The influence of thermostatic temperature on w_c content.

improve not only the hydration reaction of cement but also the reaction of GGBS or FA. (3) The order of the degrees of influence of prolonging the thermostatic time on improving w_c content of binders is the binder containing a large portion of GGBS > the binder containing a large portion of FA. The order of the degree of influence of the thermostatic time on the hydration degree is the binder containing a large portion of GGBS > the binder containing a large portion of FA. This finding is consistent with the trend for the form removal strength that is presented in Table 4.

The influence of the thermostatic temperature on w_c content of cement paste, the paste containing a large portion

of GGBS, and the paste containing a large portion of FA at the end of the steam curing process is illustrated in Figures 3(a), 3(b), and 3(c), respectively. The ordinates in Figure 3 denote the relative change rate of w_c content of pastes cured at 70°C, 80°C, and 90°C relative to the paste cured at 60°C. A positive relative change rate indicates that an increase in the thermostatic temperature from 60°C can further enhance the hydration of binder. A negative relative change rate indicates that an increase in the thermostatic temperature from 60°C would inhibit the hydration of binder.

Figure 3(a) reveals that w_c content of the hydration products of cement has not always increased with an increase

TABLE 5: Reaction degree of fly ash and GGBS/%.

	Thermostatic temperature/°C	Thermostatic time/h		
		8	12	16
Fly ash	60	2.11	4.26	6.39
	80	5.79	7.96	9.73
GGBS	60	7.86	10.15	13.18
	80	9.84	13.15	15.25

in the thermostatic temperature. When the thermostatic time is 8 h, 10 h, 12 h, and 14 h, w_c content initially increases and subsequently decreases with an increase in thermostatic temperature. When the thermostatic time is 16 h, w_c content remains unchanged with an increase in the thermostatic temperature. Therefore, when the thermostatic period is short, an increase in the thermostatic temperature from 60°C to 70°C can enhance the hydration degree of pure cement paste. For a long thermostatic period, an increase in the thermostatic temperature is not needed to enhance the hydration degree of pure cement paste.

Conclusions can be drawn from Figures 3(b) and 3(c): (1) when the thermostatic temperature increases from 60°C to 70°C, w_c content of every group increases; (2) when the thermostatic temperature increases from 70°C to 80°C, w_c content of the pastes, which have thermostatic times of 8 h, 10 h, and 12 h, increases, and w_c content of the pastes, which have a thermostatic time of 14 h and 16 h, remains unchanged or slightly decreased; (3) when the thermostatic temperature increases from 80°C to 90°C, w_c content of every group decreases. Note that the change rate of w_c content of the paste containing a large portion of FA (thermostatic period: 90°C/14 h or 16 h) relative to the change rate of the paste cured at 60°C is negative.

This result indicates that increasing the thermostatic temperature to 80°C can effectively enhance the hydration degree of paste containing a large portion of FA or GGBS. Compared with Figure 3(a), by increasing the thermostatic temperature, the hydration promoting effect of paste containing a large portion of FA or GGBS is more apparent than the hydration promoting effect of pure cement paste. This finding may be attributable to two reasons: firstly, increasing the thermostatic temperature not only improves the hydration reaction of cement but also stimulates the activity of FA or GGBS; secondly, the reaction degree of fly ash and GGBS is much lower than that of cement at normal temperature.

3.3. The Reaction Degree of Mineral Admixtures.
The influence of the thermostatic time on the reaction degree of GGBS and FA is presented in Table 5. Regardless of whether the thermostatic temperature is 60°C or 80°C, the reaction degree of GGBS or FA will increase with an extension of the thermostatic time. This result may be attributed to this reason: the longer is the thermostatic time, the longer is the time required to stimulate the activity of GGBS or FA, which will increase the reaction degree of GGBS or FA.

Table 5 also reveals that the reaction degree of FA (thermostatic period: 60°C/8 h) is only 2.1%, which demonstrates

that FA only serves a role in the microaggregate filling effect in this situation. The reaction degree of FA (thermostatic period: 60°C/16 h) is 6.4%, which is substantially higher than the reaction degree of FA after 8 h of curing during the thermostatic period. However, the reaction degree remains at a low level from the point of absolute value. Therefore, prolonging the thermostatic time at 60°C is not an effective way for the concrete containing a large portion of FA to achieve a satisfactory form removal strength. Conversely, the reaction degree of GGBS at 60°C for 8 h is 7.9%, which even exceeds the reaction degree of FA at 60°C for 16 h. The absolute value of the reaction degree of GGBS is obviously higher than that of the reaction degree of FA in an equivalent steam curing system, which suggests that the chemical effect of GGBS is obviously higher than the chemical effect of FA in the early steam curing process.

The following can also be concluded from Table 5. (1) When the thermostatic temperature increases from 60°C to 80°C, the reaction degree of GGBS and FA is improved. (2) The influence of increasing the thermostatic temperature on the reaction degree of FA is more significant from the point of growth.

From the perspective of the reaction degree of mineral admixtures as well as the hydration degree of the whole binder, it is obvious that the promoting effect of increasing the thermostatic temperature is more significant than the promoting effect of prolonging the thermostatic time on the early hydration of the binder containing a large portion of FA. Increasing the thermostatic temperature and prolonging the thermostatic time have a significant role in promoting the hydration of GGBS. This case also applies to the influence of the thermostatic temperature and thermostatic time on the form removal strength of concrete.

3.4. Strength Development of Concrete.
The comparison between the strength development of pure cement concrete after steam curing and the strength development of steam-cured concrete containing a large portion of GGBS and FA is shown in Figure 4. The strength development of pure cement concrete (thermostatic period: 60°C/9 h) is employed as the reference. Figure 4(a) shows the strength development of concrete containing FA after steam curing at 80°C. As shown in Figure 4(a), although the form removal strength of steam-cured concrete containing a large portion of FA is not lower than the form removal strength of the control group, the compressive strength of steam-cured concrete containing a large portion of FA is substantially lower than the form removal strength of the control group at the age of 28 days

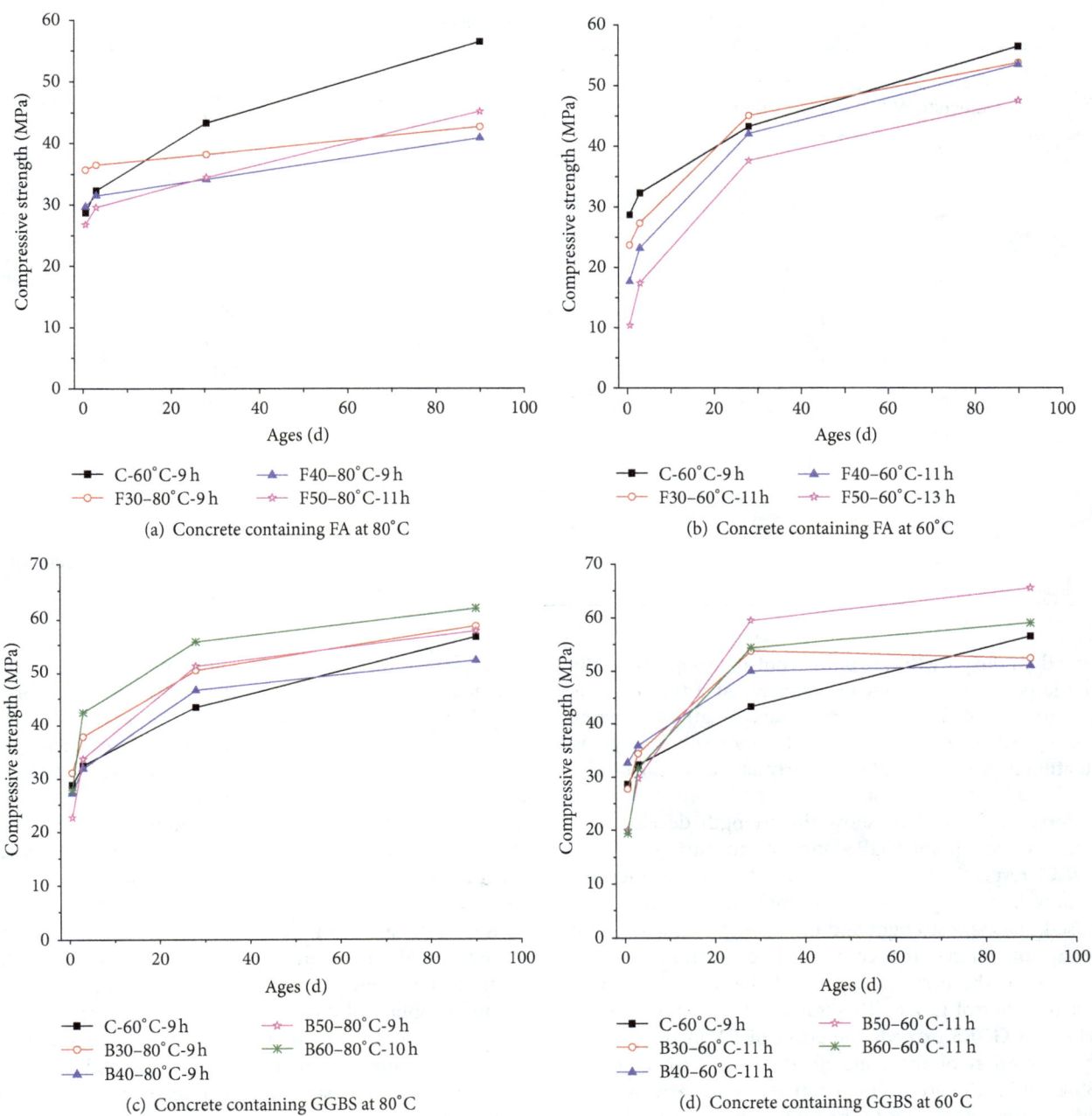

(a) Concrete containing FA at 80°C

(b) Concrete containing FA at 60°C

(c) Concrete containing GGBS at 80°C

(d) Concrete containing GGBS at 60°C

FIGURE 4: Comparison of strength development between pure cement concrete and concrete containing large portion FA or GGBS under steam curing.

and 90 days. After the age of three days, the strength growth of steam-cured concrete containing a large portion of FA is very slow, which differs from the strength growth of ordinary concrete containing FA [22–24]. The greater the content of FA, the greater the contribution of the pozzolanic reaction to the strength at a later age, and the greater the potential for subsequent growth of concrete. High temperatures enhance the form removal strength of concrete containing FA at an early age; however, it hinders the strength development of concrete at a later age. Possible reasons for this phenomenon are as follows. (1) When the thermostatic temperature is 80°C, the early reaction of cement is significant. The gel layer of

C-S-H can be formed in the particles' surfaces of cement and FA, which is disadvantage for further hydration of binder at a later age. (2) The distribution of the hydration product of cement is uneven at elevated temperatures, and a large amount of $Ca(OH)_2$ crystal exhibits orientation distribution. Thus, the contact area between FA and $Ca(OH)_2$ is reduced. The pozzolanic reaction of FA at a later age is restricted.

Figure 4(b) shows the strength development of concrete containing FA after steam curing at 60°C. As shown in Figure 4(b), although the growth rate of the late strength of concrete containing FA is similar to the growth rate of the control group, the form removal of concrete containing FA

TABLE 6: Chloride ion permeability of concrete.

Samples	Thermostatic temperature/°C	Thermostatic time/h	28 d		90 d	
			Charge passed/C	Permeability level	Charge passed/C	Permeability level
C	60	9	7111	High	4338	High
F30	60	11	1714	Low	700	Very low
	80	9	644	Very low	503	Very low
F40	60	11	1707	Low	510	Very low
	80	9	682	Very low	278	Very low
F50	60	13	1952	Low	607	Very low
	80	11	518	Very low	200	Very low
B30	60	11	2094	Moderate	1350	Low
	80	9	2628	Moderate	1709	Low
B40	60	11	2424	Moderate	1535	Low
	80	9	2765	Moderate	1650	Low
B50	60	11	1668	Low	1117	Low
	80	9	2075	Moderate	1213	Low
B60	60	11	1515	Low	813	Very low
	80	10	1150	Low	775	Very low

is lower than the growth rate of the control group, especially for the larger content of FA of concrete. At a thermostatic temperature of 60°C, the form removal strength of concrete containing FA is relatively low, which explains why the stimulation degree of 60°C for the early activity of the binder containing a large portion of FA is limited.

Figures 4(c) and 4(d) show the strength development of concrete containing GGBS after steam curing at 60°C and 80°C, respectively. Regardless of whether the thermostatic time is prolonged or the thermostatic temperature is increased, the early strength and the strength growth rate at a later age of concrete that contain a large portion of GGBS are similar to the early strength and the strength growth rate of the control group. Therefore, at high temperatures, the effect of GGBS exceeds the effect of FA in the process of the formation of the concrete strength: first, at 60°C, GGBS exhibits a relatively high activity at an early age and it substantially contributes to the form removal strength and early strength; second, GGBS can display substantial early activity at 80°C; third, even if the concrete containing a large portion of GGBS is cured at 80°C, it can obtain a satisfactory late strength. This finding suggests that GGBS serves an important role in the late strength growth of steam-cured concrete; that is, after steam curing at a high temperature, GGBS can still take a considerable pozzolanic reaction at a high reaction rate at a later age.

3.5. Resistance to Chloride Ion Permeability of Concrete. The comparison between chloride ion permeability of concrete containing a large portion of FA or GGBS and the chloride ion permeability of pure cement concrete after steam curing is illustrated in Table 6. The chloride ion permeability of pure cement concrete (thermostatic period: 60°C/9 h) at the same age is employed as the reference. According to ASTM C1202, which is related to the chloride ion permeability

grade classification, the permeability of the control group at 28 d falls in the "High" level, the permeability of concrete containing FA which is cured at 80°C falls in the "Very Low" level, and the permeability of concrete containing FA which is cured at 60°C falls in the "Low" level. At the age of 90 days, the permeability of the control group falls in the "High" level, and the permeability of concrete containing FA which is cured at 60°C and 80°C falls in the "Very Low" level. This finding suggests that the resistance to chloride ion permeability of steam-cured concrete containing FA is substantially better than the resistance of pure cement concrete, which is one of the advantages to steam-cured concrete containing a large portion of FA. FA can enhance the resistance to chloride ion permeability of concrete at a later age, which has been confirmed by a large number of experiments [25]. The main reason for this improvement is the ability of FA to improve on the pore structure of concrete due to pozzolanic reaction [26]; the secondary hydration products decrease the connected porosity of concrete. Therefore, the higher is the reaction degree of FA, the greater is the contribution to the resistance to the chloride ion permeability of concrete. High temperature curing at an early age can significantly stimulate the activity of FA and enhance the reaction degree of FA. Thus, FA enhances the resistance of steam-cured concrete to chloride ion permeability.

At the age of 28 days, the permeability of steam-cured concrete containing GGBS falls in the "Moderate" or "Low" levels. At the age of 90 days, the permeability of steam-cured concrete containing GGBS falls in the "Low" or "Very Low" levels. The greater is the mixing amount of GGBS, the better is the resistance to the chloride ion permeability of concrete. Compared with pure cement concrete, the concrete containing GGBS can achieve better resistance to chloride ion permeability. In addition, the chloride ion permeability of steam-cured concrete containing GGBS of each group

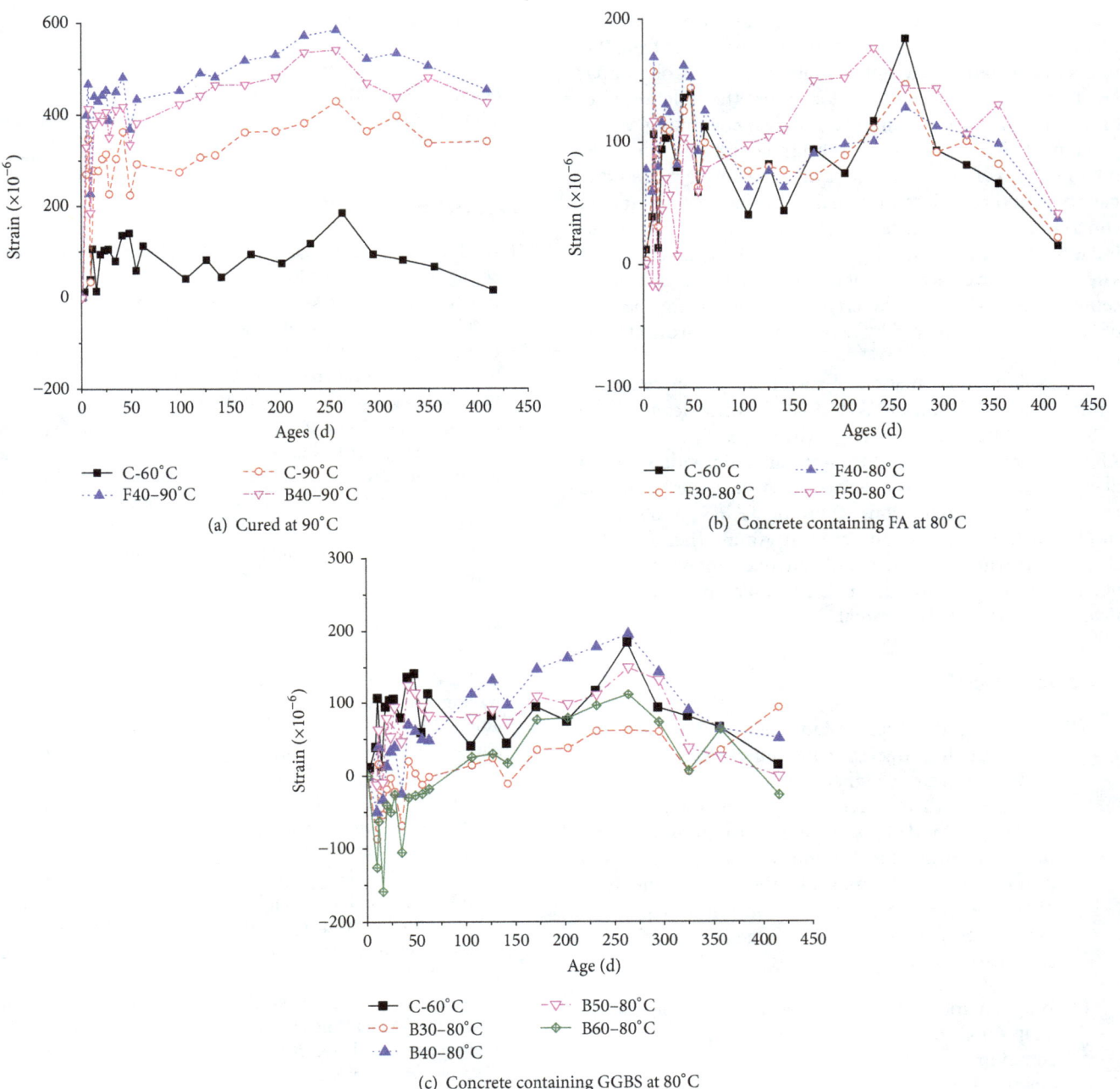

FIGURE 5: Volume deformation of steam-cured concrete.

is not substantially different due to the difference in the curing systems, as both prolonging the thermostatic time and improving the thermostatic temperature can stimulate the reaction activity of GGBS and substantially improve the pore structure of concrete. The effect of the late performance of resistance to chloride ion permeability improved by FA is better than the effect by GGBS. Although the reaction degree of GGBS is higher than the reaction degree of FA after steam curing, the pozzolanic reaction of FA can consume a mass of $Ca(OH)_2$. The amount of $Ca(OH)_2$ consumed by GGBS is minimal. Therefore, the reaction of FA plays a significant role in improving the pore structure of concrete.

3.6. *Volume Stability Analysis of Concrete.* The comparison between the volume deformation of steam-cured concrete containing a large portion of FA or GGBS and the volume deformation of pure cement concrete after steam curing is presented in Figure 5. Figure 5(a) shows the comparison between the volume deformation of concrete containing a large portion of FA or GGBS, the volume deformation of pure cement concrete with steam curing at 90°C, and the volume deformation of pure cement concrete with steam curing at 60°C. The microstrain of pure cement concrete (thermostatic period: 60°C/8 h) is employed as the reference. For steam curing at 90°C, the inflation rates for the cement concrete and the concrete containing a large portion of FA or GGBS are relatively high, of which the late microstrain varies from 200 to 600. However, the late microstrain of the control group is less than 200. These results indicate that the inflation rates of the cement concrete and the concrete containing a large portion of FA or GGBS are substantially higher than the inflation rates of the control group. A large number of studies have proven that DEF cannot weaken the late performance of concrete at thermostatic temperatures below 60°C. Thus, the inflation rate of the control group can be considered to be a safe value. Conversely, the possibility of the structure of the cement concrete and the concrete containing a large portion of FA or GGBS destroyed by DEF after steam curing at 90°C is significant.

The volume deformation of steam-cured concrete containing a large portion of FA and GGBS with steam curing at 80°C is illustrated in Figures 5(b) and 5(c), respectively. All the concrete types produce normal strain with a certain degree at a later age. The strains of steam-cured concrete containing a large portion of FA or GGBS at 80°C are similar to the strains of the control group. Therefore, the risk of destroying the structure of concrete containing a large portion of FA or GGBS due to DEF when specimens were steam-cured at 80°C is minimal.

4. Conclusions

(1) Improving the thermostatic temperature is more favourable for concrete containing FA, as indicated by the significantly higher form removal strength and higher growth of reaction degree of FA compared with the method of prolonging the thermostatic time. Both improving the thermostatic temperature and prolonging the thermostatic time contribute to a distinct enhancement of the form removal strength of concrete that incorporates a large portion of GGBS and the reaction degree of GGBS.

(2) With an increase in the thermostatic temperature from 60°C to 90°C, the hydration degree of binder containing FA or GGBS initially increases and subsequently decreases.

(3) Concrete containing FA can obtain satisfactory form removal strength with steam curing at 80°C; however, the late strength growth rate of concrete containing FA is low for the same curing conditions.

(4) The effect of late performance of resistance to chloride ion permeability improved by FA is better than the same effect achieved by GGBS.

(5) The risk of destroying the structure of concrete containing FA or GGBS due to DEF when specimens were steam-cured at 80°C is minimal.

Conflicts of Interest

The authors declare that they have no conflicts of interest.

Acknowledgments

The authors would like to acknowledge National Natural Science Foundation of China (no. 51478248) and the Tsinghua University Initiative Scientific Research Program (20131089239).

References

[1] Z.-M. He, G.-C. Long, and Y.-J. Xie, "Influence of subsequent curing on water sorptivity and pore structure of steam-cured concrete," *Journal of Central South University of Technology (English Edition)*, vol. 19, no. 4, pp. 1155–1162, 2012.

[2] H.-K. Choi, Y.-C. Choi, and C.-S. Choi, "Development and testing of precast concrete beam-to-column connections," *Engineering Structures*, vol. 56, pp. 1820–1835, 2013.

[3] H. B. Jiang, Q. Cao, A. R. Liu, T. L. Wang, and Y. Qiu, "Flexural behavior of precast concrete segmental beams with hybrid tendons and dry joints," *Construction and Building Materials*, vol. 110, pp. 1–7, 2016.

[4] Z. M. He, J. Z. Liu, and K. W. Zhu, "Influence of mineral admixtures on the short and long-term performance of steam-cured concrete," *Energy Procedia*, vol. 16, pp. 836–841, 2012.

[5] G. Long, Z. He, and A. Omran, "Heat damage of steam curing on the surface layer of concrete," *Magazine of Concrete Research*, vol. 64, no. 11, pp. 995–1004, 2012.

[6] A. M. Ramezanianpour, K. Esmaeili, S. A. Ghahari, and A. A. Ramezanianpour, "Influence of initial steam curing and different types of mineral additives on mechanical and durability properties of self-compacting concrete," *Construction and Building Materials*, vol. 73, pp. 187–194, 2014.

[7] J. L. García Calvo, M. C. Alonso, L. Fernández Luco, and M. Robles Velasco, "Durability performance of sustainable self compacting concretes in precast products due to heat curing," *Construction and Building Materials*, vol. 111, pp. 379–385, 2016.

[8] I. Won, Y. Na, J. T. Kim, and S. Kim, "Energy-efficient algorithms of the steam curing for the in situ production of precast concrete members," *Energy and Buildings*, vol. 64, pp. 275–284, 2013.

[9] F. Cassagnabère, G. Escadeillas, and M. Mouret, "Study of the reactivity of cement/metakaolin binders at early age for specific use in steam-cured precast concrete," *Construction and Building Materials*, vol. 23, no. 2, pp. 775–784, 2009.

[10] G. Long, M. Wang, Y. Xie, and K. Ma, "Experimental investigation on dynamic mechanical characteristics and microstructure of steam-cured concrete," *Science China Technological Sciences*, vol. 57, no. 10, pp. 1902–1908, 2014.

[11] S. J. Barnett, M. N. Soutsos, S. G. Millard, and J. H. Bungey, "Strength development of mortars containing ground granulated blast-furnace slag: effect of curing temperature and determination of apparent activation energies," *Cement and Concrete Research*, vol. 36, no. 3, pp. 434–440, 2006.

[12] M. Paul and F. P. Glasser, "Impact of prolonged warm (85°C) moist cure on Portland cement paste," *Cement and Concrete Research*, vol. 30, no. 12, pp. 1869–1877, 2000.

[13] G. Escadeillas, J.-E. Aubert, M. Segerer, and W. Prince, "Some factors affecting delayed ettringite formation in heat-cured mortars," *Cement and Concrete Research*, vol. 37, no. 10, pp. 1445–1452, 2007.

[14] K. Tosun, "Effect of SO_3 content and fineness on the rate of delayed ettringite formation in heat cured Portland cement

mortars," *Cement and Concrete Composites*, vol. 28, no. 9, pp. 761–772, 2006.

[15] R. Yang, C. D. Lawrence, C. J. Lynsdale, and J. H. Sharp, "Delayed ettringite formation in heat-cured Portland cement mortars," *Cement and Concrete Research*, vol. 29, no. 1, pp. 17–25, 1999.

[16] L. Lam, Y. L. Wong, and C. S. Poon, "Degree of hydration and gel/space ratio of high-volume fly ash/cement systems," *Cement and Concrete Research*, vol. 30, no. 5, pp. 747–756, 2000.

[17] C. S. Poon, L. Lam, and Y. L. Wong, "Study on high strength concrete prepared with large volumes of low calcium fly ash," *Cement and Concrete Research*, vol. 30, no. 3, pp. 447–455, 2000.

[18] C. C. Castellano, V. L. Bonavetti, H. A. Donza, and E. F. Irassar, "The effect of w/b and temperature on the hydration and strength of blastfurnace slag cements," *Construction and Building Materials*, vol. 111, pp. 679–688, 2016.

[19] K. Luke and F. P. Glasser, "Selective dissolution of hydrated blast furnace slag cements," *Cement and Concrete Research*, vol. 17, no. 2, pp. 273–282, 1987.

[20] B. A. Suprenant and G. Papadopoulos, "Selective dissolution of portland-fly-ash cements," *Journal of Materials in Civil Engineering*, vol. 3, no. 1, pp. 48–59, 1991.

[21] Q. Wang, M. Shi, and D. Wang, "Contributions of fly ash and ground granulated blast-furnace slag to the early hydration heat of composite binder at different curing temperatures," *Advances in Cement Research*, vol. 28, no. 5, pp. 320–327, 2016.

[22] T. Hemalatha and A. Ramaswamy, "A review on fly ash characteristics—towards promoting high volume utilization in developing sustainable concrete," *Journal of Cleaner Production*, vol. 147, pp. 546–559, 2017.

[23] F. U. A. Shaikh and S. W. M. Supit, "Compressive strength and durability of high-volume fly ash concrete reinforced with calcium carbonate nanoparticles," in *Fillers and Reinforcements for Advanced Nanocomposites*, chapter 11, pp. 275–307, Woodhead Publishing, 2015.

[24] A. M. Rashad, "A brief on high-volume Class F fly ash as cement replacement—a guide for Civil Engineer," *International Journal of Sustainable Built Environment*, vol. 4, no. 2, pp. 278–306, 2015.

[25] J. Liu, F. Xing, B. Dong, H. Ma, and D. Pan, "Study on water sorptivity of the surface layer of concrete," *Materials and Structures*, vol. 47, no. 11, pp. 1941–1951, 2014.

[26] H. Ma, *Multi-scale modeling of the microstructure and transport properties of contemporary concrete [Ph.D. thesis]*, The Hong Kong University of Science and Technology, Hong Kong, 2013.

19

Flexural Behavior of Concrete Beam Strengthened by Near-Surface Mounted CFRP Reinforcement Using Equivalent Section Model

Woo-tai Jung, Jong-sup Park, Jae-yoon Kang, and Moon-seoung Keum

Structural Engineering Research Institute, Korea Institute of Civil Engineering and Building Technology, Goyang, Republic of Korea

Correspondence should be addressed to Woo-tai Jung; woody@kict.re.kr

Academic Editor: Carlo Santulli

FRP (fiber reinforced polymer) has found wide applications as an alternative to steel rebar not only for the repair and strengthening of existing structures but also for the erection of new structures. Near-surface mounted (NSM) strengthening was introduced as an alternative of externally bonded reinforcement (EBR) but this method also experiences early bond failure, which stresses the importance of predicting accurately the bond failure behavior in order to evaluate precisely the performance of NSM reinforcement. This study proposes the equivalent section model assuming monolithic behavior of the filler and CFRP reinforcement. This equivalent section model enables establishing a bond failure model applicable independently of the sectional shape of the CFRP reinforcement. This so-derived bond failure model is then validated experimentally by means of beams flexure-strengthened by NSM CFRP reinforcements with various cross-sections. Finally, analytical analysis applying the bond failure model considering the equivalent section and defined failure criteria is performed. The results show the accuracy of the prediction of the failure mode as well as the accurate prediction of the experimental results regardless of the sectional shape of the CFRP reinforcement.

1. Introduction

Despite its excellence as construction material, concrete degrades with time and the serviceability deteriorates following the ongoing performance loss of the concrete member and the material itself caused by environmental actions. The width of the cracks developed in concrete enlarges gradually with time and favors the penetration of impurities, which accelerate the degradation of the structure. Steel plate bonding, external tendons prestressing, and FRP (fiber reinforced polymer) bonding are some of the common strengthening methods for degraded concrete structures. Steel plate bonding offers poor workability on site due to its heavy weight and is susceptible to corrosion. External tendons prestressing can improve the serviceability by controlling the cracks and restoring the deflection but requires anchoring devices that are designed to have substantial size to meet the tensile strength of the reinforcement. All three methods present the inconvenience of being vulnerable to external

damage because of external reinforcements factual exposure at the surface of the reinforced section. Besides, FRP bonding improves the performance of the structure by bonding the CFRP (carbon fiber reinforced polymer) sheet or plate on the tensile face of the concrete member. Owing to the lightweight, corrosion-resistant, and high strength properties of CFRP, FRP bonding is gaining wider applications as an alternative to steel plate bonding [1–6].

The structure strengthened by externally bonded CFRP sheet or plate experiences bond failure of the interface between the FRP reinforcement and concrete, which makes the CFRP reinforcement unable to develop 100% of its capacity. The risk of bond failure increases with shorter reinforced length and larger amount of reinforcement. Also, this risk depends sensitively on the skill of the technician executing the strengthening work [2]. Another drawback of the external bonding is the exposure of the reinforcement, which increases the risk of damage caused by vehicle impact or fire and makes it difficult to achieve strengthening effect without sufficient protection. Accordingly, Near-Surface Mounted (NSM) strengthening was introduced to minimize such problems and improve the utilization of the reinforcing material [2]. NSM strengthening using CFRP reinforcement started since 1990s and designates the strengthening method embedding the CFRP reinforcement and the filler inside the slot dug in the tensile zone of the concrete cover. The NSM strengthening is known to reduce the tasks executed on site compared to the external bonding and to decrease the effects of early bond failure [3, 4].

Similarly to external bonding, NSM strengthening also experiences bond failure. This topic is being studied until now and necessitates deeper investigation since different bond failure models were proposed according to the sectional shape of the reinforcement or stochastic data. There is a need is thus to provide a bond failure model for each shape of CFRP reinforcement to predict the flexural behavior of the member strengthened by NSM reinforcement (NSMR) [5–8]. Bond failure may theoretically occur in diverse manners like the interfacial failure between the CFRP reinforcement and filler, the failure of the filler-concrete interface, and so on. Because the bond failure induced experimentally may be different than that predicted theoretically, the bond failure model should rely on actual failure patterns. Accordingly, the proposed bond failure model should be applicable regardless of the sectional shape of the CFRP reinforcement based upon the close analysis of the bond failure occurring actually in the flexure-strengthened behavior.

This study intends to propose a bond failure model considering the equivalent section through the analysis of the failure patterns of specimens flexure-strengthened by NSM CFRP reinforcement and to validate this model with regard to the flexural behavior of NSM-strengthened specimens with various CFRP reinforcement shapes. Other studies have deficiencies such as complicated analysis and various failures with FRP shapes because they have used the data which is not flexural test, but pull-out test. This study offers new failure model with equivalent section which results from failure mode of flexural test.

2. Major Failure Modes of Flexure-Strengthened Specimens by NSMR

The common failure modes experienced by the RC (reinforced concrete) member strengthened by NSM CFRP are various like failure at the interface FRP-filler, cohesive failure on filler, failure at the interface filler-concrete, and cohesive failure on concrete [8]. Since bond failure is an important factor affecting the performance of the member, studies were carried out to predict experimentally and analytically the bond failure characteristics. Even if bond failure by pull-out test occurs in various ways like failure of CFRP reinforcement-filler interface, inside the filler, or of the filler-concrete interface, the bond failure of the flexure-strengthened member by NSM reinforcement may lack diversity compared to that of the pull-out test [4]. Accordingly, there is a need is to examine the major failure modes through

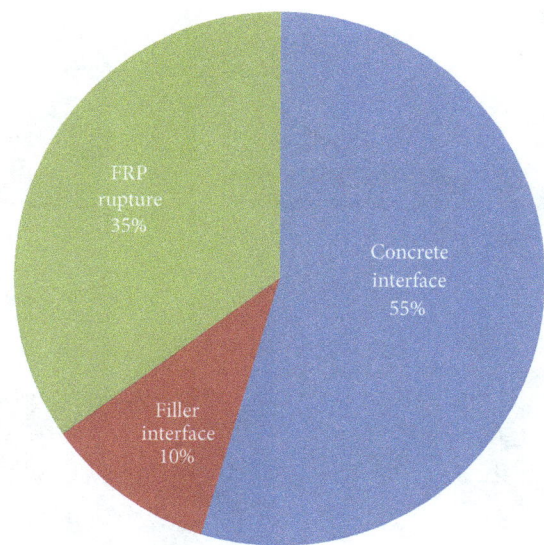

FIGURE 1: Distribution of the types of failure mode of flexure-strengthened concrete beams (2004–2016, 106 beams strengthened with NSM reinforcement) [5–8, 10, 23–40].

bending test in order to evaluate the flexural behavior of the NSM reinforcement.

The investigated results until 2016 on the failure modes observed in 106 concrete beams flexure-strengthened by NSM FRP reinforcement and shown in Figure 1 reveal that failure occurred mainly by rupture of the FRP reinforcement and failure of the filler-concrete interface (including the concrete cover). This indicates that about 90% of the failures experienced by the flexure-strengthened members by NSM FRP occur through the rupture of FRP or the failure of the concrete interface, which represent the major failure modes determining the strengthened performance of the flexural members.

This means that the bond failure in NSM strengthening depends on cohesive concrete around the slot because strength of cohesive filler is better than that of cohesive concrete generally. Therefore the filler-CFRP reinforcement ensemble behaves like a unique reinforcement. Accordingly, this study assumes the filler-CFRP reinforcement ensemble as an equivalent section to propose the bond failure model. The equivalent section illustrated in Figure 2 is defined as the section including the sections of the filler and reinforcement and behaving monolithically. Under this assumption, the stress developed at the filler-reinforcement interface is transferred to the filler-concrete interface by means of the filler, which will enable us to propose a bond failure model independent of the sectional shape of the reinforcement.

3. Proposal of Bond Failure Model

The following assumptions are adopted to establish the bond failure model applying the equivalent section derived above. First, bond failure occurs at the concrete-equivalent section interface. Second, the shear force that occurred at the CFRP-filler interface is transferred to the filler-concrete interface.

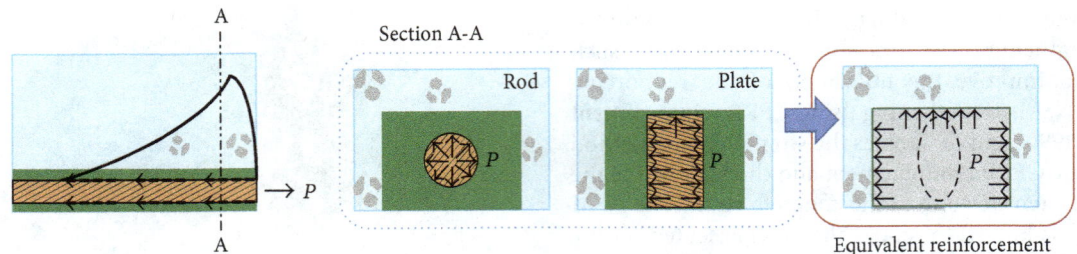

FIGURE 2: Concept of equivalent section.

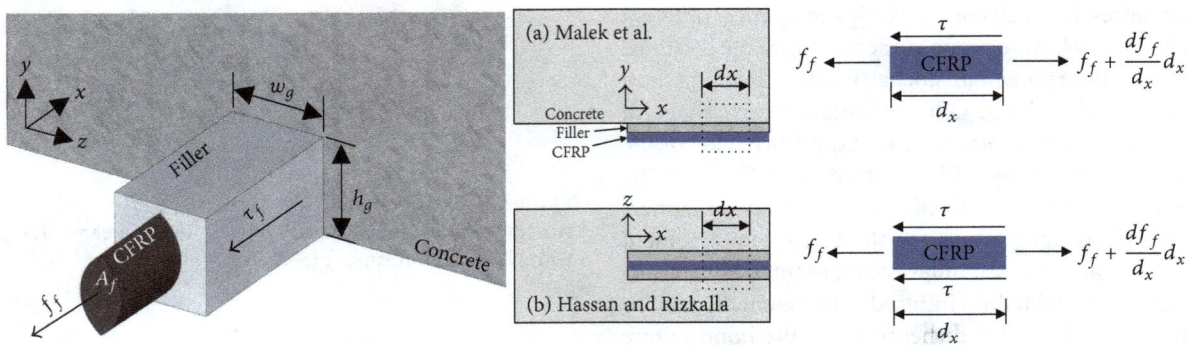

FIGURE 3: Shear stress developed at the interface [5, 9].

Third, the filler is considered only as a mean to transfer the shear force and the deformation of the filler is ignored (see (1)).

Based upon these assumptions, the equilibrium of the forces acting on the infinitesimal element dx of Figure 3(b) can be expressed as follows:

$$A_f df_f = \tau_c \left(2h_g + w_g\right) dx, \tag{1}$$

where τ_c is shear stress developed at the epoxy-concrete interface; h_g is depth of slot; w_g is width of slot; and A_f is cross-sectional area of CFRP reinforcement.

Rearranging (1) with respect to τ_c gives (2). The formulations derived by Malek et al. [9] and Hassan and Rizkalla [5] from the infinitesimal element shown in Figure 3 are expressed in (3) and (4), respectively.

$$\tau_c = \frac{A_f}{\left(2h_g + w_g\right)} \frac{df_f}{dx} \tag{2}$$

$$\tau_c = \frac{df_f}{dx} t_f \ [9] \tag{3}$$

$$\tau_c = \frac{1}{2} \frac{df_f}{dx} t_f \ [5], \tag{4}$$

where t_f is thickness of CFRP plate and f_f is tensile stress of CFRP plate.

The shear stress τ_c proposed in this study for the infinitesimal element is expressed in terms of the dimensions of the slot and the cross-sectional area of the CFRP reinforcement since the equivalent section is assumed. The formulations

for τ_c provided in (3) and (4) are expressed in terms of the dimensions of the CFRP reinforcement. Equation (5) reformulates τ_c of (2) proposed in this study for the 4-point loading condition shown in Figure 4. The derivation of (5) is skipped here since it is similar to that reported by Hassan and Rizkalla [5]:

$$\tau_c = \frac{A_f}{\left(2h_g + w_g\right)} \left(\frac{nPl_0 y_e}{I_e} \phi e^{-\phi x} + \frac{nPy_e}{I_e} \right),$$

$$\phi^2 = \frac{\left(2h_g + w_g\right) G_a}{A_f t_a E_f}, \tag{5}$$

$$n = \frac{E_f}{E_c},$$

where P is applied load; l_0 is distance from support to end of reinforcement; x is distance from end of reinforcement; y_e is distance from reinforcement to neutral axis; I_e is effective moment of inertia; E_f is elastic modulus of CFRP reinforcement; E_c is elastic modulus of concrete; t_a is thickness of epoxy; and G_a is shear elastic modulus of epoxy.

3.1. Failure Criteria.

It is primordial to establish failure criteria to simulate bond failure using the analytical model. In particular, the bond failure of the flexure-strengthened specimens by NSM reinforcement is similar to that of the external bonding since it occurs at the concrete-filler interface and not in the filler nor at the reinforcement-filler interface [5, 10].

A survey of the research on the bond characteristics of External Bonding Reinforcement (EBR) reveals that the

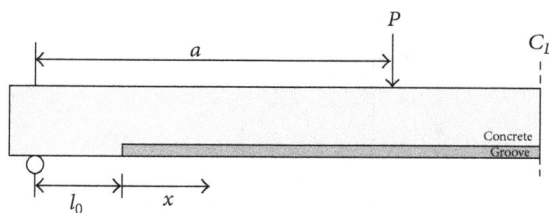

FIGURE 4: Four-point loading.

bond failure criterion at the end of the reinforcement is set as the concrete failure coefficient (flexure-tensile strength), which when exceeded means that failure has occurred [9, 11, 12]. Moreover, since bond failure occurs along the concrete surface, many researchers define the strength of the concrete surface as the interfacial bond failure criterion [13, 14]. The bond failure occurring at the concrete interface is closely related to the cracking strength of concrete and is assumed to occur when the shear stress developed at the concrete interface exceeds the cracking strength of concrete [12].

Accordingly, since cracking at the concrete interface has decisive effect on the propagation of the cracks and on the bond failure, this study adopts the concrete flexure-tensile strength (failure coefficient) proposed by numerous researchers like Malek et al. [9] and Tumialan et al. [12] as failure criterion. The cracking stress of normal concrete is calculated as the following failure coefficient (flexure-tensile strength) [15] and it is assumed that bond failure occurs when the value of the cracking stress exceeds the failure criterion of

$$\tau_{max} = f_r = 0.63\sqrt{f_{ck}} \ (MPa) = 7.5\sqrt{f_{ck}} \ (psi). \quad (6)$$

3.2. Concrete Material Model. The stress-strain curve of concrete is represented using the parabolic material model shown in Figure 5(a). This model is assumed to exhibit nonlinear elastic behavior with a parabolic shape until the ultimate compressive strain (ε_{cu}) [16].

$$f_c = f_{ck}\left[\frac{2\varepsilon_c}{\varepsilon_c'} - \left(\frac{\varepsilon_c}{\varepsilon_c'}\right)^2\right], \quad (7)$$

where f_c is stress of concrete; f_{ck} is compressive strength of concrete; ε_c is strain of concrete; ε_c' is strain when the concrete stress f_c reaches the compressive strength f_{ck}. The value of ε_c' is assumed to be 0.002 in this study. Note that the ultimate compressive strain of concrete is assumed to 0.003.

3.3. Material Model of Steel Reinforcement. As shown in Figure 5(b), the stress-strain curve of the steel reinforcement is assumed to be bilinear exhibiting linear elastic behavior before the yield stress (f_y) and perfect plastic behavior beyond the yield stress. The elastic modulus in the plastic region is set as $E_s' = 0.01E_s$ [17].

$$\begin{aligned} f_s &= E_s\varepsilon_s \quad \text{for } \varepsilon_s \le \varepsilon_y \\ f_s &= f_y + E_s'\left(\varepsilon_s - \varepsilon_y\right) \quad \text{for } \varepsilon_s \ge \varepsilon_y, \end{aligned} \quad (8)$$

where f_s is stress of steel rebar; ε_s is strain of steel rebar; ε_y is yield strain of steel rebar; E_s is elastic modulus of steel rebar; and E_s' is elastic modulus in plastic zone.

3.4. Material Model of CFRP Reinforcement. The stress-strain curve of CFRP reinforcement shown in Figure 5(c) assumes linear elastic behavior until rupture. In Figure 5(c), f_{fu} designates the rupture stress of the CFRP reinforcement and ε_{fu} is the rupture strain of the CFRP reinforcement.

$$f_f = E_f\varepsilon_f, \quad (9)$$

where f_f is stress of CFRP reinforcement; ε_f is strain of CFRP reinforcement; and E_f is elastic modulus of CFRP reinforcement.

4. Computation of Deflection

The deflection of an ordinary RC member can be classified into three zones delimited by the initiation of cracks and the yielding of the tensile rebar. The effective moment of inertia (I_e) is calculated according to this classification and applied in the formula computing the deflection [18]. In general, the whole section is considered to be effective prior to the initiation of cracks and the corresponding moment of inertia of the whole section (I_g) is applied. For the zone after cracking and before yielding of the tensile rebar, the effective moment of inertia of Branson [19] is adopted, and, for the zone after yielding of the tensile rebar, the cracked section moment of inertia (I_{cr}) is used. In this study, the flexural rigidity is calculated by applying the moment-curvature curve from early loading and used to compute the deflection per loading stage [19, 20]. The curvature (ϕ) at an arbitrary loading stage is computed as follows using the concrete strain (ε_c) and neutral axis (c).

$$\phi = \frac{\varepsilon_c}{c}. \quad (10)$$

Here, the flexural rigidity $(EI)_e$ at any loading stage can be expressed as follows from the moment-curvature (M-ϕ) relation:

$$(EI)_e = \frac{M}{\phi}. \quad (11)$$

Once the flexural rigidity is calculated, the deflection (Δ) at midspan can be computed as follows according to the loading condition that is 4-point loading here:

$$\Delta = \frac{Pa}{24(EI)_e}\left(3L^2 - 4a^2\right). \quad (12)$$

5. Section Analysis of Beam Strengthened by NSMR

Section analysis is conducted using the strain compatibility condition and equilibrium of the forces to predict the flexural behavior of the reinforced specimens. The moment-curvature distribution of ordinary RC specimens can be classified into

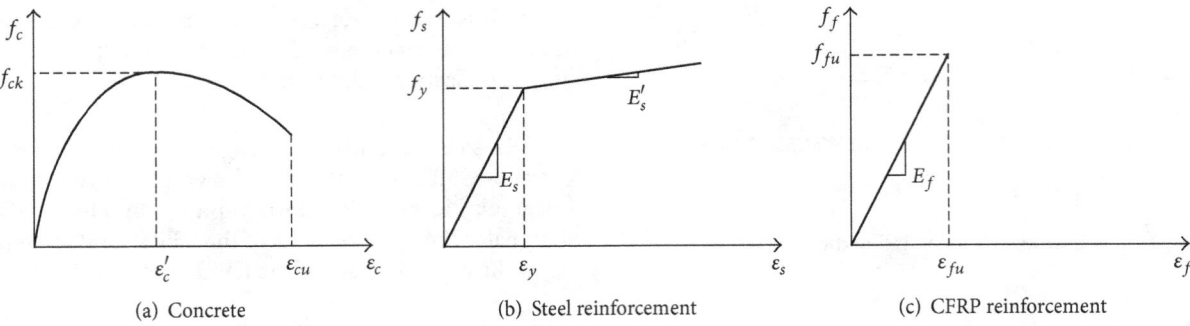

Figure 5: Material models.

three zones delimited by the initiation of cracks in concrete and the yielding of rebar. Therefore, the section analysis is performed considering three different cases that are as follows: the precracking stage when the acting moment M_a due to the load is smaller than the cracking moment; the preyielding stage after cracking and before yielding of the rebar; and the postyielding case [14, 20]. The calculation for each case uses the equilibrium of the forces and moments. Figure 6 describes the strain and stress distributions in each case.

(a) Precracking stage: $0 \leq M_a \leq M_{cr}$.

$$\frac{1}{2} f_c bc + A'_s f'_s - \frac{1}{2} (h - c) f_{ct} b - A_s f_s - A_f f_{fe} = 0$$

$$\begin{aligned} M_a \\ = \frac{1}{3} f_{ct} bh (h - c) + A_s f_s \left(d_s - \frac{1}{3} c \right) \\ + A_f f_{fe} \left(d_f - \frac{1}{3} c \right) - A'_s f'_s \left(\frac{1}{3} c - d'_s \right). \end{aligned} \tag{13}$$

(b) Preyielding stage: $M_{cr} \leq M_a \leq M_y$.

$$\alpha_1 \beta_1 f_{ck} bc + A'_s f'_s - A_s f_s - A_f f_{fe} = 0$$

$$\begin{aligned} M_a \\ = A_s f_s \left(d_s - \frac{1}{2} \beta_1 c \right) + A_f f_{fe} \left(d_f - \frac{1}{2} \beta_1 c \right) \\ - A'_s f'_s \left(\frac{1}{2} \beta_1 c - d'_s \right). \end{aligned} \tag{14}$$

(c) Postyielding stage: $M_y \leq M_a \leq M_u$.

$$\alpha_1 \beta_1 f_{ck} bc + A'_s f'_s - A_s \left[f_y + E'_s (\varepsilon_s - \varepsilon_y) \right] - A_f f_{fe} = 0$$

$$\begin{aligned} M_a \\ = A_s \left[f_y + E'_s (\varepsilon_s - \varepsilon_y) \right] \left(d_s - \frac{1}{2} \beta_1 c \right) \\ + A_f f_{fe} \left(d_f - \frac{1}{2} \beta_1 c \right) - A'_s f'_s \left(\frac{1}{2} \beta_1 c - d'_s \right). \end{aligned} \tag{15}$$

5.1. Determination of Bond Failure. The common failure modes in NSMR with 1 line of reinforcement are known to be the rupture of the CFRP reinforcement and the bond failure of the concrete-filler interface (bond model I). When NSMR is done with more than 1 line of reinforcement, the failure mode varies according to the interval between the slots or the distance to the corners. If the space of slots is sufficiently wide or the distance from slot to the corners is larger than a definite value, rupture of the CFRP reinforcement or bond failure of the concrete interface may occur similarly to the 1-line NSMR (bond model II-1). However, when it is not the case, failure may occur not only through the concrete interface but also by failure of concrete between the slots or spalling of the concrete between the slots and the corners (bond model II-2). Accordingly, this study simplifies the failure pattern as shown in Figure 7 to simulate analytically such failures. In Figure 7, even if different from the bond failure of the corner concrete only, the failure of two lines of reinforcement is considered to occur through the bond failure between the slots. In such case, this simplified pattern appears to remain applicable if failure of the corners occurs after the bond failure between the slots. During the section analysis, since it is difficult to model the amount of reinforcement in the 2 lines, the simplification assumes that the 2 lines of reinforcement and the filler behave as one unique reinforcement and filler ensemble as shown in Figure 7. Accordingly, in case of NSMR with 2 lines of reinforcement, the width of the slot, the amount of reinforcement, and the width of the filler are doubled to compute the shear stress developed at the end of the reinforcement.

5.2. Calculation Procedure. In order to predict the nonlinear behavior of the specimens strengthened by NSMR using CFRP, the analysis is conducted in three different stages. Figure 8 presents the flowchart of the calculation procedure adopted in the analysis. The calculation procedure determines the load, the position of the neutral axis, the strain in each material and the deflection using the strain compatibility condition, the constitutive equations of each material, and the equilibrium of the internal forces by increasing the strain according to the material model of the steel rebar. Final failure is decided at bond failure using the bond failure model proposed in this study when the shear stress developed at the end of the CFRP reinforcement exceeds the maximum

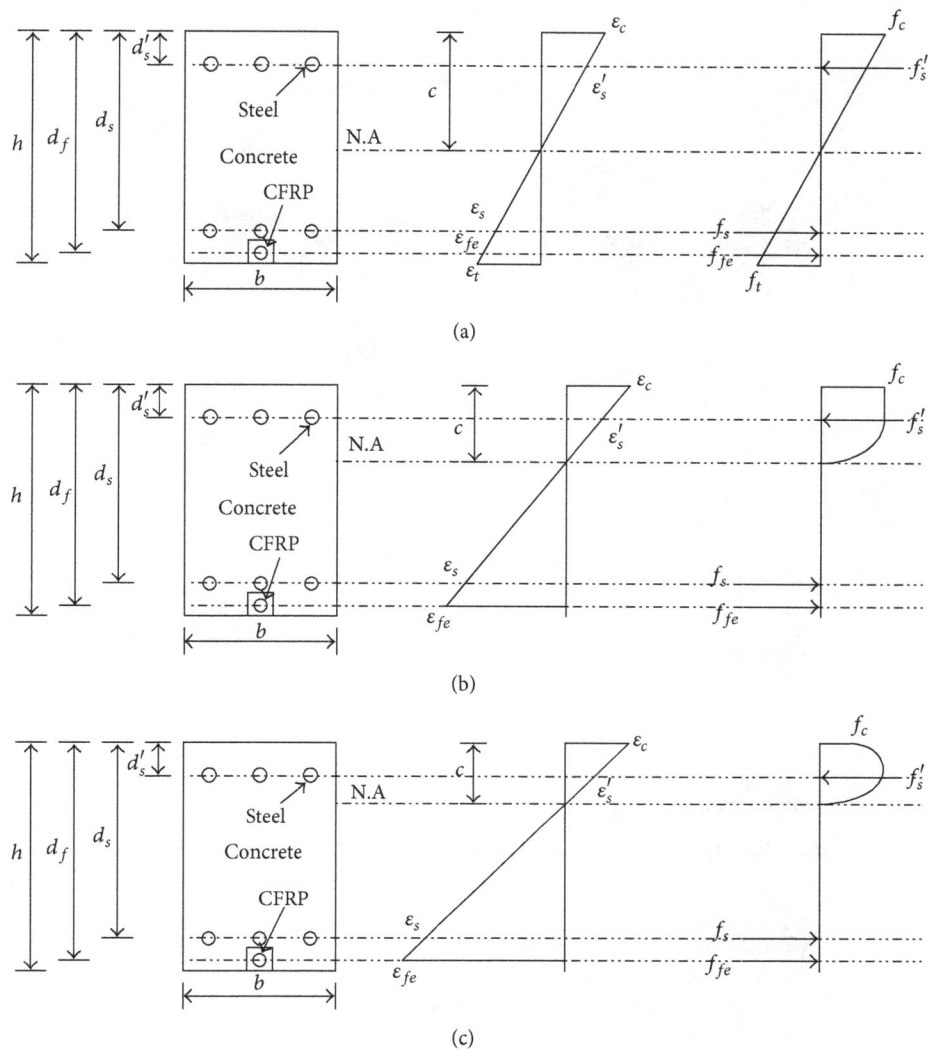

FIGURE 6: Strain and stress distributions per analysis stage.

shear strength, at compressive failure of concrete when the compressive strain at the top of concrete exceeds 0.003 and at rupture of the reinforcement when the rupture strain of the CFRP reinforcement is exceeded.

6. Validation Test of Bond Failure Model for NSMR

6.1. Test Variables. When the filler and FRP reinforcement behave as an equivalent reinforcement, the equivalent reinforcement can be modeled using the equivalent section during the section analysis. In such case, it is acceptable to skip the consideration of the concrete-filler interface and filler-FRP reinforcement interface. Therefore, it is necessary to check the behavior of this equivalent reinforcement by examining the failure mode of beams flexure-strengthened by FRP reinforcements exhibiting various sectional shapes. To that goal, a series of tests were conducted to validate the behavior of the equivalent section using CFRP reinforcements presenting rectangular, round, and trapezoidal shapes.

It is necessary to understand the arrangement details of the reinforcement in the reinforced section to mount it at the bottom of the beam. To that goal, two lines of reinforcement are arranged as shown in Figure 9 and the test variables are set as the spacing between the two lines (S = 60 mm and $2S$ = 120 mm) and as the embedded depth (25 mm and 15 mm) to evaluate their influence on the section failure mode and strengthening effect. The same reinforced length of 2,700 mm is applied in all the specimens. Table 1 summarizes the designation of the specimens with the test variables. Figure 9 depicts the cross-section of each of the specimens designated in Table 1 and corresponding reinforced areas (A_f).

6.2. Fabrication of Specimens Strengthened by NSMR. A total of 9 specimens were fabricated as shown in Figure 10 to verify the reliability of the bond failure model used in the analysis. The fabrication used ready-mixed concrete with design strength of 27 MPa. D10 SD40 tensile rebar were arranged with reinforcement ratio of 0.0041 and three D13

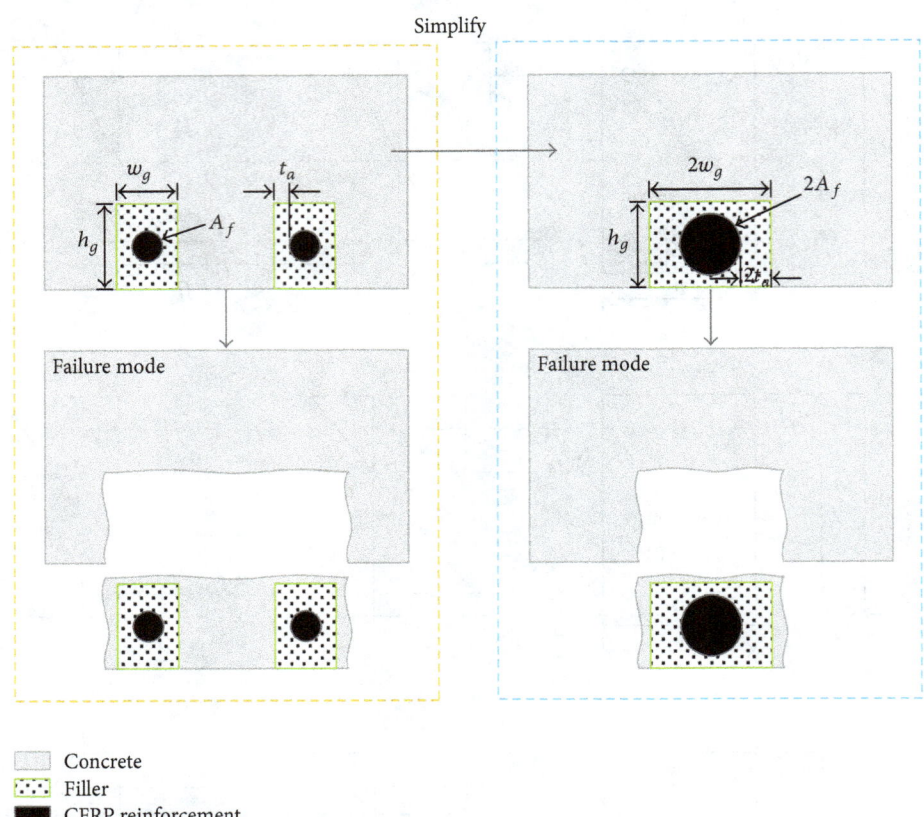

FIGURE 7: Simplification of failure mode of 2-line reinforcement [bond model II-2].

TABLE 1: Designation of specimens and corresponding test variables.

Specimen	CFRP area (mm^2)	CFRP type	Slot depth (mm)	CFRP cross-sectional shape	ρ_f
CONTROL	—	—	—	Nonstrengthened	—
R-TR-10	35.0	Rod	25	Trapezoidal	0.0003
R-PL-15	21.0	Strip	15	Rectangular	0.0003
R-PL-25	35.0	Strip	25	Rectangular	0.0005
R-RD-9	63.6	Rod	25	Round	0.0007
R-PL-25*2-S	70.0	Strip	25	Rectangular	0.0010
R-PL-25*2-2S	70.0	Strip	25	Rectangular	0.0010
R-RD-9*2-S	127.2	Rod	25	Round	0.0014
R-RD-9*2-2S	127.2	Rod	25	Round	0.0014

compressive reinforcements were disposed. D10 shear reinforcements were disposed at spacing of 100 mm to prevent shear failure. Strain gauges (T1, T2, B1, and B3) were bonded at midspan on the upper and lower steel reinforcements and at the quarters (B2, B4) on the lower rebar to measure the strain of the reinforcements according to loading (Figure 10).

The reinforcement ratio of the CONTROL beam was set to 0.0041 between the minimum ratio of 0.0032 and maximum ratio of 0.033 to induce ductile failure and not compressive failure of concrete and to increase the difference in the reinforcing performance. The equivalent reinforcement ratio of FRP is calculated using (16) for the conversion of the

FRP reinforcement ratio into steel reinforcement ratio and the corresponding values are listed in Table 1.

$$\rho_f = \frac{A_f}{bd_f} \times \frac{E_f}{E_s},\tag{16}$$

where A_f is cross-sectional area of CFRP reinforcement (mm^2); b is width of section (mm); d_f is effective depth (mm); E_f is elastic modulus of CFRP reinforcement (MPa); and E_s is elastic modulus of steel reinforcement (MPa).

Similarly to the CONTROL beam, the flexure-strengthened specimens are reinforced by an amount of

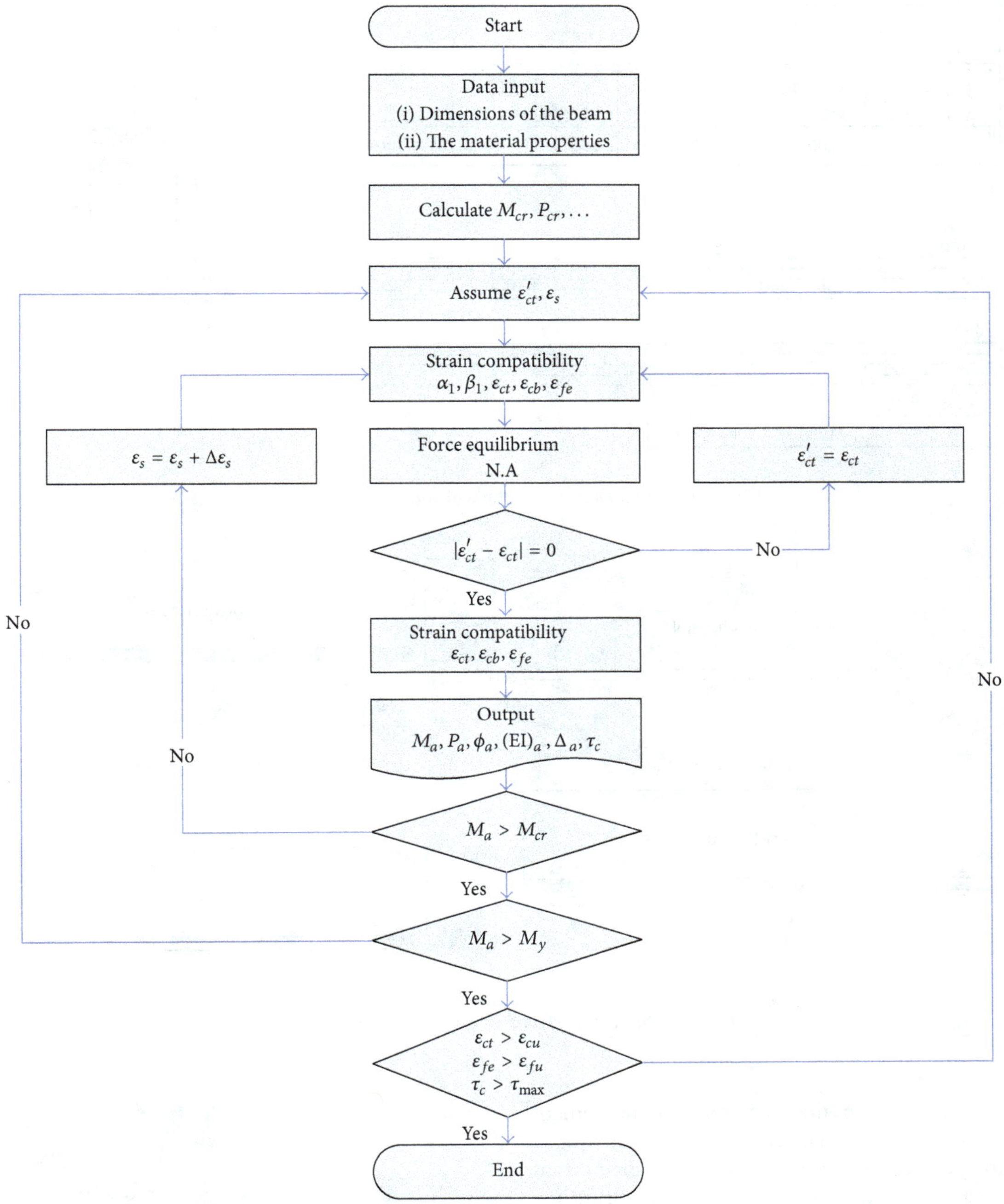

Start
Data input
(i) Dimensions of the beam
(ii) The material properties
Calculate $M_{cr}, P_{cr}, \ldots$
Assume $\varepsilon'_{ct}, \varepsilon_s$
Strain compatibility
$\alpha_1, \beta_1, \varepsilon_{ct}, \varepsilon_{cb}, \varepsilon_{fe}$
$\varepsilon_s = \varepsilon_s + \Delta\varepsilon_s$
Force equilibrium
N.A
$\varepsilon'_{ct} = \varepsilon_{ct}$
$|\varepsilon'_{ct} - \varepsilon_{ct}| = 0$
No
Yes
Strain compatibility
$\varepsilon_{ct}, \varepsilon_{cb}, \varepsilon_{fe}$
No
No
Output
$M_a, P_a, \phi_a, (EI)_a, \Delta_a, \tau_c$
No
$M_a > M_{cr}$
Yes
$M_a > M_y$
Yes
$\varepsilon_{ct} > \varepsilon_{cu}$
$\varepsilon_{fe} > \varepsilon_{fu}$
$\tau_c > \tau_{max}$
Yes
End

FIGURE 8: Flowchart of calculation procedure.

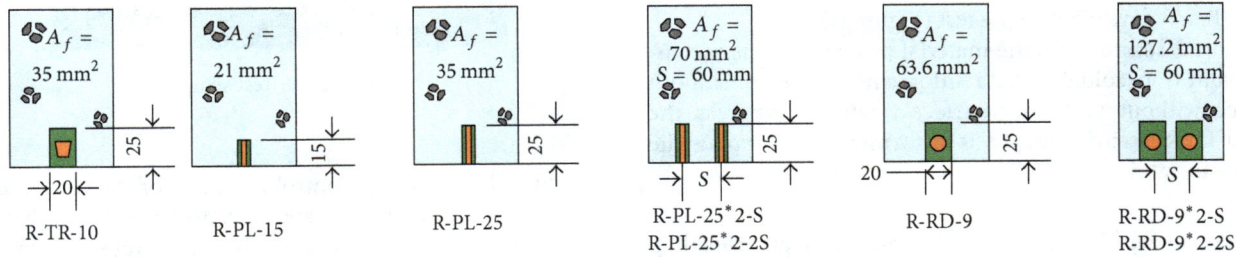

A_f = 35 mm^2
25
20
R-TR-10
A_f = 21 mm^2
15
R-PL-15
A_f = 35 mm^2
25
R-PL-25
A_f = 70 mm^2
S = 60 mm
25
S
R-PL-25*2-S
R-PL-25*2-2S
A_f = 63.6 mm^2
25
20
R-RD-9
A_f = 127.2 mm^2
S = 60 mm
S
R-RD-9*2-S
R-RD-9*2-2S

FIGURE 9: Cross-section of specimens strengthened by NSMR [unit: mm].

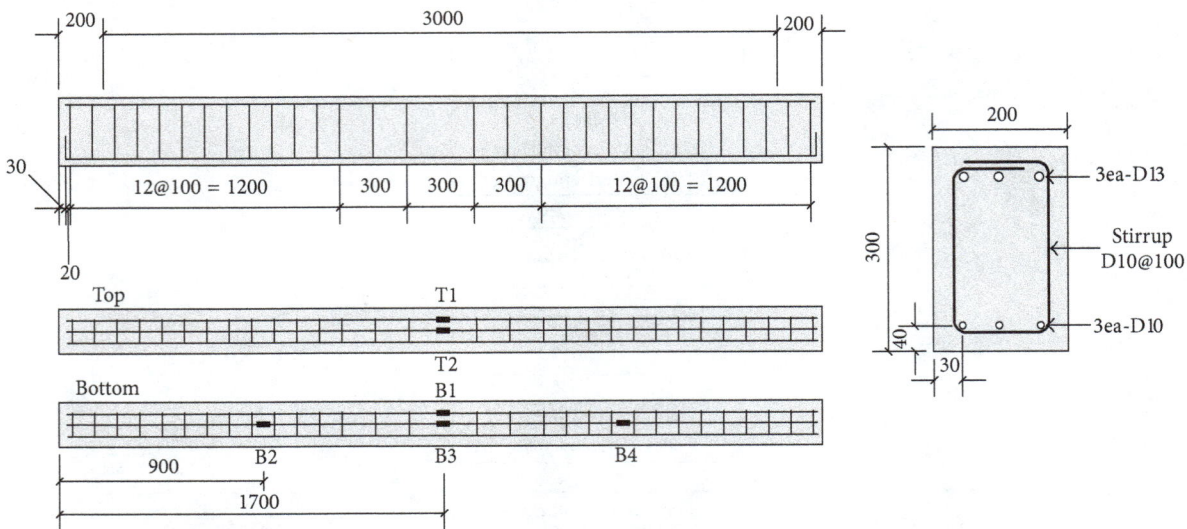

FIGURE 10: Dimensions and details of specimens [unit: mm].

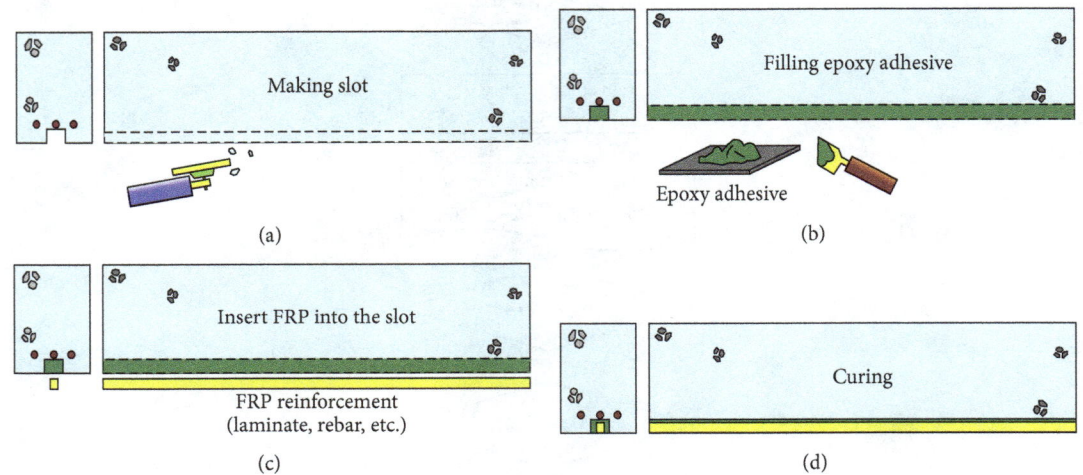

FIGURE 11: Specimen strengthening process by NSMR.

reinforcement smaller than the maximum reinforcement ratio to induce ductile failure. For the NSM strengthening, a slot was grinded using a grinder at midspan on the bottom of the beam and the fine dust was removed by air-brushing. Then, primer was applied in the slot and cured for more than 1 day. The filler for NSMR adopted an epoxy matrix mixed with hardener at a proportion of 2 : 1 and was applied in the slot. The reinforcement was finally embedded with respect to the considered test variable and curing was conducted for more than 3 days prior to the test (Figure 11).

Table 2 summarizes the material properties. The rectangular and trapezoidal CFRP reinforcements present smooth surface without particular surface treatment whereas the round CFRP reinforcement is deformed by wrapping the surface by fabrics.

6.3. Loading and Measurement Methods. Four-point loading test was conducted using UTM (Universal Testing Machine) with capacity of 980 kN (Figure 12). Loading was applied

FIGURE 12: Test setup.

through displacement control at speed of 0.02 mm/s until a displacement of 15 mm and speed of 0.05 mm/s beyond that displacement until failure. The data were measured at intervals of 1 s and recorded using a static data logger and a computer.

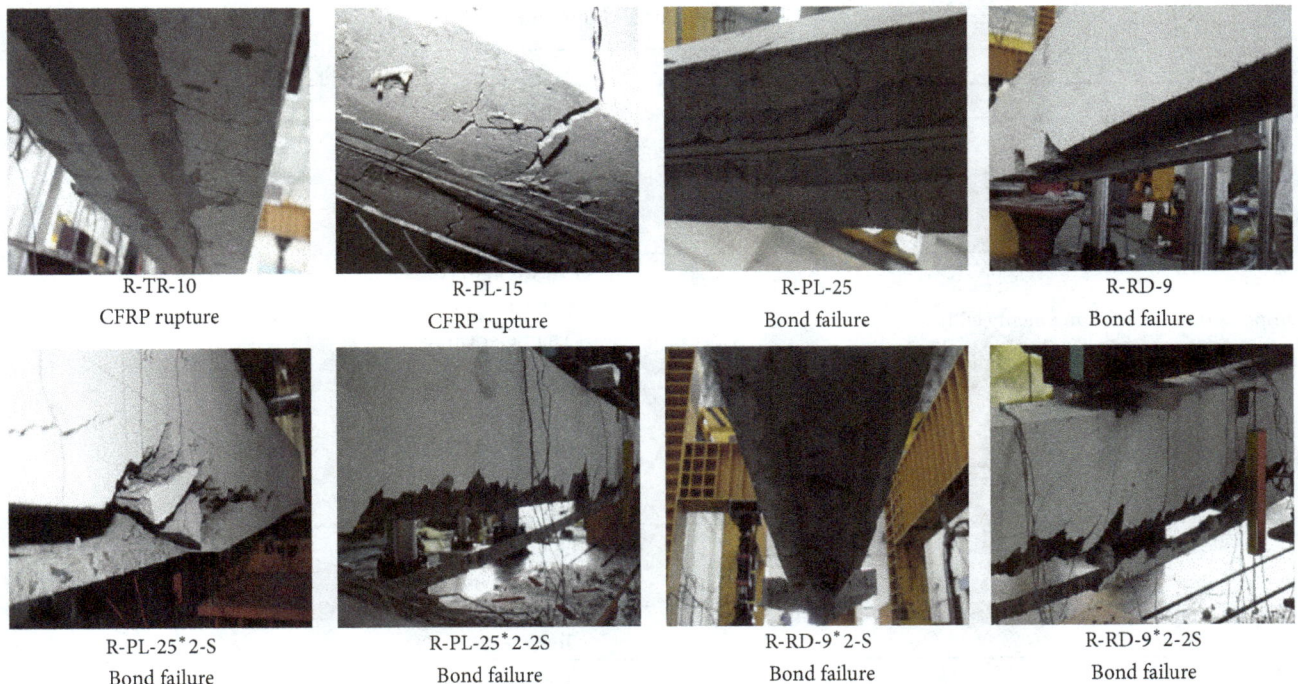

| R-TR-10 | R-PL-15 | R-PL-25 | R-RD-9 |
| CFRP rupture | CFRP rupture | Bond failure | Bond failure |

| R-PL-25*2-S | R-PL-25*2-2S | R-RD-9*2-S | R-RD-9*2-2S |
| Bond failure | Bond failure | Bond failure | Bond failure |

FIGURE 13: Failure patterns of specimens strengthened by NSMR.

The load-displacement curves were drawn using the load given by the actuator and the displacements measured by LVDTs disposed at the center and quarter points of the specimens.

7. Results of Validation Test

7.1. Failure Modes. Figure 13 shows the failure pattern of the specimens strengthened by NSMR. Failure occurred through rupture of CFRP or bond failure. First, the specimens that experienced CFRP rupture pertain to the specimens with relatively smaller amount of reinforcement and the specimens with trapezoidal reinforcement. In this failure mode, cracks perpendicular to the slots were observed without failure of epoxy, reinforcement, or concrete. Second, for the bond failure, cracks parallel to the slots were observed and failure of the epoxy and reinforcement occurred around the slots entraining the concrete around the slots. The occurrence of bond failure at the epoxy-concrete interface was observed not only in this study but also in most of the experimental studies related to the specimens flexure-strengthened by NSMR. The epoxy and the reinforcement are seen to behave monolithically as a unique reinforcement similarly to the failure mode (equivalent section behavior) suggested above.

In case of narrow spaced NSMR, the P-PL-25*2-S specimen strengthened by strip reinforcement experienced bond failure of the reinforcement and filler together with failure not only of the concrete between the slots but also of the corners. Besides, the P-RD-9*2-S specimen strengthened by rod reinforcement experienced failure of the concrete between the slots only.

In case of widely spaced NSMR, the P-PL-25*2-2S specimen strengthened by strip reinforcement experienced bond failure of the reinforcement and filler together with the quasi-disappearance of the corners. On the other hand, the P-RD-9*2-2S specimen strengthened by rod reinforcement experienced partial failure of the concrete corners due to the shock provoked by the bond failure of the reinforcement and filler.

In case of identical spacing, the different failure patterns observed according to the reinforcement can be attributed to the effect of the amount of reinforcement and moment of inertia of the cross-sectional shape.

7.2. Analysis of Strengthening Effect. Table 3 summarizes the experimental results for the specimens strengthened by NSMR. The moment, load, and deflection are represented for each instance of early cracking, rebar yielding, and maximum load. P/P_{con}, the load ratio of strengthened specimen to the nonstrengthened specimen (CONTROL), indicates the strengthening effect. The efficiency of CFRP is expressed in terms of the ratio of the maximum strain measured in the CFRP reinforcement of the specimen to the strain at rupture ($\varepsilon_{max}/\varepsilon_u$); as a matter of fact a ratio closer to 100% means that the reinforcement develops its best performance. In strengthening effect analysis, since the load ratio does not give information on the cross-sectional area and physical properties of the CFRP reinforcement, this indicator is valid only when the same reinforcement and strengthening method are used and not good when they are different. Accordingly, this study adopts the dimensionless strengthening effect per axial rigidity obtained by dividing the maximum load by the stiffness of the reinforcement to compare the strengthening

TABLE 2: Material properties.

Material	Properties	
Concrete[1]	Compressive strength (MPa)	31.3
Tension steel reinforcement (D10)[1]	Yield strength (MPa)	436
	Tensile strength (MPa)	562
	Diameter (mm)	9.53
	Area (mm^2)	71.33
Compression steel reinforcement (D13)[1]	Yield strength (MPa)	481
	Tensile strength (MPa)	608
	Diameter (mm)	12.7
	Area (mm^2)	126.7
CFRP strip (PL, smooth surface)[1]	Thickness (mm)	1.4
	Tensile strength (MPa)	2482.5
	Elastic modulus (GPa)	167
	Ultimate strain (%)	1.48
CFRP rod (TR, smooth surface)[1]	Design thickness (mm)	10
	Tensile strength (MPa)	1500
	Elastic modulus (GPa)	100
	Ultimate strain (%)	1.5
CFRP rod (RD, deformed surface)[2]	Diameter (mm)	9
	Tensile strength (MPa)	1878
	Elastic modulus (GPa)	121.42
	Ultimate strain (%)	1.55

1: from supplier; 2: from tests performed by authors according to [22].

effect brought by CFRP reinforcements exhibiting different sections and physical properties. Since the strengthening effect per axial rigidity (P_u/EA $\times$ 10^3) considers not only the efficiency of FRP but also the physical properties of the reinforcement, this indicator becomes effective when analyzing the strengthening effect according to the strengthening method and sectional details. If the value of this indicator is high, it means that the strengthening effect is good.

The experimental results show that the yield load and maximum load increase with larger cross-sectional area of the reinforcement. However, specimens R-PL-25 and R-TR-10 present identical reinforced area but different flexural behaviors. This can be explained by the fact that these two specimens have different stiffness of the reinforcements. Therefore, it appears that the comparison of the strengthening effect should consider not only the reinforced area but also the effect of the reinforcement stiffness. Accordingly, the observation of the strengthening effect per axial rigidity reveals that specimens R-TR-10 and R-PL-15 experiencing CFRP rupture have better strengthening effect than specimens R-PL-25 and R-RD-9 featured by bond failure. In case of NSMR using reinforcements with identical shape, the strengthening effect per axial rigidity tends to reduce with larger amount of reinforcement. The reduction of the strengthening effect per axial rigidity even if the reinforced area increases as much as the increase of the amount of reinforcement can be attributed to the fact that the reinforcement could not develop 100% of its performance due to bond failure. Such tendency is more apparent in the specimens strengthened by 2 lines of reinforcement and seems to be caused by the premature

occurrence of bond failure following the overlapping of the failed surfaces.

The comparison of the strengthening effects brought by 1-line and 2-line reinforcement shows that the strengthening effects of specimens R-PL-25*2-S and R-PL-25*2-2S reach, respectively, 69% and 75% which is lower than the 83% of specimen R-PL-25 with 1-line reinforcement. Moreover, specimen R-PL-25*2-S with smaller spacing than specimen R-PL-25*2-2S is more vulnerable to failure. Specimen R-RD-9*2-S with round reinforcement and narrow spacing shows reduced strengthening effect compared to the specimens with 1-line reinforcement and indicates that the failure of the slot spacing affects the bond failure. The strengthening effect of specimen R-RD-9*2-2S with large spacing reaches 91%, which is higher than the specimens with 1-line reinforcement and indicates that this specimen is less influenced by the spacing. Failure modes are affected by the position of slots when the distance between slots or the distance from slot to corner is less than 40 mm [21]. R-PL-25*2-S/2S specimens have 40 mm space between slots and corner and therefore one slot's failure affects the other slots' failure and corners. The comparison of strengthening effects per axial rigidity shows that R-PL-25*2-S specimen is larger than R-RD-9*2-S specimen because the distance between slots of R-PL-25*2-S specimen is wider than that of R-RD-9*2-S specimen.

7.3. Comparison of Experimental and Analytical Results.
Figure 14 plots concurrently the load-displacement curve measured at midspan and the analytical values derived from the analytic model of this study together with the failure

TABLE 3: Experimental results.

Specimens		CONTROL	R-TR-10	R-PL-15	R-PL-25	R-RD-9	R-PL-25*2-S	R-PL-25*2-2S	R-RD-9 2-S	R-RD-9*2-2S
Cracking	M_{cr} (N·mm*10^3)	8930	8705	8122	8663	8652	9466	11240	10164	10883
	P_{cr} (kN)	15.98	16.58	15.47	16.5	16.48	18.03	21.41	19.36	20.73
	Δ_{cr} (mm)	1.38	1.74	1.72	1.98	1.66	1.88	2.84	1.66	1.84
	$P_{cr}/P_{cr,con}$	1	1.04	0.97	1.03	1.03	1.13	1.34	1.21	1.3
Yielding	M_y (N·mm*10^3)	24512	28859	30172	32387	32855	37779	37007	40724	40562
	P_y (kN)	46.69	54.97	57.47	61.69	62.58	71.96	70.49	77.57	77.26
	Δ_y (mm)	12.48	13.36	15.5	16.06	15.36	16.46	14.16	15.22	15.4
	$P_y/P_{y,con}$	1	1.18	1.23	1.32	1.34	1.54	1.51	1.66	1.65
Maximum	M_u (N·mm*10^3)	29500	41533	41160	45245	48631	57572	56191	59036	78556
	P_u (kN)	56.19	79.11	78.40	86.18	92.63	109.66	107.03	112.45	149.63
	Δ_u (mm)	71.68	58.68	58.94	53.98	43.88	46.92	44.44	39.5	54.19
	$P_u/P_{u,con}$	1	1.41	1.4	1.53	1.65	1.95	1.9	2	2.66
Ductility	Δ_u/Δ_y	5.74	4.39	3.8	3.36	2.86	2.85	3.14	2.6	3.52
CFRP efficiency	$\varepsilon_{max}/\varepsilon_u$ (%)	—	100	100	83	84	69	75	70	91
Strength effect per axial rigidity	$P_u/EA \times 10^3$	—	22.6	22.32	14.72	12	9.36	9.14	7.28	9.69

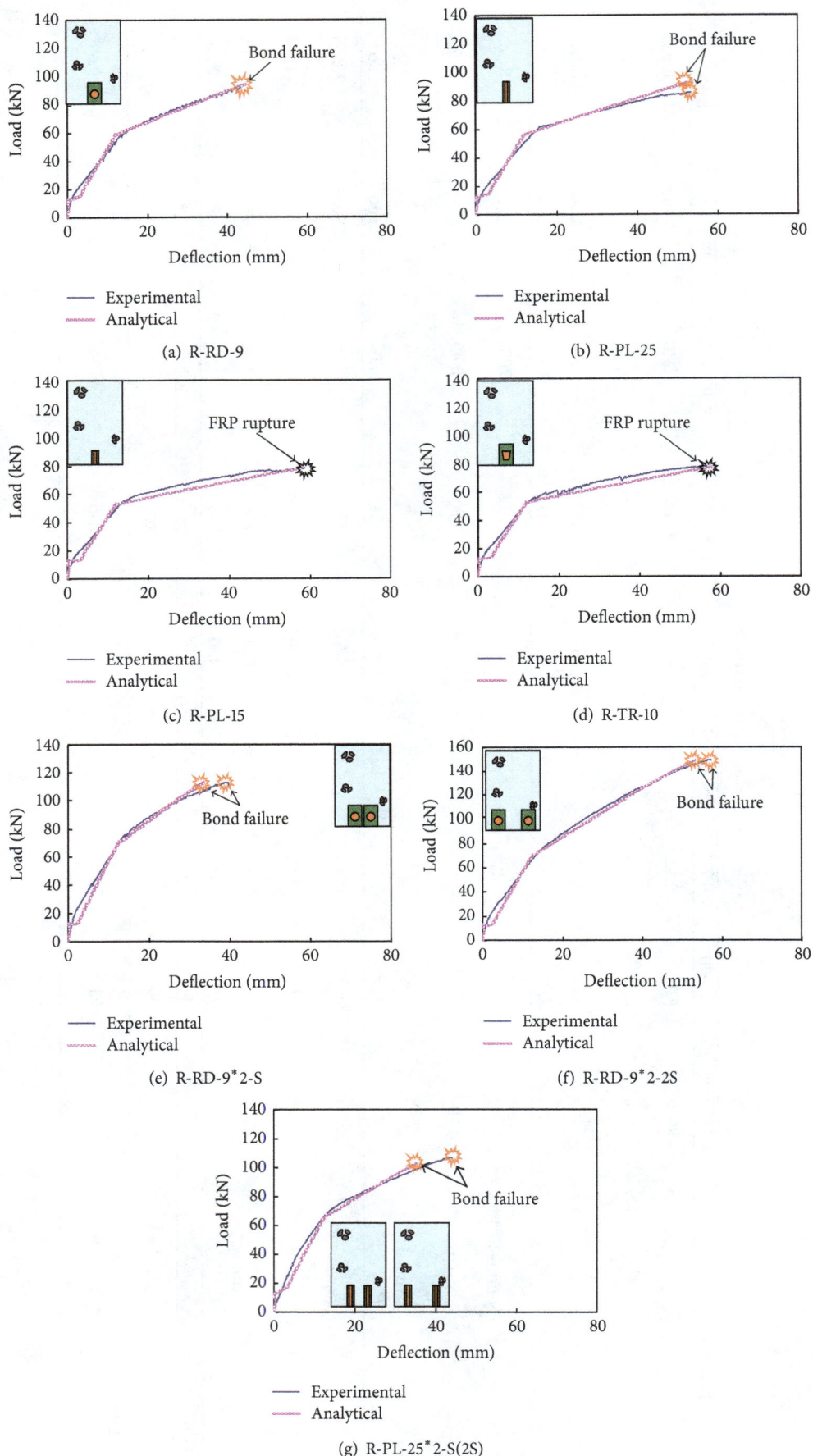

FIGURE 14: Comparison of experimental and analytical load-deflection curves.

TABLE 4: Comparison of experimental and analytical results.

Specimens		R-TR-10	R-PL-15	R-PL-25	R-RD-9	R-PL-25*2-S	R-PL-25*2-2S	R-RD-9*2-S	R-RD-9*2-2S
Moment at yielding (kN·mm)	Test	28859	30172	32855	32387	37779	37007	40724	40562
	Analysis	27698	27783	30821	29667	34198	34198	36480	36480
	Analysis/test	1.04	1.09	1.07	1.09	1.10	1.08	1.12	1.11
Maximum moment (kN·mm)	Test	41533	41160	48631	45245	57572	56191	59036	78556
	Analysis	40613	41054	49652	48584	54259	54259	59089	78372
	Analysis/test	1.02	1	0.98	0.93	1.06	1.04	1	1
Deflection at yielding (mm)	Test	13.36	15.5	15.36	16.06	16.46	14.16	15.22	15.4
	Analysis	11.96	11.96	12.1	12.04	12.24	12.24	12.35	12.35
	Analysis/test	1.12	1.3	1.27	1.33	1.34	1.16	1.23	1.25
Maximum deflection (mm)	Test	58.68	58.94	43.88	53.98	46.92	44.44	39.5	54.19
	Analysis	57.83	57.85	44.78	52.56	35.16	35.16	33.53	53.39
	Analysis/test	1.01	1.02	0.98	1.03	1.33	1.26	1.18	1.02

mode. Table 4 compares the experimental and analytical values of the moment and displacement at each stage. The analytic model proposed in this study is seen to predict the experimental maximum load within an error of 8% for all strengthened beams and shows perfect concordance with the final failure mode. The analysis of the specimens strengthened with 1-line reinforcement applied the bond failure model I. Among the specimens strengthened with 2-line reinforcement, for the cases where failure occurred between the slots (Figure 14(e)) and the cases experiencing failure between the slots and additional failure of the concrete corners (Figure 14(g)), analysis was performed using the bond failure model II-2 simplifying the failure mode. For the specimens that experienced failure of the concrete corners (Figure 14(f)), the bond failure model II-1 was applied in the analysis similarly to the 1-line failure mode.

In view of the analytical results, the bond failure model proposed in this study appears to predict accurately the failure mode of the 1-line NSMR as well as of the 2-line NSMR specimens and to simulate appropriately the load-displacement behavior.

8. Conclusions

This study examined the failure modes reported in previous studies to propose a bond failure model considering the equivalent section. Tests were performed on specimens strengthened by NMSR with various shapes of CFRP reinforcement. The following conclusions can be drawn.

The survey of the failure modes observed in former specimens flexure-strengthened by NSMR revealed that failure occurred mainly through rupture of CFRP reinforcement, failure of filler-concrete interface, and failure of filler-CFRP reinforcement interface with larger occurrence of the failure of filler-concrete interface than the failure of filler-CFRP reinforcement interface.

Based upon this inventory of the major failure modes of the flexure-strengthened specimens, the failed surface was assumed as an equivalent section representing the filler-concrete interface and a bond failure model was derived so as to be applicable regardless of the sectional shape of the CFRP reinforcement.

Tests were then conducted on flexure-strengthened specimens by NSMR to validate the proposed bond failure model. The experimental results showed that the yield load and maximum load increased with larger cross-sectional area of the CFRP reinforcement and that the strengthening effect per axial rigidity reduced even if the bonded area enlarged as much as the increase of the amount of reinforcement. This was explained by the fact that the reinforcement could not develop 100% of its performance due to bond failure.

In the case of the specimens strengthened with 2-line reinforcement, premature bond failure occurred due to the overlapping of the failed surface caused by the reinforcement spacing. Accordingly, further studies should be conducted on the details of the reinforced section to compute the minimum spacing enabling the slots to behave independently during the NSMR design.

Analytical analysis was finally carried out by setting failure criteria for the bond failure model considering the equivalent section proposed in this study. The results validated the failure mode in which the CFRP reinforcement and filler behaved monolithically and regardless of the sectional shape of the CFRP reinforcement. Moreover, the failure mode and flexure-strengthened behavior could be predicted within an error of 8%.

Competing Interests

The authors declare that they have no competing interests.

Acknowledgments

This research was supported by a grant from "A Strategic Research Project Funded by the Korea Institute of Civil Engineering and Building Technology" and "Upgrading

Technology of Prestressed NSM Reinforcement System with CFRP Tendon."

References

[1] W. T. Jung, *Flexural behavior of reinforced concrete beams strengthened with NSM CFRP reinforcements considering the equivalent section [Ph.D. thesis]*, Myongji University, Seoul, South Korea, 2009.

[2] R. EI-Hacha, "Effectiveness of near surface mounted FRP reinforcement for flexural strengthening of reinforced concrete beams," in *Proceedings of the 4th International Conference on Advanced Composite Materials in Bridges and Structures*, Alberta, Canada, July 2004.

[3] KICT, *Development of Strengthening Methods for Deteriorated Concrete Bridges*, Report, 2004.

[4] L. De Lorenzis and J. G. Teng, "Near-surface mounted FRP reinforcement: an emerging technique for strengthening structures," *Composites Part B: Engineering*, vol. 38, no. 2, pp. 119–143, 2007.

[5] T. Hassan and S. Rizkalla, "Investigation of bond in concrete structures strengthened with near surface mounted carbon fiber reinforced polymer strips," *Journal of Composites for Construction*, vol. 7, no. 3, pp. 248–257, 2003.

[6] T. Hassan and S. H. Rizkalla, "Bond mechanism of near-surface-mounted fiber-reinforced polymer bars for flexural strengthening of concrete structures," *ACI Structural Journal*, vol. 101, no. 6, pp. 830–839, 2004.

[7] I. Del Prete, A. Bilotta, L. Bisby, and E. Nigro, "Bond tests on NSM FRP strengthening using cementitious matrices for concrete structures," in *Proceedings of the 4 Contributo in Atti di Convegno, FRPRCS-12 & APFIS-2015 Joint Conference*, 2015.

[8] M. R. F. Coelho, J. M. Sena-Cruz, and L. A. C. Neves, "A review on the bond behavior of FRP NSM systems in concrete," *Construction and Building Materials*, vol. 93, pp. 1157–1169, 2015.

[9] A. M. Malek, H. Saadatmanesh, and M. R. Ehsani, "Prediction of failure load of R/C beams strengthened with FRP plate due to stress concentration at the plate end," *ACI Structural Journal*, vol. 95, no. 2, pp. 142–152, 1998.

[10] I. A. Sharaky, L. Torres, J. Comas, and C. Barris, "Flexural response of reinforced concrete (RC) beams strengthened with near surface mounted (NSM) fibre reinforced polymer (FRP) bars," *Composite Structures*, vol. 109, no. 1, pp. 8–22, 2014.

[11] S. A. Mirza, M. Hatzinikolas, and J. G. MacGregor, "Statistical descriptions of strength of concrete," *ASCE J Struct Div*, vol. 105, no. 6, pp. 1021–1037, 1979.

[12] G. Tumialan, P. Serra, A. Nanni, and A. Belarbi, "Concrete cover delamination in RC beams strengthened with FRP sheets, SP-188," in *Proceedings of the 4th International Symposium on FRP for Reinforcement of Concrete Structures (FRPRCS4 '99)*, pp. 725–735, American Concrete Institute, Baltimore, Md, USA, November 1999.

[13] U. Neubauer and F. S. Rostasy, "Design aspects of concrete structures strengthened with externally bonded CFRP plates," in *Proceedings of the 7th International Conference on Structural Faults and Repairs*, vol. 2, pp. 109–118, ECS Publications, Edinburgh, UK, 1997.

[14] J. S. Park, *Flexural behavior of RC beam strengthened with prestressed CFRP plates considering bond characteristics [Ph.D. thesis]*, Myongji University, Seoul, Republic of Korea, 2007.

[15] ACI Committee 318, *Building Code Requirements for Structural Concrete (318-05) and Commentary-(318R-05)*, American Concrete Institute, 2005.

[16] M. P. Collins and D. Mitchell, *Prestressed Concrete Structures*, Prentice Hall, 1991.

[17] H. E.-D. M. Sallam, A.-A. M. Saba, H. H. Shahin, and H. Abdel-Raouf, "Prevention of peeling failure in plated beams," *Journal of Advanced Concrete Technology*, vol. 2, no. 3, pp. 419–429, 2004.

[18] S. B. Lee, M. Y. Park, S. Y. Jang, K. S. Kim, and S. S. Kim, "Investigation on the effective moment of inertia of reinforced concrete flexural members under service load," *Journal of the Korea Concrete Institute*, vol. 20, no. 3, pp. 393–404, 2008.

[19] D. E. Branson, "Instantaneous and time-dependent deflections of simple and continuous reinforced concrete beams," HPR Report 7, part 1, Alabama Highway Department; Bureau of Public Roads, 1963.

[20] J. Y. Park, H. D. Cho, and S. H. Han, "Prediction of non-linear behavior for RC beam flexural strengthened with CFRP plate," *KSCE Journal of Civil Engineering*, vol. 24, no. 1, pp. 9–16, 2004.

[21] J. Y. Kang, Y. H. Park, J. S. Park, Y. J. You, and W. T. Jung, "Analytical evaluation of RC beams strengthened with near surface mounted CFRP laminates," in *Proceedings of the 7th International Symposium on Fiber-Reinforced Polymer Reinforcement for Reinforced Concrete Structures (FRPRCS '05)*, November 2005.

[22] ACI 440.3R-04, "Guide test methods for fiber-reinforced polymer(FRPs) for reinforcing and strengthening concrete structures," American Concrete Institute, 2004.

[23] J. A. O. Barros and A. S. Fortes, "Flexural strengthening of concrete beams with CFRP laminates bonded into slits," *Cement and Concrete Composites*, vol. 27, no. 4, pp. 471–480, 2005.

[24] R. El-Hacha and S. H. Rizkalla, "Near-surface-mounted fiber-reinforced polymer reinforcements for flexural strengthening of concrete structures," *ACI Structural Journal*, vol. 101, no. 5, pp. 717–726, 2004.

[25] A. Carolin, B. Täljsten, and A. Hejll, "Concrete beams exposed to live loading during carbon fiber reinforced polymer strengthening," *Journal of Composites for Construction*, vol. 9, no. 2, pp. 178–186, 2005.

[26] J. G. Teng, L. De Lorenzis, B. Wang, R. Li, T. N. Wong, and L. Lam, "Debonding failures of RC beams strengthened with near surface mounted CFRP strips," *Journal of Composites for Construction*, vol. 10, no. 2, pp. 92–105, 2006.

[27] J. A. O. Barros, S. J. E. Dias, and J. L. T. Lima, "Efficacy of CFRP-based techniques for the flexural and shear strengthening of concrete beams," *Cement and Concrete Composites*, vol. 29, no. 3, pp. 203–217, 2007.

[28] J. R. Yost, S. P. Gross, D. W. Dinehart, and J. J. Mildenberg, "Flexural behavior of concrete beams strengthened with near-surface-mounted CFRP strips," *ACI Structural Journal*, vol. 104, no. 4, pp. 430–437, 2007.

[29] H. Nordin and B. Täljsten, "Concrete beams strengthened with prestressed near surface mounted CFRP," *Journal of Composites for Construction*, vol. 10, no. 1, pp. 60–68, 2006.

[30] H. T. Choi, J. S. West, and K. A. Soudki, "Partially bonded near-surface-mounted CFRP bars for strengthened concrete T-beams," *Construction and Building Materials*, vol. 25, no. 5, pp. 2441–2449, 2011.

[31] A. M. Khalifa, "Flexural performance of RC beams strengthened with near surface mounted CFRP strips," *Alexandria Engineering Journal*, vol. 55, no. 2, pp. 1497–1505, 2016.

[32] S. E. El-Gamal, A. Al-Nuaimi, A. Al-Saidy, and A. Al-Lawati, "Efficiency of near surface mounted technique using fiber reinforced polymers for the flexural strengthening of RC beams," *Construction and Building Materials*, vol. 118, pp. 52–62, 2016.

[33] S.-Y. Seo, K.-B. Choi, Y.-S. Kwon, and K.-S. Lee, "Flexural strength of rc beam strengthened by partially de-bonded near surface-mounted FRP Strip," *International Journal of Concrete Structures and Materials*, vol. 10, no. 2, pp. 149–161, 2016.

[34] S. Foubert, K. Mahmoud, and E. El-Salakawy, "Behavior of prestressed hollow-core slabs strengthened in flexure with near-surface mounted carbon fiber-reinforced polymer reinforcement," *ASCE Composite Construction Journal*, vol. 20, no. 6, 2016.

[35] A. Bilotta, F. Ceroni, E. Nigro, and M. Pecce, "Efficiency of CFRP NSM strips and EBR plates for flexural strengthening of RC beams and loading pattern influence," *Composite Structures*, vol. 124, pp. 163–175, 2015.

[36] A. A. Khalil, E. E. Etman, and A. H. El-Masry, "Behavior of reinforced concrete continuous beams strenghened with near surface mounted reinforcement," in *Proceedings of the International Conference on Advances in Structural and Geotechnical Engineering*, Hurghada, Egypt, April 2015.

[37] H. Zhang, L. He, and G. Li, "Bond failure performances between near-surface mounted FRP bars and concrete for flexural strengthening concrete structures," *Engineering Failure Analysis*, vol. 56, pp. 39–50, 2015.

[38] B. Almassri, A. Kreit, F. A. Mahmoud, and R. François, "Mechanical behaviour of corroded RC beams strengthened by NSM CFRP rods," *Composites Part B: Engineering*, vol. 64, pp. 97–107, 2014.

[39] W. Ibrahim, W. A. Fattah, A. Kotb, and M. A. Mjeed, "Flexural behavior of RC beams strengthened with CFRP strips," in *Proceedings of the 7th International Conference on FRP Composites in Civil Engineering (CICE '14)*, Canada, August 2014.

[40] A. Laraba, N. Chikh, H. Mesbah, N. Djebbar, and A. Belouar, "Experimental evaluation of damaged rectangular reinforced concrete beams repaired with near surface mounted carbon fibre reinforced polymer," *Materials Research Innovations*, vol. 18, no. 6, pp. S6–S13, 2014.

Assessment of Mechanical Properties and Damage of High Performance Concrete Subjected to Magnesium Sulfate Environment

Sheng Cang,[1,2] **Xiaoli Ge,**[3] **and Yanlin Bao**[2]

[1]*Ningbo City College of Vocational Technology, Ningbo 315100, China*
[2]*Department of Mechanics and Engineering Science, Ningbo University, Ningbo 315211, China*
[3]*Jiangsu Testing Center for Quality of Construction Engineering Co., Ltd., Nanjing 210028, China*

Correspondence should be addressed to Sheng Cang; cangsheng@nbcc.cn

Academic Editor: Frederic Dumur

Sulfate attack is one of the most important problems affecting concrete structures, especially magnesium sulfate attack. This paper presents an investigation on the mechanical properties and damage evolution of high performance concrete (HPC) with different contents of fly ash exposure to magnesium sulfate environment. The microstructure, porosity, mass loss, dimensional variation, compressive strength, and splitting tensile strength of HPC were investigated at various erosion times up to 392 days. The ultrasonic pulse velocity (UPV) propagation in HPC at different erosion time was determined by using ultrasonic testing technique. A relationship between damage and UPV of HPC was derived according to damage mechanics, and a correlation between the damage of HPC and erosion time was obtained eventually. The results indicated that (1) the average increasing amplitude of porosity for HPCs was 34.01% before and after exposure to magnesium sulfate solution; (2) the damage evolution of HPCs under sulfate attack could be described by an exponential fitting; (3) HPC containing 20% fly ash had the strongest resistance to magnesium sulfate attack.

1. Introduction

Nowadays, high performance concrete (HPC) is increasingly being used in the field of construction engineering. However, these engineering structures are regularly subjected to aggressive environment. Sulfates, as highly soluble salts, are considered to be one of the main problems affecting concrete structures. Sulfate attack is generally leading to the volume change and cracking, which results in concrete deterioration. In addition, sulfates are widespread in environment, such as underground water, soil, sea water, or industrial waste water [1]. What is more, magnesium sulfate ($MgSO_4$) has the fastest, most severe effects on concrete [2]. Hekal et al. [3] have testified that magnesium sulfate is more severe than sodium sulfate concerning the influence on the properties of concrete. The chemical reactions between concrete and magnesium sulfate are summarized as follows: (1) the sulfate ion (SO_4^{2-}) reacts with the aluminate and portlandite to form gypsum and ettringite, respectively; (2) the magnesium ion (Mg^{2+}) may react with hydroxyl ion (OH^-) to form brucite [$Mg(OH)_2$] or the magnesium can partly replace the calcium in calcium silicate hydrate (C—S—H). Therefore, the magnesium silicate hydrate (M—S—H) that has no binding properties can be formed in the second reaction, and thus the hydrated paste becomes soft and incoherent [4-6].

Up to now, the effects of magnesium sulfate on ordinary concrete have been studied extensively, and there are a lot of investigations conducted into these effects [7-11]. The study on the impact of fly ash on the sulfate resistance properties by Sumer [8] has indicated that the addition of fly ash can significantly increase the resistance to magnesium sulfate attack. The investigation by Chousidis et al. [9] has shown that the durability of concrete can be improved by adding fly ash with high SO_3 and free CaO contents, but the mechanical properties of concrete can be reduced due to the addition of fly ash with significant amounts of clay minerals. Furthermore, the use of fly ash is conducive to resisting the magnesium sulfate attack into concrete, because the microstructure of concrete becomes more compact due to the pozzolanic reaction [12]. Mostofinejad et al. [10] have carried out a research on the influence of magnesium sulfate concentration on durability of concrete. The results suggested that the 5% magnesium sulfate solution is considered to be the most deteriorating environment from the compressive strength reduction viewpoint, while the 14.7% magnesium sulfate solution is observed as the most severe environment from expansion aspect. However, the composition of HPC is usually different from that of ordinary concrete, and then the behavior of HPC under magnesium sulfate attack may be different from that of ordinary concrete.

As a new construction and building material with various excellent properties, HPC has been widely used in modern concrete structures, while information on the behavior of HPC under magnesium sulfate attack is rather limited and needs to be updated. Uysal and Sumer [13] have presented an experimental investigation on the influence of fly ash on durability of self-consolidating concrete under magnesium sulfate attack, and the results indicated that the addition of fly ash can substantially improve the resistance of self-consolidating concrete against magnesium sulfate attack. Another study concerning the long-term performance of self-consolidating concrete under magnesium sulfate attack by Siad et al. [1] has shown that the resistance of self-consolidating concrete under magnesium sulfate attack can be improved due to the addition of natural pozzolan. Silica fume is also used as supplementary cementitious material for the production of HPC, because of its pozzolanic reactivity and microfiller effect. In addition, the fluidity of fresh concrete can be improved due to the addition of silica fume, which may be caused by the dispersing power of the plasticizer coupled with silica fume particle packing between cement grains displacing water or by a ball-bearing effect of silica spheres [14]. However, Behfarnia and Farshadfar [15] have pointed out that silica fume has negative effect on durability of self-consolidating concrete subjected to magnesium sulfate attack, and the debility in specimens containing silica fume is because of formation of magnesium silicate hydrate (M—S—H), a nonadherent substance, in such kind of concrete. Harbec et al. [16] have carried out an investigation on mechanical and durability properties of HPC, and the findings demonstrated that the pozzolanic reaction of glass fume contributes to controlling the expansion of HPC due to sulfate attack. In addition, it has been testified that the deterioration of concrete under sulfate attack is affected by many parameters, such as sulfate concentration [10], water to binder ratio [17, 18], participant cation [3, 19–21], temperature [22], pH [23], and even specimen's size [24].

The study aims to design a new kind of HPC that has characteristics of both high strength concrete and self-consolidating concrete, based on which the mechanical properties and damage evolution of HPC with different contents of fly ash under magnesium sulfate attack were comprehensively investigated. To this end, the porosity, mass loss, dimensional variation, compressive strength, and splitting tensile strength of HPC were studied at various erosion time up to 392 days.

TABLE 1: Chemical composition and physical properties of cement and mineral admixtures.

Materials	Cement	Fly ash	Silica fume
Chemical composition	Weight percentage (%)		
CaO	64.70	8.38	0.77
SiO_2	20.40	47.96	96.18
Al_2O_3	4.70	30.46	0.96
Fe_2O_3	3.38	5.91	0.85
MgO	0.87	2.60	0.74
SO_3	1.88	1.32	0.50
K_2O	0.83	1.61	
Na_2O		1.76	
Loss on ignition	3.24		
Physical properties			
Specific gravity	3.15	2.31	2.22
Specific surface (m^2/kg)	362.20		2.74×10^4
28 d Compressive strength (MPa)	62.8		

In addition, the ultrasonic pulse velocity (UPV) propagation in HPC at different erosion time was determined via ultrasonic testing technique. According to damage mechanics, a relationship between damage and UPV of HPC was derived, and a correlation between the damage of HPC and erosion time was obtained eventually.

2. Materials

2.1. Cement and Mineral Admixtures. Fly ash and silica fume were used as mineral admixtures in the study. The chemical composition and physical properties of cement and mineral admixtures are shown in Table 1. Note that the fly ash can be graded Class I (equivalent to ASTM C 618 Class F), according to the Chinese standard GB/T 1596-2005. The added silica fume was mainly used to improve the fluidity of the fresh HPC in the work.

2.2. Aggregates. Natural river sand with a maximum size of 4 mm was used as fine aggregate, and the specific gravity and water absorption of the aggregate were 2.63 and 1.36%, respectively. Crushed limestone aggregate with a maximum size of 16 mm was utilized as coarse aggregate, and the specific gravity and water absorption of the aggregate were 2.82 and 0.45%, respectively.

2.3. Superplasticizer. A superplasticizer of polycarboxylate obtained from local supplier was used to gain a satisfactory fluidity of HPC. Solid content and water-reducing rate of the superplasticizer were 36.0% and 33.9%, respectively.

2.4. Mix Proportions of HPC. The mixtures of HPC used in the study are shown in Table 2.

According to the mix proportions shown in Table 2, the cubic (size: 100 × 100 × 100 mm) and prismatic (size: 70 × 70 × 280 mm) specimens of HPC were cast. The specimens were covered with plastic sheets after casting and were cured 24

TABLE 2: Mix proportions and fresh properties of HPC.

Mixture	HPC0	HPC1	HPC2	HPC3
Cement (kg/m^3)	530	477	424	371
Fly ash (kg/m^3)	0	53	106	159
Silica fume (kg/m^3)	20	20	20	20
Water (kg/m^3)	160	160	160	160
Fine aggregate (kg/m^3)	477	477	477	477
Coarse aggregate (kg/m^3)	1213	1213	1213	1213
Superplasticizer (kg/m^3)	7.50	7.00	6.85	6.80
Air content (%)	4.6	4.4	4.2	4.1
Slump flow (mm)	585	598	607	616

hours at ambient condition, after which point the specimens were demolded. And then the specimens were placed into a curing room for curing over a span of 28 days with a temperature range of $21 \pm 1°C$ and relative humidity of above 95%. 90 cubic and 3 prismatic specimens were prepared for each mixture.

3. Experimental

3.1. Wetting-Drying Test. The medium used for immersion was 10% magnesium sulfate solution, and the temperature of the solution was maintained at $21 \pm 1°C$. The pH of the sulfate solution was holding in the range of 6.0–8.0 by adding an appropriate amount of sulfate acid solution (0.1 N H_2SO_4). In addition, the magnesium sulfate solution was renewed every 56 days. The wetting-drying test was conducted in accordance with Chinese standard GB/T 50082-2009, but the variation of the experimental method was arranged in order to reduce the effects of high temperature drying. The revised wetting-drying test was shown as follows: the specimens were immersed in 10% magnesium sulfate solution for 7 days, and then they were taken out from their solution tank for 7-day natural drying (at ambient temperature). After this drying process the specimens were immersed in sulfate solution again. This wetting-drying cycle was repeated for 392 days.

3.2. Microstructure. In order to study the microstructural evolution in HPCs before and after exposure to sulfate solution, the microstructure of HPCs was detected by utilizing a FEI 3D environmental scanning electronic microscopy (ESEM), and the micrographs of HPC2 were taken at different erosion times.

3.3. Porosity. To investigate the porosity of HPCs before and after exposure to sulfate solution, the porosity of HPC with different contents of fly ash was determined quantitatively by a Micromeritics AutoPore IV 9510 mercury intrusion porosimetry (MIP). It should be highlighted that the specimen used for MIP test was mortar of HPC, and the coarse aggregate of HPC was removed during the process of specimen preparation.

3.4. Mass Loss. In order to study the mass evolution of HPCs before and after exposure to sulfate solution, the mass of HPC with different contents of fly ash was measured by a high precision electronic balance with a sensitivity of 0.001 g at various erosion time. The mass loss ratio α of HPC can be calculated by the equation as follows:

$$\alpha = \frac{(m_0 - m_e)}{m_0} \times 100, \tag{1}$$

where m_0 is the initial mass and m_e is the mass of HPC after exposure to sulfate solution.

3.5. Dimensional Variation. To investigate the dimensional variation of HPCs before and after exposure to sulfate solution, the length of HPC with different contents of fly ash was determined via a high accuracy digital length comparator with a precision of 0.001 mm. The expansion ratio β of HPC can be calculated by

$$\beta = \frac{(l_e - l_0)}{l_0} \times 100, \tag{2}$$

where l_0 is the initial length and l_e is the length of HPC after exposure to sulfate solution.

3.6. Mechanical Strength

3.6.1. Compressive Strength. A universal testing machine was utilized to measure the compressive strength of HPC with different contents of fly ash at various erosion times. The loading rate was set as 0.30 MPa/s in the compressive test.

3.7. Splitting Tensile Strength. The splitting tensile strength of each mixture at different erosion time was determined by the universal testing machine equipped with a splitting tensile setup at a loading rate of 0.03 MPa/s.

It should be noted that the compressive strength and splitting tensile strength of HPC were measured in accordance with the Chinese standard GB/T 50081-2002, and the cubic specimens were also used in recent published paper for splitting tensile strength of sacrificial concrete [25].

3.8. UPV Test. The experiment on UPV in the study was conducted according to literature [26]. In the actual measuring procedure, the experiment on UPV was implemented according to CECS02-88 [27]. The specific parameters of ultrasonic inspection instrument are presented in Table 3.

3.9. Damage. When HPC is exposed to sulfate solution, damage is initiated in it. The damage leads to a variation in UPV propagation through HPC. According to damage mechanics, the relationship between damage and Young's modulus can be expressed as [28]

$$D = 1 - \frac{E_e}{E_0}, \tag{3}$$

Assessment of Mechanical Properties and Damage of High Performance Concrete Subjected to Magnesium...

205

TABLE 3: Specific parameters of ultrasonic inspection instrument.

Transmitting voltage	Transmitting pulse width	Amplifier gain	Sampling period
1000 V	0.08 ms	82 dB	0.4 ms

where D is damage, E_e is Young's modulus at different erosion time, and E_0 is initial Young's modulus.

Supposing that HPC is homogeneous material, Young's modulus E correlates its UPV V, and their relationship is shown as follows:

$$V = \sqrt{\frac{E(1-v)}{[\rho(1-2v)(1+v)]}}, \tag{4}$$

where ρ is the density of HPC and v is Poisson's ratio.

Assuming that sulfate attack has a negligible effect on the Poisson's ratio of HPC. According to (4), then

$$E_0 = \frac{[\rho_0(1-2v)(1+v)V_0^2]}{(1-v)},$$

$$E_e = \frac{[\rho_e(1-2v)(1+v)V_e^2]}{(1-v)}. \tag{5}$$

Substituting (5) into (3), the damage expression of HPC can be obtained:

$$D = 1 - \left(\frac{\rho_e}{\rho_0}\right) \times \left(\frac{V_e}{V_0}\right)^2, \tag{6}$$

where ρ_0 and ρ_e are density of HPC before and after exposure to sulfate solution, respectively. V_0 and V_e are UPV of HPC before and after exposure to sulfate solution, respectively.

The density of HPC before exposure to sulfate solution ρ_0 was determined according to the Chinese standard GB/T 50080-2002 and can be expressed as

$$\rho_0 = \frac{m_0}{t_0} = \frac{m_0}{a^3}, \tag{7}$$

where m_0 is the initial mass, t_0 is the initial volume, and $a = 100$ mm.

Thus, the density of HPC after exposure to sulfate solution ρ_e can be calculated from the following equation:

$$\rho_e = \frac{m_e}{t_e} = \frac{(1-\alpha)m_0}{[a(1+\beta)]^3} = \frac{(1-\alpha)\rho_0}{(1+\beta)^3}. \tag{8}$$

Consequently, the damage of HPC after exposure to sulfate solution can be calculated through (6), since the density and UPV of HPC before and after exposure to sulfate solution can be obtained via experiment.

It should be stressed that the cubic (size: $100 \times 100 \times 100$ mm) specimens were used for mass loss, compressive strength, splitting tensile strength, and UPV experiments, and the prismatic (size: $70 \times 70 \times 280$ mm) specimens were utilized for dimensional variation experiment. In addition, 3 replicate measurements were carried out on the porosity,

mass loss, dimensional variation, compressive strength, splitting tensile strength, and UPV experiments at various erosion time, and only the average values were reported so as to improve the accuracy of experimental results. The ESEM test of HPCs was carried out at 0 day, 84 days, 224 days, and 392 days, but the porosity, mass loss, dimensional variation, compressive strength, splitting tensile strength, and UPV experiments were performed at 0 day and every 28 days until 392 days.

4. Results and Discussion

4.1. Microstructure. ESEM investigations illustrated vivid variations in the microstructure of HPC at different erosion time. Figure 1 presents the ESEM micrographs of HPC2 before and after exposure to sulfate solution. Note that the green square in Figure 1 was enlarged to show details and was presented in its right side and that energy dispersive spectrometer (EDS) analysis was carried out on the blue square in right micrographs.

As shown in Figure 1, the matrix of HPC2 before exposure to sulfate solution presented a continuous microstructure with few microcracks, and the microcracks were due to sample making. A small amount of microcracks was observed at 84 days, and ettringite was found in the microcracks. The main elemental compositions of EDS area for HPC2 at 84 days were C, O, Mg, Si, S, Ca, and Pt, and their weight percent was 6.93%, 17.77%, 0.41%, 4.04%, 1.06%, 17.00%, and 50.91%, respectively, the result of which indicated that the material in the blue square was ettringite. The ettringite filled the cracks in HPC2 at this erosion time, which resulted in the decrease of its porosity. A fairly large number of microcracks emerged with the erosion time up to 224 days. The main elemental compositions of EDS area for HPC2 at 224 days were C, O, Mg, Si, S, Ca, and Pt, and their weight percent was 8.24%, 16.56%, 0.10%, 15.59%, 0.78%, 13.80%, and 43.62%, respectively, the result of which suggested that the material in the blue square was also ettringite. The ettringite was so much as to cause the expansion of HPC2 at 224 days. Connected cracks spread all over the specimen at 392 days, and the ettringite led to the cracking of HPC2. The main elemental compositions of EDS area for HPC2 at 392 days were C, O, Mg, Si, S, Ca, and Pt, and their weight percent was 21.70%, 22.54%, 0.48%, 3.20%, 0.31%, 11.10%, and 39.39%, respectively, the result of which indicated that the material in the blue square was also ettringite. The microstructure evolution of HPC2 under sulfate attack was in line with the results of Siad et al. [1].

4.2. Porosity. The porosity of HPCs at different erosion time is shown in Figure 2.

As seen in Figure 2, the porosity of HPCs decreased slowly before a sharp rise with the increase of erosion time,

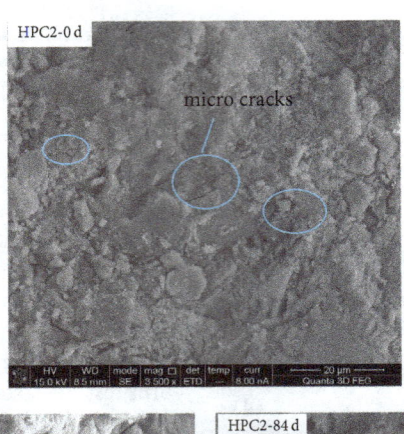

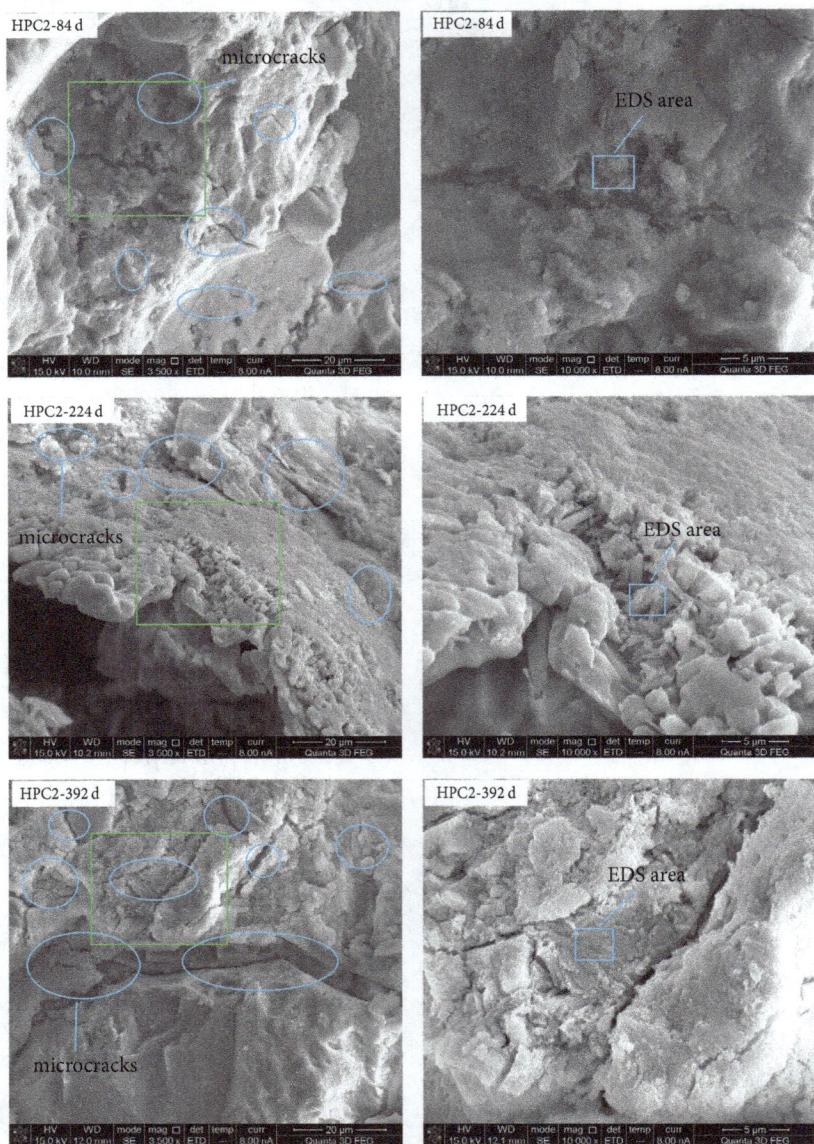

FIGURE 1: ESEM micrographs of HPC2 before and after exposure to sulfate solution.

and the porosity reached the minimum value at 84 (HPC3) or 112 (HPC0, HPC1, and HPC2) days, which meant that sulfate attack contributed to improving the microstructure of HPCs in the early stage (from 0 to 84 or 112 days). The decrease of porosity could be due to the fact that the HPCs absorbed water which resulted in further hydration of the cement and that a small amounts of reaction products between sulfate solution and HPCs filled cracks. The increased porosity could be attributed to the inner expansion of HPCs under sulfate attack. Between 0 and 392 days, the porosity increasing

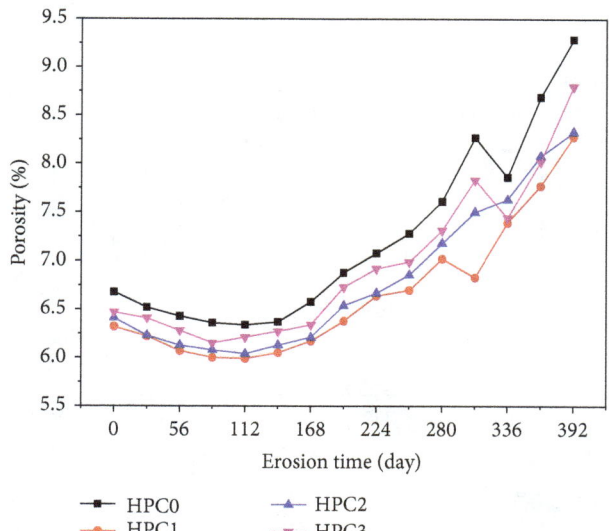

FIGURE 2: Porosity of HPCs at different erosion time.

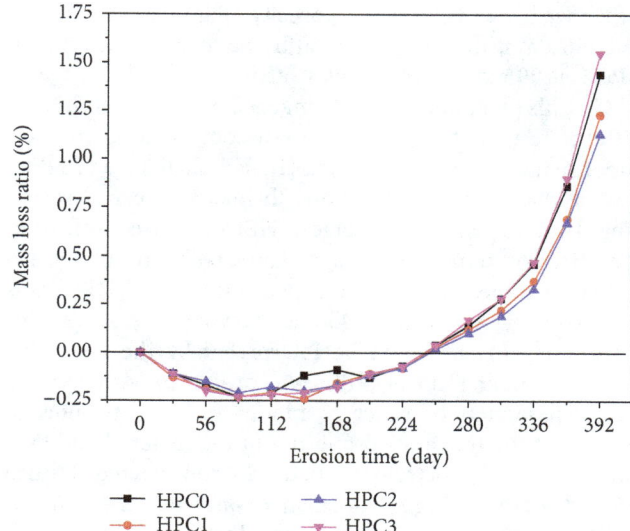

FIGURE 3: Mass loss ratio of HPCs before and after exposure to sulfate solution.

amplitudes of HPC0, HPC1, HPC2, and HPC3 were 39.07%, 31.01%, 29.95%, and 36.01%, respectively, which indicated that the average increasing amplitude of porosity for HPCs was 34.01% before and after exposure to 10% magnesium sulfate solution.

In general, the porosity of HPCs with fly ash (HPC1, HPC2, HPC3) was lower than that of HPC without fly ash (HPC0), which was mainly due to pozzolanic effect and microfiller effect of fly ash [29]. On the one hand, the microheads in fly ash could disperse well in concrete, and combined firmly with gel produced in cement hydration, and hence the matrix of concrete could be improved. On the other hand, the gel produced from pozzolanic action could fill in the capillary pore in concrete, which could effectively reduce the porosity of concrete. In addition, the porosity of HPC0 was the highest, followed by HPC3, HPC2, and HPC1, respectively, which suggested that HPC1 had the strongest resistance to magnesium sulfate attack in terms of porosity.

4.3. Mass Loss. The mass loss ratio of HPCs before and after exposure to sulfate solution is presented in Figure 3.

As illustrated in Figure 3, the mass loss ratio of HPCs decreased firstly and then increased with the increment of erosion time, and a minimum value was reached at 84 days. This changing trend was accordant well with that of the porosity (Figure 2). Between 0 and 224 days, the mass loss ratio was negative, which indicated that the mass of HPCs was increased during this time. The phenomenon of mass gain for HPCs is in line with the results of high strength self-consolidating concrete under magnesium sulfate attack [1]. The mass gain could be due to water filling cracks, and to the water used to precipitate the hydrated phase, such as ettringite [30]. The mass loss ratio of HPCs, as a whole, was about between −0.25% and 1.60%.

After 336 days, the mass loss ratio of HPCs was increased significantly, which suggested that there was a sharp rise in the degree of deterioration in HPCs. Furthermore, the

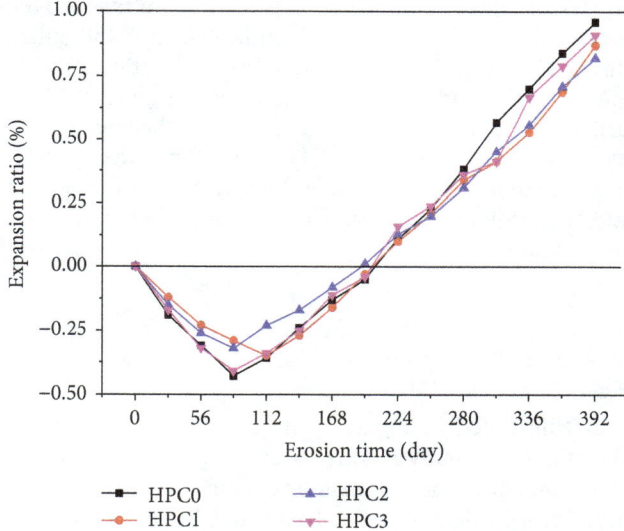

FIGURE 4: Expansion ratio of HPCs at various erosion time.

mass loss ratio of HPC2 was the lowest after 252 days, which indicated that this kind of HPC had the strongest resistance to magnesium sulfate attack in terms of mass loss ratio. It was worth noting that the mass loss ratio of HPC3 was higher than that of HPC0 after 280 days, which demonstrated that the sulfate attack resistance of HPC3 was even weaker than that of HPC0. That is, with this content (30%) of fly ash, the sulfate attack resistance of HPC could be decreased slightly.

4.4. Dimensional Variation. The expansion ratio of HPCs at various erosion time is shown in Figure 4.

As seen in Figure 4, the expansion ratio of HPCs decreased quickly and then increased rapidly with the increase of erosion time, and the expansion ratio reached the minimum value at 84 (HPC0, HPC2, and HPC3) or 112 (HPC1) days, the changing trend of which was matched rather well with that of the porosity (Figure 2). Between 0 and 196 days, the expansion ratio was negative except for HPC2 at 196 days, which was attributed to shrinkage of HPCs. This phenomenon of shrinkage for HPCs is consistent with the results of high strength self-consolidating concrete under magnesium sulfate attack [1]. The shrinkage of HPCs may be due to their autogenous shrinkage at early erosion time. The autogenous shrinkage of HPCs is caused by further hydration of cement and pozzolanic reactions of fly ash and silica fume. Overall, the expansion ratio of HPCs was approximately between −0.45% and 0.96%. The absence or the low expansion might be interpreted by the low value water to binder ratio of HPCs. According to Maltais et al. [31], a reduction of water to binder ratio contributes to decreasing the transport properties of the material and thus limits the extent of deterioration. In addition, Neville [32] also pointed out that in high performance concrete there is limited pore space to accommodate the products of reactions with sulfate, namely, M—S—H, and gypsum that partly result in expansion.

After 280 days, the expansion ratio of HPCs with fly ash (HPC1, HPC2, and HPC3) was lower than that of HPC without fly ash (HPC0), which indicated that the sulfate attack resistance of HPC could be improved by adding fly ash. And this result was in line with published literature [9, 13]. In particular, the expansion ratio of HPC2 was the lowest at 392 days. That meant this kind of HPC (HPC2) had the strongest resistance to magnesium sulfate attack in terms of expansion ratio, the result of which was accordant with that of mass loss ratio (Figure 3).

4.5. Compressive Strength.

The compressive strength of HPCs before and after exposure to magnesium sulfate solution is presented in Figure 5.

As illustrated in Figure 5, the compressive strength of HPCs increased firstly and then decreased with the increment of erosion time, and a maximum value was reached at 84 days. An investigation by Acharya and Patro [11] has also found that the compressive strength of concrete under sulfate attack increases firstly and then decreases with the increase of erosion time, although the compressive strength of concrete is only determined at 28, 91, and 180 days in the literature. Between 0 and 84 days, there was a slow increase in the compressive strength of HPCs, after that the compressive strength decreased at a slower speed (compared to its increase) until 168 days, while the compressive strength was falling rapidly after 196 days. This result was in line with the changing trend of porosity during the same erosion time. The increase of compressive strength could be attributed to the fact that the HPCs absorbed water which resulted in further hydration of the cement and that a small amounts of reaction products between sulfate solution and HPCs filled cracks. The decrease of compressive strength was due to expansion, cracking, and the formation of M—S—H. Expansion and cracking generally resulted from expansive forces generated by sulfate reacting with the calcium aluminum hydrates to form ettringite [33]. The formation of M—S—H was because the magnesium could partly replace the calcium in C—S—H [6].

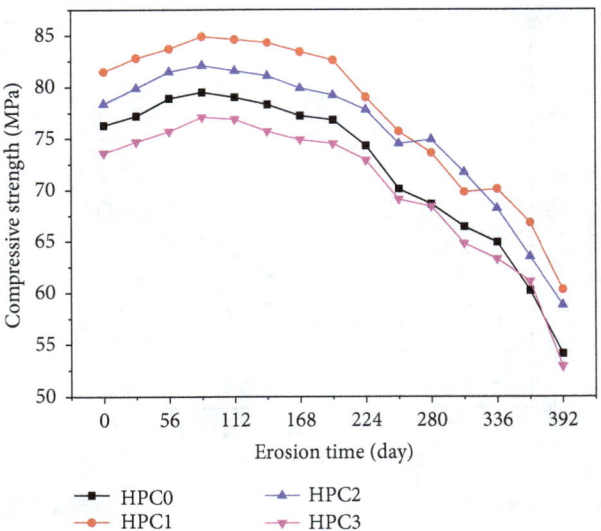

FIGURE 5: Compressive strength of HPCs before and after exposure to sulfate solution.

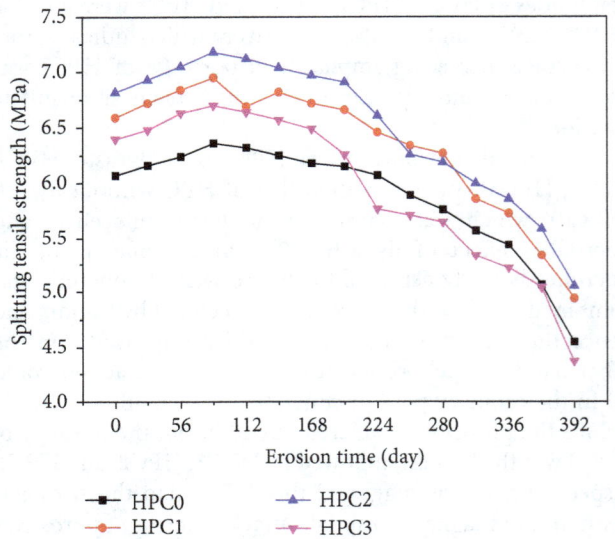

FIGURE 6: Splitting tensile strength of HPCs at different erosion time.

The cracking, in turn, made it easy to transport sulfate ions into concrete, which accelerated the deterioration of HPCs.

The compressive strength of HPC1, on the whole, was the highest before and after exposure to magnesium sulfate solution, which indicated that this kind of HPC had the strongest resistance to magnesium sulfate attack in terms of compressive strength. In addition, the compressive strength of HPC3 was lower than that of HPC0, which demonstrated that the sulfate attack resistance of HPC3 was lower than that of HPC0. That is, with this content (30%) of fly ash, the compressive strength of HPC could be reduced slightly, which was consistent with that of mass loss ratio (Figure 3).

4.6. Splitting Tensile Strength.

The splitting tensile strength of HPCs at different erosion time is shown in Figure 6.

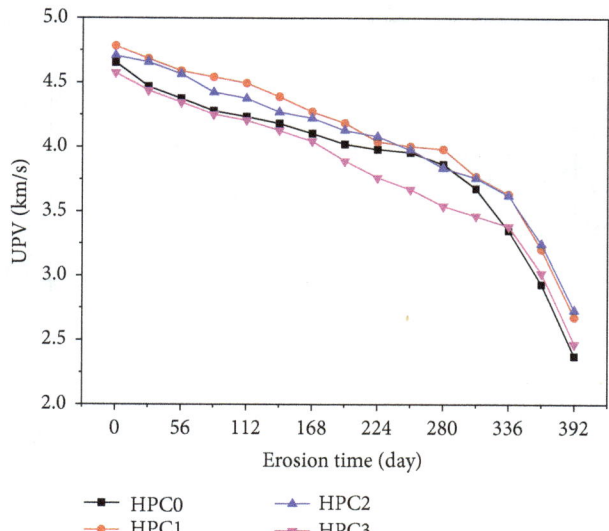

FIGURE 7: UPV of HPCs before and after exposure to sulfate solution.

As shown in Figure 6, the splitting tensile strength of HPCs increased firstly and then decreased with the increase of erosion time, and a maximum value was reached at 84 days. This changing trend was in agreement with that of flexural strength in previous study [11], although the flexural strength of concrete is simply measured at 28, 91, and 180 days in the paper. There was a slow increase in the splitting tensile strength of HPCs between 0 and 84 days, and then the splitting tensile strength decreased at a slower speed (compared to its increase) until 168 days, yet the splitting tensile strength dropped rapidly after 196 days. This result was accordant with the changing trends of porosity (Figure 2) and compressive strength (Figure 5) during the same erosion time. The reasons for the increase and decrease of splitting tensile strength were the same as those of compressive strength mentioned above.

The splitting tensile strength of HPC2, in general, was the highest before and after exposure to magnesium sulfate solution, which suggested that this kind of HPC had the strongest resistance to magnesium sulfate attack in terms of splitting tensile strength. In addition, the splitting tensile strength of HPC3 was lower than that of HPC0 after 224 days, which indicated that the sulfate attack resistance of HPC3 was lower than that of HPC0. In other words, with this content (30%) of fly ash, the splitting tensile strength of HPC could be decreased slightly, the result of which was accordant with those of mass loss ratio (Figure 3) and compressive strength (Figure 5).

4.7. UPV. The UPV of HPCs before and after exposure to sulfate solution is presented in Figure 7.

As seen in Figure 7, the UPV of HPCs decreased continually with the increase of erosion time, and there was an accelerating process in the decrease of UPV after 308 days. This phenomenon corresponded broadly with the changing trends of the compressive strength (Figure 5) and splitting

tensile strength (Figure 6), the result of which indicated that the deterioration of HPCs in terms of mechanical strengths could be evaluated by the means of UPV test. The decrease of UPV with the increase of erosion time was due to the damaged microstructure and the induced sulfate attack damage in HPCs.

After 308 days, the UPV of HPCs was decreased significantly, which suggested that there was a sharp rise in the degree of deterioration in HPCs. Furthermore, the UPV of HPC2 was the highest after 308 days, which indicated that this kind of HPC had the strongest resistance to magnesium sulfate attack in terms of UPV. It was worth noting that the UPV of HPC3 was lower than that of HPC0 before 308 days, which demonstrated that the sulfate attack resistance of HPC3 was lower than that of HPC0 during this period.

4.8. Relationships between Strengths and UPV. The relationships between strengths and UPV for HPCs are shown in Figure 8.

As illustrated in Figure 8(a), the compressive strength of HPCs increased quickly and then there was a slight fall with the increase of UPV, and their compressive strength evolution was accordant, the result of which could be described by a polynomial fitting, as the full line in Figure 8(a). The equation of the polynomial was shown as follows:

$$f_{cs} = 197.28883 - 142.24795V + 44.76914V^2 \\ - 4.20657V^3, \tag{9}$$

where f_{cs} is the compressive strength of HPCs and V is the UPV propagation in HPCs.

Similar to compressive strength, the splitting tensile strength of HPCs also increased sharply and then decreased slightly with the increase of UPV, as shown in Figure 8(b). Their splitting tensile strength evolution was consistent, and the results suggested that a polynomial relationship could provide an approximation to assess the splitting tensile strength of HPCs via UPV. The following was the expression of the polynomial:

$$f_{sts} = 18.56919 - 13.69795V + 4.24942V^2 \\ - 0.39801V^3, \tag{10}$$

where f_{sts} is the splitting tensile strength of HPCs.

The R-Square of the polynomial fitting for compressive strength and splitting tensile strength was 0.9587 and 0.9684, respectively, which indicated that the fitting results matched rather well with the experimental data and that (9) and (10) could be used to evaluate the compressive strength and the splitting tensile strength of HPCs by UPV. In addition, the relationships between mechanical strengths and UPV of HPCs were different from those of ferro-siliceous concretes, and the compressive strength-UPV and splitting tensile strength-UPV relationships of ferro-siliceous concretes were Weibull distribution and exponential form, respectively [34]. Maybe that was because the deterioration of ferro-siliceous concretes was due to high temperature.

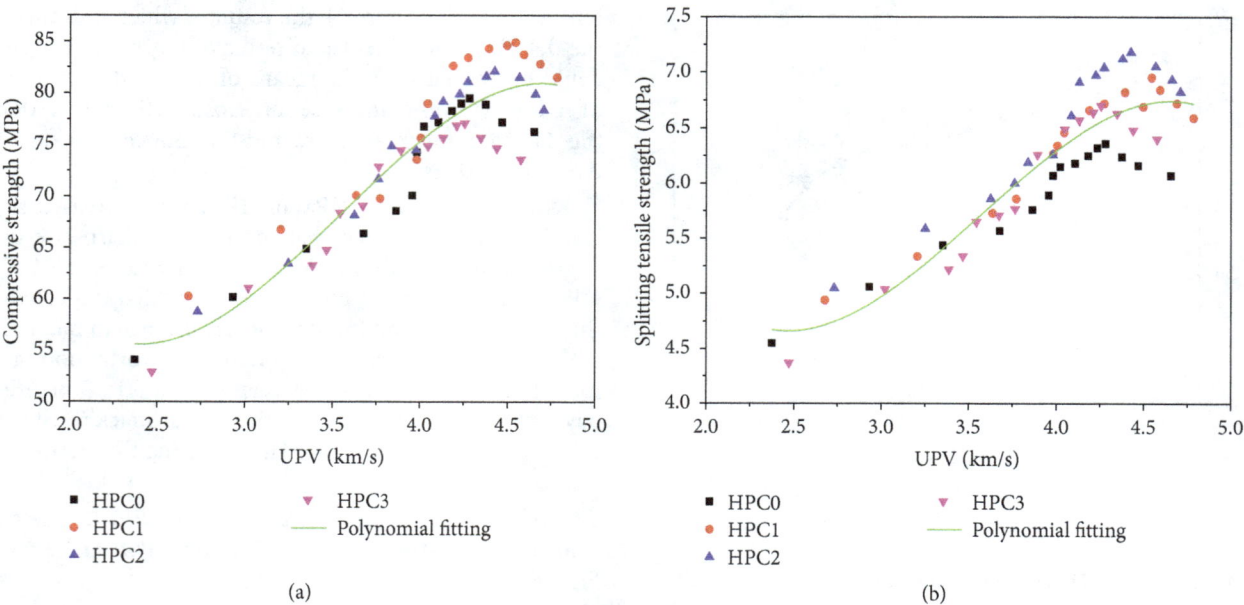

FIGURE 8: Relationships between strengths and UPV for HPCs: (a) compressive strength versus UPV and (b) splitting tensile strength versus UPV.

It should be emphasized that the empirical formulas might not precisely predict the mechanical strengths (both compressive and splitting tensile strength) of HPCs with different compositions, such as different mineral admixtures, different aggregate types, and different water to binder ratio. On the whole, however, the models established in the paper were remarkably accurate to assess the mechanical strengths of HPCs under magnesium sulfate attack. In practice, the empirical models could provide a convenient way to evaluate mechanical strengths of HPCs subjected to magnesium sulfate environment.

4.9. Damage.
According to Figures 3 and 4, both the mass loss ratio and expansion ratio of HPCs could be obtained. The initial density of HPCs could be calculated from (7). Based on these results, the density of HPCs at different erosion time could be determined, as shown in Figure 9. Therefore, the damage of HPCs at various erosion time could be calculated through (6) and is plotted in detail in Figure 10.

As illustrated in Figure 9, the density of HPCs increased firstly and then decreased with the increment of erosion time, and a maximum value was reached at 84 days. This changing trend was accordant well with those of the porosity (Figure 2), the mass loss ratio (Figure 3), the compressive strength (Figure 5), and the splitting tensile strength (Figure 6). The phenomenon of density gain for HPCs is in line with the results of high strength self-consolidating concrete under magnesium sulfate attack [1]. The density gain could be attributed to water filling cracks and to the water used to precipitate the hydrated phase, such as ettringite [29]. In addition, the density of HPCs was important input data for the calculation of their damage.

As shown in Figure 10, the damage of HPCs increased sharply with the increase of erosion time. The damage of

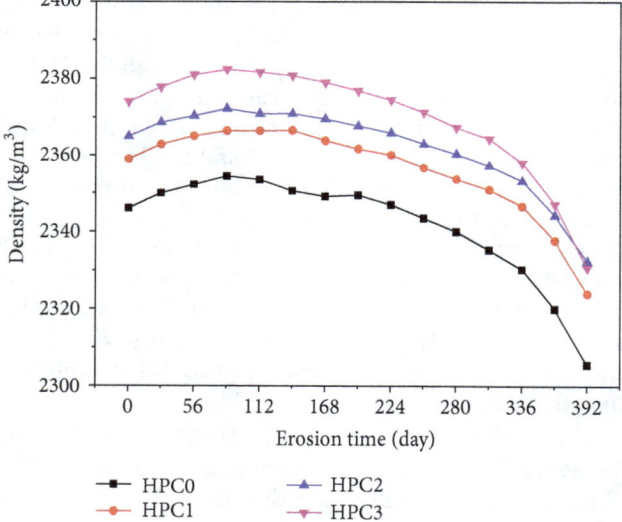

FIGURE 9: Density of HPCs at different erosion time.

HPC2 was broadly the lowest before and after exposure to magnesium sulfate solution, which suggested that this kind of HPC had the strongest resistance to magnesium sulfate attack in terms of damage. When HPCs were exposed to magnesium sulfate solution, their damage evolution was accordant and could be described by an exponential fitting (the full curve in Figure 10). The exponential model could be expressed by an equation as follows:

$$D = 0.11881e^{0.00464T} - 0.07298, \tag{11}$$

where D is the damage of HPCs and T is the erosion time of HPCs.

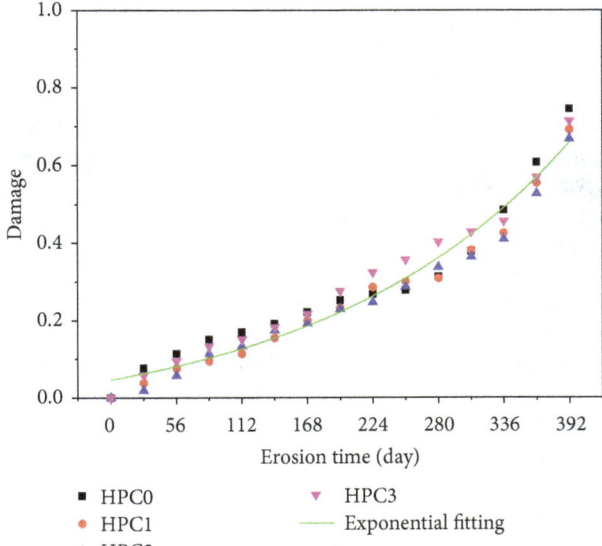

FIGURE 10: Damage of HPCs at various erosion time.

The *R*-Square of the exponential fitting was 0.9629, which indicated that the fitting result agreed very well with the experimental data and that (11) could be applied to characterize the damage evolution of HPCs under magnesium sulfate attack. It should be stressed that the damage of HPCs increased slowly between 0 and 84 days, which matched well with the changing trend of UPV (Figure 7) in the same term, but accorded not so well with the changing trends of porosity (Figure 2), compressive strength (Figure 5), and splitting tensile strength (Figure 6) during the same period. And different form of damage evolution was also reported by other researchers, such as Weibull distribution model [35]. These results might slightly jeopardize the damage assessment. On the whole, however, the model established in the study was reasonably precise to describe the damage evolution HPCs under magnesium sulfate attack. In practice, the established model could be utilized in damage assessment of HPCs and evaluation of magnesium sulfate attack.

In summary, HPC1 and HPC2 had the stronger resistance to magnesium sulfate attack than that of HPC0 and HPC3. Furthermore, compared to HPC1, HPC2 was shown to has higher porosity, lower mass loss ratio, lower expansion ration, lower compressive strength, higher splitting tensile strength, and lower damage. Consequently, the HPC2, on the whole, had the strongest resistance to magnesium sulfate attack.

5. Conclusions

In the paper, a new kind of HPC that has characteristics of both high strength concrete and self-consolidating concrete is designed, and the mechanical properties and damage evolution of HPC with different contents of fly ash under magnesium sulfate attack are comprehensively investigated. The main conclusions drawn in this study are as follows:

(1) The porosity of HPCs decreases slowly before a sharp rise with the increase of erosion time, and the porosity reaches the minimum value at 84 (HPC3) or 112 (HPC0, HPC1, and HPC2) days. The average increasing amplitude of porosity for HPCs is 34.01% before and after exposure to 10% magnesium sulfate solution. HPC1 has the strongest resistance to magnesium sulfate attack in terms of porosity.

(2) The mass loss ratio of HPCs decreases firstly and then increases with the increment of erosion time, and a minimum value is reached at 84 days. The mass loss ratio of HPCs, as a whole, is about between −0.25% and 1.60%. HPC2 has the strongest resistance to magnesium sulfate attack in terms of mass loss ratio.

(3) The expansion ratio of HPCs decreases quickly and then increases rapidly with the increase of erosion time, and the expansion ratio reaches the minimum value at 84 (HPC0, HPC2, and HPC3) or 112 (HPC1) days. Overall, the expansion ratio of HPCs is approximately between −0.45% and 0.96%. HPC2 has the strongest resistance to magnesium sulfate attack in terms of expansion ratio.

(4) The compressive strength of HPCs increases firstly and then decreases with the increment of erosion time, and a maximum value is reached at 84 days. HPC1 has the strongest resistance to magnesium sulfate attack in terms of compressive strength.

(5) The splitting tensile strength of HPCs increases firstly and then decreases with the increase of erosion time, and a maximum value is reached at 84 days. HPC2 has the strongest resistance to magnesium sulfate attack in terms of splitting tensile strength.

(6) The UPV of HPCs decreases continually with the increase of erosion time, and there is an accelerating process in the decrease of UPV after 308 days. HPC2 has the strongest resistance to magnesium sulfate attack in terms of UPV.

(7) The relationships between mechanical strengths and UPV for HPCs can be described by 2 polynomial fittings. In practice, the empirical models can provide a convenient way to evaluate mechanical strengths of HPCs subjected to magnesium sulfate environment.

(8) The damage of HPCs increases sharply with the increase of erosion time and can be described by an exponential fitting. In practice, the established model can be utilized in damage assessment of HPCs and evaluation of magnesium sulfate attack. HPC2 has the strongest resistance to magnesium sulfate attack in terms of damage.

Conflicts of Interest

The authors declare that there are no conflicts of interest regarding the publication of this paper.

Acknowledgments

The research is financially supported by Scientific Research Project of the Education Department of Zhejiang Province of China (no. Y201327450), the Zhejiang Open Foundation of the Most Important Subjects (nos. zj1208 and zj1209), and Scientific Research Project of Jiangsu Provincial Department of Housing and Urban-Rural Development (no. 094828867208), which are gratefully appreciated.

References

[1] H. Siad, M. Lachemi, S. K. Bernard, M. Sahmaran, and A. Hossain, "Assessment of the long-term performance of SCC incorporating different mineral admixtures in a magnesium sulphate environment," *Construction and Building Materials*, vol. 80, pp. 141–154, 2015.

[2] K. Yildirim and M. Sumer, "Effects of sodium chloride and magnesium sulfate concentration on the durability of cement mortar with and without fly ash," *Composites Part B: Engineering*, vol. 52, pp. 56–61, 2013.

[3] E. E. Hekal, E. Kishar, and H. Mostafa, "Magnesium sulfate attack on hardened blended cement pastes under different circumstances," *Cement and Concrete Research*, vol. 32, no. 9, pp. 1421–1427, 2002.

[4] D. D. Higgins, "Increased sulfate resistance of ggbs concrete in the presence of carbonate," *Cement and Concrete Composites*, vol. 25, no. 8, pp. 913–919, 2003.

[5] D. Bonen and M. D. Cohen, "Magnesium sulfate attack on portland cement paste-I. Microstructural analysis," *Cement and Concrete Research*, vol. 22, no. 1, pp. 169–180, 1992.

[6] D. Bonen and M. D. Cohen, "Magnesium sulfate attack on portland cement paste - II. Chemical and mineralogical analyses," *Cement and Concrete Research*, vol. 22, no. 4, pp. 707–718, 1992.

[7] C. Jaturapitakkul, K. Kiattikomol, W. Tangchirapat, and T. Saeting, "Evaluation of the sulfate resistance of concrete containing palm oil fuel ash," *Construction and Building Materials*, vol. 21, no. 7, pp. 1399–1405, 2007.

[8] M. Sumer, "Compressive strength and sulfate resistance properties of concretes containing Class F and Class C fly ashes," *Construction and Building Materials*, vol. 34, pp. 531–536, 2012.

[9] N. Chousidis, I. Ioannou, E. Rakanta, C. Koutsodontis, and G. Batis, "Effect of fly ash chemical composition on the reinforcement corrosion, thermal diffusion and strength of blended cement concretes," *Construction and Building Materials*, vol. 126, pp. 86–97, 2016.

[10] D. Mostofinejad, F. Nosouhian, and H. Nazari-Monfared, "Influence of magnesium sulphate concentration on durability of concrete containing micro-silica, slag and limestone powder using durability index," *Construction and Building Materials*, vol. 117, pp. 107–120, 2016.

[11] P. K. Acharya and S. K. Patro, "Acid resistance, sulphate resistance and strength properties of concrete containing ferrochrome ash (FA) and lime," *Construction and Building Materials*, vol. 120, pp. 241–250, 2016.

[12] J. A. Jain and N. Neithalath, "Chloride transport in fly ash and glass powder modified concretes—influence of test methods on microstructure," *Cement and Concrete Composites*, vol. 32, no. 2, pp. 148–156, 2010.

[13] M. Uysal and M. Sumer, "Performance of self-compacting concrete containing different mineral admixtures," *Construction and Building Materials*, vol. 25, no. 11, pp. 4112–4120, 2011.

[14] H. Vikan and H. Justnes, "Rheology of cementitious paste with silica fume or limestone," *Cement and Concrete Research*, vol. 37, no. 11, pp. 1512–1517, 2007.

[15] K. Behfarnia and O. Farshadfar, "The effects of pozzolanic binders and polypropylene fibers on durability of SCC to magnesium sulfate attack," *Construction and Building Materials*, vol. 38, pp. 64–71, 2013.

[16] D. Harbec, A. Zidol, A. Tagnit-Hamou, and F. Gitzhofer, "Mechanical and durability properties of high performance glass fume concrete and mortars," *Construction and Building Materials*, vol. 134, pp. 142–156, 2017.

[17] J.-K. Chen and M.-Q. Jiang, "Long-term evolution of delayed ettringite and gypsum in Portland cement mortars under sulfate erosion," *Construction and Building Materials*, vol. 23, no. 2, pp. 812–816, 2009.

[18] M. Sahmaran, O. Kasap, K. Duru, and İ. Ö. Yaman, "Effects of mix composition and water-cement ratio on the sulfate resistance of blended cements," *Cement and Concrete Composites*, vol. 29, no. 3, pp. 159–167, 2007.

[19] O. S. B. Al-Amoudi, "Attack on plain and blended cements exposed to aggressive sulfate environments," *Cement and Concrete Composites*, vol. 24, no. 3-4, pp. 305–316, 2002.

[20] R. S. Gollop and H. F. W. Taylor, "Microstructural and microanalytical studies of sulfate attack. I. Ordinary portland cement paste," *Cement and Concrete Research*, vol. 22, no. 6, pp. 1027–1038, 1992.

[21] Y.-S. Park, J.-K. Suh, J.-H. Lee, and Y.-S. Shin, "Strength deterioration of high strength concrete in sulfate environment," *Cement and Concrete Research*, vol. 29, no. 9, pp. 1397–1402, 1999.

[22] M. H. Rarani, M. R. M. Aliha, M. M. Shokrieh, and M. R. Ayatollahi, "Mechanical durability of an optimized polymer concrete under various thermal cyclic loadings—an experimental study," *Construction and Building Materials*, vol. 64, pp. 308–315, 2014.

[23] Q. Zhou, J. Hill, E. A. Byars, J. C. Cripps, C. J. Lynsdale, and J. H. Sharp, "The role of pH in thaumasite sulfate attack," *Cement and Concrete Research*, vol. 36, no. 1, pp. 160–170, 2006.

[24] X. Brunetaud, M.-R. Khelifa, and M. Al-Mukhtar, "Size effect of concrete samples on the kinetics of external sulfate attack," *Cement and Concrete Composites*, vol. 34, no. 3, pp. 370–376, 2012.

[25] H. Chu, J. Jiang, W. Sun, and M. Zhang, "Effects of graphene sulfonate nanosheets on mechanical and thermal properties of sacrificial concrete during high temperature exposure," *Cement and Concrete Composites*, vol. 82, pp. 252–264, 2017.

[26] H.-Y. Chu and J.-K. Chen, "The experimental study on the correlation of resistivity and damage for conductive concrete," *Cement and Concrete Composites*, vol. 67, pp. 12–19, 2016.

[27] China Association for Engineering Construction Standardization, Beijing: China Architecture & Building (in Chinese), 2005.

[28] S. Murakami, *Continuum Damage Mechanics: A Continuum Mechanics Approach to The Analysis of Damage And Fracture*, Springer, Berlin, Germany, 2012.

[29] C. Cao, W. Sun, and H. Qin, "Analysis on strength and fly ash effect of roller-compacted concrete with high volume fly ash," *Cement and Concrete Research*, vol. 30, no. 1, pp. 71–75, 2000.

[30] A. Skaropoulou, K. Sotiriadis, G. Kakali, and S. Tsivilis, "Use of mineral admixtures to improve the resistance of limestone cement concrete against thaumasite form of sulfate attack," *Cement and Concrete Composites*, vol. 37, no. 1, pp. 267–275, 2013.

[31] Y. Maltais, E. Samson, and J. Marchand, "Predicting the durability of Portland cement systems in aggressive environments - Laboratory validation," *Cement and Concrete Research*, vol. 34, no. 9, pp. 1579–1589, 2004.

[32] A. Neville, "The confused world of sulfate attack on concrete," *Cement and Concrete Research*, vol. 34, no. 8, pp. 1275–1296, 2004.

[33] B. Tian and M. D. Cohen, "Does gypsum formation during sulfate attack on concrete lead to expansion?" *Cement and Concrete Research*, vol. 30, no. 1, pp. 117–123, 2000.

[34] H. Chu, J. Jiang, W. Sun, and M. Zhang, "Mechanical and physicochemical properties of ferro-siliceous concrete subjected to elevated temperatures," *Construction and Building Materials*, vol. 122, pp. 743–752, 2016.

[35] H.-Y. Chu, J.-Y. Jiang, W. Sun, and M. Zhang, "Thermal behavior of siliceous and ferro-siliceous sacrificial concrete subjected to elevated temperatures," *Materials and Design*, vol. 95, pp. 470–480, 2016.

21

Evaluating the Performance of Absolute RSSI Positioning Algorithm-Based Microzoning and RFID in Construction Materials Tracking

M. Truijens,[1] X. Wang,[2,3] H. de Graaf,[4] and J. J. Liu[5]

[1] *Woodside Energy Ltd., School of Built Environment, Curtin University, Perth, WA 6845, Australia*
[2] *Curtin-Woodside Oil, Gas & LNG Construction and Project Management, Australasian Joint Research Centre for Building Information Modelling (BIM), Curtin University, Perth, WA 6845, Australia*
[3] *Department of Housing and Interior Design, Kyung Hee University, Seoul 130-701, Republic of Korea*
[4] *Industrial Automation Group MD, Joondalup Business Park, Perth, WA 6027, Australia*
[5] *College of Science, China University of Petroleum, Beijing 102249, China*

Correspondence should be addressed to X. Wang; xiangyu.wang@curtin.edu.au and J. J. Liu; liujj@cup.edu.cn

Academic Editor: Changzhi Wu

High accuracy of construction materials tracking with radio frequency identification technology (RFID) is challenging to achieve. The microzoning method consists essentially of an absolute received signal strength indication (RSSI) positioning algorithm on the basis of measuring the distance of tag from antennas base. In this paper, we analyse and examine the effects of microzoning method on the performance of RFID tags. A system was set up whereby RFID tags and antennas with the microzoning method were developed and studied. The performance of the tag antennas was studied with the practical read-range measurements. The study results showed that this absolute algorithm worked reliably and was suitable for RFID applications requiring identification of positions of onsite materials and components. The results also showed that the algorithm achieved a large read range and high accuracy. The study investigates the RFID solutions for Australian LNG (liquefied natural gas) industry and was initiated by the collaboration between Woodside Energy, Curtin University, and Industrial Automation Group Pty Ltd.

1. Introduction

Accurate location of LNG (liquefied natural gas) plant assets like piping, valve, and equipment is a major issue in Australian LNG industry. The current methods used to pinpoint the location of set-in-place assets are useful in varying degrees. Thus far, unfortunately, none of these methods gives the degree of accuracy. That is, the current methods are not able to provide accurate and comprehensive data on the location of each asset. For example, in LNG supply chain, it is normally a challenging task to efficiently identify, track, and position pipes as many of the present-day applications are still using the paper-based methods. The problems encountered in terms of the paper-based material tracking are getting late deliveries, missing components, and incorrect installations, which result in additional labor and material costs. In the current LNG practice, pipe installation activities at the construction site normally need workers to search paper instructions, on which the destination and delivery information are stored. Once the instructions are produced, they are transferred to the storage area and archived as the layout plan for assembly workers to retrieve information from. If the workers cannot locate the relevant information in time, an extended search needs to be carried out at the construction site to locate the missing component [1].

To address these issues, radio frequency identification technology (RFID) which uses radio frequency waves to transmit data between readers and tags is widely adopted. RFID automatically streamlines the identification and acquisition of data, without the need for direct contact between readers and tags. A typical RFID system includes an antenna, a transceiver (RFID reader), and a transponder (radio frequency tag) [2]. The antenna generates an electromagnetic zone where the tag detects the activation signal and responds by sending the stored data from its memory through radio frequency waves.

Types of RFID tags can be active, passive, or battery assisted passive (BAP) [3]. The active tags are equipped with built-in batteries and work on a "tag-talks-first" principle. Thus, their activation is not dependent on the RFID reader, and the tags periodically transmit their ID signal. The active tags can be detected over a long distance (>100 m) [4]. The passive tags, however, require an RFID reader to generate the activated electromagnetic field while the tags can only be detected in a short range (around 10–15 m maximum) [4]. The RFID tags used in the study belong to BAP, a new classification of RFID. BAP is recognized in a new international standard of RFID technology: ISO/IEC 18000-6:2010 Class 3 [5]. The BAP tags do not transmit their ID, but the tags "backscatter" the reader's signal with their ID. The battery in BAP is used only for powering the chip (and hence amplifying the backscattered signal). The BAP tags provide the same performance (>100 m) as the active tags but closer to the cost of the passive tags. Moreover, within 3-4 years the BAP technology does not need the battery to be replaced inside the tags, whereas in case of the active tags, the battery needs to be replaced frequently (order of days-weeks). Therefore the BAP tags are scalable in terms of cost and can be used throughout the entire LNG plant to cater for millions of materials and components search. Apart from storing tag IDs, BAP tags also support 64 kb rewritable memory in which users can read/write data.

The RFID reader can be portable or fixed. The portable readers come along with an antenna and can be held like a hand gun. The portable readers are also equipped with a wireless module through which they can communicate the tag information to the server. The fixed readers need to be connected to external patch antennas that can be mounted on walls or clamped on poles. The fixed readers communicate through an Ethernet connection to the server. Fiber optic cabling or Wi-Fi can be used for connecting hundreds of readers to a server. This study uses both portable and fixed readers to increase giving the flexibility of the RFID system.

2. Relevant Work

RFID has been widely deployed in variety of applications, such as logistics [6], mining [7], air cargo [8], hospitals [9], museums [10], retailing [11–13], and waste management [14]. RFID has recently attracted significant attention in construction areas such as material tracking (e.g., structural steel members), quality control, equipment monitoring and inspection, and maintenance [15–18].

Ren et al. [19] identify that poor materials management typically incurs low construction productivity, cost overrun, and delays. They further specify major contributors to such problems, namely, lack of active, accurate, and integrated information flow. Motamedi et al. [20] use permanently

attached tags to allow different users to share and handle lifecycle information.

Most of the research in investigating the technical capacity of RFID is focused upon the hardware study of tags and antennas such as RFID tags and antennae [21, 22]. There has been very few research on investigating the algorithm, methods, and theories on RFID implementation and evaluation that can be applied in practice for practitioners. Zhou and Shi [23] propose an algorithm based on a signal propagation model to get the accurate location information of objects. A multilateration method is formulated as the kernel of the algorithm to maximize the accuracy of distance measurements between RFID tags and readers. Song et al. [24] present a method to extend the use of current RFID technology in tracking the precise location of tagged materials on construction sites. The combination of RFID and GPS technologies, as evaluated through experimentation, presents the opportunity of densely deploying low cost RFID tags with GPS-supported RFID readers to track construction materials. Park and Hashimoto [25] present a novel method using the read time of only a few number of passive HD RFID tags without any external sensors, signal strength measurement, or a vision system. The experimental results show that their method offers a modular and cost-effective way for servicing mobile robots in indoor environment, enabling the synchronized locational and orientational estimation of robots. Li and Becerik-Gerber [26] emphasize the significance of indoor location information in improving the utilization and maintenance of facilities. They review 21 research projects where RFID-based indoor location sensing (ILS) solutions were applied in the context of algorithm design, devices, test setup, and performance evaluation. A summary on the use of the proximity method and the underlying rationales in RFID-based ILS is made, which indicates that no single solution satisfies the widespread deployment of RFID-based ILS.

Based on the RSSI, there are three better known localization algorithms, the lateration algorithm, the minimum maximum algorithm, and the ring overlapping circles algorithm [27, 28]. Papamanthou et al. [29] examine the RSSI measurement model for location estimation and provide the first detailed formulation of the probability distribution of the position of a sensor node. On the other hand, range-based algorithms make use of the RSSI to estimate the distance between nodes. Then, different techniques, such as lateration [30], triangulation [31], or statistical inference [32], are used to estimate the position of strayed nodes with respect to the beacons. Unfortunately, RSSI-based ranging is severely affected by errors due to the unpredictable radio propagation behavior. Hence a RSSI positioning algorithm with microzoning is presented in the next section.

3. RSSI Positioning Algorithm Based on Microzoning Method

This section presents and examines the microzoning method which consists essentially of absolute received signal strength indication (RSSI) positioning algorithm on the basis of measuring the distance of tag from antennas base. In this

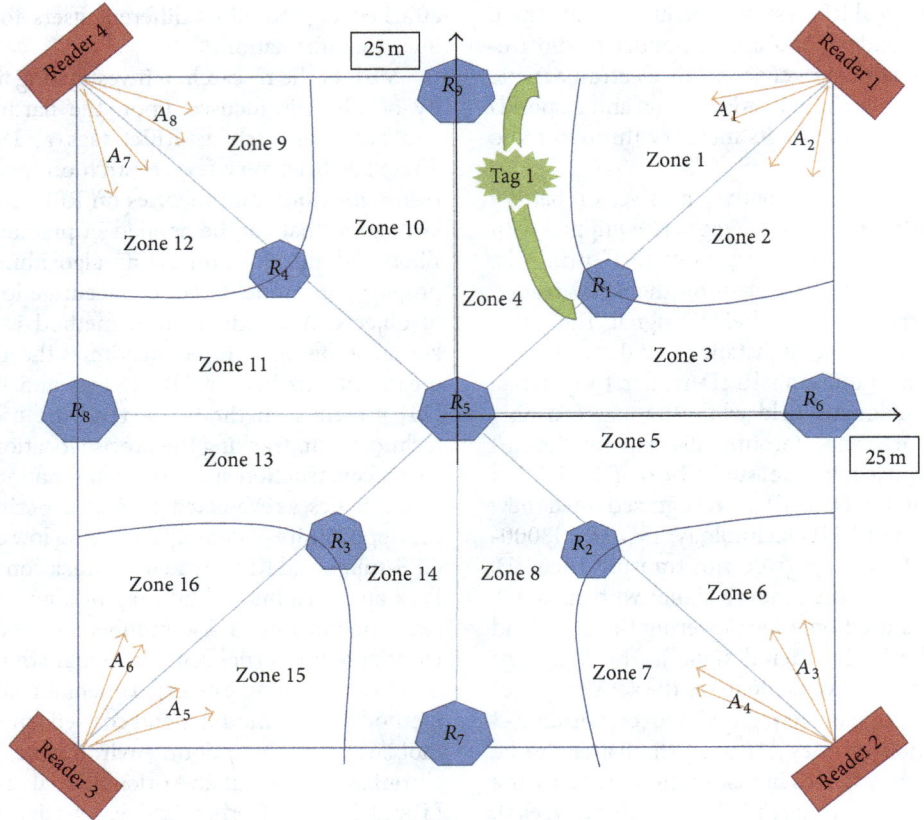

FIGURE 1: Demonstration of RSSI positioning algorithm based on microzones.

paper, we analyse and examine the effects of microzoning method on the performance of RFID tags. A system was set up whereby the RFID tags and antennas with the microzoning method were developed and studied. The performance of the tag antennas was studied with the practical read-range measurements. The aim of microzoning method is to locate the position of an RFID tag in the given area based on RSSI signals. In other words, this method was to measure the distance of tag from antennas using absolute RSSI. Within the algorithm, the position of the tag is estimated down to predefined microzones based on the reference tags. The signal strength of the tag is compared with that of the referenced tags, and therefore the relative position of the tag can be estimated. Figure 1 depicts the design of method and the layout of study using microzoning. A square region is divided into eight subzones, Zone ($i = 1, 2, \ldots 12$). Nine reference tags R_j ($i = 1, 2, \ldots 9$) are set in the square and four readers Reader_k ($k = 1, 2, 3, 4$) are fixed at four canners. A tag which needs to be located is set in one of the microzones (see Figure 1).

The algorithm will be described using following parameters and variables.

(A_{2k-1}, A_{2k}): antenna pair, $k = 1, 2, 3, 4$,

R_j: reference tag, $j = 1, 2, \ldots, 9$,

RA_j: RSSI of the (target) tag from antenna A_j,

$RA_{i,j}$: RSSI of the ith reference tag from antenna A_j,

(X, Y): position of the tag,

(x_i, y_i): position of the ith reference tag,

(x'_j, y'_j): position of the jth antenna,

dA_j, ($j = 1, 2, \ldots, 8$): Euclidean distance of the tag from the antennas using absolute RSSI

$$dA_j = \sqrt{\left(X - x'_j\right)^2 + \left(Y - y'_j\right)^2}, \tag{1}$$

dR_j, ($j = 1, 2, \ldots, 9$): Euclidean distance of the reference tag from the antennas using absolute RSSI

$$dR_j = \sqrt{\left(X - x_j\right)^2 + \left(Y - y_j\right)^2}. \tag{2}$$

The proposed algorithm is carried out in the following 4 stages.

Stage 1 (identify the quadrant). Identify which quadrant the tag belongs to by finding the reader with the best RSSI value from both the antenna pair by the following:

$$\text{RSSI}_k = \max\left\{RA_{2k-1} + RA_{2k}\right\}, \tag{3}$$

where $k = 1, 2, 3, 4$ and is corresponded to Zone k, respectively, for example, locating "Tag 1" as shown in Figure 1. The calculation of RSSI values of antennas is listed in Table 1. It is found that the tag is in Quadrant 1 (which includes

TABLE 1: The calculation of locating "Tag 1."

k	1	2	3	4
RA_{2k-1}	−30	−60	−100	−70
RA_{2k}	−40	−70	−90	−60
$RA_{2k-1} + RA_{2k}$	−70	130	190	−130
Quadrant	**1**	**2**	**3**	**4**

Zone 1, Zone 2, Zone 3, and Zone 4). Therefore, we depend on 2 antenna pairs to conclude if the tag is in one quadrant.

Stage 2 (identify subquadrant). Based on the differential RSSI readings from two antenna pairs in the identified quadrant, it can be found to which subquadrant the tag belongs to. For example, two antenna pairs in Quadrant 1 are A_1 and A_2. Since A_1's RSSI is greater than A_2's, the tag is in the subquadrant of A_1 (which includes Zone 1 and Zone 4).

Stage 3 (identify zone). Based on the relative readings between Tag 1 and the reference tags, it can be found which zone the tag belongs to. For example, suppose the RSSI of reference tags from antenna A_1 are as follows $RA_{r1,1} = −25$, $RA_{r1,9} = −40$, and $RA_{r1,5} = −45$; then it can be inferred that the tag is in Zone 4, that is, between R_1 and R_9–R_5.

Stage 4 (identify approximate distance). Based on the relative readings between the reference tags and the actual tag using the distance of the reference tags from the antenna, the approximate distance of a tag from the antenna can then be determined by (4)

$$d = 10^{(-B-RA)/10n}, \tag{4}$$

where RA denotes the RSSI value of the tag from antenna A, n is a signal propagation constant or exponent, d is a distance from the tag to the reference tag, and B is the received signal strength at 1 m distance.

For example, the RSSI of Tag 1 from antenna A_1 is −35, and the reference tag R_1 is at a distance 6 m. Then it can be inferred that Tag 1 is in Zone 4 at a distance of approximately 10 m from the antenna A_1 (green patch).

Then the approximate position coordinate (X, Y) of the tag can be identified by (1) and (2).

It should be noted that the accuracy of results from Stage 3 and Stage 4 is subject to the issues caused by null effect of radio signals. The null effect can be explained that due to wave nature of radio frequencies, a tag can show a low RSSI value than what it is expected to show at a certain distance. For example, if a tag is supposed to show −30 dbm for a distance of 20 meters, due to surface reflection, it might show −60 dbm for the same 20 meters. A potential solution to overcome the problem due to the null effect of radio signals can be by placing a second reader (antenna pair) a few meters above the planned reader (antenna pair) in the same direction; we could take the best RSSI signal out of two readers for the same tag. The following devices are used in the experiment.

(i) Intelleflex FMR reader × 4,

(ii) Intelleflex antenna 12″ patch antenna × 8 pair,

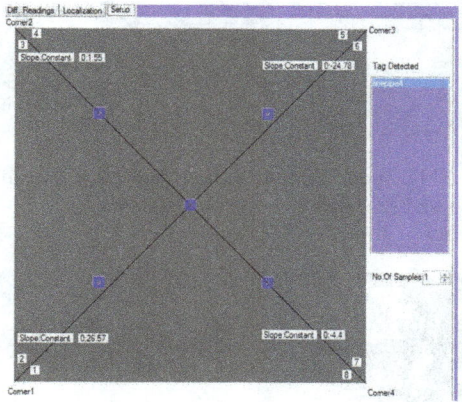

FIGURE 2: Graphical interface of the tracked items.

Tag	Location X	Location Y	Timestamp	Corner
linepipe4	18	6	05/04/2012 08:59:01 AM	corner1
linepipe4	18	6	05/04/2012 08:59:01 AM	corner2
linepipe4	18	6	05/04/2012 08:59:01 AM	corner3
linepipe4	18	6	05/04/2012 08:59:01 AM	corner4

FIGURE 3: RFID application on the computer that lists the location (X, Y) and timestamp of the tracked item.

(iii) Intelleflex general purpose tags × 10 (9 reference + 1 blind). Suggested tags for reference tags: nonmetal tags,

(iv) some snapshots: the following figures demonstrate an example of the graphical interface of the tracked items (Figure 2), the RFID application on the computer that lists the location (X, Y) and timestamp of the tracked item (Figure 3), and RFID application on the computer that lists differential RSSI, angle, and timestamp of the tracked item (Figure 4).

4. Experimentation

Field experimentation was conducted in this paper. The test setup, configuration, and conclusion for the test are discussed in this section. The aim of the experiment is to find the maximum range of tag detection and to find the relationship between absolute RSSI (signal strength) and distance of tag from antennas. Various factors affecting the relation between signal and distance are listed as follows:

(i) height of tag,

(ii) angle of tag,

(iii) multipath loss or gain.

Testing sites are as follows:

(1) Packard Street, Joondalup,

(2) Elcar Park (see Figure 5),

(3) Heathridge Park (see Figure 6).

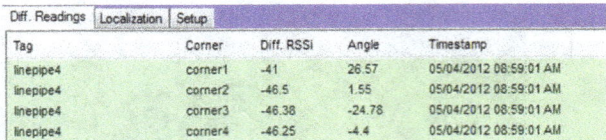

Diff. Readings	Localization	Setup			
Tag		Corner	Diff. RSSI	Angle	Timestamp
linepipe4		corner1	-41	26.57	05/04/2012 08:59:01 AM
linepipe4		corner2	-46.5	1.55	05/04/2012 08:59:01 AM
linepipe4		corner3	-46.38	-24.78	05/04/2012 08:59:01 AM
linepipe4		corner4	-46.25	-4.4	05/04/2012 08:59:01 AM

FIGURE 4: RFID application on the computer that lists differential RSSI, angle, and timestamp of the tracked items.

FIGURE 5: Elcar Park.

Equipment used is as follows:

(i) Intelleflex reader: 1,

(ii) 12″ patch antenna: 2 pairs,

(iii) metal Mounting Tag: 1 for location purpose,

(iv) custom application software that would query two antenna ports and log the results in a csv file with the below given format "tagid, description, port, signal strength."

Test setup is as follows:

(i) two pairs of antennas are connected to a single reader. The first pair of antennas is fixed at 2.5 m above the ground, and the second pair of antenna is fitted 3.5 m above the ground,

(ii) all the four antennas are pointed in a single plane, in same direction, perpendicular to the ground,

(iii) the metal tag used for location is mounted on a metal plate and held at a height of 1.5 m on a steel pole.

5. Data Collection and Results Analysis

In the test site 1, the tag is read along over a linear distance every 10 m by two pairs of antennas with one pair at 2.5 m height and the other at 3.5 m height. According to Figure 7, this graph shows that the tags were detected up to a range of 130 m in the centre line. When the bottom antenna goes to a null at 30 m, the top antenna provides a reliable reading; when the top antenna goes to a null at 40 m, the bottom antenna provides a reliable reading. Thus a reliable detection and signal strength can be obtained from nearly all the points by using spatial diversity.

FIGURE 6: Heathridge Park.

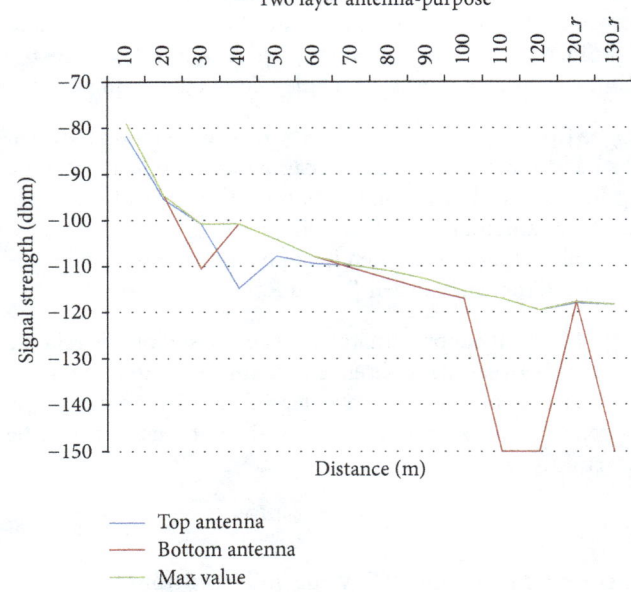

FIGURE 7: Long range detection at test site 1.

This test was conducted over a sealed road with objects (cars) on the sides. The RFID tags were detected up to a range of 120 m in the centre line. It is possible that the objects on the side of the road could serve as a magnifying tunnel and help increase the range of detection. Signal magnification caused due to objects presented along the signal path may boost the detection range.

In the test site 2 with dual layer antenna test, tag moved along a straight line from 10 m to 40 m, 2 m at a time. According to Figure 8, the signal strength does not vary consistently with the distance for one antenna pair. Also, there are various nulls, which indicate that the tags may not be detected in many instances. However, by considering the maximum of two antenna values, tags can be detected and valid signal strength be measured at all instances.

Tag moved along centre line from 10 m to 100 m, 10 m at a time. According to Figure 9, it is observed that the tags were not detected between 50 m and 90 m. After analyzing the park with the help of a spectrum analyzer, the major reason for this is traced back to the presence of strong radio interference in the region, due to a cell phone tower in the very close

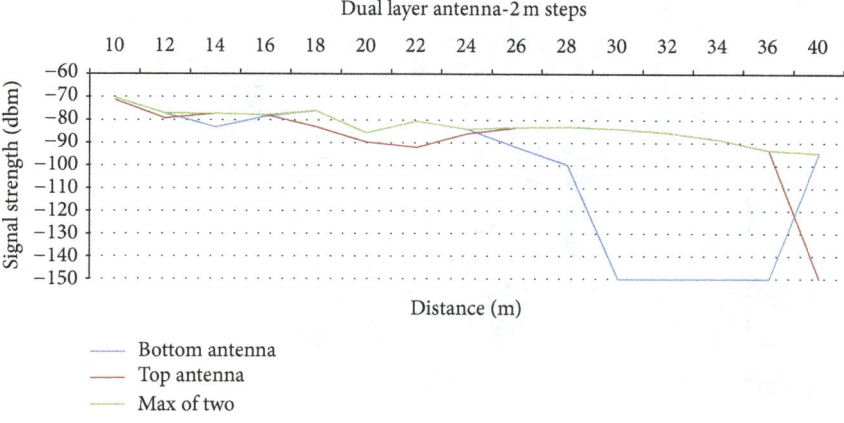

FIGURE 8: Dual layer antenna test −2 m step.

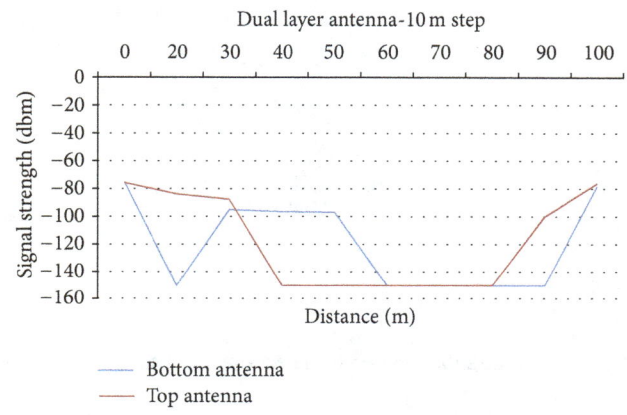

FIGURE 9: Long range detection at venue 2.

vicinity of the test site. Since the cell phone tower was sending signals in 950 MHz frequency range whose strength was as strong as −30 dBm, they were causing interference to the RFID system.

In test site 3, according to Figure 10, the bottom antenna shows reliable readings; however, the top antenna does not show a reliable reading due to the presence of multiple nulls. Thus spatial diversity of antennas will help reliably detect tags at all distances. Location 1 had object influence and Location 2 had radio frequency interference; therefore Location 3 was chosen to be free from both influences. An open park was verified free of radio interferences with the help of a spectrum analyzer and used for rest of the study. In this location, a range of 110 m was achieved.

6. Conclusions

This paper analysed and examined the effects of micro-zoning method on the performance of RFID tags. This experimental design aimed at demonstrating the proof of concept in a specified area of zone. The experimentation was conducted based on the proposed design and the test results were promising and showing that the proposed tag positioning (triangulation) concept is feasible (e.g., tags were detected up to a range of 130 m in the centre line). It is observed that the error is much less for tags closer to the centre and slightly higher for tags closer to the edge. Still the error is found to be considerably less than anticipated. The test results were positive with very low error rate. It is proved that spatial diversity of antennas (two layers of antenna, at different heights) could help reliably detect tags at all distances (up to the maximum range) to overcome the loss in signal strength due to multipath loss.

A significant discovery from Trials 1, 2, and 3 is the presence of "nulls" in the radio propagation, due to multipath loss. As a result of nulls, the RSSI value drops considerably even in short distances. After analyzing the test results, it was confirmed that absolute RSSI cannot be used for calculating the distance of a tag from an antenna.

Further tests need to be conducted to test different tag positions at different locations in square area. Moreover, tests should also be conducted with tags at various heights from the ground. As all these studies have been conducted with no obstacles in place at the current stage, the major work in future may focus on studying how this concept is applicable when obstacles are introduced in place. A lot of challenges

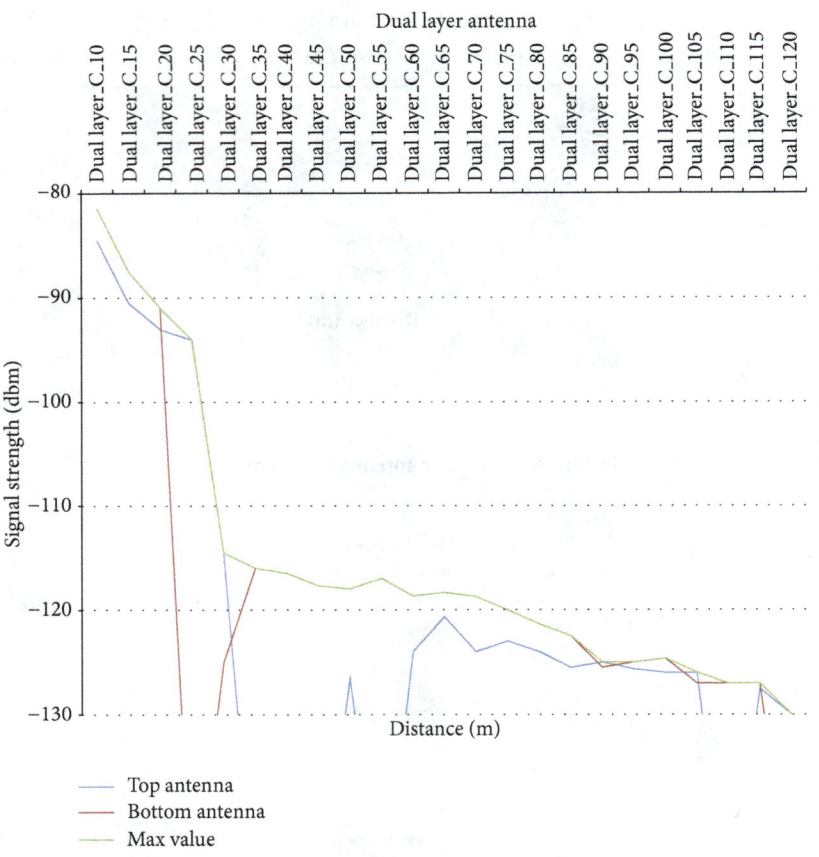

FIGURE 10: Long range detection at test site 3.

are expected in a larger zone as the accuracy of RFID system is expected to drop considerably with the distance increase. Moreover, a larger region may have a considerable path loss and hence it may be necessary to use spatial diversity (two layers of antennas). For integrating the four X and Y coordinates of a tag that were derived from four different corners (rather than averaging), more sophisticated methods like statistical analysis or information theories can be used to get better results.

Conflict of Interests

The authors declare that there is no conflict of interests regarding the publication of this paper.

Acknowledgments

The work presented in this paper was initiated by Mr. Martijn Truijens, the Lean Construction Technology Advisor, Woodside Energy Limited (WEL), as part of the Project Echo Construction Research initiative. Also, part of this research was supported under Australian Research Council Linkage Project scheme (project name: Transforming LNG Plant Construction Productivity through Mobile Computing, Project no. LP130100451).

References

[1] G. Demiralp, G. Guven, and E. Ergen, "Analyzing the benefits of RFID technology for cost sharing in construction supply chains: a case study on prefabricated precast components," *Automation in Construction*, vol. 24, pp. 120–129, 2012.

[2] J. F. Alicot, S. R. Maitin, R. F. Devoe, and D. E. Jones, "Distributed Radio Frequency Identification Reader," EP Patent 2, 070, 000, 2011.

[3] A. M. Barker, G. R. Hanson, A. K. Sexton, J. Jones, E. B. Freer, and A. L. Sjoreen, "An active RFID Accountability System (RAS) for constrained wireless environments," in *Proceedings of the Future of Instrumentation International Workshop (FIIW '11)*, pp. 25–27, IEEE, 2011.

[4] S. Nainan, R. Parekh, and T. Shah, "RFID technology based attendance management system," *Evolution*, vol. 10, p. 11, 2013.

[5] Intelleflex Corporation, "White Paper on 9 Key Points of Differentiation, Comparing Intelleflex Technology and Active RFID," 2011.

[6] R. Angeles, "RFID technologies: supply-chain applications and implementation issues," *Information Systems Management*, vol. 22, no. 1, pp. 51–65, 2005.

[7] T. M. Ruff and D. Hession-Kunz, "Application of radio-frequency identification systems to collision avoidance in metal/nonmetal mines," *IEEE Transactions on Industry Applications*, vol. 37, no. 1, pp. 112–116, 2001.

[8] Y. S. Chang, M. G. Son, and C. H. Oh, "Design and implementation of RFID based air-cargo monitoring system," *Advanced Engineering Informatics*, vol. 25, no. 1, pp. 41–52, 2011.

[9] X. Qu, L. T. Simpson, and P. Stanfield, "A model for quantifying the value of RFID-enabled equipment tracking in hospitals," *Advanced Engineering Informatics*, vol. 25, no. 1, pp. 23–31, 2011.

[10] S. Hsi and H. Fait, "RFID enhances visitors' museum experience at the Exploratorium," *Communications of the ACM*, vol. 48, no. 9, pp. 60–65, 2005.

[11] J.-L. Hou and T.-G. Chen, "An RFID-based Shopping Service System for retailers," *Advanced Engineering Informatics*, vol. 25, no. 1, pp. 103–115, 2011.

[12] P. Jones, C. Clarke-Hill, D. Comfort, D. Hillier, and P. Shears, "Radio frequency identification in retailing and privacy and public policy issues," *Management Research News*, vol. 27, pp. 46–56, 2004.

[13] B. Eckfeldt, "What does RFID do for the consumer?" *Communications of the ACM*, vol. 48, no. 9, pp. 77–79, 2005.

[14] S. Abdoli, "RFID application in municipal solid waste management system," *International Journal of Environmental Research*, vol. 3, no. 3, pp. 447–454, 2009.

[15] W. C. Stone, L. Pfeffer, and K. Furlani, "Automated part tracking on the construction job site," in *Proceedings of the 4th International Conference and Exposition/Demonstration on Robotics for Challenging Situations and Environments*, pp. 96–103, March 2000.

[16] E. J. Jaselskis and T. El-Misalami, "Implementing radio frequency identification in the construction process," *Journal of Construction Engineering and Management*, vol. 129, no. 6, pp. 680–688, 2003.

[17] J. Yagi, E. Arai, and T. Arai, "Parts and packets unification radio frequency identification (RFID) application for construction," *Automation in Construction*, vol. 14, no. 4, pp. 477–490, 2005.

[18] X. Wang, P. E. D. Love, M. J. Kim, C. S. Park, C.-P. Sing, and L. Hou, "A conceptual framework for integrating building information modeling with augmented reality," *Automation in Construction*, vol. 34, pp. 37–44, 2013.

[19] Z. Ren, C. J. Anumba, and J. Tah, "RFID-facilitated construction materials management (RFID-CMM)—a case study of water-supply project," *Advanced Engineering Informatics*, vol. 25, no. 2, pp. 198–207, 2011.

[20] A. Motamedi, R. Saini, A. Hammad, and B. Zhu, "Role-based access to facilities lifecycle information on RFID tags," *Advanced Engineering Informatics*, vol. 25, no. 3, pp. 559–568, 2011.

[21] C. Cho, H. Choo, and I. Park, "Broadband RFID tag antenna with quasi-isotropic radiation pattern," *Electronics Letters*, vol. 41, no. 20, pp. 1091–1092, 2005.

[22] J. R. Smith, K. P. Fishkin, B. Jiang et al., "RFID-based techniques for human-activity detection," *Communications of the ACM*, vol. 48, no. 9, pp. 39–44, 2005.

[23] J. Zhou and J. Shi, "A comprehensive multi-factor analysis on RFID localization capability," *Advanced Engineering Informatics*, vol. 25, no. 1, pp. 32–40, 2011.

[24] J. Song, C. T. Haas, and C. H. Caldas, "A proximity-based method for locating RFID tagged objects," *Advanced Engineering Informatics*, vol. 21, no. 4, pp. 367–376, 2007.

[25] S. Park and S. Hashimoto, "An intelligent localization algorithm using read time of RFID system," *Advanced Engineering Informatics*, vol. 24, no. 4, pp. 490–497, 2010.

[26] N. Li and B. Becerik-Gerber, "Performance-based evaluation of RFID-based indoor location sensing solutions for the built environment," *Advanced Engineering Informatics*, vol. 25, no. 3, pp. 535–546, 2011.

[27] X. Nguyen and T. Rattentbury, "Localization algorithms for sensor networks using RF signal strength," Tech. Rep., University of California at Berkeley, 2003.

[28] I. F. Akyildiz, W. Su, Y. Sankarasubramaniam, and E. Cayirci, "Wireless sensor networks: a survey," *Computer Networks*, vol. 38, no. 4, pp. 393–422, 2002.

[29] C. Papamanthou, F. P. Preparata, and R. Tamassia, "Algorithms for location estimation based on rssi sampling," in *Algorithmic Aspects of Wireless Sensor Networks*, pp. 72–86, Springer, Berlin, Germany, 2008.

[30] A. Savvides, H. Park, and M. B. Srivastava, "The n-hop multilateration primitive for node localization problems," *Mobile Networks and Applications*, vol. 8, no. 4, pp. 443–451, 2003.

[31] A. Savvides, H. Park, and M. B. Srivastava, "The bits and flops of the n-hop multilateration primitive for node localization problems," in *Proceedings of the 1st ACM International Workshop on Wireless Sensor Networks and Applications*, pp. 112–121, ACM, September 2002.

[32] N. Patwari, A. O. Hero III, M. Perkins, N. S. Correal, and R. J. O'Dea, "Relative location estimation in wireless sensor networks," *IEEE Transactions on Signal Processing*, vol. 51, no. 8, pp. 2137–2148, 2003.

22

Comparative Study on the Performance of Blended and Nonblended Fly Ash Geopolymer Composites as Durable Construction Materials

Debabrata Dutta ⓘ **and Somnath Ghosh**

Department of Civil Engineering, Jadavpur University, 188 Raja S.C. Mallick Road, Kolkata 700032, India

Correspondence should be addressed to Debabrata Dutta; ddebabrata83@gmail.com

Academic Editor: Robert Černý

This article represents that the mechanical and microstructural properties and durability of fly ash-based geopolymers blended with silica fume and borax are better than those of conventional fly ash-based geopolymers. Fly ash itself contains the sources of silica and alumina which are required for geopolymerisation. But a sufficient amount of high-reactive silica is able to rapidly initiate geopolymerisation with activation. Pure potassium hydroxide pellets and sodium silicate solution were used for preparation of alkaline activator solution. Fly ash geopolymer paste exhibited better mechanical properties in the presence of silica fume with slight portion of borax. The effect of silica fume-blended geopolymer paste on temperature fluctuation (heating and cooling cycle at certain temperatures) showed better performance than nonblended fly ash-based specimens. Durability property was evaluated by immersion of geopolymer specimens in 10% magnesium sulfate solution for a period of one year. The change in weight, strength, and microstructure was studied and compared. In the magnesium sulfate solution, a significant drop of strength to around 37.26% occurred after one year for nonblended fly ash-based specimens. It is evident that specimens prepared incorporating silica fume had the best performance in terms of their properties.

1. Introduction

Numerous studies have already been done on fly ash-based geopolymers, as fly ash contains large amounts of silica and alumina. It is also clear that the reaction process basically involves alkaline activation of the source material by alkali hydroxide and sodium silicate solution followed by heat curing [1]. Again, this material exhibits good strength and durability when compared to conventional concrete [2]. The geopolymer chemistry elaborates the formation of 3D polymeric chain by Si–O–Al–O bond during the reaction between the source material rich in silica and alumina [3]. But the chemistry of geopolymers is altered for different combinations of raw material and alkalis [4]. The higher viscosity of the activator interferes with the rheological characteristics of the geopolymers. Thus, the porosity of geopolymers increases for finer precursor materials [5]. Inclination is needed towards several disposals over fly ash as the supplemental material in

geopolymers like silica fume. Silica fume is a very fine material (particle size ranging from $1\,\mu$m to $15\,\mu$m). The basic problem in using silica fume as a base material for geopolymers is the absence of aluminium (trivalent in character). This phenomenon may be overcome with two different alternatives. The primary one is the choice of an activator, and the second one is the intrusion of secondary material. The choice of an activator is subjected to specific chemistry. Higher presence of monomers and dimers which exists during dissolution of Al–Si appreciates sodium hydroxide for better stabilization. Again, for larger silicate oligomers, potassium hydroxide is preferred for better coordination [6]. Latest research exhibits typical geopolymer varieties from simple to multiple phases. In the present research, silica fume was incorporated in fly ash-based geopolymers. Commercial borax (like substitute of aluminium) was used as another secondary input in the mixture. Borax arises in nature as evaporated dump formed by the continual evaporation of seasonal lakes. The effectiveness of the pioneer composite was studied by comparing typical parameters like strength and durability with traditional fly ash-based geopolymers.

2. Chemistry of the Study

Fly ash activated with sodium hydroxide sometimes creates cracks with aging [7]. It is because of continuous pore pressure developed within the hardened composite by the late precipitated alkali compound [7]. However, at the initial level, it shows successive strength gaining for the time being [7]. On the contrary, potassium hydroxide is better for developing stable structures only under higher concentration of sodium silicate [8]. Earlier studies depict that despite having the same electric charges, the Na^+ and K^+ act in a different way because of dissimilar size. The smaller cation favors the ion-pair reaction with the smaller silicate oligomers, like silicate monomers, dimers, and trimers [9–11]. Xu et al. observed that the smaller silicate oligomers like monomers and dimer are better stabilized by Na^+ (sodium ion) and result in higher extent of dissolution, whereas more silicate solution emphasizes larger silicate oligomers which can be better coordinated by K^+ cation of larger size [6]. The previous study showed stable compressive strength for fly ash-based geopolymers activated with potassium hydroxide with higher concentration of sodium silicate [12]. The major drawback of this geopolymer composite is the rise in water content in the mixture. The higher amount of sodium silicate in fact allows the presence of additional water embedded within it. Consequently, additional water makes the structure more porous, lesser in weight, and mainly permeable to some extent. Based on the discussed theory, a suitable compensator of sodium silicate has been evaluated in this study where silica fume has been introduced as the primary source of reactive silica to initiate the faster reaction. The presence of a higher oligomer in the chemical environment is favorable for the purpose of using potassium as the charge compensator or system stabilizer. Geopolymer is a Si–O–Al–O tetrahedral frame structure. But rise in the ratio of Si to Al may insist the formation of a chain-like structure over frame structure. So, another secondary input borax (B) has been typically applied in a manner to maintain the Si/X ratio, where X = Al or B, considering that both have three valence electrons. Another cause of choosing borax in this study is specifically to enhance the X content without allowing further incorporation of additional silica.

3. Experimental Procedure

3.1. Materials. The materials used in this research were class F fly ash (ASTM C618) produced by Kolaghat Thermal Power Plant, India; silica fume supplied by Oriental Trexim Pvt. Ltd., India; and commercial borax collected from DRD Educational & Consultancy Pvt. Ltd., India. In fly ash, 78% of the particles were finer than 45 microns with Blaine's specific surface area equal to 380 m^2/kg. The BET surface area of silica fume was 18,900 m^2/kg. Commercial borax had specific gravity and BET surface area equal to 1.7 and 557 m^2/kg, respectively. Potassium hydroxide pellets and

sodium silicate solution were collected from Loba Chemie Limited, India.

3.2. Specimen Preparation. Alkaline activator was prepared by dissolving potassium hydroxide pellets directly into water. The dissolved hydroxide pellets were left at room temperature for 24 hrs. After that, predetermined quantity of sodium silicate solution was added 3 hours before being used for mixing as done earlier [13]. In the activator solution, the K_2O content was maintained at 6% of the total base material plus supplementary material (fly ash, silica fume, and borax). Apparent silicate modulus (SiO_2/K_2O) was kept as 1 and 0.1. Originally, silicate modulus in the activator has to be calculated from the formulation ($SiO_2/(K_2O + Na_2O)$). In this case, apparent silicate modulus is taken earlier for easier calculation. It is already described that Na^+ is essentially required to promote monomers, dimers, and trimers in the starting of activation. Here, a slight amount of sodium silicate solution was added in some cases only to confirm the presence of Na^+ in the mixture. The mixture was prepared in the model P660 Hobart mixer of capacity 600 cc and speed 60 cps. Fly ash with activator solution was mixed for 5 minutes. For blended mixtures, fly ash with supplementary material was mixed with activator solution for 5 minutes. The mixture was then transferred into a cubical mould and subjected to a table vibration for 2 minutes to expel entrapped air. After 60 minutes of rest period in open air, the cubes were cured in a hot air oven for a period of 48 hours at 85°C and allowed to cool inside the oven after that. The scanning electron micrographs and chemical composition of fly ash, silica fume, and borax are represented in Figure 1 and Table 1, respectively. The mix proportions of six sample mixes are elaborated in Table 2.

3.3. Test Details

3.3.1. Strength Test. Compressive strength of the specimens was tested by digital compression testing machine of model number EM500 supplied by Enkay Enterprise. The least count was 0.001 kN. The compressive strength of cube specimens was done as per ASTM C109.

3.3.2. Exposure Test in Sulfate Solution. Experimental setup was prepared to investigate the performance of blended and non-blended fly ash-based geopolymer samples in 10% concentration of magnesium sulfate solutions for one year. The test specimens (cubical) were immersed vertically in a glass pan. After immersion, the water level was maintained at 4 cm-5 cm over specimens. Throughout the exposure, regular investigations on physical appearance, residual strength, and weight changes were monitored at preselected intervals.

3.3.3. Physical Changes and Optical Microscopy. At preset intervals, the exposed geopolymer specimens were removed from the solutions and observed for any remarkable changes in its physical appearance. A crack detection microscope WF 10x, manufactured by C&D (Micro services) Ltd. (U.K.) was

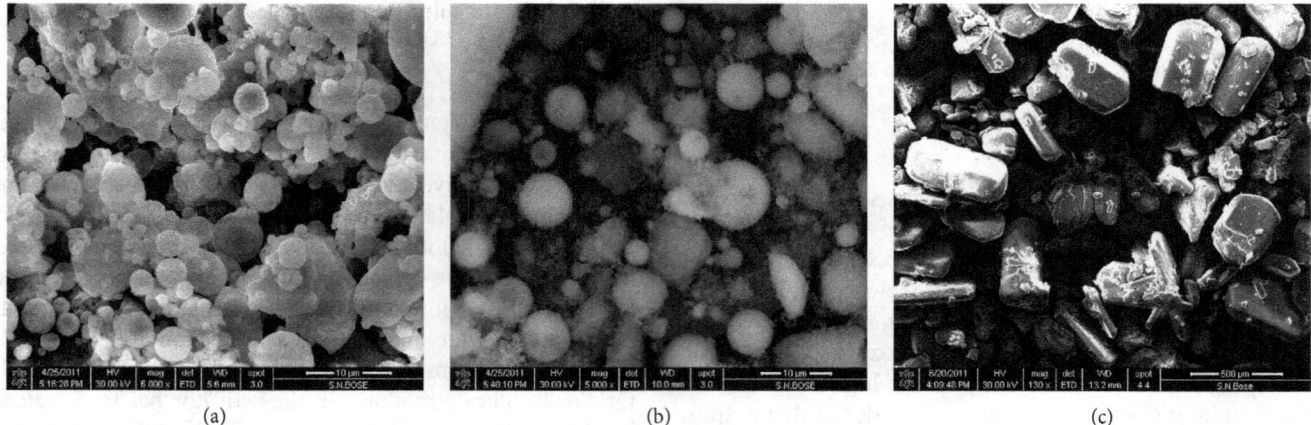

FIGURE 1: Scanning electron microscopic images of (a) fly ash, (b) silica fume, and (c) borax.

TABLE 1: Chemical composition of different raw materials (quantity in %).

Chemical composition	SiO_2	Al_2O_3	Fe_2O_3	TiO_2	CaO	MgO	K_2O	Na_2O	SO_3	P_2O_5	B_2O_3	LOI (%)
Fly ash	56.01	29.8	3.58	1.75	2.36	0.30	0.73	0.61	Nil	0.44	Nil	0.44
Silica fume	92.00	0.46	1.60	Nil	0.29	0.28	0.61	0.51	0.19	Nil	Nil	1.00
Borax	Nil	Nil	1.0	Nil	Nil	Nil	Nil	17.0	1.0	Nil	59.00	22.08

TABLE 2: Mix proportions of geopolymer mixes.

Sample ID	Fly ash* (%)	Silica fume* (%)	Borax* (%)	K_2O content in activator (%)	SiO_2 content in activator (%)	Water* (%)
SMF1	100	0	0	6	6	33
SMF2	100	0	0	6	0.6	33
SMS1	90	10	0	6	6	33
SMSB1	90	7.5	2.5	6	6	33
SMSB2	90	5.0	5.0	6	6	33
SMSB3	90	5.0	5.0	6	0.6	33

*This is the percentage amount with respect to the total weight of fly ash, silica fume, and borax. Calculated as the % of (fly ash + silica fume + borax); measured by weight.

used to detect the surface changes of the specimens at preset intervals throughout the exposure period. Observations on unexposed samples were conducted for making comparison with exposed samples.

3.3.4. Change in Weight. Before immersion in saline water, every specimen was kept submerged in potable water for 1 hour. After that, the samples were taken out and weighed. This measured weight value was indicated as the primary weight for individual. Specimens were weighed after every preselected interval. Before weighing, the specimens were clothed and brushed out to remove the free water and needle-like outcome (if any). Here, simple scrub brushes and cotton cloth were used for cleaning. Every specimen was brought to a saturated surface dry condition by applying mild airflow over the specimens for 5 minutes. Digital electronic balance of least count equal to 0.001 gm was used to conduct weighing. Change in weight indicates the percentage change (increment or decrement) with respect to the primary data at different intervals.

3.3.5. Change in Strength and Residual Strength. The average strength of any defined series of sample before the first immersion in the setup was treated as the primary strength value for specimens. A set of ten samples of different series were subjected to strength test at preselected intervals. Before testing, the specimens were kept at room temperature under airflow for 24 hrs. Digital electronic balance of least count equal to 0.001 gm was used to conduct weighing. The percentage change in strength (+/−) with respect to the primary strength value at different intervals is evaluated. The residual strength indicates the percentage of strength achieved at different intervals to the primary strength value. The residual strength at any interval is expressed as "100 − percentage change in strength."

3.3.6. Thermal Fluctuation Setup. A chest freezer of Model No. VT3-NUCAB 400L with glass-top configuration supplied by Hindustan Unilever, India, has been used in this study. The temperature fluctuating domain is limited from 8°C to −20°C. Each specimen was subjected to 20 consecutive

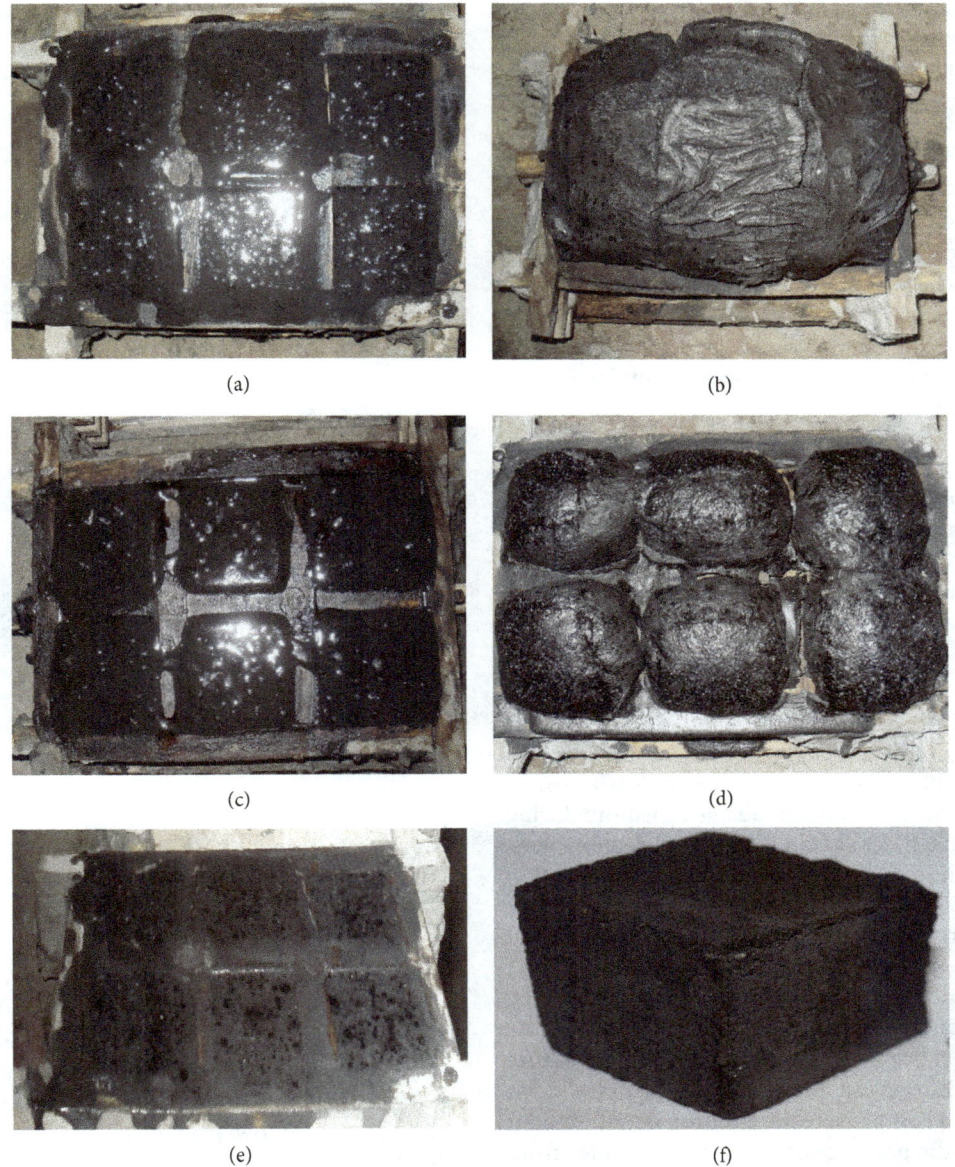

(a)

(b)

(c)

(d)

(e)

(f)

FIGURE 2: Physical appearance of the samples: (a, b) SMS1, (c, d) SMSB1, and (e, f) SMSB2.

cycles for a program of almost 85 hours. At the initial phase, the specimens were allowed to cool at 8°C for 2 hours. After that, it was brought to −20°C for the immediate next 2 hours. The temperature change rate was reported almost as 2°C/minute.

3.3.7. Scanning Electron Microscopy.
Scanning electron microscopy was conducted by QUANTA 2000 with a capacity of 2.4 nm at 30 kV at high vacuum condition. The geopolymer sample was collected as a small scrap form to conduct this study. Samples with irregular shape were collected at the time of crushing. For the research, scraps were collected from the inner part of the specimen under consideration. No further grinding or polishing was done for SEM analysis.

3.3.8. Mercury Intrusion Porosimetry.
MIP samples were made by cutting a cylinder of ¼ in. diameter to ½ in. height,

having a bulk sample volume of 1.00 cc, which were tested on Micromeritics Autopore II (Central Glass and Ceramic, India) from 0 to 60,000 psi, with Hg surface tension 480.00 erg/cm^2 and contact angles (I) 140.00° and (E) 140.00°. MIP was used to examine a statistical comparison of the tested samples in terms of mean and median pore size, pore distribution, total porosity, bulk density, and apparent density. The bulk volume of each test specimen was 1 cc, and the maximum applied intrusion pressure during the test was about 53,500 psi. In this method, mercury is intruded under pressure in an evacuated sample, and volume of intruded mercury is monitored against pressure. In the feature of this very test, the pore volume distribution over pore diameter is presented as a distribution function $F = -dV/d\log D$, where V is the collective pore volume and D the diameter of pores. It indicates that the part under

TABLE 3: Compressive strength at 3 days after heat curing.

Number	Sample ID	Compressive strength (MPa) at 3 days after heat curing
1	SMF1	14.25
2	SMF2	0.0
3	SMS1	—
4	SMSB1	18.62
5	SMSB2	29.05
6	SMSB3	28

a function of any pore diameter range capitulates pore volume of pores in that range.

3.3.9. Energy Dispersive X-Ray Analysis. Energy dispersive X-ray analysis was carried on a point of selected samples for SEM under QUANTA FEG 250. The geopolymer sample was collected as a small scrap form to conduct this study.

4. Results and Discussion

4.1. Physical Appearance. Excessive volumetric increment was observed for sample SMS1 (Figure 2). As silica fume is very reactive, it emphasizes the formation of dihydrogen by oxidation of free silicon by water of alkaline medium during synthesis [14]. In alkali-activated fly ash-based geopolymers, the purpose of sodium silicate is to start the polymerization at the earlier stage [13]. But, for blended geopolymers, silica fume helps the formation of in situ inorganic foam itself at the initial stage [14]. The volumetric enlargement of SMS1 was due to the increment in dihydrogen production by water in the presence of reactive silica fume. Further, the addition of external borax can better stabilize the structure macroscopically. Borax can play the role of alumina in a similar manner to compensate the additional requirement of alumina due to the presence of much reactive silica from silica fume.

4.2. Compressive Strength. At hardened state, compressive strength is considered as the characteristic material value for the identification of the structural feature. Table 3 presents the mean compressive strength of the six samples measured in digital compressive strength testing equipment. In this study, the noticeable feature was visualized in absence of sodium silicate for fly ash-based geopolymer SMF2. Figure 3 shows the disintegrated part of partially reacted geopolymer SMF2. Sample SMSB1 had almost a shape of fungus mushroom, where a swelled top has risen upon the perfect cube base. The swelled portion was cut out by using an electrically operated low-speed concrete saw cutter to obtain a perfect cube. Excessive swelling was observed for sample SMS1. Extrication of any right cubical unit from the honeycombed, asymmetrical, and distorted outcome of sample SMS1 was not possible. Because of that, sample SMS1 was not supportive for the measurement of compressive strength in test setup.

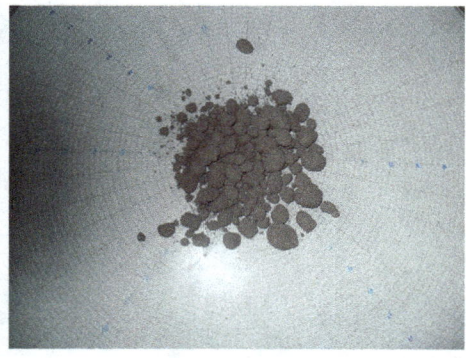

(a)

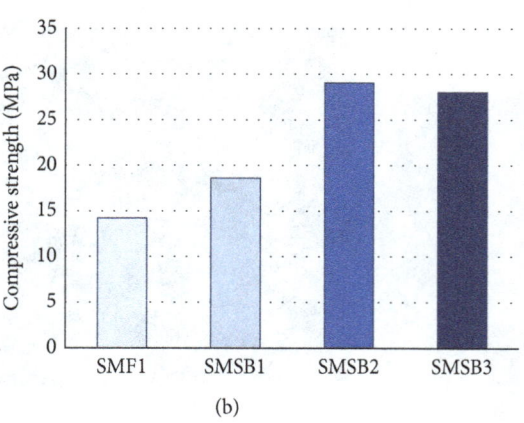

(b)

FIGURE 3: (a) Sample SMF2 after mixing and (b) compressive strength at the third day after curing.

Fly ash activation in the presence of external silica fume and borax sources yields a better geopolymer. Successive betterment in compressive strength was observed with the incorporation of silica fume and borax in certain percentage in the composition. Another important finding was that lowering of sodium silicate in mixing had little to no effect in polymerization or condensation process of silica fume-blended geopolymers (as treated for SMSB3). Maximum compressive strength was obtained for SMSB2 (29.05 MPa).

4.3. Thermal Fluctuation Effect. For a hardened geopolymer, it can be assumed that the most thermal sensitive part within the geopolymer composite is the crystalline alkali precipitation. In Figures 4(a) and 4(b), the regular crystal structure within the pore indicates the presence of alkali. Point A indicates alkaline surplus within the pores of the geopolymer body for fly ash-based geopolymers. In Figure 5, EDAX shows the existence of sodium-based alkaline precipitation, at point A. This entrapped crystal compound is basically due to the late and less reaction. In fact, alkali hydroxide is essential for the dissolution of silica and alumina. The metallic cations (Na, K, and others) maintain the structure neutrality as aluminium has fourfold coordination. But, alkali metal hydroxide which acts as a catalyst expelled out during the hardening of the gel phase with the progression of reaction [15]. In this research, it is supposed that incorporation of

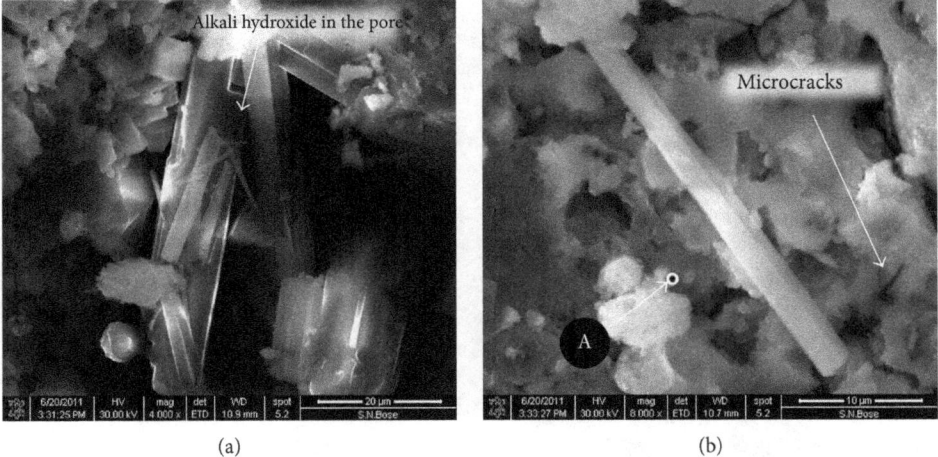

(a)   (b)

FIGURE 4: (a) Crystal structure within SMF1. (b) After fatigue exposure, microcracks in SMF1.

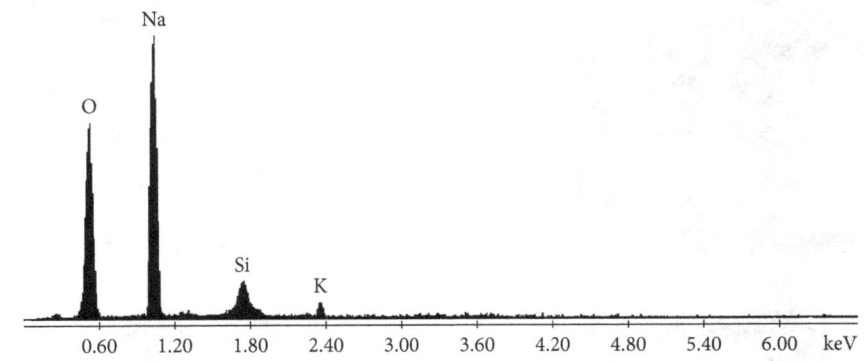

FIGURE 5: EDAX analysis at the position near A.

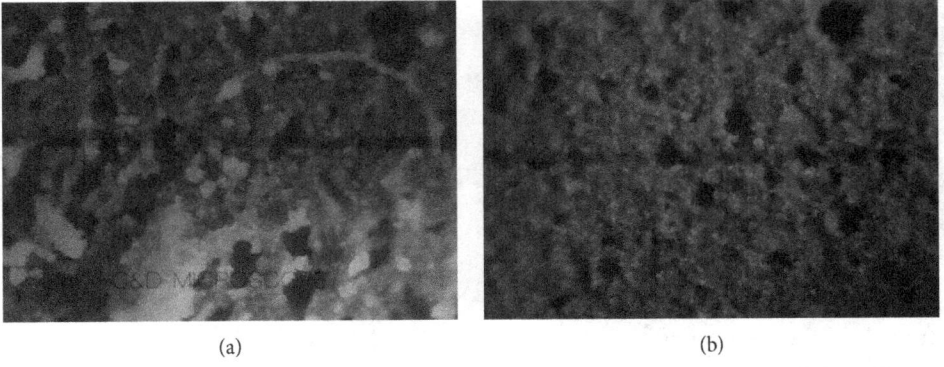

(a)   (b)

FIGURE 6: Optical microscopy (10x magnification). Sample SMF1 (a) and sample SMSB2 (b) after 30 days.

silica fume emphasizes the speed of reaction which allows the expulsion of metal hydroxide instantaneously. The existence of the entrapped alkaline entity in the nonblended composites after hardening possibly resulted from the slower rate of reaction. Now, the temperature fluctuation affects the volumetric change of alkali solution within the pore. This can cause remarkable fatigue on the geopolymer skeleton and successive crack formation (as shown in Figure 4(b)). The result shows failure for sample SMF1 at the end of temperature fluctuating cycles.

4.4. Efflorescence Behavior (Study at Micro- and Macrolevel). Geopolymer efflorescence is the outcome from the geopolymer body which is highly alkaline. The charge compensator alkali hydroxide basically extrudes from the pores with the process of synthesis or the formation of Si–O–Al. Basically, alkali metal hydroxide acts as a catalyst, but almost the same amount which is added during synthesis is leached out from the hardened structure [15]. Moderate amount of late leaching was noticeable for sample SMF1 under optical microscopy (as shown in Figure 6(a)). But silica

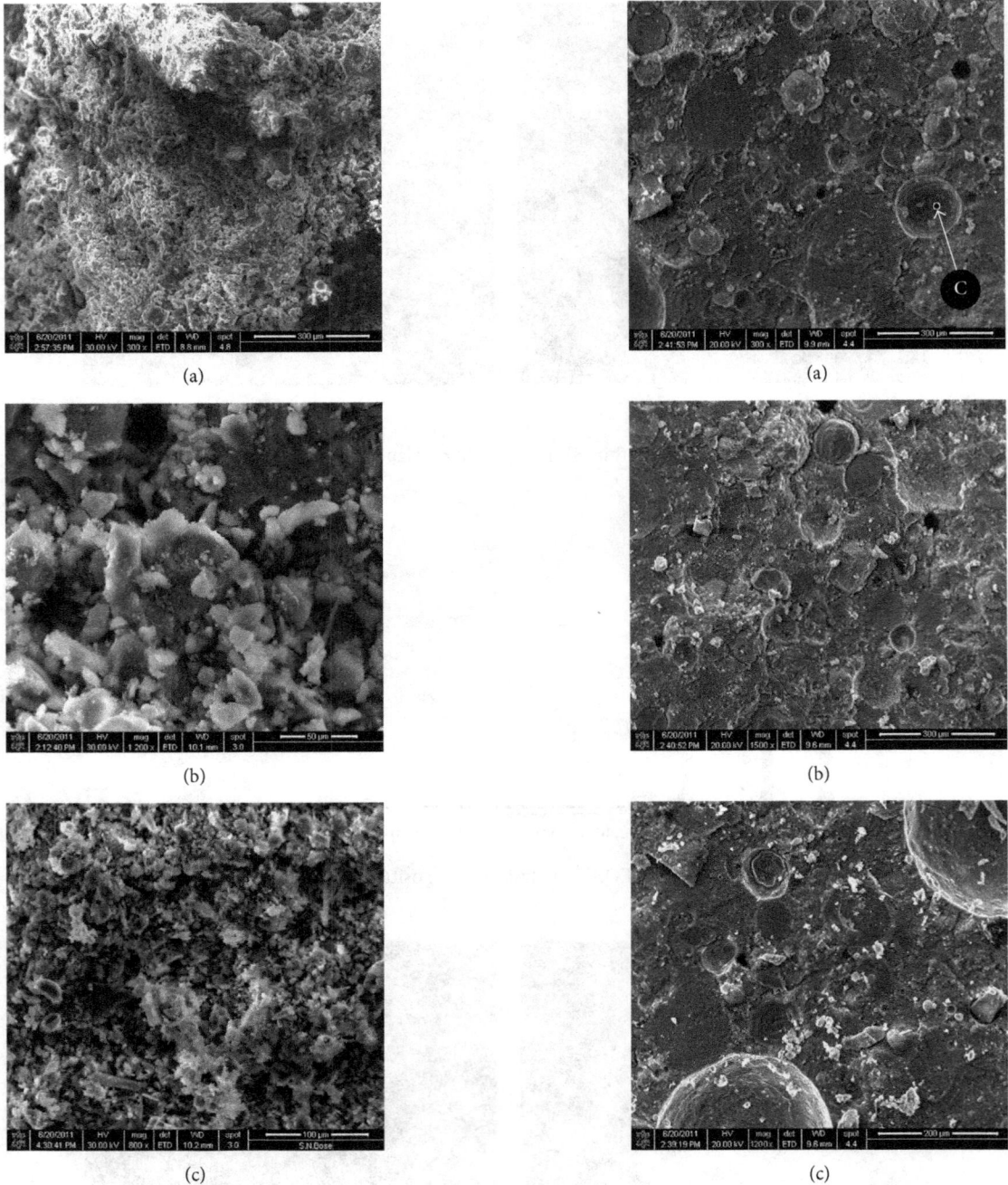

FIGURE 7: (a) Outer surfaces of sample SMF1 at 3 days (a), 30 days (b), and 60 days (c) from the end of heat curing.

FIGURE 8: (a) Outer surfaces of sample SMSB2 at 3 days (a), 30 days (b), and 60 days (c) from the end of heat curing.

fume-blended geopolymers did not exhibit such characteristics (as shown in Figure 6(b)). It is because of the presence of silica fume which makes a very reactive complex and compact structure. Generally, the application of sodium silicate initializes the primary polymerization, but for the blended mix, this role is almost taken by the reactive silica fume [16]. Excessive leaching was observed under scanning electron microscopy for sample SMF1 with aging. Figures 7(a)–7(c) represent the SEM images of sample SMF1 at the age of 3, 30, and 60 days. The sample SMSB2 did not show excessive leaching with aging. Figures 8(a)–8(c) show compact and intact structure

supporting strong alkali silicate reaction product as defined by point C in Figure 8(a).

4.5. Durability Study in Magnesium Sulfate Exposure.
Samples SMF1, SMS1, SMSB1, SMSB2, and SMSB3 were immersed in 10% magnesium sulfate solution for 12 months. The methodology was the same as Thokchom et al. taken earlier [17]. At a preselected interval, the physical appearance of geopolymer specimens was examined. Elongated needle-like crystal formation began appearing on the surfaces of sample SMF1 after few weeks (as shown in Figure 9(a)).

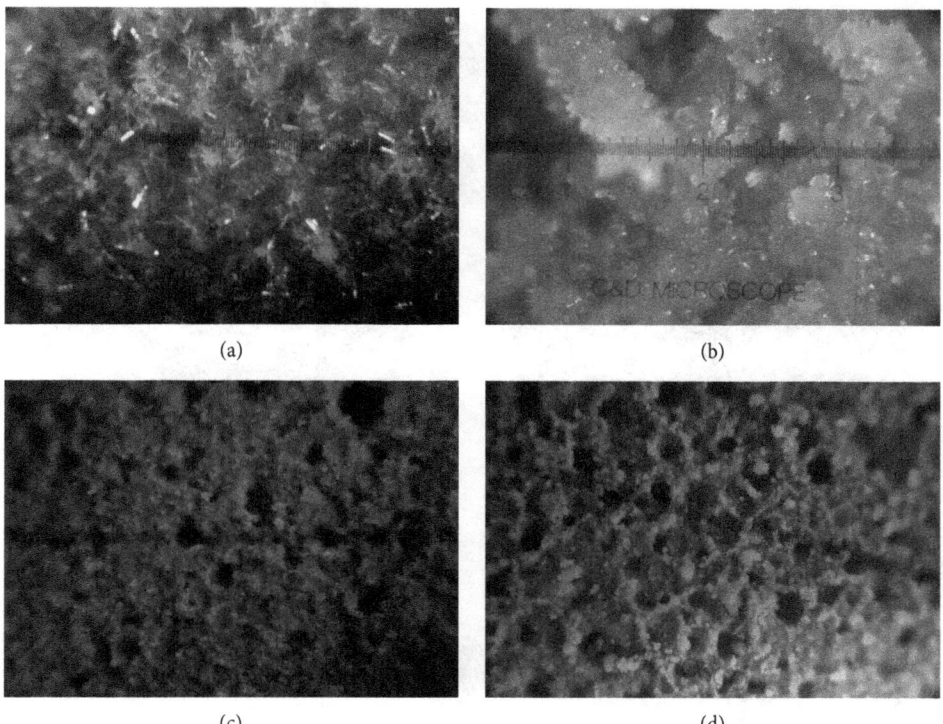

FIGURE 9: (a) Sample SMF1 showing a needle-like structure after 1 month of exposure to magnesium sulfate solution (optical microscope at 10x magnification). (b) Sample SMF1 showing white precipitant after 6 months of exposure to magnesium sulfate solution (optical microscope at 10x magnification). (c) Sample SMSB2 showing the fresh surface after 1 month of exposure to magnesium sulfate solution (optical microscope at 10x magnification). (d) Sample SMSB2 showing very little precipitants on the surface after 6 months of exposure to magnesium sulfate solution (optical microscope at 10x magnification).

These images of surfaces of the specimens were observed under an optical microscope with a magnification of 10x. After 12 weeks of exposure in 10% magnesium sulfate solution, sample SMF1 showed needle-like elongated crystal formation. It is due to the reaction of alkali hydroxide (within the pores) with magnesium sulfate, which forms less-soluble magnesium hydroxide with precipitation of alkali sulfate. White precipitant in larger quantity was observed for sample SMF1 after 6 months (as shown in Figure 9(b)), which may be magnesium hydroxide with sodium sulfate. Again, it was confirmed through scanning electron microscopy (as shown in Figures 10(a) and 10(b)) and EDAX analysis (as shown in Figure 10(c)) of scrap taken from the inner part of the sample. This outcome was the basic cause behind the drop in weight and strength after 9 months for sample SMF1. Bakharev [18] reported that the loss of strength is due to the migration of alkalis from the specimens and also due to the diffusion of calcium and sulfur near the surface region. But samples SMSB2 and SMSB3 did not show any formation at the outer surface for the time being during sulfate exposure (as shown in Figures 9(c) and 9(d)).

4.6. Change in Weight and Residual Strength.

The result shows a remarkable increment in the weight of the non-blended samples exposed to magnesium sulfate solution at room temperature up to 3 months (Figure 11(a)). The similar trend was observed for compressive strength (Figure 11(b)). But materials blended with silica fume and borax did not show any remarkable change in connection with weight and strength. As mentioned earlier, the entrapped alkali within the pores of geopolymers participates in the ionic transaction in the presence of magnesium sulfate. This phenomenon has a great impact on the change in weight and strength for the time being during sulfate exposure. Again, the volumetric change within the pores enhances the pore pressure. This may be treated as the initial cause of strength increment. But later on, this continuous change in volume (compound within the pore) deteriorates the polymer structure. Due to the lack of the presence of the untreated hydroxide within the pore, silica fume-based geopolymers did not exhibit in this manner. The drop in compressive strength of sample SMF1 was around 37.26%. But there was almost no such change for SMSB2. The detail of change in weight and strength at different times of exposure to sulfate solution (as a percentage of primary value) is plotted in Figures 11(a) and 11(b), respectively. The figures indicate a noticeable drop from the primary values in connection with weight and strength for the SMF1 specimen.

The results of volume intrusion of the MIP of geopolymers are shown in Figure 12. The volume intrusion down to 1 micrometer was indicated for sample SMSB2. The results of volume intrusion of samples clearly reflect the change in chemistry with the change in the constituent materials. The

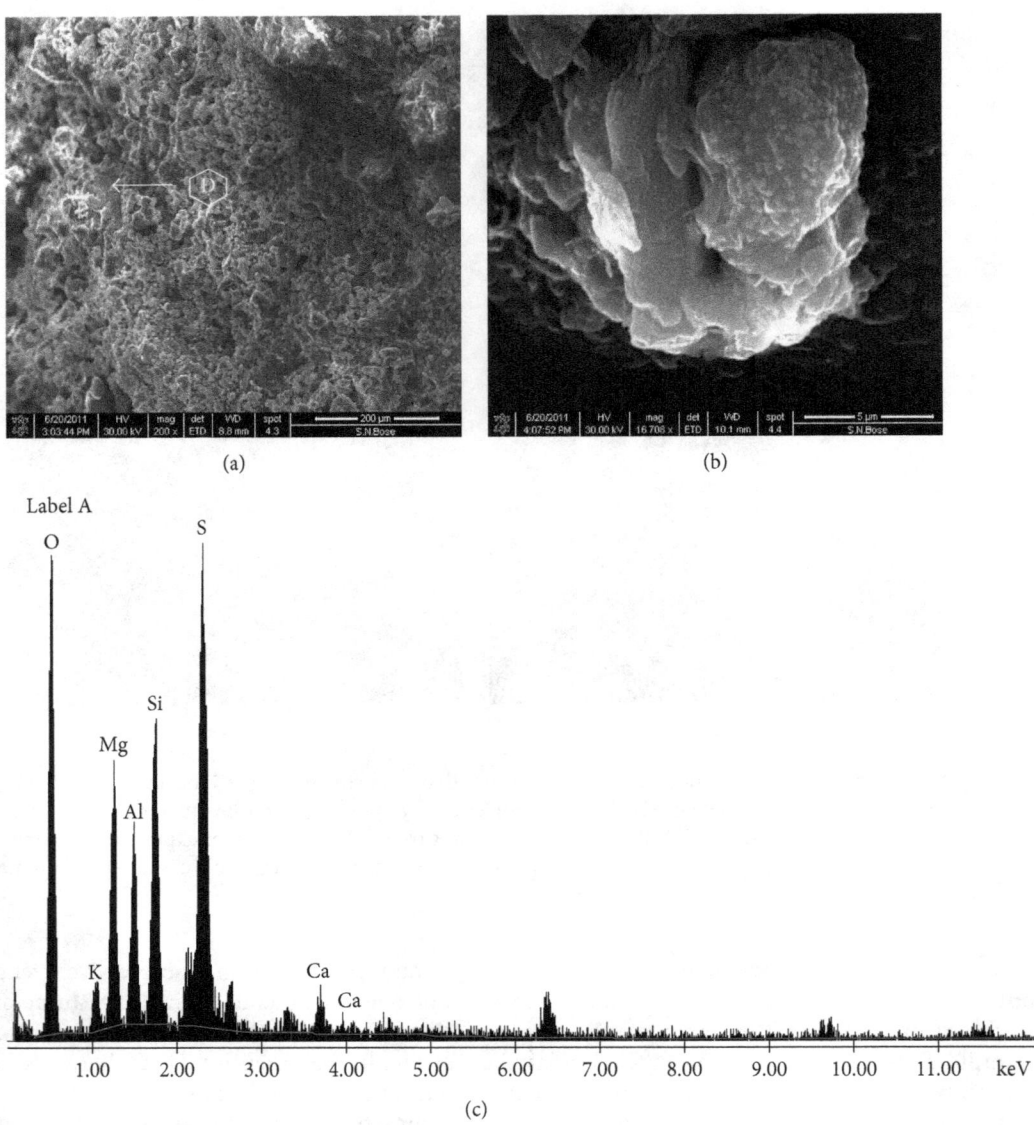

FIGURE 10: (a) Sample SMF1 showing white precipitant at the inner surface after 3 months at 200x magnifications. (b) SEM analysis near point D at 16708x magnification. (c) EDAX analysis at point D.

fine silica fume with high surface area eased the dissolution of silica which resulted in higher rate of reaction and produced denser matrix with higher compressive strength. Reactive silica fume significantly produced higher percentage of larger pores with a lower limit greater than 10 micrometers (the specimens SMS1 and SMSB1). The study reveals that geopolymer is a tetrahedral aluminosilicate structure. This structure (Al–O–Si) consisting of aluminium and silica is interlinked tetrahedrally by allotting alternate oxygen atoms. But based on the Si/Al atomic ratio, the geopolymeric aluminosilicate structure diverges in different families from amorphous to semicrystalline frameworks like polysialate type (Si–O–Al–O–), polysialate-siloxo type (Si–O–Al–O–Si–O–), and polysialate-disiloxo type (Si–O–Al–O–Si–O–Si–O–) [19]. For samples SMS1 and SMSB1, the excessive rise in Si/Al ratio (due to the incorporation of reactive silica supplements) may insist the possible formation of Si–O–Si chain structure rather than any framework structure. However, sample SMSB2 again tends

to form framework structure because of the presence of borax containing boron (B). Here, the role of aluminium (Al) is compensated by boron (another fourfold). The difference in pore morphology must signify the change in structural formation.

Figure 12(d) illustrates the curve located within a small area for SMSB2. This indicates that the silica fume-blended geopolymers in the presence of borax reduces the mean pore size. Also, the threshold volume intruded values for the nonblended geopolymer SMF1 are larger than other silica-blended specimens. Thus, the reduced total volume of porosity and better pore size distribution contribute to the strength development.

5. Conclusions

Based on the experimental investigation, it is concluded that silica fume-blended fly ash-based geopolymers leads

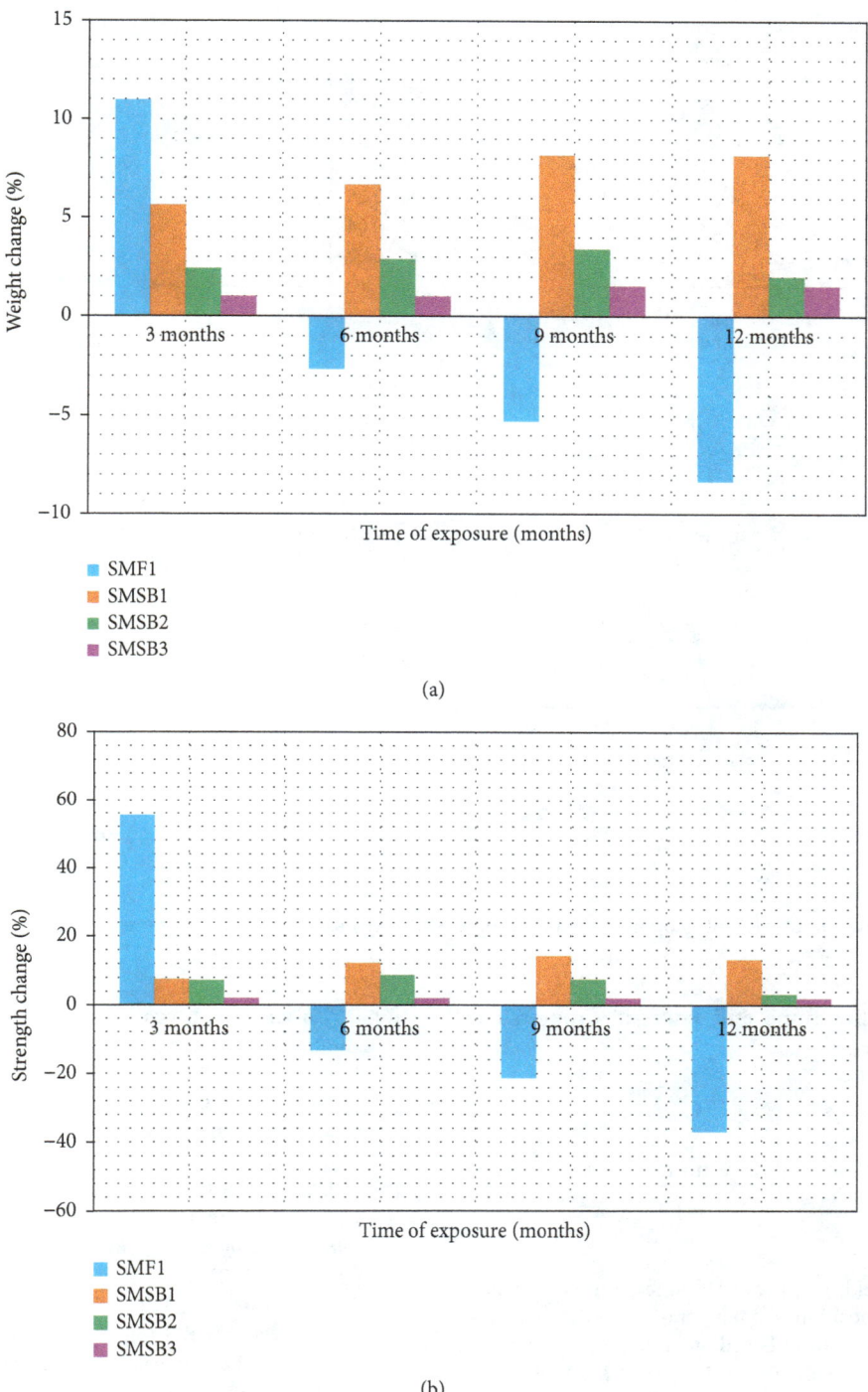

FIGURE 11: Percentage gain or loss in (a) weight and (b) strength with time of exposure to magnesium sulfate.

to a new trend which appears to generate more amorphous products and accelerates the rate of reaction in lower alkalinity. Sample SMSB3 (activated with very low value of sodium silicate) showed the compressive strength of 28 MPa after just 3 days, which is 96.49% greater than the compressive strength of sample SMF1. The maximum compressive strength (29.05 MPa) was obtained for SMSB2 (90% fly ash, 10% silica fume, and 5% borax-made specimens). Almost no efflorescence was observed for sample

blended with silica fume in the presence of borax. Temperature fluctuation has little to no effect in microstructure and mechanical properties of silica fume and borax-blended geopolymers when compared to the geopolymer produced from only fly ash. The resistance to sulfate attack for silica fume and borax-blended geopolymers was excellent and almost stable in connection with weight and strength change. The threshold values of the intruded volume pore sizes and distribution confirm dense

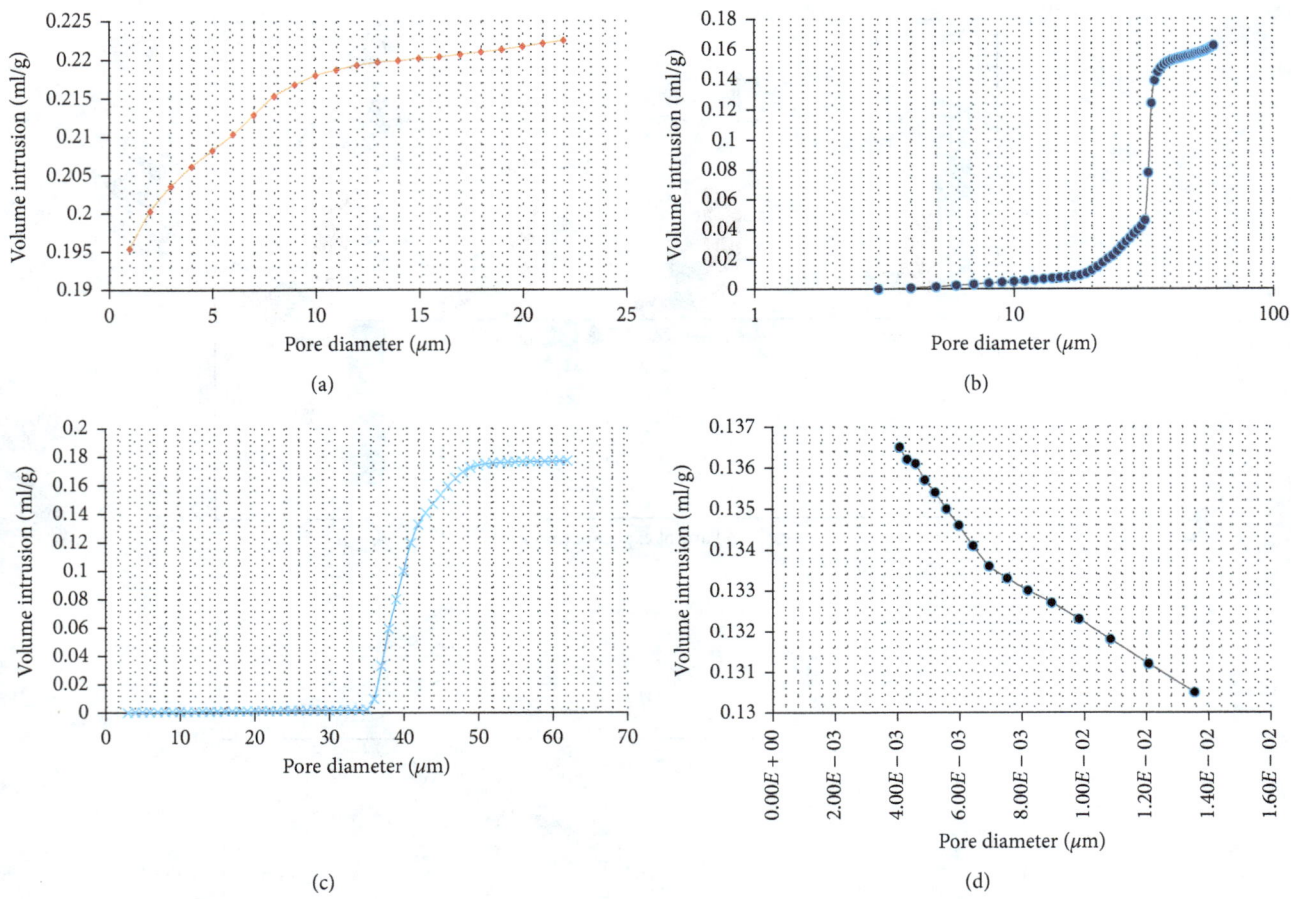

FIGURE 12: Intruded volume of MIP of geopolymers. (a) Sample SMF1, (b) sample SMS1, (c) sample SMSB1, and (d) sample SMSB2.

microstructure for silica fume and borax-incorporated fly ash-based geopolymers.

Conflicts of Interest

The authors declare that they have no conflicts of interest.

References

[1] J. G. S. van Jaarsveld, J. S. J. van Deventer, and G. C. Lukey, "The effect of composition and temperature on the properties of fly ash- and kaolinite-based geopolymers," *Chemical Engineering Journal*, vol. 89, no. 1–3, pp. 63–73, 2002.

[2] T. Bakharev, J. G. Sanjayan, and Y. B. Cheng, "Resistance of alkali-activated slag concrete to acid attack," *Cement and Concrete Research*, vol. 33, no. 10, pp. 1607–1611, 2003.

[3] E. I. Diaz, E. N. Allouche, and S. Eklund, "Factors affecting the suitability of fly ash as source material for geopolymers," *Fuel*, vol. 89, no. 5, pp. 992–996, 2010.

[4] E. P. Marinho, *Desenvolvimento de Pastas Geopoliméricas Para Cimentação de Postos de Petróleo*, Universidade Federal do Rio Grande do Norte, BrazilUniversidade Federal do Rio Grande do Norte, 2004 Ph.D. thesis, Thesis.

[5] P. H. R. Borges, L. F. Fonseca, V. A. Nunes, T. H. Panzera, and C. C. Martuscelli, "Andreasen particle packing method on the development of geopolymer concrete for civil engineering," *Journal of Materials in Civil Engineering*, vol. 26, no. 4, pp. 692–697, 2014.

[6] H. Xu and J. S. J. van Deventer, "The geopolymerisation of aluminosilicate minerals," *International Journal of Mineral Processing*, vol. 59, no. 3, pp. 247–266, 2000.

[7] D. Dutta and S. Ghosh, "Microstructure of fly ash geopolymer paste with blast furnace slag," *Current Advances in Civil Engineering*, vol. 2, no. 3, pp. 95–101, 2014.

[8] D. Dutta and S. Ghosh, "Parametric study of geopolymer paste with the different combination of activators," *International Journal of Engineering Innovation & Research*, vol. 3, no. 6, pp. 786–793, 2014.

[9] A. V. Mc Cormick, A. T. Bell, and C. J. Radke, "Evidence from alkali metal NMR spectroscopy for ion pairing in alkaline silicate solution," *Journal of Physical Chemistry*, vol. 93, no. 5, pp. 1733–1737, 1989.

[10] W. M. Hendricks, A. T. Bell, and C. J. Radke, "Effect of organic and alkali metal cations on the distribution of silicate anions in aqueous solutions," *Journal of Physical Chemistry*, vol. 95, no. 23, pp. 9513–9518, 1991.

[11] T. W. Swaddle, J. Salerno, and P. A. Tregloan, "Aqueous aluminates, silicates and aluminosilicates," *Chemical Society Reviews*, vol. 23, no. 5, pp. 319–325, 1994.

[12] D. Dutta, S. Chakrabarty, C. Bose, and S. Ghosh, "Comparative study of geo-polymer paste prepared from different activators," *STM Journals*, vol. 2, pp. 1–10, 2012.

[13] S. Thokchom, P. Ghosh, and S. Ghosh, "Durability of fly ash geopolymer mortars in nitric acid–effect of alkali (Na_2O) content," *Journal of Civil Engineering and Management*, vol. 17, no. 3, pp. 395–399, 2011.

[14] E. Prud'homme, P. Michaud, E. Joussein et al., "Silica fume as porogent agent in geo-materials at low temperature," *Journal of the European Ceramic Society*, vol. 30, no. 7, pp. 1641–1648, 2010.

[15] J. S. G. van Jaarsveld, J. S. J. van Deventer, and L. Lorenzen, "Factors affecting the immobilization of metals in geo-polymerized fly ash," *Metallurgical and Materials Transactions, B*, vol. 29, no. 1, pp. 283–291, 1998.

[16] J. T. Gourley, "Geopolymers, opportunities for environ-mentally friendly construction materials," in *Proceedings of Conferenc in adaptive materials for a modern society*, vol. 49, 15–26, pp. 1455–1461, Institute of Materials Engineering Australia, Sydney, Australia, October 2003.

[17] S. Thokchom, D. Dutta, and S. Ghosh, "Effect of incorporating silica fume in fly ash geopolymers," *World Academy of Science, Engineering and Technology*, vol. 60, pp. 245–247, 2011.

[18] T. Bakharev, "Durability of geopolymer materials in sodium and magnesium sulfate solutions," *Paste and Concrete Research*, vol. 35, no. 6, pp. 1233–1246, 2005.

[19] J. Davitovits, "Geopolymers: inorganic polymeric new materials," *Journal of Thermal Analysis*, vol. 37, no. 8, pp. 1633–1656, 1991.

Permissions

All chapters in this book were first published in AMSE, by Hindawi Publishing Corporation; hereby published with permission under the Creative Commons Attribution License or equivalent. Every chapter published in this book has been scrutinized by our experts. Their significance has been extensively debated. The topics covered herein carry significant findings which will fuel the growth of the discipline. They may even be implemented as practical applications or may be referred to as a beginning point for another development.

The contributors of this book come from diverse backgrounds, making this book a truly international effort. This book will bring forth new frontiers with its revolutionizing research information and detailed analysis of the nascent developments around the world.

We would like to thank all the contributing authors for lending their expertise to make the book truly unique. They have played a crucial role in the development of this book. Without their invaluable contributions this book wouldn't have been possible. They have made vital efforts to compile up to date information on the varied aspects of this subject to make this book a valuable addition to the collection of many professionals and students.

This book was conceptualized with the vision of imparting up-to-date information and advanced data in this field. To ensure the same, a matchless editorial board was set up. Every individual on the board went through rigorous rounds of assessment to prove their worth. After which they invested a large part of their time researching and compiling the most relevant data for our readers.

The editorial board has been involved in producing this book since its inception. They have spent rigorous hours researching and exploring the diverse topics which have resulted in the successful publishing of this book. They have passed on their knowledge of decades through this book. To expedite this challenging task, the publisher supported the team at every step. A small team of assistant editors was also appointed to further simplify the editing procedure and attain best results for the readers.

Apart from the editorial board, the designing team has also invested a significant amount of their time in understanding the subject and creating the most relevant covers. They scrutinized every image to scout for the most suitable representation of the subject and create an appropriate cover for the book.

The publishing team has been an ardent support to the editorial, designing and production team. Their endless efforts to recruit the best for this project, has resulted in the accomplishment of this book. They are a veteran in the field of academics and their pool of knowledge is as vast as their experience in printing. Their expertise and guidance has proved useful at every step. Their uncompromising quality standards have made this book an exceptional effort. Their encouragement from time to time has been an inspiration for everyone.

The publisher and the editorial board hope that this book will prove to be a valuable piece of knowledge for researchers, students, practitioners and scholars across the globe.

List of Contributors

Daegeon Kim
Architecture Engineering, Dongseo University, Busan, Republic of Korea

Nafeesa Shaheen and Siraj Ud din
NUST Institute of Civil Engineering (NICE), School of Civil and Environmental Engineering (SCEE), National University of Sciences and Technology (NUST), Sector H-12, Islamabad 44000, Pakistan

Rao Arsalan Khushnood
Department of Structural, Geotechnical and Building Engineering (DISEG), Politecnico di Torino, Corso Duca degli Abruzzi 24, Torino 10129, Italy

Sheng-bo Zhou and Wei-an Xuan
Guangxi Key Laboratory of Road Structure and Materials, Guangxi Transportation Research Institute Co., Ltd., Nanning 530007, China

Jun-lin Liang
School of Civil Engineering & Architecture, Guangxi University, Nanning 530004, China

Ye Qiu
College of Civil and Transportation Engineering, Hohai University, Nanjing 210098, China

Nasim Shatarat
Civil Engineering Department, The University of Jordan, Amman 11942, Jordan

Mutasem Shehadeh
Department of Mechanical Engineering, American University of Beirut, Beirut, Lebanon

Mohammad Naser
The Hashemite University of Jordan, Zarqa, Jordan

Woong Kim
Department of Bio-Industry Mechanical Engineering, Kongju National University, Yesan 32439, Republic of Korea

Jong-Chan Jeon, Byung-Hwan An and Chan-Gi Park
Department of Rural Construction Engineering, Kongju National University, Yesan 32439, Republic of Korea

Joo-Ha Lee
Department of Civil Engineering, University of Suwon, Hwaseong 18323, Republic of Korea

Hae-Do Kim
Rural Research Institute, Korea Rural Community Corporation, Ansan 15634, Republic of Korea

Guojin Tan, Zhiqing Zhu, Yafeng Gong and Ziyu Liu
College of Transportation, Jilin University, Changchun, Jilin 130025, China

Chenglin Shi
Jilin Provincial Transport Scientific Research Institute, Changchun, Jilin 130012, China

Tereza Komárková, Jaromír Láník and Ondřej Anton
Faculty of Civil Engineering, Brno University of Technology, Veveří 95, Brno, Czech Republic

Lijun Niu and Wenfang Zhang
College of Architecture and Civil Engineering, Taiyuan University of Technology, Taiyuan 030024, China

Syed Ishtiaq Ahmad and Mohammad Anwar Hossain
Department of Civil Engineering, Bangladesh University of Engineering and Technology, Dhaka, Bangladesh

Wei Tian and Nv Han
School of Civil Engineering, Chang'an University, Xi'an 710061, China

Roman Fediuk, Aleksey Smoliakov and Aleksandr Muraviov
Far Eastern Federal University, Vladivostok 690950, Russia

Lei Xu and Yefei Huang
College of Water Conservancy and Hydropower Engineering, Hohai University, Nanjing 210098, China

Shanshan Luo, Xinyan Guo and Xiaohong Zheng
School of Civil Engineering and Transportation, South China University of Technology, Guangzhou 510640, China

Peiyan Huang
School of Civil Engineering and Transportation, South China University of Technology, Guangzhou 510640, China
State Key Laboratory of Subtropical Building Science, South China University of Technology, Guangzhou 510640, China

Seungho Cho
Architectural Engineering Department, Seoul National University of Science and Technology, Seoul, Republic of Korea

Seunguk Na
Architectural Engineering Department, College of Architecture, Dankook University, Yongin-si, Gyeonggi-do, Republic of Korea

Humphrey Danso
Department of Construction and Wood Technology, University of Education Winneba, Kumasi, Ghana

Zhixin Li, Kaidong Xu, Jina Wang, Xianwei Ma and Jishou Niu
School of Material and Chemistry Engineering, Henan University of Urban Construction, Pingdingshan 467036, China

Jiahui Peng
College of Materials Science and Engineering, Chongqing University, Chongqing 400045, China

Ki Yong Ann, Min Jae Kim, Jun Pil Hwang and Ki Hwan Kim
Department of Civil and Environmental Engineering, Hanyang University, Ansan 15588, Republic of Korea

Chang-geun Cho
School of Architecture, Chosun University, Gwangju 61452, Republic of Korea

Mengyuan Li, Qiang Wang and Jun Yang
Department of Civil Engineering, Tsinghua University, Beijing, China

Woo-tai Jung, Jong-sup Park, Jae-yoon Kang and Moon-seoung Keum
Structural Engineering Research Institute, Korea Institute of Civil Engineering and Building Technology, Goyang, Republic of Korea

Sheng Cang
Ningbo City College of Vocational Technology, Ningbo 315100, China
Department of Mechanics and Engineering Science, Ningbo University, Ningbo 315211, China

Xiaoli Ge
Jiangsu Testing Center for Quality of Construction Engineering Co., Ltd., Nanjing 210028, China

Yanlin Bao
Department of Mechanics and Engineering Science, Ningbo University, Ningbo 315211, China

Index

9 781639 890170